AF867395

Celestial Mechanics and Astrodynamics

Progress in ASTRONAUTICS and AERONAUTICS

(a continuation of Progress in Astronautics and Rocketry)

A series of volumes sponsored by

American Institute of Aeronautics and Astronautics

1290 Avenue of the Americas, New York, New York 10019

Progress Series Editor

Martin Summerfield

Princeton University, Princeton, New Jersey

Titles in the Series

Volume 1. SOLID PROPELLANT ROCKET RESEARCH. 1960
Editor: MARTIN SUMMERFIELD, *Princeton University, Princeton, New Jersey*

Volume 2. LIQUID ROCKETS AND PROPELLANTS. 1960
Editors: LOREN E. BOLLINGER, *The Ohio State University, Columbus, Ohio;* MARTIN GOLDSMITH, *The RAND Corporation, Santa Monica, California;* AND ALEXIS W. LEMMON JR., *Battelle Memorial Institute, Columbus, Ohio*

Volume 3. ENERGY CONVERSION FOR SPACE POWER. 1961
Editor: NATHAN W. SNYDER, *Institute for Defense Analyses, Washington, D. C.*

Volume 4. SPACE POWER SYSTEMS. 1961
Editor: NATHAN W. SNYDER, *Institute for Defense Analyses, Washington, D. C.*

Volume 5. ELECTROSTATIC PROPULSION. 1961
Editors: DAVID B. LANGMUIR, *Space Technology Laboratories, Inc., Canoga Park, California;* ERNST STUHLINGER, *NASA George C. Marshall Space Flight Center, Huntsville, Alabama;* AND J. M. SELLEN JR., *Space Technology Laboratories, Inc., Canoga Park, California*

Volume 6. DETONATION AND TWO-PHASE FLOW. 1962
Editors: S. S. PENNER, *California Institute of Technology, Pasadena, California;* AND F. A. WILLIAMS, *Harvard University, Cambridge, Massachusetts*

Volume 7. HYPERSONIC FLOW RESEARCH. 1962
Editor: FREDERICK R. RIDDELL, *Avco Corporation, Wilmington, Massachusetts*

Volume 8. GUIDANCE AND CONTROL. 1962
Editors: ROBERT E. ROBERSON, *Consultant, Fullerton, California;* AND JAMES S. FARRIOR, *Lockheed Missiles and Space Company, Sunnyvale, California*

ACADEMIC PRESS · NEW YORK AND LONDON

Progress in ASTRONAUTICS and AERONAUTICS

(a continuation of Progress in Astronautics and Rocketry)

Titles in the Series (continued)

Volume 9. ELECTRIC PROPULSION DEVELOPMENT. 1963
Editor: ERNST STUHLINGER, *NASA George C. Marshall Space Flight Center, Huntsville, Alabama*

Volume 10. TECHNOLOGY OF LUNAR EXPLORATION. 1963
Editors: CLIFFORD I. CUMMINGS AND HAROLD R. LAWRENCE, *Jet Propulsion Laboratory, California Institute of Technology, Pasadena, California*

Volume 11. POWER SYSTEMS FOR SPACE FLIGHT. 1963
Editors: MORRIS A. ZIPKIN AND RUSSELL N. EDWARDS, *Space Power and Propulsion Section, Missiles and Space Division, General Electric Company, Cincinnati, Ohio*

Volume 12. IONIZATION IN HIGH-TEMPERATURE GASES. 1963
Editor: KURT E. SHULER, *National Bureau of Standards, Washington, D.C.*
Associate Editor: JOHN B. FENN, *Princeton University, Princeton, New Jersey*

Volume 13. GUIDANCE AND CONTROL — II. 1964
Editors: ROBERT C. LANGFORD, *General Precision Inc., Little Falls, New Jersey;* AND CHARLES J. MUNDO, *Institute of Naval Studies, Cambridge, Massachusetts*

Volume 14. CELESTIAL MECHANICS AND ASTRODYNAMICS. 1964
Editor: VICTOR G. SZEBEHELY, *Yale University Observatory, New Haven, Connecticut*

Volume 15. HETEROGENEOUS COMBUSTION. 1964 *(in preparation)*
Editors: HANS G. WOLFHARD, *Research and Engineering Support Division, Institute for Defense Analyses, Washington, D. C.;* IRVIN GLASSMAN, *Guggenheim Laboratories for Aerospace Propulsion Sciences, Department of Aerospace and Mechanical Sciences, Princeton University, Princeton, New Jersey;* AND LEON GREEN JR., *Research and Technology Division, Air Force Systems Command, Washington, D. C.*

(Other volumes are planned)

ACADEMIC PRESS · NEW YORK AND LONDON

Celestial Mechanics and Astrodynamics

Edited by

Victor G. Szebehely
Yale University Observatory,
New Haven, Connecticut

A Selection of Technical Papers
based mainly on
the American Institute of Aeronautics and Astronautics
Astrodynamics Conference
held at New Haven, Connecticut
August 19-21, 1963

ACADEMIC PRESS · NEW YORK · LONDON · 1964

ACADEMIC PRESS INC.
111 FIFTH AVENUE
NEW YORK, NEW YORK 10003

United Kingdom Edition
Published by
ACADEMIC PRESS INC. (LONDON) LTD.
BERKELEY SQUARE HOUSE, LONDON W. 1

Library of Congress Catalog Card Number: 64-25707

PRINTED IN THE UNITED STATES OF AMERICA

AIAA ASTRODYNAMICS CONFERENCE
New Haven, Conn., August 19-21, 1963

The AIAA Astrodynamics Conference of 1963 was conceived and initiated by the 1962 Astrodynamics Committee of the American Rocket Society and was carried out under the auspices of the 1963 Astrodynamics Committee of the American Institute of Aeronautics and Astronautics.

THE ARS ASTRODYNAMICS COMMITTEE 1962

Victor G. Szebehely, Chairman
Yale University Observatory, New Haven, Conn.

Merle Andrew
Air Force Office of Scientific Research, Washington, D. C.

Robert M. L. Baker
University of California, Los Angeles, Calif.

G. M. Clemence
U. S. Naval Observatory, Washington, D. C.

Edward R. Dyer
National Academy of Sciences, Washington, D. C.

Theodore N. Edelbaum
United Aircraft Corporation, East Hartford, Conn.

C. R. Gates
Jet Propulsion Laboratory, Pasadena, Calif.

F. T. Geyling
Bell Telephone Laboratories, Whippany, N. J.

Rudolf F. Hoelker
NASA George C. Marshall Space Flight Center,
Huntsville, Ala.

George Leitmann
University of California, Berkeley, Calif.

Eugene Levin
Aerospace Corporation, El Segundo, Calif.

Samuel Pines
Analytical Mechanics Associates, Inc., Massapequa, N. Y.

Joseph Siry
NASA Goddard Space Flight Center, Greenbelt, Md.

Eberhard W. Wahl
Air Force Systems Command, Bedford, Mass.

THE AIAA ASTRODYNAMICS COMMITTEE 1963

Robert M. L. Baker Jr., Chairman
Lockheed-California Company, Los Angeles, Calif.

John V. Breakwell
Lockheed Missiles and Space Company, Palo Alto, Calif.

Arthur E. Bryson Jr.
Harvard University, Cambridge, Mass.

Carl B. Cox
The Boeing Company, Seattle, Wash.

J. M. A. Danby
Yale University, New Haven, Conn.

R. L. Duncombe
U. S. Naval Observatory, Washington, D. C.

Theodore N. Edelbaum
United Aircraft Corporation, East Hartford, Conn.

Paul Herget
Cincinnati College, Cincinnati, Ohio

Robert C. Langford
General Precision, Inc., Little Falls, N. J.

George Leitmann
University of California, Berkeley, Calif.

Eugene Levin
Aerospace Corporation, El Segundo, Calif.

Gordon J. F. MacDonald
University of California, Los Angeles, Calif.

B. H. Paiewonsky
Princeton University, Princeton, N. J.

Paul R. Peabody
Jet Propulsion Laboratory, Pasadena, Calif.

Joseph W. Siry
NASA Goddard Space Flight Center, Greenbelt, Md.

Victor G. Szebehely
Yale University Observatory, New Haven, Conn.

William T. Thomson
University of California, Los Angeles, Calif.

THE ASTRODYNAMICS CONFERENCE COMMITTEE 1963

PREFACE

This interdisciplinary book combines the astronomical and the engineering approaches to those questions of space research which are known as orbit and trajectory problems. The new word Astrodynamics (no reference to stellar dynamics) intends to represent a field which emphasizes the engineering aspects of dynamical astronomy.

The application of a highly developed mathematical science which is soundly embedded in hundreds of years of tradition to the newest engineering problems is one of the most challenging tasks to representatives of both fields. This book intends to meet this challenge by covering the most significant and recent developments in a systematic, though not at all textbook like, manner. It is prepared for the worker in the field with background in celestial mechanics and with familiarity with the engineering problems.

The chapters are organized according to the major functional subjects of space dynamics rather than along operational lines. Indeed, it is aimed at the discussion, first, of general ideas, and then these are interspersed with examples.

The reader finds in the first chapter some of the most advanced analytical methods, recently proposed and developed, that attack new and old problems of celestial mechanics. An impressive amount of careful mathematics is included in this chapter, and it should be realized that these papers describe some of the methods that will continue to be used probably five years or more hence.

It was with regret that I did not include papers on the subject of artificial earth satellite motion without drag. This is one of the most important and popular subjects in space dynamics, and, ever since its analytical solution by Brouwer in 1959, a large number of papers have been contributed to the literature. Some of these solutions are ingenious indeed, but it seems to be hard to compete with results obtained by the powerful methods of general perturbations. Satellite motion with drag, on the other hand, is one of the problems where general perturbation methods have not yet yielded significant, useful results, and therefore this problem still is "not solved."

Several years ago I was present at the initial careful lectures describing the subjects of the first three papers of this chapter. The authors were feeling their way, and the reader may notice that we still do not really know whether these papers will open new avenues, and, if they do, where these new directions will lead us. They are sound, new, and exciting! In the several hundred years old subject of celestial mechanics, a reference to a seemingly new idea almost always can be located in the early literature. The point is, of course, what happens to the development of the idea. Laplace's sphere of activity unquestionably influenced the authors of the first paper, and it is well known that Euler solved completely the two fixed force center problem mentioned in the third paper. Nevertheless, Tisserand did not propose the asymptotic expansions introduced by the first paper, and Euler did not apply his solution to the problem of the third. The five papers of this chapter represent the major problems in analytical space dynamics – i.e., the restricted problem of three bodies, the problem of artificial satellite motion with drag, and the problem of developing new perturbation theories.

The second chapter is devoted to the physics of the solar system from the point of view of celestial mechanics. Although this might be considered a narrow view, it is probably the most exacting and precise approach to physics. Constants entering the equations of motion of celestial mechanics represent the physical world. Solutions of these equations cannot be more meaningful than the accuracy of the constants, which in turn are determined from observing the motions described by the equations. The problem is vexing for the mathematician because of the lack of uniqueness of the solution, and so it gives true enjoyment to the astronomer. Solution comes from the combination of fields, as so often it does. The constants of celestial mechanics are discussed in the first two papers in this chapter. The third paper attacks the inverse of the problem treated by the last paper of Chapter 1, viz., the determination of the density of the atmosphere from satellite observations. The density, of course, is not constant, nor is it the subject of conventional celestial mechanics, but still this is one of the major problems of space dynamics at present.

Brouwer's paper on the question of astronomical constants will truly delight the reader, independently of his background, since he will see a strictly logical approach to a problem which seldom has been approached in the past in a "straightforward" manner.

I am also glad to be able to include the JPL Mariner II paper in this chapter since I feel that some of the very important and very sound results reported in this paper show what high-level competence can accomplish quietly among noisy and spectacular space activities.

One of the major engineering problems of space dynamics, orbit selection, is the subject of the third chapter. The papers in this chapter are systems oriented, written mostly by engineers for those who are concerned with the preselection of the most suitable trajectories for certain specified missions. Emphasis was put on the methods of orbit selection and not necessarily on the results which are obtained when the methods are being applied. In this chapter the reader will find papers on the systems engineering aspects of artificial earth satellites, including two outstanding contributions to guidance problems treated from the point of view of trajectory analysis, three papers on launch window determination for lunar missions, and three on trajectories from the moon to the earth. These last five papers, all on lunar trajectories, constitute the "pot of gold" at the end of the rainbow, the arch of which started in Chapter 1. Few will be the readers who will enjoy thoroughly both ends of the rainbow; nevertheless, the main purpose of this book is to bridge the gap between the above-mentioned two groups of papers.

It is true that the restricted problem of three bodies, treated in the first chapter, is an idealized version of lunar orbit mechanics, and even the most avid proponents of analysis will admit this. On the other hand, it is also true that a digital computer is a very poor systems engineer, and that, to understand lunar orbit mechanics, it is not enough to compute hundreds of "reasonable looking" orbits. The request from the systems engineer of Chapter 3 to the analyst of Chapter 1 is to produce simple, useful, and accurate analytical results, applicable to lunar trajectory problems.

The analyst's understanding of this request and the systems engineer's understanding of the difficulties involved in trying to satisfy this request, and finally, his appreciation of the basic analytical work, are the keys to advance in space dynamics. This is the reason, I believe, that this book serves a most useful purpose.

Chapter 4 applies the general systems engineering aspects of the previous chapter to a specific mission. The three papers prepared by the three participating groups are aimed at a systematic presentation of the Ranger and Mariner missions. The over-all mission aspects and the post injection phase are described in the first paper, the boost phase trajectory is treated in a second, and the guidance problems and the complete trajectory simulation are covered in the third contribution. That this trilogy of systems engineering will be a classic in the literature, I have but little doubt.

Probably the most popular subjects of space dynamics, orbit transfer, and optimization problems are treated in the fifth chapter. The common distinguishing feature of these rather analytical papers is that their subject does not appear in classical celestial mechanics, where neither optimization nor modification of orbits is possible. The optimization aspects of space dynamics will have a long and distinguished future, and Kelly's paper in this volume always will be considered an important step. Orbit transfer questions under idealized two-body conditions are treated so thoroughly in this volume that not very much new can be expected along these lines in the future.

Inasmuch as the basic science behind the subjects treated in this volume is mechanics, realization of a very strong relation between Chapter 5 and the rest of the book is of some importance. The variational method applied to trajectory optimization is, of course, one of the fundamental formulations of dynamics. Hamilton's principle, its consequence, the Euler-Lagrange equations, and other happy schooltime memories of the reader are re-examined and utilized today.

The last chapter deals with a classical subject of celestial mechanics, viz., the determination of orbits of objects already in motion. This chapter, therefore, closes the circle of subjects that starts with perturbation problems of celestial mechanics and ends with the inverse problem, i.e., orbit determination.

The principal aim of general perturbation methods is, in general, to establish orbits when initial conditions are given. Orbit determination deals with an already established orbit, and its aim is to find the initial conditions or orbital parameters that will describe the future orbit. The orbit, of course, is established experimentally, and only its nominal initial conditions are known. Some of the techniques of orbit determination go way back to Laplace; on the other hand, new observational methods, the use of high-speed data processing machines, and requirements for real time orbit prediction necessitate the development of new techniques.

The long preface to this volume also serves as a very short introductory course in celestial mechanics and astrodynamics. The more than two-hundred-year long history of celestial mechanics and the less than ten-year-old literature of astrodynamics will, of course, soon be indistinguishable, since from the point of view of mechanics these fields are identical, and only the emphasis might differ from time to time. This volume is offered to the profession as a step in this fusion process.

With few exceptions, the papers in this volume were presented at the AIAA Astrodynamics Conference at Yale University, August 19–21, 1963. Although I was the general chairman of the Conference, the various session chairmen enjoyed complete freedom in selecting the papers to be presented. The success of the conference and of this volume, therefore, depended largely on them. Drs. R. M. L. Baker Jr., D. Brouwer, G. M. Clemence, C. R. Gates, G. Leitmann, E. Levin, and A. B. Mickelwait performed a magnificent service to the profession.

"Astrodynamics in Space Research" was the title of the banquet lecture delivered by Professor Samuel Herrick, Department of Engineering, University of California, Los Angeles. The history of astrodynamics in the AIAA is closely associated with him, since he was the initiator of the first committee on this subject and has maintained an active role in this field.

My appreciation and thanks go to Dr. Martin Summerfield, Series Editor, for his many valuable advices, and to Miss Ruth F. Bryans, Managing Editor, Scientific Publications Department of the AIAA, and members of her staff. She made my work of putting this volume together a most enjoyable experience.

Victor G. Szebehely
Yale University Observatory
New Haven, Conn.

June 1964

ASTRODYNAMICS IN THE AIAA

Samuel Herrick
University of California, Los Angeles, Calif.

In all aspects and areas of space research, interest became avid with the spectacular advent of Sputnik I in 1957. This development was especially dramatic to those of us who were attending the International Astronautical Congress in Barcelona in the next few days, or who found their mail swamped in the ensuing months with requests for previously developed materials and information, or both.

Nowhere was resurgent interest more evident than in the field of celestial mechanics. It was so even in the most abstract phases of the subject, which derive, along with explanations of observed phenomena, largely from the crystalline insight of the French school of celestial mechanics, founded or carried forward by Lagrange, Laplace, Clairaut, and Poincaré, to name but a few. It was even more evident in the severely practical aspects, in the determination of orbits, which were put on a firm foundation by the German school of Gauss, Encke, Oppolzer, Bauschinger, and Stracke, and brought to this country by Leuschner. In this school we find the emphasis not merely on the explanation, or on the existence of a solution, but on the utilization and comparison of all practical solutions. No present-day "celestial mechanic" would confess to being solely partisan to either of these schools. All are vitally interested in the ever-refreshing, ever-surprising mathematical basis, and all, except perhaps a very few very pure mathematicians, derive inspiration from the problems presented to us by artificial vehicles in space. Some of those problems require entirely new solutions, and all of them require at least revisions or adaptations of our predecessors' perspicuous solutions to problems presented by nature.

With Sputnik I the practical determination of orbits may be said to have entered upon an active phase, with the possibility of thrust guidance and control added to its previously passive "navigation" or path-prediction of objects in space. This extension and enrichment of celestial mechanics and orbit determination, together with the use of new observation and control devices and data and the increasing emphasis upon nongravitational forces, have tended to crystallize and develop under the term Astrodynamics.

Recognizing these trends, the American Rocket Society asked me to form in early 1959 the technical committee that soon came to be known as the Astrodynamics Committee. From 1961 to 1964 this committee has flourished under the chairmanships of Dr. Robert M. L. Baker Jr. and Dr. Victor G. Szebehely, who skillfully arranged its amalgamation with at least a portion of the Institute of Aerospace Sciences' Panel on Flight Mechanics, chaired by Dr. Arthur E. Bryson, and its rebirth as one of the original 32 technical committees of the AIAA. By its nature it has conscious overlap and liaison with the Atmospheric Flight Mechanics Committee, which also grew out of Dr. Bryson's panel, and the Guidance and Control Committee, with which it will hold its first joint specialist conference in August 1964, largely because of the energetic efforts of Dr. Robert C. Langford, chairman of the Guidance and Control Committee and liaison member of the Astrodynamics Committee.

The planning and development of specialist conferences and of sessions at meetings of AIAA and its predecessor societies has been a major function of the Astrodynamics Committee. I regret that I have not space to salute the many sessions, their devoted chairmen, the contributions of authors, and the notable attendance. Typical was Dr. Joseph W. Siry's excellent session in New York in January of 1964 which attracted an attendance of 400.

The first of our specialists conferences was held as an "astrodynamics colloquium" in conjunction with the Stockholm International Astronautical Congress on August 16, 1960. With an attendance of about 70, it filled a day with 16 accepted papers. The following year the Congress was held in Washington, D. C., and in part because of my chairmanship thereof and in part because the host society was the American Rocket Society, which had sponsored the previous year's colloquium, the latter was integrated into the general sessions of the Congress, where it has since remained. The Astrodynamics Committee, in endeavoring to reduce the number of meetings in the field, on several occasions gave its active support to specialist conferences initiated by groups without a specific society affiliation, but has not itself organized another specialist conference until Dr. Szebehely's energetic efforts resulted in the present one.

CONTENTS

III. Orbit Selection

IV. Orbit Mechanics of the Ranger and Mariner Missions

I.

ORBIT PREDICTION AND GENERAL PERTURBATIONS

NUMERICAL ASPECTS OF UNIFORMLY VALID ASYMPTOTIC APPROXIMATIONS FOR A CLASS OF TRAJECTORIES IN THE RESTRICTED THREE-BODY PROBLEM

P. A. Lagerstrom* and J. Kevorkian†

California Institute of Technology, Pasadena, Calif.

Abstract

In previous papers the authors have discussed the theoretical aspects of uniformly valid asymptotic approximations for the motion of a particle in a trajectory originating near a body of relatively large mass (the "earth") and passing close to a body of relatively small mass (the "moon"). It was shown that determination of the orbit during and after moon passage to the lowest order requires in principle the determination of a correction term to the Keplerian solution for the approach orbit. This correction term was expressed in terms of certain integrals. The present paper summarizes these theoretical results. The relevant definite integrals are evaluated numerically so that the motion during and after moon passage can be explicitly related to the initial conditions near the earth. Typical results are exhibited in various graphs. The Keplerian integrals for the motion during moon passage obtained by the present first-order theory are compared with exact values calculated numerically on an IBM 7094 for 27 orbits corresponding to a wide range of initial conditions and three values of the mass ratio. As expected, the comparisons indicate that the accuracy of the present theory improves

---

Presented as Preprint 63-389 at the AIAA Astrodynamics Conference, New Haven, Conn., August 19-21, 1963. This work was jointly sponsored by the U. S. Air Force under Grant No. AF-AFOSR-62-256, and the Douglas Aircraft Co., Inc. under Independent Research and Development Fund No. 80225-400 52673.

*Professor of Aeronautics; also Consultant, Douglas Aircraft Co., Inc., Santa Monica, Calif.

†Research Fellow in Aeronautics; also Consultant, Douglas Aircraft Co., Inc., Santa Monica, Calif.

as the mass ratio becomes smaller. The largest value of the mass ratio used is that for the actual earth-moon system. Even in this case the results of the asymptotic theory are satisfactory as long as the assumption of small initial perigee height is maintained.

1. Introduction

In Refs. 1-3, the authors discussed the theoretical aspects of the motion of a particle of negligible mass ("space-vehicle") whose trajectory originates near a body of relatively large mass (the "earth") and passes close to a body of relatively small mass (the "moon"). In Ref. 1 it was assumed that the two primaries were fixed in an inertial frame; this case is, in principle, integrable and was used as a mathematical model for illustrating the use of certain asymptotic methods. These methods were then applied in Refs. 2 and 3 to the restricted three-body problem. In Ref. 1 it was suggested that there was no mathematical justification for the direct "matching of conics" in deriving solutions for earth-to-moon trajectories; the conic before moon passage does not provide sufficient information for defining the hyperbolic motion during moon passage. In Refs. 2 and 3, a detailed study of the matching problem indeed indicated that, with the exception of the energy, the integrals during moon passage contain terms contributed by consideration of the corrections to order μ of the orbit before moon passage.

The intuitive physical justification for this hinges on the fact that the angular momentum relative to the moon is, by definition, a parameter of order μ in the natural variables (because for the class of trajectories we consider the orbit approaches to within a distance of order μ of the moon with a velocity of order unity) and can only be defined if the orbit before moon passage is known to order μ.

In the present paper, the theoretical results of Refs. 2 and 3 are supplemented by detailed numerical data. In particular, the definite integrals occuring in the first-order solution are tabulated, and it is shown that these integrals tend to the values predicted by the theory of Ref. 3 for minimal energy. In addition, 27 orbits corresponding to a wide range of initial values are computed by exact numerical integration. Three different values of the mass-ratio μ are used; the largest one, namely, $\mu = 0.01215$, corresponds to the actual earth-moon system. The resulting integrals during moon passage are then compared with the theoretical values to provide guidelines for the range of applicability of the results.

Most of the theoretical results presented here apply in general for the restricted three-body problem. It is only in deriving the relations between specific dimensional initial conditions and the dimensionless parameters used in the theory that we need to introduce the numerical constants for the actual system in consideration.

2. Resume of Theoretical Background

The material in this section is a resume of the formulation of the problem and basic results of Refs. 2 and 3 to which the reader is referred for detailed discussion.

2.1 Formulation of the Problem

We use the assumptions of the planar, circular restricted three-body problem. With the center of the earth as the origin of a nonrotating Cartesian coordinate system, the equations of motion of the particle of negligible mass in terms of the conventional dimensionless variables become

$$\frac{d^2x}{dt^2} + (1-\mu)\frac{x}{r^3} = \mu f \tag{2.1a}$$

$$\frac{d^2y}{dt^2} + (1-\mu)\frac{y}{r^3} = \mu g \tag{2.1b}$$

where (cf. Fig. 1 for the geometry)

$$f = \frac{\xi_m - x}{r_m^3} - \xi_m \qquad g = \frac{\eta_m - y}{r_m^3} - \eta_m \tag{2.2a}$$

μ = mass of moon/mass of earth and moon (2.2b)

ξ_m = x coordinate of moon = $\cos(t-T)$ (2.2c)

η_m = y coordinate of moon = $\sin(t-T)$ (2.2d)

T = phase constant of moon's motion (2.2e)

$r_m^2 = (x-\xi_m)^2 + (y-\eta_m)^2 \qquad r^2 = x^2 + y^2$ (2.2f)

The terms $-\xi_m$, $-\eta_m$, which are added to the gravitational attraction components in f and g, represent the apparent force due to the fact that the x-y coordinate system is noninertial. It is convenient to specify the initial conditions by prescribing the values of the Keplerian integrals obtained by setting $f = g = 0$ in (2.1). These integrals are

$$h_e = \frac{1}{2}\left[\left(\frac{dx}{dt}\right)^2 + \left(\frac{dy}{dt}\right)^2\right] - \frac{1-\mu}{r} \tag{2.3a}$$

$$\ell_e = x\frac{dy}{dt} - y\frac{dx}{dt} \tag{2.3b}$$

$$p_e = \ell_e\frac{dx}{dt} + (1-\mu)\frac{y}{r} \tag{2.3c}$$

$$q_e = -\ell_e\frac{dy}{dt} + (1-\mu)\frac{x}{r} \tag{2.3d}$$

We consider trajectories whose perigee distance relative to the earth for the initial part of the orbit is of order μ (for the actual earth-moon system this is numerically reasonable for space vehicles launched from a vicinity of the surface of the earth or, say, a parking orbit). As a consequence, ℓ_e, the angular momentum relative to the earth, is of order $\mu^{\frac{1}{2}}$. We shall only consider orbits that initially are elliptic or parabolic relative to the earth (although the hyperbolic case presents no special problems) and that approach to within a distance of order μ of the moon. Thus we assume that $-1 \leqslant h_e \leqslant 0$. The theoretical discussion for the minimal energy case $h_e = -1$ is given in Ref. 3, and will be summarized separately at the end of this section. For the present purposes, we set at $x = 0$

$$h_e = -\rho^2 \quad 0 \leqslant \rho^2 < 1 \quad \ell_e = \mu^{\frac{1}{2}}\lambda \quad p_e = 0 \quad q_e > 0 \quad t = 0 \tag{2.4}$$

Here ρ and λ are constants independent of μ. We note that $\theta = \tan^{-1}(p_e/q_e)$ is the apse angle; hence, with $p_e = 0$, the elliptic orbit relative to the earth is oriented symmetrically along the x axis. Since the orientation of our coordinate system relative to the line joining the earth to the moon is prescribed by the phase-constant T, we can insure the approach of the space-vehicle to within a distance of order μ of the moon by an appropriate choice of T, to be given later.

The initial conditions given by (2.4) are equivalent to the following initial values of the coordinates and velocities at perigee

$$x = -\frac{(1-\mu)}{2\rho^2}\left\{1 - \left[1 - \frac{2\mu\rho^2\lambda^2}{(1-\mu)^2}\right]^{\frac{1}{2}}\right\} \quad y = 0 \tag{2.5a}$$

$$\frac{dx}{dt} = 0 \quad \frac{dy}{dt} = -\frac{2\mu^{\frac{1}{2}}\lambda\rho^2}{1-\mu}\left[1 - \left(1 - \frac{2\mu\rho^2\lambda^2}{(1-\mu)^2}\right)^{\frac{1}{2}}\right]^{-1} \tag{2.5b}$$

or the following Keplerian elements for the orbit relative to the earth

$$a = \text{semimajor axis} = (1-\mu)/2\rho^2 \tag{2.6a}$$

$$\epsilon = \text{eccentricity} = \{1 - [2\mu\rho^2\lambda^2/(1-\mu)^2]\}^{\frac{1}{2}} \tag{2.6b}$$

$$\theta = \text{longitude of perigee (measured from x axis)} = \pi \tag{2.6c}$$

$$\tau = \text{time of perigee passage} = 0 \tag{2.6d}$$

2.2 Outer Expansion before Moon Passage

The motion before lunar passage may be regarded as a Keplerian ellipse perturbed by the small force corresponding to μf and μg. This leads to the following "outer expansion" for y and t in terms of x:

$$y = \mu^{\frac{1}{2}} y_{\frac{1}{2}}(x, \mu) + \mu y_1(x) + O(\mu^2) \tag{2.7a}$$

$$t = t_0(x, \mu) + \mu t_1(x) + O(\mu^2) \tag{2.7b}$$

The leading terms of Eq. (2.7), which are the exact solutions of Eq. (2.1) with f = g = 0, and the initial conditions (2.5) or (2.6) are given by

$$\mu^{\frac{1}{2}} y_{\frac{1}{2}}(x, \mu) = -2^{-\frac{1}{2}}\rho^{-1}\mu^{\frac{1}{2}}\lambda\nu \tag{2.8a}$$

$$t_0(x, \mu) = 2^{-3/2}\rho^{-3}(1-\mu)(\sin^{-1}\nu - \epsilon\nu) \tag{2.8b}$$

where
$$\nu = [1 - (\epsilon - xa^{-1})^2]^{\frac{1}{2}} \tag{2.8c}$$

Note that $t_0(x, \mu)$ may be further expanded in the form

$$t_0(x, \mu) = t_{00}(x) + \mu t_{01}(x) + O(\mu^2) \tag{2.9a}$$

where

$$t_{00}(x) = \frac{1}{2^{3/2}\rho^3}[\sin^{-1} 2\rho(x-\rho^2x^2)^{\frac{1}{2}} - 2\rho(x-\rho^2x^2)^{\frac{1}{2}}] \tag{2.9b}$$

$$t_{01}(x) = -t_{00}(x) + \frac{2(1-\rho^2\lambda^2)x^2 + 3x\lambda^2}{2[2x(1-\rho^2x)]^{\frac{1}{2}}} \tag{2.9c}$$

Letting $\rho \to 0$ in the preceding formulas gives the solution for parabolic initial orbits.

With the phase constant defined in the form

$$T = T_0 + \mu^{\frac{1}{2}} T_{\frac{1}{2}} + \mu T_1 + O(\mu^2) \qquad (2.10a)$$

the necessary and sufficient condition that the particle pass within a distance of order μ of the moon is

$$T_0 = t_{00}(1) = \frac{1}{2^{3/2}\rho^3}\left[\sin^{-1} 2\rho(1-\rho^2)^{\frac{1}{2}} - 2\rho(1-\rho^2)^{\frac{1}{2}}\right] \qquad (2.10b)$$

$$T_{\frac{1}{2}} = -y_{\frac{1}{2}}(1, 0) = \lambda[2(1-\rho^2)]^{\frac{1}{2}} \qquad (2.10c)$$

In the preceding equation, T_1 may be chosen arbitrarily. Since this choice corresponds to a rotation of the apse by an angle of order μ, one can prescribe the relative positions of the moon and particle arbitrarily at the onset of lunar passage.

The perturbation terms t_1 and y_1 given in Ref. 2 may be expressed in the convenient integral form,

$$t_1(x) = \int_0^x A(x;\xi)\, f_0(\xi)\, d\xi \qquad (2.11a)$$

$$y_1(x) = x \int_0^x B(x;\xi)\, \xi t'_{00}(\xi)\, g_0(\xi)\, d\xi \qquad (2.11b)$$

where

$$A(x;\xi) = -\int_\xi^x [t'_{00}(z)]^3\, dz \qquad (2.11c)$$

$$= -\frac{1}{2^{\frac{1}{2}}4}\left[\frac{z^{\frac{1}{2}}}{2}(1-\rho^2 z)^{\frac{1}{2}} + \frac{z^{\frac{1}{2}}}{(1-\rho^2 z)^{\frac{1}{2}}} - \frac{3}{2\rho}\sin^{-1}\rho z^{\frac{1}{2}}\right]_{z=\xi}^{z=x}$$

$$B(x;\xi) = \int_\xi^x \frac{t'_{00}(z)dz}{z^2} = \frac{1}{t'_{00}(\xi)} - \frac{1}{t'_{00}(x)} \qquad (2.11d)$$

$$f_0(x) = \frac{\cos[t_{00}(x)-T_0] - x}{\left\{1+x^2 - 2x\cos[t_{00}(x)-T_0]\right\}^{3/2}} - \cos[t_{00}(x)-T_0] \qquad (2.11e)$$

$$g_0(x) = \frac{\sin[t_{00}(x)-T_0]}{\left\{1+x^2 - 2x\cos[t_{00}(x)-T_0]\right\}^{3/2}} - \sin[t_{00}(x)-T_0] \qquad (2.11f)$$

As x tends to unity (i. e. near the moon), Eqs. (2. 11a) and (2. 11b) have the following behavior:

$$t_1(x) = (3-2\rho^2)^{-3/2}\log(1-x) + \gamma(\rho) + o(1) \tag{2.12a}$$

$$y_1(x) = (3-2\rho^2)^{-3/2}\log(1-x) + \delta(\rho) + o(1) \tag{2.12b}$$

where

$$\gamma(\rho) = \int_0^1 \left[A(1;\xi)f_0(\xi) + \frac{1}{(3-2\rho^2)^{3/2}(1-\xi)}\right] d\xi \tag{2.13a}$$

$$+ \frac{1}{(3-2\rho^2)^{3/2}}$$

$$\delta(\rho) = \int_0^1 \left[B(1;\xi)\xi t_{00}'(\xi)\, g_0(\xi) + \frac{1}{(3-2\rho^2)^{3/2}(1-\xi)}\right] d\xi \tag{2.13b}$$

$$+ \frac{1}{(3-2\rho^2)^{3/2}}$$

The fundamental theme of Refs. 1 and 2 was the fact that, because the terms of order μ affect the value of the angular momentum for the moon passage, the motion during moon passage can be calculated only after the approach orbit has been determined to order μ. As we shall see presently, the constants γ and δ play a crucial role in determining this value of the angular momentum.

2. 3 Moon Passage (Inner Expansion)

The motion during moon passage is hyperbolic in terms of the "inner" variables

$$\bar{x} = \frac{x-\xi_m}{\mu} \qquad \bar{y} = \frac{y-\eta_m}{\mu} \qquad \bar{t} = \frac{t-T_0-\mu\bar{\tau}}{\mu} \tag{2.14}$$

and is given in the standard parametric form

$$\bar{x} = \bar{a}(\bar{\epsilon} - \cosh u)\cos\bar{\theta} \mp \bar{a}(\bar{\epsilon}^2-1)^{\frac{1}{2}}\sinh u \sin\bar{\theta} \tag{2.15a}$$

$$\bar{y} = \bar{a}(\bar{\epsilon} - \cosh u)\sin\bar{\theta} \pm \bar{a}(\bar{\epsilon}^2-1)^{\frac{1}{2}}\sinh u \cos\bar{\theta} \tag{2.15b}$$

$$\bar{t} = \bar{a}^{3/2}(\bar{\epsilon}\sinh u - u) \tag{2.15c}$$

where the upper signs in the square roots in (2. 15) correspond to positive angular momentum relative to the moon.

The elements of the hyperbola are

$\bar{a}$ = semimajor axis

$\bar{\epsilon}$ = eccentricity

$\bar{\theta}$ = counter-clockwise angle from $\bar{x}$ axis to apse

$T_0 + \mu\bar{\tau}$ = time of pericenter passage.

The Keplerian integrals for the hyperbola

$$\bar{h} = \tfrac{1}{2}[(d\bar{x}/d\bar{t})^2 + (d\bar{y}/d\bar{t})^2] - (1/\bar{r}) \qquad \bar{r}^2 = \bar{x}^2 + \bar{y}^2 \tag{2.16a}$$

$$\bar{\ell} = \bar{x}(d\bar{y}/d\bar{t}) - \bar{y}(d\bar{x}/d\bar{t}) \tag{2.16b}$$

$$\bar{p} = \bar{\ell}(d\bar{x}/d\bar{t}) + (\bar{y}/\bar{r}) \tag{2.16c}$$

$$\bar{q} = -\bar{\ell}(d\bar{y}/d\bar{t}) + (\bar{x}/\bar{r}) \tag{2.16d}$$

are related to the elements by

$$\bar{a} = 1/2\bar{h} \qquad \bar{\epsilon}^2 = 1 + 2\bar{h}\,\bar{\ell}^2 \tag{2.17a}$$

$$\bar{p} = -\bar{\epsilon}\sin\bar{\theta} \qquad \bar{q} = -\bar{\epsilon}\cos\bar{\theta} \tag{2.17b}$$

If we denote the $\bar{x}$ and $\bar{y}$ components of velocity along the approach asymptote by U_I and V_I, respectively, we have

$$\begin{aligned} U_I &= (1/\bar{\epsilon}\,\bar{a}^{\frac{1}{2}})[\cos\bar{\theta} \mp (\bar{\epsilon}^2 - 1)^{\frac{1}{2}}\sin\bar{\theta}] \\ &= [1/(1+2\bar{h}\,\bar{\ell}^2)][2\bar{h}\bar{p}\bar{\ell} - (2\bar{h})^{\frac{1}{2}}\bar{q}] \end{aligned} \tag{2.18a}$$

$$\begin{aligned} V_I &= (1/\bar{\epsilon}\,\bar{a}^{\frac{1}{2}})[\sin\bar{\theta} \pm (\bar{\epsilon}^2 - 1)^{\frac{1}{2}}\cos\bar{\theta}] \\ &= [-1/(1+2\bar{h}\,\bar{\ell}^2)][2\bar{h}\,\bar{q}\,\bar{\ell} + (2\bar{h})^{\frac{1}{2}}\bar{p}] \end{aligned} \tag{2.18b}$$

The corresponding components for the second asymptote are

$$\begin{aligned} U_{II} &= (-1/\bar{\epsilon}\,\bar{a}^{\frac{1}{2}})[\cos\bar{\theta} \pm (\bar{\epsilon}^2 - 1)^{\frac{1}{2}}\sin\bar{\theta}] \\ &= [1/(1+2\bar{h}\bar{\ell}^2)][2\bar{h}\,\bar{p}\,\bar{\ell} + (2\bar{h})^{\frac{1}{2}}\bar{q}] \end{aligned} \tag{2.19a}$$

$$\begin{aligned} V_{II} &= (-1/\bar{\epsilon}\,\bar{a}^{\frac{1}{2}})[\sin\bar{\theta} \mp (\bar{\epsilon}^2 - 1)^{\frac{1}{2}}\cos\bar{\theta}] \\ &= [1/(1+2\bar{h}\,\bar{\ell}^2)][-2\bar{h}\,\bar{q}\,\bar{\ell} + (2\bar{h})^{\frac{1}{2}}\bar{p}] \end{aligned} \tag{2.19b}$$

2.4 Matching

Matching the outer expansion for the approach orbit to order μ with the inner expansion determines all the elements of the hyperbola in the form

$$\bar{h} = \tfrac{1}{2}(3-2\rho^2) = \tfrac{1}{2}(1+U^2) \qquad U = 1/t'_{00}(1) = [2(1-\rho^2)]^{\frac{1}{2}} \tag{2.20a}$$

$$\bar{\ell} = 1 + \lambda^2/2 - UT_1 - UT_0 + U(\gamma-\delta) \tag{2.20b}$$

$$\bar{p} = U\bar{\ell} + (2\bar{h})^{-\frac{1}{2}} \qquad \bar{q} = \bar{\ell} - U(2\bar{h})^{-\frac{1}{2}} \tag{2.20c}$$

$$\bar{\tau} = (1+U^2)^{-3/2}\log\mu + U^{-1}\bar{\ell}(1-\bar{a}) + \delta + T_1$$
$$+ \bar{a}^2 - \bar{a}^{3/2}\log 2/\bar{\epsilon}\, U\bar{a}^{3/2} \qquad \bar{a} = (1+U^2)^{-1} \tag{2.20d}$$

With the exception of $\bar{h}$, which can be calculated from first-order considerations, all the integrals for the motion during moon passage are affected by the inclusion of the correction to order μ for the approach orbit. This effect is exhibited explicitly in $\bar{\ell}$ and $\bar{\tau}$ by the added terms $U(\gamma-\delta)$ and δ, respectively, and implicity in $\bar{p}$, $\bar{q}$ as well as $\bar{\tau}$. The numerical values of this difference as a function of the initial energy ρ will be discussed in Sec. (3.2).

2.5 First-Order Orbit after Moon Passage

The first-order orbit after moon passage is again a Keplerian conic with the following initial conditions at $t = T_0$

$$x = 1 \qquad y = 0 \tag{2.21a}$$

$$dx/dt = U_{II} \qquad dy/dt = V_{II} + 1 \tag{2.21b}$$

with U_{II} and V_{II} defined by Eq. (2.19), and the values for the Keplerian integrals given by (2.20).

The Keplerian integrals for the motion relative to the earth after moon passage are

$$h_{e_{II}} = \tfrac{1}{2}[U_{II}^2 + (V_{II}+1)^2] - 1 = \bar{h} + V_{II} - \tfrac{1}{2} \tag{2.22a}$$

$$\ell_{e_{II}} = V_{II} + 1 \tag{2.22b}$$

$$p_{e_{II}} = (V_{II}+1)\,U_{II} \tag{2.22c}$$

$$q_{e_{II}} = -(V_{II} + 1)\, V_{II} + 1 \qquad (2.22d)$$

The motion is either elliptic or hyperbolic depending on the sign of $h_{e_{II}}$. With (2.22), the elements of this conic can readily be calculated.

2.6 Uniformly Valid Composite Expansion

A composite expansion for the part of the motion which consists of the earth-to-moon passage and the passage around the moon, which is uniformly valid to order μ, is

$$y = \mu^{\frac{1}{2}} y_{\frac{1}{2}}(x, \mu) + \mu y_1(x) + \mu[\bar{y}(\bar{x}) - \alpha^*(\bar{x})] + O(\mu^2) \qquad (2.23a)$$

$$t = t_0(x, \mu) + \mu t_1(x) + \mu[\bar{t}(\bar{x}) - \beta^*(\bar{x})] + O(\mu^2) \qquad (2.23b)$$

where $\bar{y}(\bar{x})$ and $\bar{t}(\bar{x})$ are the expressions obtained from (2.15) for the motion during moon passage, and α^* and β^* are the following limiting values of $\bar{y}$ and $\bar{t}$ when $\bar{x} \to -\infty$:

$$\alpha^*(\bar{x}) = (V_I/U_I)\bar{x} - \bar{a}\bar{\epsilon}(V_I \cos\bar{\theta} - U_I \sin\bar{\theta})/U_I \qquad (2.24a)$$

$$\beta^*(\bar{x}) = (\bar{x} - \bar{a}\bar{\epsilon} \cos\bar{\theta})/U_I \qquad (2.24b)$$

It is pointed out that the logarithmic singularities in y_1 and t_1 at $x = 1$ are cancelled out in Eqs. (2.23) by virtue of the behavior of the "boundary-layer" correction terms at this point.

2.7 Minimal Energy Trajectories

As $\rho \to 1$, the formulas for the behavior of y_1 and t_1 near $x = 1$ given in (2.12) are no longer valid, since for this value of ρ the singularities of the outer expansions at $x = 1$ change type. In fact, for this case it is shown in Ref. 3 that the perturbation terms t_1 and y_1 have the following behavior as $x \to 1$:

$$t_1 = \alpha[2(1-x)]^{-\frac{1}{2}} + \gamma_1 + O[(1-x)^{\frac{1}{2}}] \qquad (2.25a)$$

where α and γ_1 are the constants,

$$\alpha = -\int_0^1 f_0(\xi)d\xi \doteq 0.42 \qquad (2.25b)$$

$$\gamma_1 = \int_0^1 [A(0;\xi) + 3\pi/2^{5/2}] \, f_0(\xi)\, d\xi \tag{2.25c}$$

and

$$y_1 = \tfrac{1}{2} \log (1-x) + \delta_1 + O[(1-x) \log (1-x)] \tag{2.26a}$$

where

$$\delta_1 = 1 + \int_0^1 [\xi g_0(\xi) + 1/2(1-\xi)]\, d\xi \doteq 1.547 \tag{2.26b}$$

The numerical value of γ_1 will not be needed for the calculations in the present paper; it contributes a term of order μ to the energy relative to the moon [cf. Eqs. (20a) and (34) of Ref. 3].

Matching the outer expansions to order μ with the inner expansion to order unity gives the following elements for the hyperbolic motion during the moon passage:

$$\bar{h} = \tfrac{1}{2} \tag{2.27a}$$

$$\bar{\ell} = 1 + \lambda^2/2 + \alpha \tag{2.27b}$$

$$\bar{p} = 1 \qquad \bar{q} = \bar{\ell} \tag{2.27c}$$

$$\bar{\tau} = \log \mu + \delta_1 + 1 - (3/2) \log 2 + \log [1 + \bar{\ell}^2]^{\frac{1}{2}} + T_1 \tag{2.27d}$$

If we let ρ tend to unity in Eqs. (2.20), the values for $\bar{h}$, $\bar{p}$, and $\bar{q}$ tend continuously to the values given previously. We actually expect all the elements of the hyperbolic motion relative to the moon to vary continuously as $\rho \to 1$. Continuity of $\bar{\ell}$ implies

$$\lim_{U \downarrow 0} U(\gamma - \delta) = \alpha \doteq 0.42 \tag{2.28}$$

and continuity of $\bar{\tau}$ implies

$$\lim_{U \downarrow 0} [\delta + \log U] = \delta_1 - \tfrac{1}{2} \log 2 \doteq 1.201 \tag{2.29}$$

The numerical verification of the above conditions will be discussed in Sec. 3.3.

3. Discussion of Theoretical Results

The main problem studied here is to determine the elements of the moon passage and conditions after moon passage as functions of the initial conditions. As pointed out in Ref. 2 (and Sec. 2 of the present paper) $\bar{\ell}$, the angular momentum relative to the moon, is especially important since it is needed to determine the first-order orbit after moon passage. In its turn, $\bar{\ell}$ depends on γ and δ. Thus the crucial computations involve evaluating these two constants in terms of ρ (cf. Table 1). Then $\bar{\ell}$ may be used as the fundamental parameter defining the motion after moon passage.

The calculations referred to previously apply in general for the restricted three-body problem. Specific numerical results associated with any two primaries, e. g., the actual earth-moon system, need only be used in establishing initial conditions and in imposing practical limitations on the solution, e. g., the finite size of the two primaries.

3. 1 Initial Conditions

A motion will be specified by prescribing the perigee distance P(km) measured from the center of the earth, the speed W (km/hr) at perigee and the phase constant T.

The dimensional initial values P and W are related to the dimensionless parameters ρ and λ by [cf. Eqs. (2. 4) and (2.5)]

$$\rho = [(1-\mu)D/P - W^2 D/2G(m_e+m_m)]^{\frac{1}{2}} \tag{3. 1a}$$

$$\lambda = [\mu DG(m_e + m_m)]^{-\frac{1}{2}} PW \tag{3. 1b}$$

where

D = earth-moon distance

G = universal gravitational constant

m_e, m_m = masses of earth and moon respectively

For the magnitude of the various constants for the actual earth-moon system we shall use the values given in Ref. 4. In particular,

$D = 3.844 \times 10^5$ km

$\mu = 0.01215$

Period of earth-moon system $= 2\pi[D^3/G(m_e+m_m)]^{\frac{1}{2}}$

$= 27.321661$ days.

If these values are used, Eqs. (3.1) reduce to

$$\rho = (3.7972 \times 10^{5}\ P^{-1} - 3.6779 \times 10^{-8}\ W^{2})^{\frac{1}{2}} \qquad (3.2a)$$

$$\lambda = 6.40758 \times 10^{-9}\ PW \qquad (3.2b)$$

For any given ρ and λ, the phase constant must be prescribed to order $\mu^{\frac{1}{2}}$ according to Eqs. (2.10b) and (2.10c). This quantity, denoted by $T_0 + \mu^{\frac{1}{2}} T_{\frac{1}{2}}$, is the required initial angle between the apse and the line joining the earth-moon centers in order to insure that the space vehicle comes within a distance of order μ of the moon. The general value for the phase constant T contains, in addition, the arbitrary term μT_1, which can be chosen at will to achieve any desired lunar interaction. The choice of T_1 will be discussed when we consider the motion during moon passage.

3.2 Outer Expansion for Motion before Moon Passage

The motion before moon passage is to first order a Keplerian conic relative to the earth and is defined by Eq. (2.8a) for $\mu^{\frac{1}{2}} y_{\frac{1}{2}}$ and (2.8b) for t_0. The formulas for the perturbation terms y_1 and t_1 are given in (2.11) in integral form.‡ We note that these integrals contain only the initial parameter ρ, and can readily be evaluated numerically for the range $0 \leq x < 1$, $0 \leq \rho < 1$.

We now return to consideration of the behavior of y_1 and t_1 near $x = 1$. The functions γ and δ defined by (2.12) and (2.13) have been obtained by numerical evaluation of the definite integrals for values of ρ sufficiently smaller than unity. In fact, for $0 \leq \rho < 0.9$ these integrals were computed by using Simpson's rule with 100 equal intervals in the range $0 \leq x \leq 1$ on an IBM 7094 computer. The results are given in Table 1.

It was pointed out in Ref. 3 that γ and δ are singular at $\rho = 1$. An equivalent representation for these functions which exhibits the singularities, and hence provides an accurate means for their evaluation near $\rho = 1$ was also given in Ref. 3. Thus the entries in Table 1 for values of $\rho \geq 0.9$ were obtained by first computing all the finite parts of γ and δ (by

‡In Ref. 2 approximate formulas in a closed form were derived for those integrals for the case $\rho = 0$. A comparison with the present more accurate results shows that the values of $\gamma(0)$ and $\delta(0)$ given in Ref. 2 are, as was claimed there, correct within less than 2%.

the same scheme as was used for $\rho < 0.9$) then adding the appropriate singular terms for values of ρ up to 0. 999. The reader is referred to Eqs. (49) and (50) of Ref. 3 and the subsequent discussion for the details of this analysis.

Table 1 Numerical values of the functions γ and δ

ρ	γ	δ	ρ	γ	δ
0. 0	0. 344309	0. 336751	0. 60	0. 545123	0. 537267
0. 05	0. 345176	0. 337617	0. 65	0. 600401	0. 592408
0. 10	0. 348049	0. 340483	0. 70	0. 671596	0. 663294
0. 15	0. 352928	0. 345349	0. 75	0. 765630	0. 756582
0. 20	0. 359951	0. 352355	0. 80	0. 894272	0. 883318
0. 25	0. 369328	0. 361712	0. 85	1. 079670	1. 063419
0. 30	0. 381352	0. 373713	0. 90	1. 372556	1. 336503
0. 35	0. 396425	0. 388760	0. 95	1. 933565	1. 816922
0. 40	0. 415087	0. 407396	0. 97	2. 398922	2. 152756
0. 45	0. 438072	0. 430354	0. 99	3. 659764	2. 798566
0. 50	0. 466383	0. 458633	0. 999	9. 034912	3. 981138
0. 55	0. 501409	0. 493620			

The logarithmic singularities of y_1 and t_1 at $x = 1$ are a direct consequence of the nonvalidity of the outer expansions at this point. It can be verified by direct calculation that these singularities cancel identically in the formulas for the uniformly valid solution given in (2. 23).

3. 3 Moon Passage

The coordinate system to which the motion is referred during moon passage is also nonrotating as seen by Eqs. (2. 14). In addition, distances and the time measured from pericenter passage are magnified by the order μ. The term $\mu\bar{\tau}$ in the definition of the inner time variable represents a correction to the value T_0 for the time elapsed until moon passage.

The integrals for the hyperbolic motion during moon passage obtained by matching are summarized in Eqs. (2. 20) and (2. 27) for the case $\rho = 1$.

The simple result for the energy $\bar{h}$ given in Eq. (2. 20a) could have been obtained without consideration of y_1 and t_1. To first order, the outer solution gives the velocity $(U, 0)$ with respect to the earth at $x = 1$, where U is determined from t_{00}. When the particle is near the moon, its velocity relative to the moon is then $(U, -1)$. This is to be taken as the velocity along the approach asymptote for the hyperbola defining the moon passage from which (2. 20a) follows.

The angular momentum $\bar{\ell}$ given in (2. 20b) explicitly involves $U(\gamma-\delta)$, a term arising from consideration of the corrections to order μ for y and t. In a sense, this term measures the error in the "patching of conics" to first order, and should be studied in detail.

In addition to Fig. 2, we give below the tabulated values of $U(\gamma-\delta)$ for values of ρ ranging from zero to unity.

Table 2 Numerical values of the angular momentum correction $U(\gamma-\delta)$

ρ	$U(\gamma-\delta)$	ρ	$U(\gamma-\delta)$
0. 0	0. 010695	0. 60	0. 008888
0. 05	0. 010677	0. 65	0. 008590
0. 10	0. 010646	0. 70	0. 008384
0. 15	0. 010597	0. 75	0. 008464
0. 20	0. 010525	0. 80	0. 009295
0. 25	0. 010429	0. 85	0. 012114
0. 30	0. 010306	0. 90	0. 022224
0. 35	0. 010154	0. 95	0. 051508
0. 40	0. 009969	0. 97	0. 084632
0. 45	0. 009747	0. 99	0. 171808
0. 50	0. 009491	0. 999	0. 319549
0. 55	0. 009200	1. 0	0. 417021

In Table 2 the value of $U(\gamma-\delta)$ for $\rho = 1$ is the limiting value of $U\gamma$ as $\rho \to 1$, since $U\delta \to 0$ as $\rho \to 1$ [cf. Ref. 3, Eqs. (49) and (50)].

Returning to Fig. 2, we note that for $\bar{\ell} = 1$ (which as will be shown presently corresponds to grazing the surface of the moon approximately for all values of ρ), $U(\gamma-\delta)$ is the percentage error committed by neglecting the corrections of order μ to the outer expansion. The striking feature of this graph is the negligible magnitude of the error for values of ρ not close to unity. This undoubtedly accounts for the reputed success of the patching of conics and its failure near the minimal energy value where the higher-order terms become extremely significant.

It should be borne in mind that even when $U(\gamma-\delta)$ is numerically small, its effect is still very significant when we are considering moon-collision trajectories for which $\bar{\ell} = 0$.

Equation (2. 20b), when used in conjunction with the formulas relating the various properties of the motion during and after moon passage to $\bar{\ell}$, can be used to choose a value of T_1 to yield any desired value for $\bar{\ell}$. Figure 3 illustrates the dependence of $[\bar{\ell} + UT_1 - (\lambda^2/2)]$ on ρ and may be used as a quick approximation for formula (2. 20b). Since $\bar{\ell}$ is the central parameter for the motion during and hence after moon passage, all results will henceforth be referred to this quantity and ρ instead of the three original parameters ρ, λ and T_1. For example, Eq. (2. 20c) gives the value for the integral of motion $\bar{p}$ as a function of $\bar{\ell}$ and terms which depend on ρ only. It would be inconvenient to substitute the explicit dependence of $\bar{\ell}$ on ρ, λ, and T_1 in this and subsequent formulas.

Some of the interesting aspects of the motion during moon passage can now be derived in terms of $\bar{\ell}$ and ρ. The distance of closest approach to the moon $\bar{d}$ is simply related to $\bar{\ell}$ and $\bar{h}$ by

$$\bar{d} = (1/2\bar{h})[(1 + 2\,\bar{h}\,\bar{\ell}^{\,2})^{\frac{1}{2}} - 1] \tag{3. 3}$$

The dimensional value of $\bar{d}$ is plotted as a function of $\bar{\ell}$ for various values of ρ in Fig. 4. When the value for $\bar{d}$ corresponding to the surface of the moon (1, 738 km) is used in (3.3) we obtain a lower bound on $|\,\bar{\ell}\,|$ at which the trajectories graze the surface. This bound is approximately unity as can be seen from Fig. 4, and is plotted as a function of ρ in Fig. 5, which shows two other interesting loci for $\bar{\ell}$ which will be discussed in the next section.

3. 4 First-Order Orbit after Moon Passage

The elements of the first-order orbit after moon passage are most conveniently expressed in terms of U_{II} and V_{II}, the velocity components relative to the moon along the second asymptote. Since at lunar encounter the velocity of the moon relative to the earth is (0, 1) we have the initial conditions given in (2. 21) and the Keplerian integrals (2. 22).

Using Eqs. (2. 19) for U_{II} and V_{II} and expressing $\bar{p}$, $\bar{q}$, and $\bar{h}$ in terms of $\bar{\ell}$ and ρ, we obtain

$$U_{II} = \frac{[(3-2\rho^2)\,\bar{\ell}^{\,2} - 1]\,[2(1 - \rho^2)]^{\frac{1}{2}} + 2(3-2\rho^2)^{\frac{1}{2}}\,\bar{\ell}}{1 + (3-2\rho^2)\,\bar{\ell}^{\,2}} \tag{3. 4a}$$

$$V_{II} = \frac{1-(3-2\rho^2)\bar{\ell}^{-2} + 2[2(3-2\rho^2)(1-\rho^2)]^{\frac{1}{2}}\ \bar{\ell}}{1 + (3-2\rho^2)\bar{\ell}^{\,2}} \tag{3.4b}$$

Using Eqs. (2.22a) and (2.22b) for h_{eII} and ℓ_{eII} with U_{II} and V_{II} as given previously, we have plotted the energy and angular momentum relative to the earth after moon passage as a function of $\bar{\ell}$ for various values of ρ in Figs. 6 and 7. We note that h_{eII} is negative for all negative $\bar{\ell}$ that clear the surface of the moon, whereas for positive $\bar{\ell}$, the orbits after moon passage escape the earth unless ρ is close to unity or $\bar{\ell}$ is very large.

The final parameter required to specify the orbit relative to the earth is the initial direction, which is simply $\psi = \tan^{-1}[(V_{II} + 1)/U_{II}]$ and is plotted in Figs. 8 and 9.

Returning to Fig. 5, the curve marked $\bar{\ell}_{\text{max-boost}}$ corresponds to those trajectories that receive the maximum increase in total energy relative to the earth due to the lunar encounter. To first order, the energy relative to the earth after moon passage is given by Eq. (2.22a). Since h_{eI} is simply $-\rho^2 = \bar{h} - (3/2)$ [cf. formula (2.20a)] we obtain

$$h_{e_{II}} - h_{e_I} = V_{II} + 1 \tag{3.5}$$

Thus, in order to maximize $h_{eII} - h_{eI}$, we must seek the value of $\bar{\ell}$ for which V_{II} is maximum. This is easily obtained from equation (3.4b):

$$\bar{\ell}_{\text{max-boost}} = U^{-1}[1-(1+U^2)^{-\frac{1}{2}}] \tag{3.6}$$

in which case Eq. (3.4a) gives $U_{II} = 0$ (i.e. the velocity along the second asymptote is in the positive $\bar{y}$ direction).

The graph for this value of $\bar{\ell}$ is plotted in Fig. 5, and we see that it is impossible to achieve maximum energy boost because of the relatively large radius of the moon.

A class of trajectories of considerable practical interest are those that correspond to the same first-order conic before and after moon passage. Since to first order the family of trajectories we are studying have zero vertical velocity relative to the earth, we seek a value of $\bar{\ell}$ such that the vertical velocity relative to the earth after moon passage, i.e. $V_{II} + 1$, is also zero. With this condition, we obtain the following requirement for $\bar{\ell}$ from Eq. (2.19b):

$$\bar{\ell} = -1/U(1 + U^2)^{\frac{1}{2}} \tag{3.7}$$

in which case Eq. (2.19a) gives $U_{II} = -U$, as expected. This graph for $\bar{\ell}$ is also shown in Fig. 5. We note that it is only possible to achieve such a "mirror trajectory" without grazing the lunar surface for values of $\rho > 0.83$.

For those trajectories that do not escape the earth, it is interesting to compute the minimum distance relative to the earth which is simply

$$d_{e_{II}} = (1/2h_{e_{II}})[1 - (1 + 2h_{e_{II}} \ell_{e_{II}}^2)^{\frac{1}{2}}] \tag{3.8}$$

The curves for d_{eII} are given as functions of $\bar{\ell}$ for those values of ρ where $h_{eII} < 0$ in Figs. 10 and 11. We note that it is possible to approach the surface of the moon (i.e. $\bar{\ell} \approx 1$) and return to a close vicinity of the surface of the earth only with values of ρ close to unity, a result which further emphasizes the importance of the minimal energy trajectories for practical considerations.

4. Accuracy of First-Order Theory

In this section we shall present the results of exact numerical solutions of the equations of motion for a wide range of initial conditions and three values of the mass ratio μ. As a measure of the accuracy of the results, we shall compare the Keplerian integrals during moon passage obtained from the theoretical and exact solutions. Since the motion during moon passage is the most critical part of the total solution, the accuracy of our results in this range provides sufficient criteria for evaluating the over-all accuracy of a given trajectory.

4.1 Numerical Solution, Choice of Initial Conditions

A numerical integration program for an IBM 7094 was prepared to compute the solution of Eqs. (2.1) with arbitrary initial conditions and arbitrary values of the mass ratio. The accuracy of the numerical integration was controlled by testing the constancy of the Jacobi integral at each integration step. For all cases that were considered, this integral remained constant to at least four significant figures. This would indicate that, barring consistent compensating errors, the coordinates and velocities throughout the trajectory are accurate to four significant figures. Since only two or three significant figures provide adequate comparisons between the theoretical and numerical results, no additional refinement of the numerical integration procedure was needed.

In the present dimensionless formulation of the problem, a trajectory is defined by the parameter μ and the three initial conditions ρ, λ, and T_1. Table 3 gives a list of the 27 cases that have been numerically computed.

Table 3 List of initial conditions

Ex.	ρ	λ	T_1	μ
1	0.4011087	1.500000	2.682633	0.00050
2	0.4011087	1.500000	2.682633	0.00100
3	0.4011087	1.500000	2.682633	0.01215
4	0.4011087	2.052384	3.440000	0.00050
5	0.4011087	2.052384	3.440000	0.00100
6	0.4011087	2.052384	3.440000	0.01215
7	0.4011087	2.500000	4.226483	0.00050
8	0.4011087	2.500000	4.226483	0.00100
9	0.4011087	2.500000	4.226483	0.01215
10	0.9000000	1.500000	5.637602	0.00050
11	0.9000000	1.500000	5.637602	0.00100
12	0.9000000	1.500000	5.637602	0.01215
13	0.9000000	2.052384	7.229222	0.00050
14	0.9000000	2.052384	7.229222	0.00100
15	0.9000000	2.052384	7.229222	0.01215
16	0.9000000	2.500000	8.882030	0.00050
17	0.9000000	2.500000	8.882030	0.00100
18	0.9000000	2.500000	8.882030	0.01215
19	1.0000000	2.168064	0	0.00050
20	1.0000000	2.168064	0	0.00100
21	1.0000000	2.168064	0	0.01215
22	1.0000000	2.168064	4.000000	0.00050
23	1.0000000	2.168064	4.000000	0.00100
24	1.0000000	2.168064	4.000000	0.01215
25	1.0000000	2.168064	8.000000	0.00050
26	1.0000000	2.168064	8.000000	0.00100
27	1.0000000	2.168064	8.000000	0.01215

In the first 18 examples, the values of ρ, λ, and T_1 are such that the similarity parameter $UT_1 - (\lambda^2/2)$ is kept at the constant value of +2.35. We note from Eq. (2.20b) that $\bar{\ell}$ depends only on ρ and this parameter. In addition to the obvious simplication of having one theoretical value of $\bar{\ell}$ for each set of nine initial values, this choice of initial conditions tests the accuracy of the predicted behavior of $\bar{\ell}$ on the initial parameters. More precisely, one would expect that for sufficiently small values of μ for which the present first-order theory is accurate, widely disparate values of λ and T_1 will result in one value of $\bar{\ell}$ as long as the similarity parameter is kept constant. We shall presently see that this is indeed the case.

Since the theoretically predicted value of $\bar{\ell} + UT_1 - (\lambda^2/2)$ is always positive, it was necessary to choose $UT_1 - (\lambda^2/2)$ positive and sufficiently large to obtain the more interesting cases of negative $\bar{\ell}$. Unfortunately, for $\rho = 1$ the similarity parameter reduces to the negative quantity $-(\lambda^2/2)$ and it is not possible to include this case in the same family of solutions considered for the non-minimal values of ρ. In this case, we have varied T_1 holding λ fixed in order to verify the theoretically predicted independence of $\bar{\ell}$ on T_1.

A more complete survey, including different values of $UT_1 - (\lambda^2/2)$, is certainly desirable; it is, however, beyond the limited scope of the present paper.

4.2 Comparison of Numerical and Theoretical Results

The theoretical values of the Keplerian integrals for the 27 cases listed in Table 3 were computed using Eqs. (2.20) and (2.27) (for $\rho = 1$). The numerical values of these integrals were obtained directly by using their definitions [cf. Eqs. (2.16) and (2.14) for $\bar{\tau}$] once the exact values of the coordinates and velocities during moon passage were known.

Table 4 gives both theoretical and exact results for the cases considered.

Table 4 Theoretical and exact values of the Keplerian integrals during moon passage

Ex.	$\bar{h}$	$\bar{h}_{exact}$	$\bar{\ell}$	$\bar{\ell}_{exact}$	$\bar{\theta}^o$	$\bar{\theta}^o_{exact}$	$\bar{\tau}$	$\bar{\tau}_{exact}$
1	1.339	1.303	-1.98	-1.90	35.3	32.7	0.379	0.410
2	1.339	1.287	-1.98	-1.97	35.3	32.6	0.539	0.518
3	1.339	1.116	-1.98	-1.73	35.3	22.9	1.109	0.959
4	1.339	1.297	-1.98	-1.89	35.3	32.0	1.137	1.136
5	1.339	1.267	-1.98	-1.87	35.3	30.8	1.297	1.254
6	1.339	1.057	-1.98	-1.54	35.3	17.2	1.867	1.574
7	1.339	1.281	-1.98	-1.85	35.3	31.1	1.923	1.884
8	1.339	1.255	-1.98	-1.81	35.3	29.4	2.083	1.991
9	1.339	1.011	-1.98	-1.32	35.3	11.0	2.653	2.206

Table 4 continued

Ex.	$\bar{h}$	$\bar{h}_{exact}$	$\bar{\ell}$	$\bar{\ell}_{exact}$	$\bar{\theta}^o$	$\bar{\theta}^o_{exact}$	$\bar{\tau}$	$\bar{\tau}_{exact}$
10	0.690	0.654	-1.76	-1.71	5.86	4.09	1.517	1.484
11	0.690	0.638	-1.76	-1.80	5.86	4.79	1.945	1.875
12	0.690	0.468	-1.76	-1.69	5.86	-4.16	3.485	3.359
13	0.690	0.642	-1.76	-1.78	5.86	4.76	3.108	3.024
14	0.690	0.620	-1.76	-1.83	5.86	4.73	3.536	3.420
15	0.690	0.407	-1.76	-1.68	5.86	-7.06	5.076	5.005
16	0.690	0.632	-1.76	-1.86	5.86	5.39	4.761	4.660
17	0.690	0.606	-1.76	-1.87	5.86	4.68	5.189	5.066
18	0.690	0.358	-1.76	-1.64	5.86	-10.3	6.729	6.757
19	0.500	0.457	3.77	3.55	14.8	16.4	-4.733	-5.358
20	0.500	0.432	3.77	3.50	14.8	17.0	-4.040	-4.911
21	0.500	0.256	3.77	2.54	14.8	27.9	-1.543	-3.362
22	0.500	0.452	3.77	3.56	14.8	16.2	-0.733	-1.228
23	0.500	0.432	3.77	3.52	14.8	16.6	-0.040	-0.610
24	0.500	0.260	3.77	2.83	14.8	21.0	2.457	1.661
25	0.500	0.452	3.77	3.58	14.8	16.1	3.267	2.970
26	0.500	0.432	3.77	3.56	14.8	16.3	3.950	3.690
27	0.500	0.267	3.77	3.39	14.8	12.4	6.457	6.819

In addition to Table 4 we show in Figs. 12 to 15 the differences between the theoretical and exact integrals as functions of the small parameter μ.

As should be expected, the discrepancy tends to zero as μ becomes small. In addition to this consistent behavior, we note that the theoretical values are more accurate for the smaller values of λ. This is again consistent with the initial assumption regarding the smallness of ℓ_e. Since for $\rho = 1$ only T_1 was varied, we note from Figs. 12 to 15 that the accuracy of the theoretical results improves for the larger values of T_1. This again is physically consistent, since the magnitude of T_1 determines the actual distance between the particle and the moon, and hence for large values of T_1 the effect of the moon's attraction becomes less significant.

For the smaller values of μ, the errors in the present theory are indeed small. Consider for example case 8, which according to Figs. 12 to 15 corresponds to the largest discrepancies for $\rho = 0.90$, $\mu = 0.001$. If the exact Keplerian integrals are converted into dimensional values of minimum distance, maximum speed, and time of moon passage, we obtain for these parameters the values: 313 km,

8210 km/hr, and 52.1 hr. The errors in the present theory then correspond to 9 km in the distance, 70 km/hr for the speed, and 2.6 sec. for the time of closest approach. The error in the orientation of the apse is given directly in Fig. 14 as 5.9°. Of course, $\mu = 0.001$ corresponds to approximately 1/8 the size of the moon, and the dimensional figures quoted do not correspond to a real situation. On the other hand, it is pointed out that for interplanetary trajectories in the solar system the largest value of μ corresponding to Jupiter is only 1.18×10^{-5}, and the accuracy of the present theory for interplanetary applications would be very satisfactory. Table 5 gives the errors in the dimensional parameters during moon passage for the nine cases considered for $\mu = 0.01215$. The error is defined as the theoretical minus the exact value.

Table 5 Error in the first-order theory for trajectories in the actual earth-moon system

Ex.	Distance, km	Speed, km/hr	Orientation, deg	Time, hr
3	453	147	12.4	0.190
6	953	- 92	18.1	0.372
9	1,569	-570	24.3	0.567
12	- 187	535	10.0	0.160
15	- 294	615	12.9	0.090
18	- 247	810	16.1	0.0354
21	3,736	332	-13.1	2.310
24	2,171	520	- 6.1	1.010
27	-1,051	795	2.4	-0.460

Examples 9, 18, 21, 24, and 27 show large discrepancies because they correspond to large values of λ. Furthermore, example 21 is particularly critical because T_1 is also small. When $\mu^{\frac{1}{2}}\lambda$ is indeed small as was assumed initially, the accuracy of the present first-order theory is satisfactory even for this relatively large value of μ.

Significant improvement in a first-order theory would be provided by not making the simplifying assumption of $\ell_e = O(\mu^{\frac{1}{2}})$. This has been done in Ref. 5, and it would be interesting to examine the numerical verification of the latter results.

It is believed that consideration of higher-order terms is unrealistic from a practical point of view, as further refinements would be of the same order of magnitude as effects not included in the present model. The purpose of the present first-order theory is to provide sufficiently accurate information about the behavior of the motion to enable the efficient choice of initial conditions for accurate numerical studies.

References

[1]Lagerstrom, P. A. and Kevorkian, J., "Matched conic approximation to the two fixed force-center problem," Astron. J. 68, 84-92 (March 1963).

[2]Lagerstrom, P. A. and Kevorkian, J., "Earth-to-moon trajectories in the restricted three-body problem," J. Mécan. 2, 189-218 (June 1963).

[3]Lagerstrom, P. A. and Kevorkian, J., "Earth-to-moon trajectories with minimal energy," J. Mécan. 2 (December 1963).

[4]Danby, J. M. A., Fundamentals of Celestial Mechanics (The Macmillan Company, New York, 1962), App. C.

[5]Perko, L. M., "Interplanetary trajectories in the restricted three-body problem," AIAA Preprint No. 64-52 (1964).

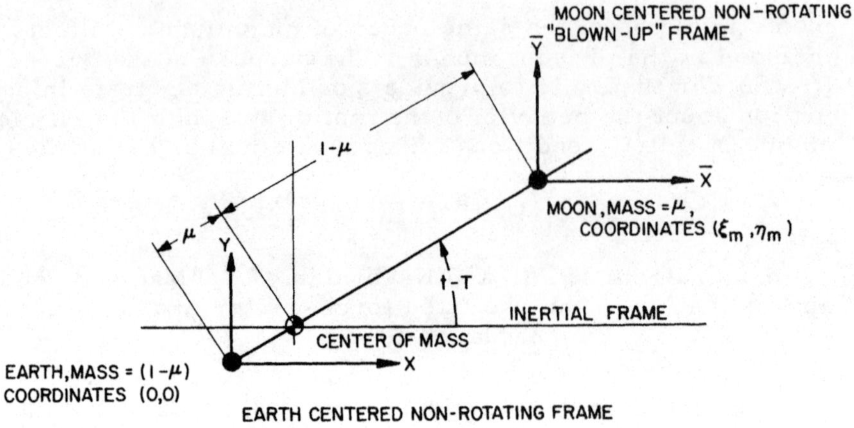

Fig. 1 Coordinate system.

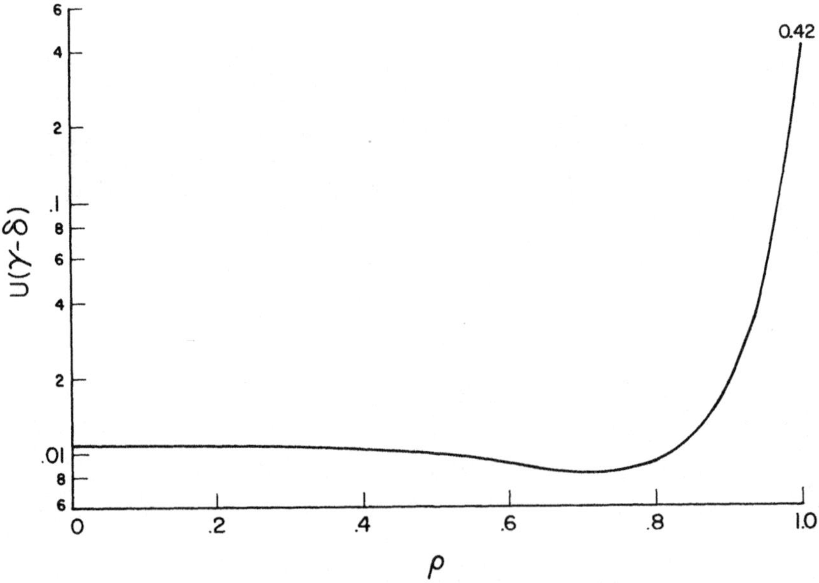

Fig. 2 Angular momentum correction, Eq. (2.12).

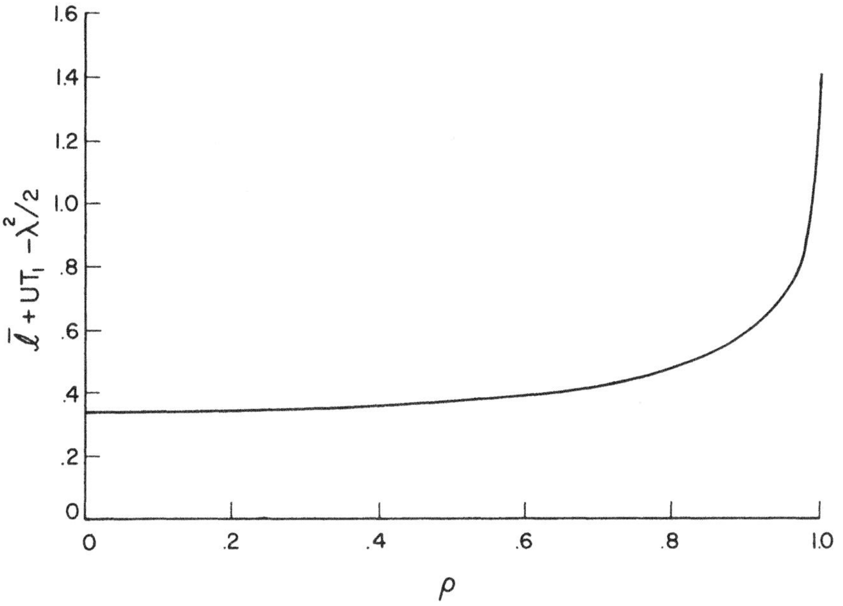

Fig. 3 Angular momentum relative to the moon, Eq. (2. 20b).

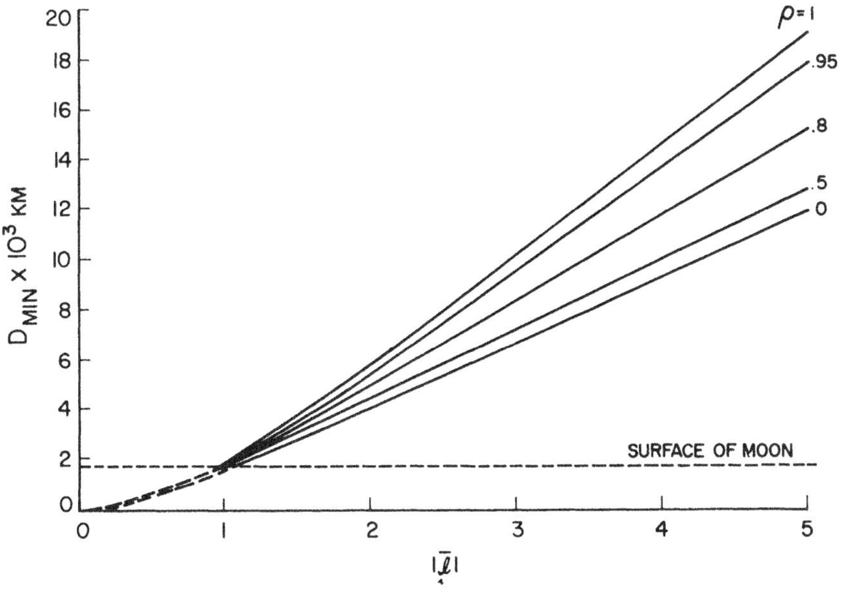

Fig. 4 Distance of minimum approach to the moon, Eq. (3. 3).

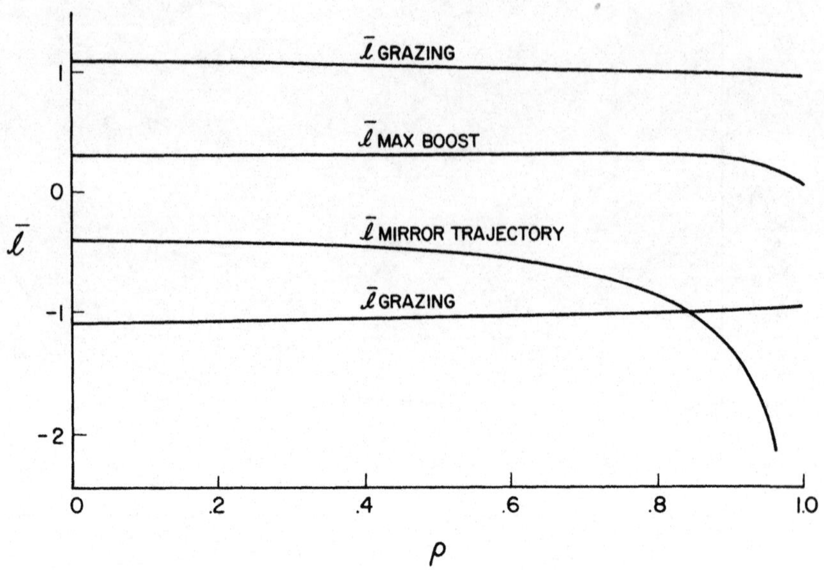

Fig. 5 Some exceptional values for the angular momentum relative to the moon, Eqs. (3. 3), (3. 6) and (3. 7).

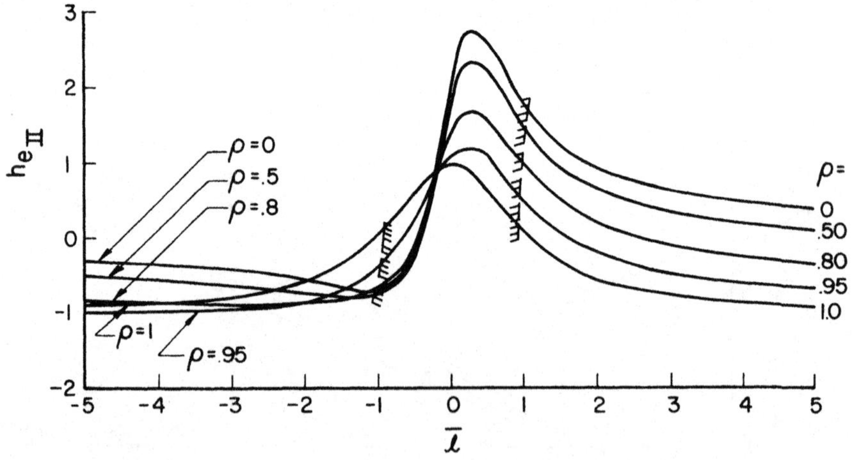

Fig. 6 Energy relative to earth after moon passage, Eq. (2. 22a).

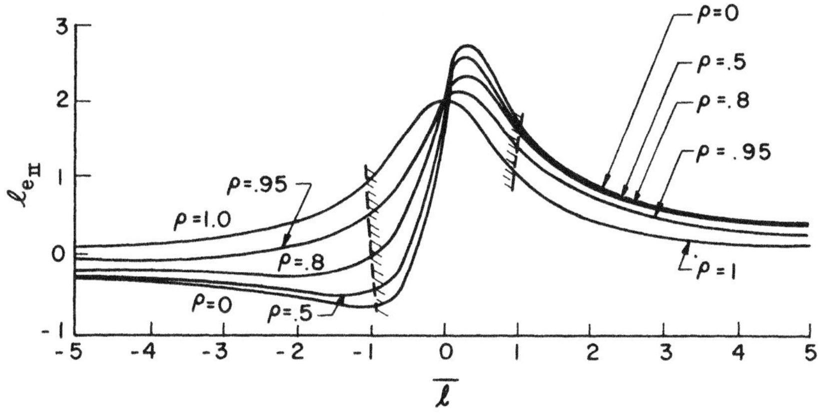

Fig. 7 Angular momentum relative to earth after moon passage, Eq. (2. 22b).

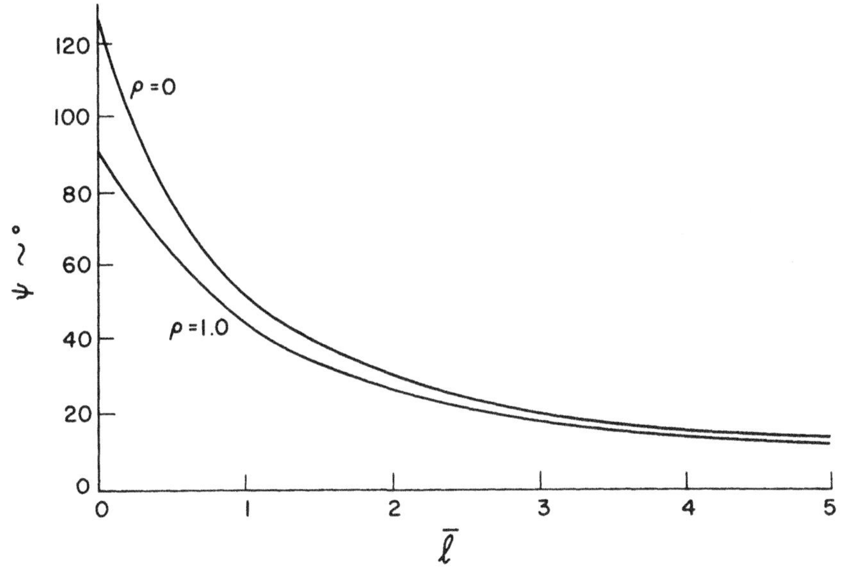

Fig. 8 Initial flight-path angle relative to earth after moon passage, $\bar{\ell} \geqslant 0$.

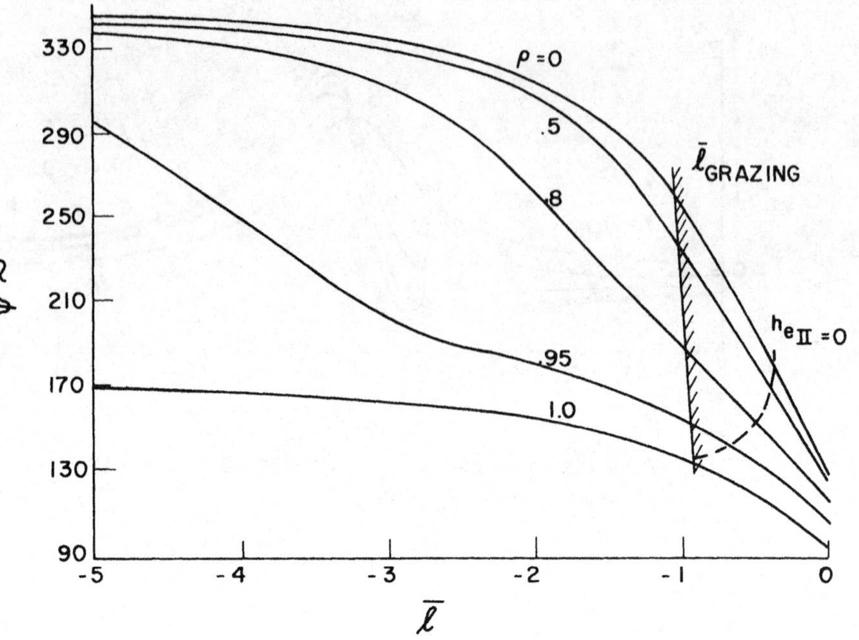

Fig. 9 Initial flight-path angle relative to earth after moon passage, $\bar{\ell} \leqslant 0$.

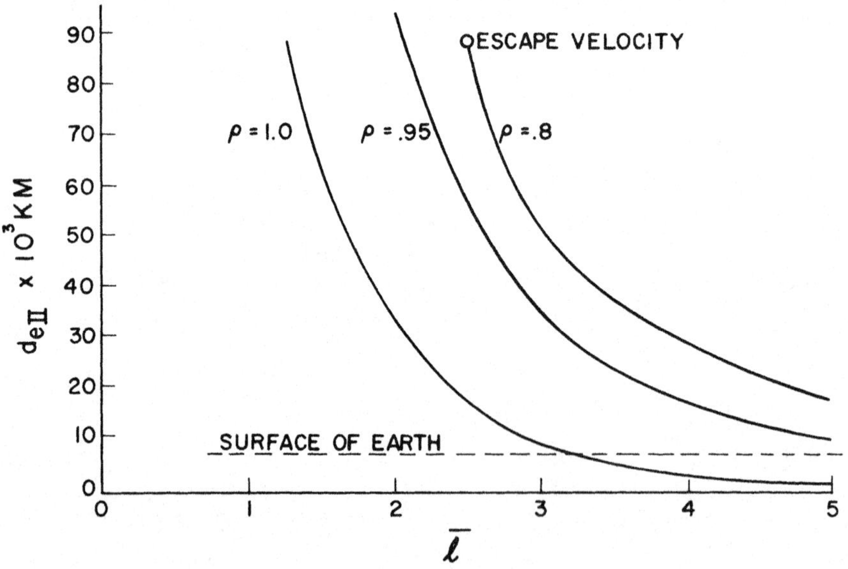

Fig. 10 Distance of minimum approach to earth after moon passage, $\bar{\ell} \geqslant 0$, Eq. (3.8).

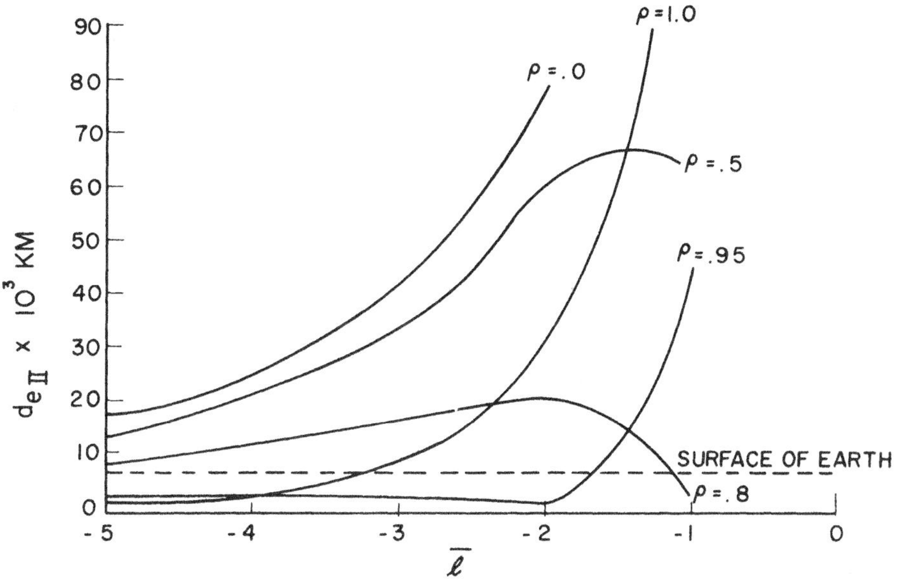

Fig. 11 Distance of minimum approach to earth after moon passage, $\bar{\ell} \leq 0$, Eq. (3.8).

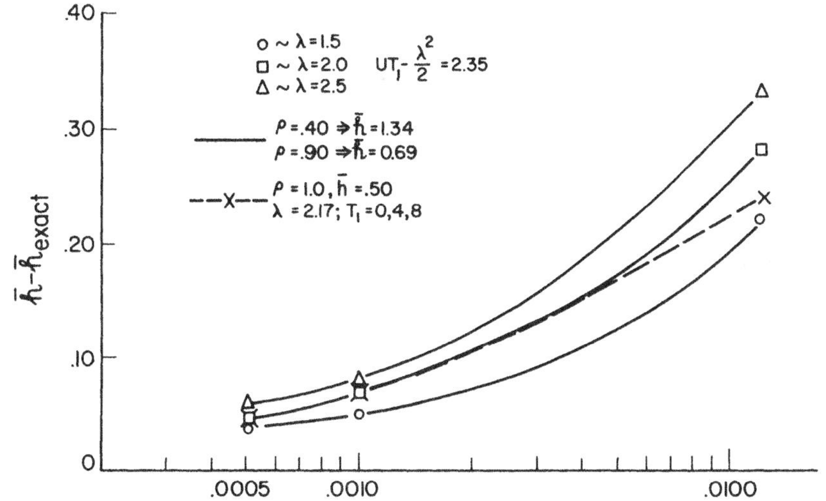

Fig. 12 Accuracy of first-order theory for energy during moon passage, Eq. (2.20a) and Table 4.

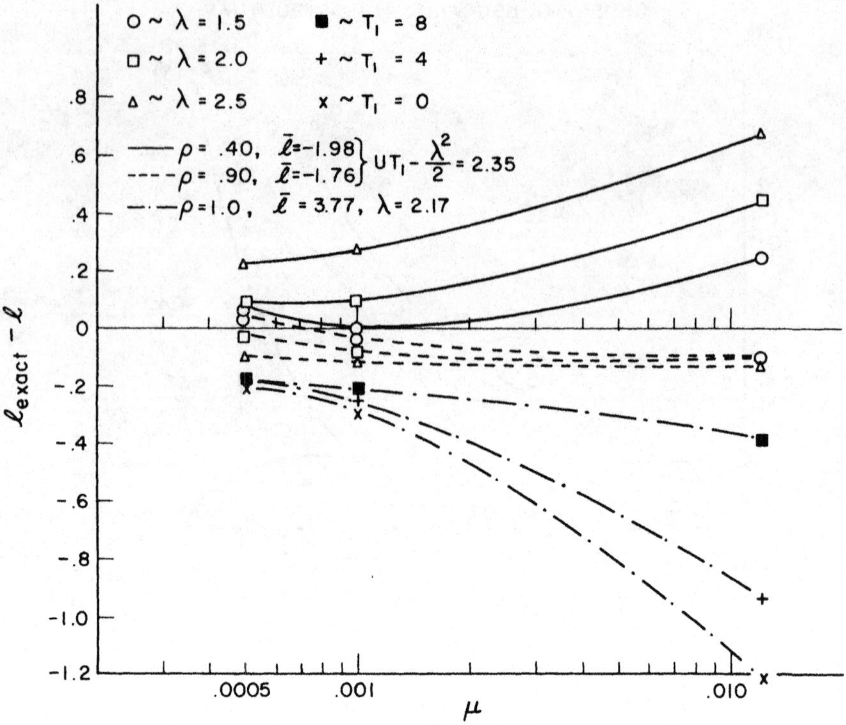

Fig. 13 Accuracy of first-order theory for angular momentum during moon passage, Eq. (2. 20b) and Table 4.

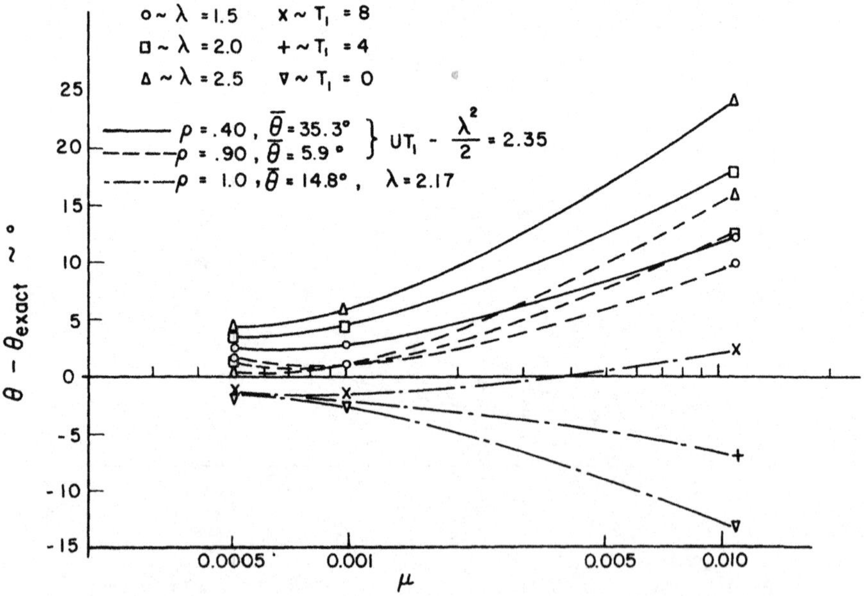

Fig. 14 Accuracy of first-order theory for apse angle during moon passage, Eq. (2. 20c) and Table 4.

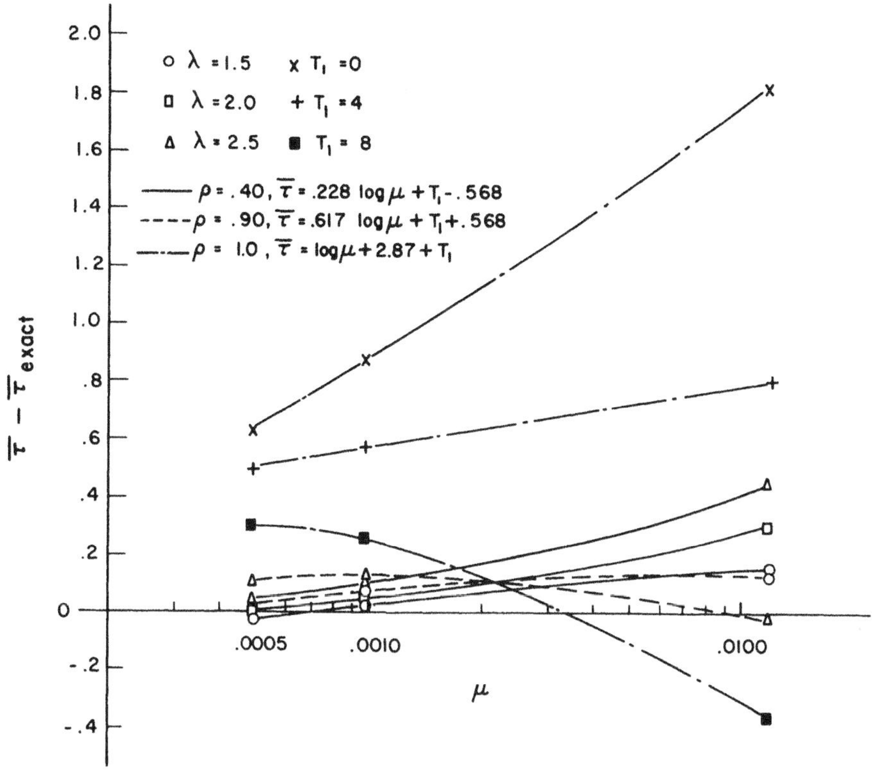

Fig. 15 Accuracy of first-order theory for time of closest approach to moon, Eq. (2. 20d) and Table 4

A GROUP OF EARTH-TO-MOON TRAJECTORIES WITH CONSECUTIVE COLLISIONS

Victor Szebehely,* David A. Pierce,+ and E. Myles Standish Jr.+

Yale University Observatory, New Haven, Conn.

Abstract

A set of trajectories connecting the earth and the moon is described. All members of this group of orbits go through, in extension, the centers of the earth and the moon, and are therefore termed orbits with consecutive collisions. The model of the planar restricted problem of three bodies is used, and the equations of motion are regularized in order to obtain solutions through the singularities. The principles of regularization are discussed in some detail. The members of the group of orbits presented in this paper are of considerable interest because they are simple and because several of them are closely related to trajectories proposed for Apollo-type missions.

I. Introduction

One of the essential differences between the problems of celestial mechanics and problems of space dynamics is the collision or close approach problem. Bodies participating in the planetary and lunar theories of celestial mechanics do not experience collisions, whereas trajectories connecting the vicinities of celestial bodies are of vital interest in space dynamics. The mathematical concept of collision between two bodies assumes two point masses and speaks about collision when at a certain time the two points have identical coordinates in space. Of

---

Received by the AIAA June 16, 1964.

The research reported in this article was partially supported by a grant from the Air Force Office of Scientific Research and by contracts with the Office of Naval Research and with NASA.

* Visiting Professor in Celestial Mechanics.

+ Graduate Student, Department of Astronomy.

course, in this sense, collision does not occur in space dynamics, since the impact of the probe with the surface of a body prevents it.

Consider now trajectories that, in extension, bring the centers of the two bodies to the same point in space simultaneously. These trajectories are called collision trajectories.

The next idealization to be introduced into these investigations is the concept of the planar restricted problem of three bodies. The foundation of this problem is as simple as its solution is difficult, and as basic as is its relation to celestial mechanics and space dynamics. Consider two point masses, m_1 and m_2, revolving in circular orbits around their mutual mass center, and introduce into this system a third body (moving in the plane of the motion of m_1 and m_2) with such small mass that it does not influence the orbits of the two primaries. When the orbit of this third (infinitesimal) mass is of interest in the gravitational field of the revolving primaries, we speak of the restricted problem of three bodies. Examples are the motion of the moon in the earth-sun system and, the subject of this paper, the motion of a space probe in the earth-moon system.

Trajectories connecting the vicinities of the earth and the moon are of great interest, and, further, we can search for orbits which actually connect the actual centers of the primaries.

The Newtonian gravitational field around a body is inversely proportional to the square of the distance (r); therefore as $r \rightarrow 0$, the force increases without limit. As a consequence the relative velocity of the bodies participating in the collision is also increasing without limit, and during very short time intervals, very large changes in the velocity can take place. In the mathematical language we speak about a singularity of the field or of the differential equations of motion.

Even when the mathematical definition of collision is not satisfied, and the point masses describe nonintersecting, close-approach trajectories, the effect of singularities can be rather serious from a computational point of view. As the relative distance between the bodies is decreasing, even without collision, very large and sudden changes in the velocity take place, so that extremely small time intervals are required in a numerical integration of the orbit.

It was originally Levi-Civita's[1] and Sundman's[2] idea to introduce a new independent variable, a "pseudo time" (τ) instead of the ordinary time(t), which would change uniformly during collision. The new and old time differentials are usually related by

$$d\tau = dt/f(r) \tag{1}$$

where f(r) is a function of the distance between the colliding particles. The generalization of this basic idea to the problem of "double" or "consecutive" collision (i.e., when the infinitesimal particle collides first with one primary and then with the other in the framework of the restricted problem) is due to Thiele and Birkhoff.[3]

Application of Birkhoff's ideas has not been attempted in the past, probably because of the conceptual, analytical, and numerical difficulties involved. The present paper treats the consecutive collision problem in order to illustrate the application of Birkhoff's basic ideas to earth-to-moon trajectories.

The terminology involved here has not crystalized, since the problem discussed is new. The expressions "double" and "triple" collision have been used in connection with the general problem of three bodies when there is no restriction regarding either the masses or the orbits of any of the participating particles. In such cases we speak of "double collision" when only two of the participating bodies collide, whereas the third one is at a different position. "Triple collision" means simultaneous collision of all three bodies. Using this terminology in the restricted problem, only double collisions are possible, since the two primaries can never collide. Therefore, strictly speaking, all collisions are double collisions in the framework of the restricted problem.

For the previous reasons, the term "consecutive collision" is introduced to describe orbits that connect the two singularities (i.e., the two primaries).

II. Conceptual Questions

The fundamental idea behind the computation of trajectories that pass through singularities is that the singularities can be eliminated from the differential equations of motion by the selection of a suitable new independent variable. This process is known as regularization. In order to illustrate its basic principles, uncomplicated by mathematical manipulations, a simple physical problem is treated first. Then consecutive collision trajectories are discussed.

Consider a simplified version of the simplest problem in celestial mechanics: the problem of two bodies. Let one of the bodies be of mass $m_1 = 1$ and the other be of much smaller mass, $m_2 \ll 1$, so that m_1 can be considered stationary while m_2 is moving in the force field of the former. The equation of motion of the small mass (m_2) is

$$\ddot{\bar{r}} = -\bar{r}/|\bar{r}|^3 \tag{2}$$

where $\bar{r}$ is the position vector of m_2, m_1 is located at the origin ($\bar{r} = 0$), and the units of mass and time are so chosen that the gravitational constant is 1. If now m_2 is given initial conditions such that its motion occurs along a straight line through m_1 (along a rectilinear orbit) the vector character of $\bar{r}$ in Eq. (2) can be ignored so we have

$$\ddot{r} = -1/r^2 \tag{3}$$

where r is now the (scalar) distance between m_1 and m_2, i.e., between the origin and m_2. The singularity of Eq. (3) is at $r = 0$, i.e. at the point where the moving (m_2) and the fixed (m_1) point masses collide.

Integrating Eq. (3) yields

$$\dot{r}^2 = 2/r - C \tag{4}$$

where C is an integration constant. At the singularity ($r = 0$) the acceleration ($\ddot{r}$) and the velocity of m_2 are infinite. In fact, close to $r = 0$, we have, from Eq. (4)

$$dr/dt = \pm\sqrt{2/r} \tag{5}$$

which says that the velocity $\to \infty$ as $1/\sqrt{r}$, when $r \to 0$.

The Levi-Civita idea of regularization is to introduce a pseudo time(τ), by the previously mentioned relation $d\tau = dt/f(r)$, which changes Eq. (5) into

$$dr/d\tau = \pm \, dt/d\tau \sqrt{2/r} \tag{6}$$

or

$$dr/d\tau = \pm \sqrt{2r} \tag{7}$$

The new velocity ($dr/d\tau$) is zero at the previous singularity ($r=0$), and Eq. (4) also becomes regular;viz.

$$dr/d\tau = \sqrt{r(2-Cr)} \tag{8}$$

The original second-order differential Eq. (3) becomes

$$d^2r/d\tau^2 + Cr = 1 \tag{9}$$

which is also devoid of singularities.

Since $\Delta r \simeq \pm \Delta\tau \sqrt{2r}$, from Eq. (7), we see that, as $r \to 0$, the integration step size ($\Delta\tau$) must be increased to maintain a constant Δr. Note that Eq. (5) gives $\Delta r \simeq \pm \Delta t \sqrt{2/r}$; therefore, as $r \to 0$, Δt has to be decreased in order to have a constant Δr. Hence, in order to maintain a constant step along the trajectory, integration in the τ system, using the transformation $f(r) = r$, requires an increasing step size(i.e., decreased number of steps), whereas integration in the original t system results in a decreasing step size (increased number of steps). Herein lies one of the keys to regularization.

A slight generalization of the previously used regularizing transformation ($d\tau = dt/r$) results in an even more elegant form. Let $d\tau = dt/\sqrt{r}$, which is certainly allowed since the introduction of the new independent variable is up to us. Eq. (7), which is valid close to collision, then becomes

$$dr/d\tau = \pm \sqrt{2} \tag{10}$$

and the complete equation for the pseudo velocity (8) becomes

$$dr/d\tau = \pm \sqrt{2 - Cr} \tag{11}$$

Now the velocity at the singularity is $\pm\sqrt{2}$, and, as $r \to 0$, we have $\Delta r \simeq \sqrt{2}\,\Delta\tau$, i.e., the integration in this pseudosystem uses a constant step size near collision.

The purpose of the two preceding illustrations is to show that different regularizing functions modify the orbit (particularly the velocity) at the singularity in different ways, and that, even though regularization removes the singularity and makes it possible to treat the problem in this vicinity, it does not guarantee that the motion in the pseudosystem is completely uniform. Hence, in the numerical integration of regularized equations of motion a variable step size may still be reguired to maintain accuracy.

The second-order differential equation of motion in the second case becomes

$$d^2r/d\tau^2 = -C \tag{12}$$

which is not only regular, but also considerably simpler than either Eq. (3) or (9).

Integration of the regularized equations of motion is simple, and a discussion of it would lend little to the basic concepts to be explained. It should be mentioned, nevertheless, that the relation between the actual and the pseudo times follows from

$$t - t_0 = \int_{\tau_0}^{\tau} f(r)\,d\tau \tag{13}$$

So, for instance, in the last case, when $f(r) = \sqrt{r}$, we have, from Eq. (12), the result that r is a quadratic function of τ, say $r = Q(\tau)$, so that the real time is given as

$$t = t_0 + \int_{\tau_0}^{\tau} \sqrt{Q(\tau)}\,d\tau \tag{14}$$

III. The Birkhoff Regularization

Several regularizing transformations have been proposed to simultaneously eliminate both singularities that occur in the potential function of the restricted problem of three bodies. Their advantages and disadvantages have been described recently[4]. For the purposes of this paper, only Birkhoff's selection of

$$z = f(w) = \frac{w^2 + \mu(1-\mu)}{2w + 1 - 2\mu} \tag{15}$$

as a transformation function is mentioned. Here $z=x+iy$ and $w = u + iv$ are the physical (original) and the transformed (or new) coordinates, respectively. The time transformation associated with the preceding geometrical transformation is given by

$$d\tau = dt/|f'(w)|^2 \tag{16}$$

Detailed properties of this transformation are given in Ref. 4, so here only three pertinent comments are necessary. Firstly, the invariant points of this transformation are $P_1(\mu, 0)$ and $P_2(\mu-1, 0)$, i.e., the primaries are not moved by the transformation. Secondly, at $w = (-1 + 2\mu)/2$, $z\rightarrow\infty$, i.e., the midpoint between the primaries in the transformed plane corresponds to infinity in the physical plane. Thirdly, the transformation is two-to-one. For any one point in the physical plane (excepting the locations of the two primaries) there are two corresponding points in the w plane, so that a single trajectory in the physical system is mapped into two trajectories, both valid for the purposes of computations, in the regularized system.

IV. Equations of Motion

Using the assumptions of the restricted problem, the equations of motion for the infinitesimal particle become

$$\begin{aligned} \ddot{x} - 2\dot{y} &= \Omega_x \\ \ddot{y} + 2\dot{x} &= \Omega_y \end{aligned} \tag{17}$$

where the conventional dimensionless, synodic coordinates (x,y) are used, dots denote derivatives with respect to the dimensionless time, and

$$\Omega = \tfrac{1}{2}\left[(1-\mu)r_1^2 + \mu r_2^2\right] + (1-\mu)/r_1 + \mu/r_2 \tag{18}$$

with

$$r_1^2 = (x-\mu)^2 + y^2$$
$$r_2^2 = (x-\mu+1)^2 + y^2 \tag{19}$$

The mass parameter μ is defined by

$$\mu = m_2/(m_1+m_2) \tag{20}$$

In complex notation, Eq. (17) can be written as

$$\ddot{z} + 2i\dot{z} = \text{grad}_z\, \Omega \tag{21}$$

where

$$\text{grad}_z\, \Omega = \Omega_x + i\Omega_y$$

After the regularization transformation indicated in the previous section [Eq. (15)], these equations become

$$w'' + 2i\,|f'(w)|^2 w' = \text{grad}_w\, \Omega^* \tag{22}$$

where primes denote derivatives with respect to the new (pseudo) time τ, and

$$\Omega^* = (\Omega - C/2)\,|f'(w)|^2 \tag{23}$$

The integral corresponding to the energy integral of the two-body problem [Eq. (4)] is named after Jacobi, and it can be derived from Eq. (17) in the form

$$\dot{x}^2 + \dot{y}^2 = 2\Omega - C \tag{24}$$

In complex notation, this integral becomes

$$|\dot{z}|^2 = 2\Omega - C \tag{25}$$

The corresponding Jacobian integral in the transformed plane is

$$|w'|^2 = 2\Omega^* \tag{26}$$

The new potential function Ω^* can be obtained without difficulty since the functions Ω and f are given by Eqs. (18) and (15).

In what follows, certain special numerical solutions of Eq. (22) are presented. These special solutions can not be obtained by integrating the original equations of motion [Eq. (17)] because all the solutions go through the singularities located at the aforementioned primaries.

In the regularized system, the velocities are finite at the singularities. The squares of the magnitudes of the regularized velocities are, from Eq. (26),

$$|w'(P_1)|^2 = 2\Omega^*(r_1=0, r_2=1) = 8(1-\mu)$$

and

$$|w'(P_2)|^2 = 2\Omega^*(r_1=1, r_2=0) = 8\mu \tag{27}$$

This means that any orbit that goes through the two primaries and that is computed in the regularized plane has several fixed conditions: the two positions of the singularities ($u_1 = \mu$, $v_1 = 0$; $u_2 = \mu - 1$, $v_2 = 0$) and the two velocity magnitudes at these points [$(u_1')^2 + (v_1')^2 = 8(1-\mu)$, $(u_2')^2 + (v_2')^2 = 8\mu$]. There is only one free (independent) parameter left in the problem: the Jacobian constant C.

The selection of a particular value of C (less than a critical value given by the Jacobian integral) allows a trajectory to be constructed, which connects P_1 and P_2 in the w plane. The totality of such trajectories, for all possible values of C, form the family of orbits with consecutive collisions. We define a group as a subset of the family in the following way. Consider a trajectory connecting P_1 and P_2 which is obtained using a value of the Jacobian constant C_1. Changing C_1 to $C_2 = C_1 + \Delta C$ results in another trajectory also connecting P_1 and P_2. If, as $\Delta C \to 0$, the second trajectory approaches the first one, **i.e.**, if the members of the set can be generated by analytic continuation, then we speak about a group of orbits with consecutive collisions.

A convenient dependent parameter is the pseudofiring angle (denoted Ψ, on Figs. 2-4) that determines the direction of the w_1' complex velocity vector at P_1. For a given value of C, this angle can be found by solving a two-point boundary value problem in the w plane, using an iteration procedure that begins with a two-body approximation.‡ Consider, for instance, a very large negative value of C which corresponds to a trajectory with a high relative energy. This follows from the Jacobian integral, which at a fixed point in the physical plane gives

$$(\text{velocity})^2 = 2\Omega - C = \text{const} - C$$

and therefore the velocity squared and the relative energy are very large when C is large and negative. If the fixed point is not a singular point, then the velocity becomes infinite as $C \to -\infty$. Consider now a rectilinear trajectory with consecutive collisions in the synodic, physical plane. This orbit is possible only when $C = -\infty$ and the velocity is infinite along the entire trajectory from one primary to the other. This orbit is just that part of the x axis which lies between $P_1(\mu, 0)$ and $P_2(\mu - 1, 0)$. Increasing C from $-\infty$ generates the group of consecutive collision trajectories that are described in the next section.

V. Numerical Results

The points of the curve of Fig. 1 represent the trajectories belonging to the group mentioned previously. The insert shows the synodic coordinate system (x,y) and the firing angle θ, in the synodic system, which is used to present the results. The mass ratio is that for the earth-moon system, $\mu = 1/82.45$. The $C = C(\theta)$ curve is asymptotic to the C axis, i.e., $\theta \to 0$, $C \to -\infty$ as mentioned before. The orbits corresponding to the maximum point of the curve ($C_o \simeq 2$, $\theta_o \simeq 60^o$) seem to represent a change in the basic characteristics of the members of the family, since, for $\theta < \theta_o$, the maximum distance of the space probe from the earth is the same as the earth-moon distance. When $\theta > \theta_o$, the maximum distance of the probe from the earth becomes larger than the earth-moon distance.

At $\theta = 360^o$, the physical situation is identical to the $\theta = 0$ case; nevertheless, Fig. 1 indicates another value of C, i.e. $C(0) \neq C(2\pi)$. The explanation of this lies in

‡ Measuring all angles counterclockwise from the line P_1P_2, a trajectory in the physical plane with a firing angle θ at P_1 is mapped into two trajectories in the regularized plane with pseudofiring angles of $\theta/2 \pm 90^o$.

the definition of a "group" of trajectories. By analytic continuation, the curve was produced that is shown on Fig. 1. At $\theta = 2\pi + \Delta\theta$, $(\Delta\theta > 0)$, another member of our group, can be obtained close to C = 1, but of course it is also true that trajectories with very large negative values of C are also available.

It is of considerable practical and historical interest to mention that one of the points on the $C(\theta)$ curve of Fig. 1 can be (and as a matter of fact is) obtained by modifying a proposed circumlunar Apollo mission trajectory[5]. The coordinates of this point are C = 1.805 and $\theta = 57^\circ\!.97$, and the original Apollo trajectory is that given in Ref. 5 as trajectory number 1. After establishing this orbit, whose transit time is 79 hr as compared to the original orbit's transit time of 76 hs, the curve can be completed by varying one of the parameters, C or θ, and differentially correcting the other. Cowell integrations of the regularized differential equations of motion [Eq. (21)] were performed, using the Gauss-Jackson method of numerical integration with seventh differences and variable integration step sizes. Computation of most of the orbits started at the center of the earth, but it makes no difference at which primary the trajectories are started if the moon moves in a retrograde direction around the earth when the orbits are started at the moon.

Although the trajectories are computed in a rotating coordinate system, as is the usual procedure for the restricted problem so that the Jacobian integral can be used for a check on the computations, it is easier to see the motion of the space probe in a fixed frame. This is because 1) as it will be shown in the next section, our trajectories are slightly perturbed conic sections in a fixed frame, 2) it is customary to study lunar trajectories in a fixed, geocentric frame, and 3) for several of the orbits, especially those with long transit times, the path of the probe in the rotating system is due essentially to the motion of the moon (i.e., of the system) and it is very difficult to separate the effects of the trivial geometry from that of the dynamics.

Figure 2 shows a typical fast trajectory with C = -2.98. The probe reaches the moon in about 36 hrs, during which time the moon has traveled 20°. The four lines of Fig.2

represent the same orbit in three different coordinate systems. The straight line with a slight curvature at its left end refers the motion to a fixed, geocentric system where the moon initially is at (-1.0,0.0). The solid curve, connecting the points E and M shows the trajectory in the rotating system. This is the way the orbit looks to an observer moving with the earth-moon system. The dots represent the two branches of the trajectory in the rotating, regularized (w) system. Mid-course times are denoted in hours along all four curves to give an idea of how the points are transformed in the coordinate systems.

As the energy is further decreased by increasing C, the space probe takes longer to reach the lunar distance, and so the trajectory must point in a direction farther ahead of the moon. This way the curve is continued up to where an orbit is obtained corresponding to a point in the vicinity of $C = 2$, $\theta = 60^o$. Near the top of the curve of Fig. 1 are such orbits, one of which is illustrated in Fig. 3. This low-energy orbit has just enough energy to travel out far enough so that the moon's attraction can pull the probe into a collision. For C much larger than 2, the probe cannot sufficiently overcome the attraction of the earth, so it falls back toward it without reaching the center of the moon.

It must be pointed out that this does not mean that a probe with $C > 2$ cannot go from the center of the earth to the center of the moon; rather, it means that such an orbit is not contained in our group of trajectories. In fact, an analysis of the zero velocity curves for $\mu = 1/82.45$ shows that if $C < 3.2$ orbits connecting the earth and the moon do exist. However, as Egorov[6] points out, the probe may have to make several hundred revolutions around the earth, in a highly eccentric ellipse with its apogee distance very slowly increased by the moon's perturbations, before reaching the moon.

The rest of the curve of Fig. 1 ($60^o < \theta \leq 360^o$) represents trajectories on which the space probe, still traveling essentially on geocentric, rectilinear ellipses, reaches the lunar radius before the moon has traveled θ^o. Since $C < 2$, the probe reaches an apogee greater than 1.0, and it collides with the moon as it falls back toward the earth. Figure 4 is a typical trajectory of this kind. With $C = 1.43$, the probe leaves the earth in a direction just opposite to that of the moon ($\theta = 180^{o}$), passes the moon's distance in about three days, and reaches apogee ($r = 1.431$) in about eight days. After this, it falls back toward the earth and reaches the center of the moon half a sidereal month after leaving the earth. Also, this figure shows, for the sophisticated reader who compares the two dotted trajectories in the regularized system, that since the Birkhoff transformation maps the region of infinity to the region midway between the two primaries ($\mu - \frac{1}{2}, 0$), one of the dotted curves goes near this point. For the other regularized trajectory, the region of infinity is unchanged by the transformation.

To the right of the point corresponding to the Fig. 4, trajectories on the C (θ) curve are higher energy orbits with lower values of C. These have greater apogee distances and must therefore start in a direction still farther in front of the moon. At the end of this group is the trajectory that starts the probe heading right for the moon, but by the time it gets to the lunar distance, the moon has moved on about 25° so the probe reaches apogee ($r = 2.1$) and then falls back to hit the moon about one month after leaving the earth.

The trajectories discussed are by no means the only ones that pass through the centers of both the primaries, and they are only a small percentage of the entire family. When the curve of Fig. 1 is continued to the right ($\theta > 2\pi$), it branches, and the trajectories for the approximate range $360^{o} < \theta < 420^{o}$ encounter strong perturbations from the moon as they pass it on the way out to apogee. If these trajectories are allowed to collide with the moon on the way out, the curve shown on Fig. 1 is, of course, just repeated. For $\theta \gg 2\pi$ there exist orbits that are associated with several revolutions of the moon around the earth before collision and also orbits that are characterized by several close encounters of the probe with the moon or the earth before collision. Discussion of these very interesting orbits and of the serious problem of the family of trajectories with consecutive collisions, the question of the totality of such orbits, will be presented in a forthcoming paper.

VI. Two-Body Approximations

Since the mass parameter in the earth-to-moon restricted problem is very small, Keplerian orbits can be expected to approximate, with reasonable accuracy, most of the trajectories presented on Fig. 1.

For straight-line orbits, we have [Eq. (4)] $\dot{r}^2 = 2/r - C$ from which we obtain by integration

$$t = C^{-3/2}\left[2\alpha - \sin 2\alpha\right]_{\alpha=\alpha_0}^{\pi/2}$$

Here t is the travel time from $r = r_0$ to apogee ($r = 2a = 2/C$) and

$$\alpha_0 = \arctan\left[Cr_0/(2-Cr_0)\right]^{1/2}$$

Neglecting the mass of the moon ($\mu = 0$), and selecting the firing angle such that the preceding evaluated travel time corresponds to the motion of the moon, results in an approximation to points on Fig. 1. Care must be taken, of course, of the two basically different kinds of orbits previously mentioned.

The time to travel from $r = 0$ to $r = 1$, on either kind of trajectory, is given by the sum and difference of

$$t_1 \text{ (from } r = 0 \text{ to } r = 2a) = \pi C^{-3/2}$$

and

$$t_2 \text{ (from } r = 2a \text{ to } r = 1) = 2C^{-3/2}\left[\pi/2 - \alpha_1 + \tfrac{1}{2}\sin 2\alpha_1\right]$$

where

$$\tan \alpha_1 = \left[C/(2-C)\right]^{1/2}$$

Hence the total time (in units of a radian of the moon's motion) to travel from $r = 0$ to $r = 1$ on the fast trajectory is

$$T_{fast} = t_1 - t_2 = 2C^{-3/2}\left[\alpha_1 - (2C-C^2)^{1/2}/2\right]$$

and the time to travel from $r = 0$ to $r = 2a$ and then back to $r = 1$ is

$$T_{lofted} = t_1 + t_2 = 2C^{-3/2}\left[\pi - \alpha_1 + (2C - C^2)^{1/2}/2\right]$$

When these travel times are computed as functions of C and plotted on a curve similar to that of Fig. 1, excellent agreements between the two-body approximations and the three-body integrations are obtained over the entire range $0 \leq \theta \leq 360^\circ$. We may conclude that a two-body analysis makes an excellent first approximation, good to about three figures, for the group of simple integrated, consecutive collision trajectories discussed in this paper.

References

[1] Levi-Civita, T., "Traiettorie singolari ed urti nel problema ristretto dei tre corpi," Ann. Mat. Ser. 3 9, 1-32 (1904).

[2] Sundman, C.F., Acta Math. 36, 105-192 (1912).

[3] Birkhoff, G.D., "The restricted problem of three bodies," Rend. Circ. Mat. Palermo 39, 1-70 (1915).

[4] Szebehely, V., "Perturbations of the regularized equations of the restricted problem of three bodies," Astron. J. 69, 309-315 (1964).

[5] Gapcynski, J.P. and Woolston, D.S., "Characteristics of three precision circumlunar trajectories for the year 1968," NASA TN D-1028 (March 1962).

[6] Egorov, V.A., "Certain problems of moon flight dynamics," The Russian Literature of Satellites (Intern. Phys. Index Inc., New York, 1958), Part I, pp. 123-132.

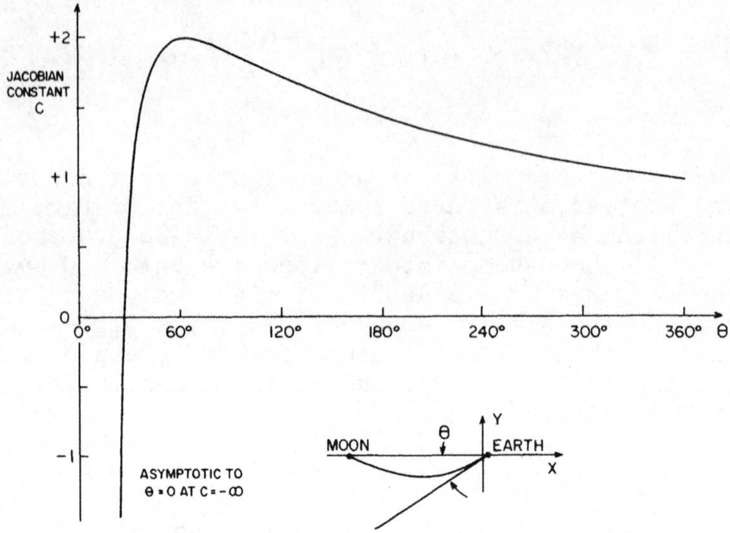

Fig. 1. The Jacobian constants and firing angles for a group of simple, consecutive collision earth-to-moon trajectories in the framework of the restricted problem of three bodies. μ=1/82.45.

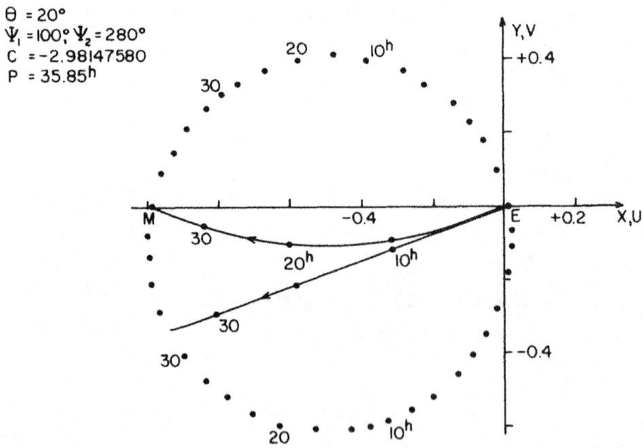

Fig. 2. A typical fast earth-to-moon trajectory with consecutive collisions. Solid curves are referred to the fixed and rotating physical (x,y) systems, dotted curves to the regularized (u,v) system. μ = 1/82.45.

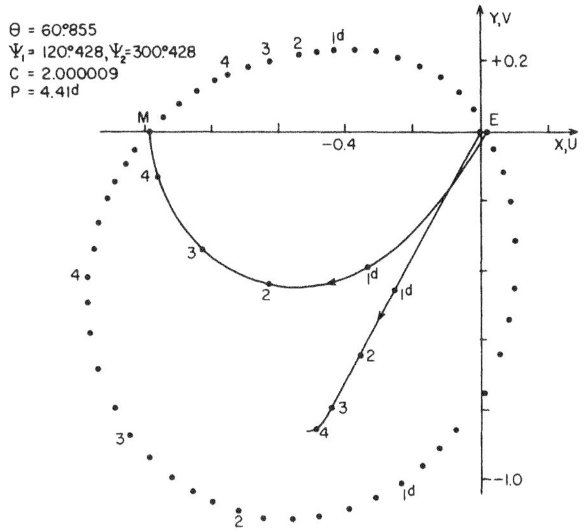

Fig. 3 A typical low energy earth-to-moon trajectory with consecutive collisions. Solid curves are referred to the fixed and rotating physical (x,y) systems, dotted curves to the regularized (u,v) system. $\mu = 1/82.45$.

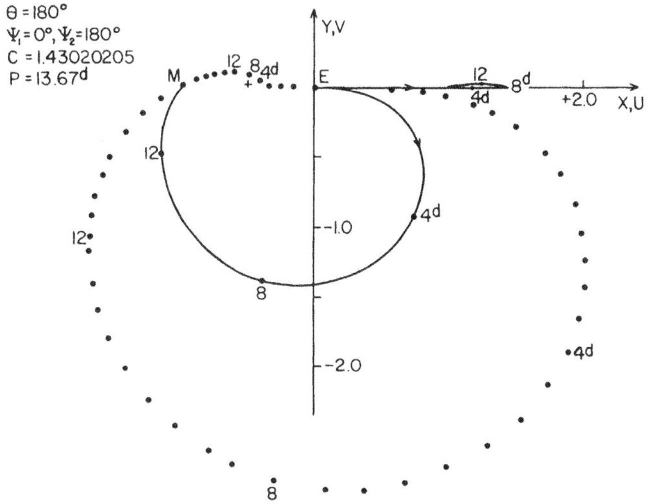

Fig. 4 A typical lofted earth-to-moon trajectory with consecutive collisions. Solid curves are referred to the fixed and rotating physical(x,y) systems, dotted curves to the regularized (u,v) system. $\mu = 1/82.45$.

TWO FIXED CENTER APPROXIMATIONS TO THE RESTRICTED PROBLEM

Mary Payne*

Republic Aviation Corporation, Farmingdale, N. Y.

Abstract

Two perturbation theories of the two-fixed-center problem in three dimensions are developed. In one of these theories perturbation equations are derived for the initial conditions in a coordinate system rotating about an accelerated origin. The origin is to be selected in such a way as to minimize the effects of the nonintegrable terms in the perturbation equations. Several procedures for the determination of the origin are discussed. The other perturbation theory is formulated in an inertial coordinate system. Again, perturbation equations are derived for the initial conditions, and fixed positions for the earth and moon are selected so as to reduce the effect of the nonintegrable terms in the perturbation equations. The results of some numerical calculations are presented showing the range of validity of the theory for typical initial conditions at launch from the earth, at closest approach to the moon, and at various midcourse positions along both the earth-to-moon and moon-to-earth portions of a lunar trajectory.

Nomenclature†

P = momentum conjugate to R in the inertial system
R = position vector of vehicle relative to the earth-moon barycenter
R_1 = position vector of the vehicle relative to the earth

Presented as Preprint 63-390 at the AIAA Astrodynamics Conference, New Haven, Conn., August 19-21, 1963. This work was performed under contract number NAS 8-2605 with the Aero-Astrodynamics Laboratory of Marshall Space Flight Center, NASA. The author wishes to express her appreciation to Mr. William E. Miner of the Astrodynamics and Guidance Theory Division, MSFC, for his support and encouragement. She also wishes to thank her colleagues Dr. George Nomicos and Mr. Maxwell Eichenwald for their valuable assistance.

*Staff Engineer, Applied Mathematics

†Capital letters denote vectors, small letters their magnitudes, and bars over vectors denote vectors in a rotating system.

r_1 = distance of vehicle from earth
R_2 = position vector of the vehicle relative to the moon
r_2 = distance of vehicle from moon
μ = gravitational constant times mass of the earth
μ' = gravitational constant times mass of the moon
L = position vector of moon relative to the earth
ℓ = distance of moon from earth
L = velocity vector of the moon relative to the earth
Ω = angular velocity vector of the moon relative to the earth
ω = magnitude of Ω

Subscripts

E = Euler problem
R = restricted problem
o = initial value
f = final value

Introduction

The two-fixed-center problem was first discussed by Euler[1], with the explicit intention of developing a perturbation theory for the restricted problem. Euler demonstrated integrability of the two-fixed-center problem, but did not develop its solution sufficiently to actually work out the perturbation theory. Legendre[2] obtained a fairly complete solution of the two-dimensional two-fixed-center problem in terms of elliptic functions. A detailed three-dimensional solution in a form suitable for computation may be found in references 3 and 4. More recently, Charlier[5] undertook the development of a perturbation theory for the two-dimensional restricted problem based on the action and angle variables for the two-fixed-center problem. This effort, too, is incomplete because of the extreme complexity of the problem. The difficulties of this approach are discussed for the three-dimensional problem in reference 6.

In this report, two general methods of obtaining approximations to the three-dimensional restricted problem in terms of the two-fixed-center problem will be discussed in detail. The first method is based on a formulation of the restricted problem in a rotating coordinate system and the second on a formulation in an inertial system. In both methods, perturbation equations are obtained for the initial conditions of the two-fixed-center problem regarded as osculating time varying parameters for the restricted problem. Both of these methods represent generalizations of a method developed by Arenstorf[7] for treating the two-dimensional restricted problem in a coordinate system rotating about the barycenter of the earth and moon.

The formulation in the rotating system involves the selection of four scalar parameters in such a way as to reduce the effects of the nonintegrable terms in the perturbation equations. Three of these parameters define the origin, to be regarded as the center of rotation, of the rotating system. The fourth allows part of one of the integrable terms to be used to reduce the effect of some of the nonintegrable terms. A method for the determination of these four parameters is presented, and a set of osculating initial conditions is obtained by an approximate integration of the perturbation equations. In addition to this set, three other sets are obtained by variations in the values of these parameters. In all of the methods developed the center of rotation is close to the center of the earth if the portion of the restricted orbit to be approximated has a close approach to the earth and no close approach to the moon. The center of rotation is close to the moon if the portion of the restricted orbit has a close approach to the moon and not to the earth. For midcourse portions, the center of rotation is somewhere between the earth and the moon. No attempt has so far been made to extend the theory to the approximation of portions containing close approaches to both the earth and the moon.

The formulation in the inertial system makes use of fictitious fixed positions for the earth and moon, so selected as to reduce the effect of the nonintegrable terms in the perturbation equations. Two sets of formulas result which differ in the approximations used in the integration of the perturbation equations.

Altogether then, six schemes are developed for approximating the restricted problem by the two-fixed-center problem. These schemes have been tested numerically for various portions of a typical lunar trajectory obtained by numerical integration. Some results of this numerical comparison are presented, following the analytical treatment. The comparison shows clearly that the formulations in the rotating system are superior and the reasons for this are discussed in the last section. Some possible improvements in the theory and its limitations are also discussed.

Theory for the Rotating System

Derivation of the Perturbation Equations

The equations of motion of the restricted problem in an inertial system with origin at the barycenter are

$$\ddot{R} = -\mu\,(R_1/r_1^{\,3}) - \mu'\,(R_2/r_2^{\,3}) \tag{1}$$

Consider a point A defined by

$$A = \alpha L + \beta \dot{\Omega} + \gamma \dot{L} \qquad A_1 = \alpha L + \gamma \dot{L} \tag{2}$$

where L and $\dot{L}$ are position and velocity vectors of the moon relative to the earth. Since the earth and the moon are assumed moving in circular orbits about the barycenter under their mutual gravitation attraction, L, A, A_1 and Ω satisfy the relations,

$$\ddot{L} = -(\mu + \mu')(L/\ell^3) = \Omega x (\Omega x L) \qquad \dot{L} = \Omega x L$$
$$\Omega = L x \dot{L}/\ell^2 \qquad \ddot{A} = \Omega x (\Omega x A) = -\omega^2 A_1 \tag{3}$$

The point A thus rotates about the barycenter with the earth and the moon as shown in Fig. 1. The location of the barycenter is not realistic; actually it is 3/4 of an earth radius from the center of the earth. In Figs. 2 and 3 coordinate systems with origin at the earth and at A are shown with the x-y plane being that of the moon's motion. In these figures the axes have fixed directions in inertial space, but since the origins are accelerated the coordinate systems are not inertial. Of the earth, moon, barycenter and A, only that one selected as the origin is at reast in such a coordinate system; the rest move in circles with angular velocity Ω about the origin. Also shown in these figures is a typical lunar trajectory as it might appear in such a coordinate system. In Fig. 4 this trajectory is shown in a coordinate system rotating with angular velocity Ω. In such a system the earth, moon, barycenter and A are all fixed and the coordinate systems defined with each as origin are equivalent.

The equations of motion for the restricted problem in an accelerated coordinate system with origin at A, but with axes always parallel to those of the inertial system, are

$$\ddot{R}_A = -\mu (R_1/r_1^3) - \mu' (R_2/r_2^3) - \ddot{A} \tag{4}$$

since

$$R = A + R_A \qquad \dot{R} = \dot{A} + \dot{R}_A \qquad \ddot{R} = \ddot{A} + \ddot{R}_A \tag{5}$$

and finally in a coordinate system rotating about A with angular velocity Ω, the equation of motion becomes

$$\ddot{\bar{R}}_A = -\mu (\bar{R}_1/r_1^3) - \mu' (\bar{R}_2/r_2^3) - \ddot{\bar{A}} - \Omega x (\Omega x \bar{R}_A) - 2(\Omega x \dot{\bar{R}}_A) \tag{6}$$

where bars denote vectors in the rotating system. We assume that, at time t = 0, the axes of the rotating system are parallel to those in the inertial system, so that the constant vectors $\bar{A}$, $\dot{\bar{A}}$, and $\ddot{\bar{A}}$ satisfy

$$\bar{A} = A(0) \qquad \dot{\bar{A}} = \dot{A}(0) \qquad \ddot{\bar{A}} = -\omega^2 \bar{A}_1 = -\omega^2 A_1(0) \tag{7}$$

It is readily verified that the Hamiltonian

$$H_A = (1/2)\bar{P}^2_A - (\mu/r_1) - (\mu'/r_2) - \omega^2 \bar{R}_A \cdot A_1 - \Omega \cdot \bar{R}_A \times \bar{P}_A \tag{8}$$

is a Hamiltonian for the problem represented by Eq. (6) with

$$\dot{\bar{R}}_A = \text{grad}_{\bar{P}_A} H_A = \bar{P}_A - \Omega \times \bar{R}_A \tag{9}$$

and

$$\dot{\bar{P}}_A = \ddot{\bar{R}}_A + \Omega \times \dot{\bar{R}}_A = -\text{grad}_{\bar{R}_A} H_A$$

$$= -\mu (R_1/r_1^3) - \mu' (R_2/r_2^3) + \omega^2 \bar{A}_1 - \Omega \times \bar{P}_A$$

$$= \mu(R_1/r_1^3) - \mu' (R_2/r_2^3) + \omega^2 \bar{A}_1 - \Omega \times \dot{\bar{R}}_A - \Omega \times (\Omega \times \bar{R}_A) \tag{10}$$

which reduces immediately to Eq. (6). A word on the relation between position in the rotating system $\bar{R}_A$ and its conjugate momentum $\bar{P}_A$ and the position R_A and velocity $\dot{R}_A$ in the nonrotating system is necessary for the interpretation of results to be obtained later. Since the rotating and nonrotating systems are assumed coincident at $t=0$,

$$\bar{R}_{A0} = R_{A0} \tag{11}$$
$$\bar{P}_{A0} = \dot{\bar{R}}_{A0} + \Omega \times \bar{R}_{A0} = \dot{R}_{A0}$$

are vector equations that are valid component by component, and since $\dot{\bar{R}}_{A0}$ is the velocity relative to A in the rotating system whereas $\Omega \times \bar{R}_{A0}$ is the velocity due to the rotation of the system, it is seen that the initial value of the momentum conjugate to $\bar{R}_A$ is just the velocity in the nonrotating system. The same statements hold for time t also, except that, to get component agreement, a rotation through ωt is necessary - that is, at time t,

$$R_A = M^{-1}(t)\, \bar{R}_A$$
$$\dot{R}_A = M^{-1}(t)\, \bar{P}_A = M^{-1}(t)\, (\dot{\bar{R}}_A + \Omega \times \bar{R}_A) \tag{12}$$

where $M^{-1}(t)$ may be regarded either as a rotation of the axes of the rotating system through an angle $-\omega t$ or as a rotation of $\bar{R}_A$ and $\bar{P}_A$ relative to the rotating axes through an angle ωt, both rotations about the vector Ω that is the same in both systems.

The Hamiltonian H_A may be written as the sum of

$$H_E = (1/2)\,\overline{P}_A^{\,2} - (\mu/r_1) - (\mu'/r_2) \tag{13}$$

the Hamiltonian for the Euler problem with Hamilton equation

$$\begin{aligned} \dot{\overline{R}}_{AE} &= \operatorname{grad}_{\overline{P}_{AE}} H_E = \overline{P}_{AE} \\ \dot{\overline{P}}_{AE} &= \ddot{\overline{R}}_{AE} = -\mu\,(R_1/r_1^{\,3}) - \mu'\,(R_2/r_2^{\,3}) \end{aligned} \tag{14}$$

and a perturbation

$$H_1 = -\omega^2\,\overline{R}_A \cdot \overline{A}_1 - \Omega \cdot \overline{R}_A \times \overline{P}_A \tag{15}$$

where the subscript E in Eqs. (14) refers to the functional forms for $\overline{R}_A$ and $\overline{P}_A$ obtained by solving Eqs. (14).

A solution of the restricted problem with Hamiltonian given by Eq. (8) and Hamiltonian Eqs. (9) and (10) is now sought in the functional form of the solution of the Euler problem with time-varying initial conditions. That is, one seeks the solution of the restricted problem, denoted by a subscript R, in the form

$$\begin{aligned} \overline{R}_{AR}\,(\overline{R}_{AR0}, \overline{P}_{AR0}, t) &= \overline{R}_{AE}\left(\overline{R}_{AE0}(t), \overline{P}_{AE0}(t), t\right) \\ \overline{P}_{AR}\,(\overline{R}_{AR0}, \overline{P}_{AR0}, t) &= \overline{P}_{AE}\left(\overline{R}_{AE0}(t), \overline{P}_{AE0}(t), t\right) \end{aligned} \tag{16}$$

with initial conditions for the restricted and Euler problems satisfying the relations

$$\begin{aligned} &\overline{R}_{AR}(\overline{R}_{AR0}, \overline{P}_{AR0}, 0) = \overline{R}_{AR0} = \overline{R}_{AE}\left(\overline{R}_{AE0}(0), \overline{P}_{AE0}(0), 0\right) = \\ &\quad \overline{R}_{AE}(0) \\ &\overline{P}_{AR}(\overline{R}_{AR0}, \overline{P}_{AR0}, 0) = \overline{P}_{AR0} = \overline{P}_{AE}\left(\overline{R}_{AE0}(0), P_{AE0}(0), 0\right) = \\ &\quad \overline{P}_{AE0}(0) \end{aligned} \tag{17}$$

It has been shown by Arenstorf[7] that the functions $\overline{R}_{AE0}(t)$ and $\overline{P}_{AE0}(t)$, necessary for the validity of Eq. (16), satisfy the differential equations

$$\begin{aligned} (d/dt)\,\overline{R}_{AE0}(t) &= \operatorname{grad}_{\overline{P}_{AE0}} \overline{H}_1 \\ (d/dt)\,\overline{P}_{AE0}(t) &= -\operatorname{grad}_{\overline{R}_{AE0}} \overline{H}_1 \end{aligned} \tag{18}$$

where

$$\overline{H}_1 = \overline{H}_1\left(\overline{R}_{AE0}(t), \overline{P}_{AE0}(t), t\right) \tag{19}$$

is obtained by substitution of $\overline{R}_{AE}\left(\overline{R}_{AE0}(t), \overline{P}_{AE0}(t), t\right)$ and $\overline{P}_{AE}\left(\overline{R}_{AE0}(t), \overline{P}_{AE0}(t), t\right)$ for $\overline{R}_A$ and $\overline{P}_A$ in H_1 [given by Eq. (15)]. To actually carry out the substitution using the solution of the Euler problem (which is known in closed form) and then to compute the gradients required in Eq. (18) would be very complex because of the extreme complexity of the closed-form solution. Even if this could be carried out, the integration of the resulting, highly nonlinear equations in $\overline{R}_{AE0}(t)$ and $\overline{P}_{AE0}(t)$ would be very difficult. Furthermore, any approximation method for integration of perturbation equations for initial conditions must be developed with great care to avoid the introduction of troublesome secular terms, which increase in order with higher order approximations.

In view of this last fundamental difficulty, only a first approximation will be attempted. This approximation will lead to some integrable terms in the perturbation equations and the point A will be selected in such a way as to reduce the effect of the nonintegrable terms, which will then be ignored. The resulting expressions for the time variation in the initial conditions and hence the solution of the restricted problem represented by Eq. (16) will thus have limited validity in time. The hardest part of the problem will be in obtaining an estimate for duration of validity. Although this might appear to restrict considerably the application of the theory, it should nevertheless be noted that from the solutions of a sequence of two-fixed-center problems, each valid for a certain time, the solution of the restricted problem may be constructed solely in terms of closed form calculations without the use of numerical integration.

Explicit Form of the Perturbation Equations

To proceed with the approximation $\overline{H}_1$ is written in the form

$$\begin{aligned}\overline{H}_1 = &-\Omega\cdot \overline{R}_{AE0}\times\overline{P}_{AE0} - \omega^2\delta\overline{R}_{AE0}\cdot\overline{A}_1 - \omega^2(1-\delta)\overline{R}_{AE}\cdot\overline{A}_1 - \\ &\int\left\{\omega^2\delta\overline{P}_{AE}\cdot\overline{A} - \Omega\cdot\overline{R}_{AE}\times\left(\mu(R_1/r_1^3) - \mu'(R_2/r_2^3)\right)\right\}dt\end{aligned} \tag{20}$$

where the integral is obtained by time differentiation of $(-\omega^2\delta\overline{R}_{AE}\cdot\overline{A}_1 - \Omega\cdot\overline{R}_{AE}\times\overline{P}_{AE})$ and use of the Hamilton equations (14) for the Euler problem. The first two terms of $\overline{H}_1$ will be shown to lead to integrable terms in the perturbation equations (18) for the initial conditions. The factor δ permits part of the $R_{AE}\cdot A_1$ term to appear with the integrable terms and part with the nonintegrable terms. This second part helps to reduce the effect of the other non-integrable terms. The third term and the integral are not written explicitly in terms of initial conditions. It is these terms for which an effort at minimization will be made by proper selection of the factor δ and the point A. To see how this may be done, one now takes the gradients of $\overline{H}_1$ with respect to $\overline{R}_{AE0}$ and $\overline{P}_{AE0}$ to obtain the perturbation equations. The differentiation of the triple product in the integral is facilitated by noting that

$$\overline{R}_1 = \overline{R}_A + \overline{A} + \mu' \overline{L}/(\mu + \mu') \qquad \overline{R}_2 = \overline{R}_A + \overline{A} - \mu \overline{L}/(\mu + \mu') \tag{21}$$

so that

$$\overline{R}_A \times \overline{R}_1 = \overline{R}_A \times \left(\overline{A} + \frac{\mu'}{\mu + \mu'}, \overline{L}\right) \qquad \overline{R}_A \times \overline{R}_2 = \overline{R}_A \times \left(\overline{A} - \frac{\mu}{\mu + \mu'}, \overline{L}\right) \tag{22}$$

The perturbation equations for the time derivatives of $\overline{R}_{AE0}(t)$ and $\overline{P}_{AE0}(t)$ are readily verified to be

$$\begin{aligned}(d/dt)\,\overline{R}_{AE0}(t) &= \text{grad}_{PAE0}\overline{H}_1 = -\overline{\Omega} \times \overline{R}_{AE0} - \omega^2(1-\delta)\Psi_{RP}\overline{A}_1 - \\ &\int \left[\omega^2 \delta \Psi_{PP}\overline{A}_1 - \Psi_{RP}(\overline{\Omega} \times M - \omega Q)\right] dt\end{aligned} \tag{23}$$

and

$$\begin{aligned}(d/dt)\,\overline{P}_{AE0}(t) &= -\text{grad}_{RAE0}\,\overline{H}_1 = -\overline{\Omega} \times \overline{P}_{AE0} + \omega^2 \delta \overline{A}_1 + \\ &\omega^2(1-\delta)\Psi_{RR}\overline{A}_1 + \int \left[\omega^2 \delta \Psi_{PR}\overline{A}_1 - \Psi_{RR}(\overline{\Omega} \times M - \omega Q)\right] dt\end{aligned} \tag{24}$$

where the M and Q are vectors given by

$$\begin{aligned} M &= \mu\left(\overline{A} + \frac{\mu'}{\mu + \mu'}, \overline{L}\right)/r_1^{\,3} + \mu'\left(\overline{A} - \frac{\mu}{\mu + \mu'}, \overline{L}\right)/r_2^{\,3} \\ Q &= \frac{3}{\omega}\left\{\mu\left[\overline{\Omega} \cdot \overline{R}_{AE} \times \left(\overline{A} + \frac{\mu'}{\mu + \mu'}, \overline{L}\right)\overline{R}_1/r_1^{\,5}\right] + \right. \\ &\quad \left. \mu'\left[\overline{\Omega} \cdot \overline{R}_{AE} \times \left(\overline{A} - \frac{\mu}{\mu + \mu'}, \overline{L}\right)\overline{R}_2/r_2^{\,5}\right]\right\} \end{aligned} \tag{25}$$

These vectors are so defined that they have the same dimension. The Ψ's are matrices given by

$$\begin{aligned} \Psi_{RR} &= \left(\partial \overline{R}_{AEj}/\partial \overline{R}_{AE0i}\right) \qquad & \Psi_{RP} &= \left(\partial \overline{R}_{AEj}/\partial \overline{P}_{AE0i}\right) \\ \Psi_{PR} &= \left(\partial \overline{P}_{AEj}/\partial \overline{R}_{AE0i}\right) \qquad & \Psi_{PP} &= \left(\partial \overline{P}_{AEj}/\partial \overline{P}_{AE0i}\right) \end{aligned} \tag{26}$$

with the ith row and j^{th} column containing the derivative of the jth component of the time varying vector in the numerator with respect to the ith component of the initial value vector in the denominator evaluated at $R_{AE0}(t)$ and $P_{AE0}(t)$. It may be noted that the transposes of these matrices constitute the transition matrix for the Euler problem with the transposes of the first two matrices forming the top three rows and the transposes of the last two matrices forming the bottom three rows.

Determination of the Origin A and the Parameter δ

The first term in the right hand side of Eq. (23) and the first two terms on the right side of Eq. (24) depend only on the initial values $R_{AE0}(t)$ and $P_{AE0}(t)$, and if these were the only terms present, Eqs. (23) and (24) would be integrable. The remaining terms all involve components of the transition matrix for the Euler problem and no attempt will be made to include them in the integration. Instead, methods will be sought for making them small, and this will be done by seeking an approximate minimization of the vectors on which the matrices operate. These vectors appear in both equations as follows:

$$N_1 = \omega^2(1-\delta)\bar{A}_1 \qquad \text{outside the integrals}$$
$$N_2 = \omega^2\delta\,\bar{A}_1 \qquad \text{inside the integrals} \qquad (27)$$

together with M and Q defined in Eqs. (25), which appear inside the integrals. It will be noted that all these vectors have the same dimension. The vectors M and Q are functions of time. Since however, they have, effectively, the cubes of r_1 and r_2 in the denominator, it is clear that they are large only for brief periods of time at approach to the earth or to the moon closer than a few earth radii.

As a first trial at minimization, δ and A were sought such that the scalar σ,

$$\sigma = N_1^{\;2} + N_2^{\;2} + M_o^{\;2} + M_f^{\;2} \qquad (28)$$

is minimized, where M_o and M_f are computed from initial and anticipated final conditions, respectively. The omission of Q is heuristically justified by an argument of the following type. Suppose the initial position is close to the earth and the final position close to the moon. Initally, the r_2 terms are small, so that to reduce the r_1 terms $\{\bar{A} + \mu'\,\bar{L}/(\mu+\mu')\}$ must nearly vanish in order to keep M_o small. It will then follow that Q_o is also small. Evidently, of course, such a procedure will mean that both M_f and Q_f will become more or less large depending on the final value of r_2. In effect, this will place a limitation on the duration of validity of the two fixed center approximation.

The minimization of Eq. (28) will now be carried out. Since M and Q are independent of δ, partial derivatives of σ with respect to δ involve only the N_1 and N_2 terms

$$\frac{\partial \sigma}{\partial \delta} = N_1 \cdot \frac{\partial N_1}{\partial \delta} + N_2 \cdot \frac{\partial N_2}{\partial \delta} = \omega^4 \bar{A}_1^{\,2}\left(2\delta - 2(1-\delta)\right) \tag{29}$$

which vanishes for $\delta = \frac{1}{2}$. It now remains to minimize

$$\sigma_1 = \frac{1}{2}\omega^4 \bar{A}_1^{\,2} + M_o^{\,2} + M_f^{\,2} \tag{30}$$

with respect to α, β, and γ. That is, the equations

$$\frac{\partial \sigma_1}{\partial x} = \omega^4 \bar{A}_1 \cdot \frac{\partial \bar{A}_1}{\partial x} + M_o \cdot \frac{\partial M_o}{\partial x} + 2M_f \cdot \frac{\partial M_f}{\partial x} = 0 \tag{31}$$

where x denotes α, β, and γ, must be solved for α, β, and γ. Making use of Eq. (2) for $\bar{A}$, Eq. (3) for $\bar{A}_1$ and Eq. (25) for M, one obtains the following partial derivatives:

$$\frac{\partial \bar{A}_1}{\partial \alpha} = \frac{\partial \bar{A}}{\partial \alpha} = \bar{L}, \quad \frac{\partial \bar{A}_1}{\partial \beta} = 0, \quad \frac{\partial \bar{A}}{\partial \beta} = \bar{\Omega}, \quad \frac{\partial A_1}{\partial \gamma} = \frac{\partial \bar{A}}{\partial \gamma} = \dot{\bar{L}} \tag{32}$$

$$\frac{\partial M}{\partial \alpha} = \bar{L}\left(\frac{\mu}{r_1^3} + \frac{\mu'}{r_2^3}\right), \quad \frac{\partial M}{\partial \beta} = \bar{\Omega}\left(\frac{\mu}{r_1^3} + \frac{\mu'}{r_2^3}\right), \quad \frac{\partial M}{\partial \gamma} = \dot{\bar{L}}\left(\frac{\mu}{r_1^3} + \frac{\mu'}{r_2^3}\right) \tag{33}$$

Substitution in Eq. (31) for x = β and γ lead to

$$\beta = \gamma = 0 \tag{34}$$

while for x = α, one obtains

$$\alpha = -\frac{\mu\mu'}{\mu + \mu'} \, \frac{\sum\limits_{i=0,f} \left(\frac{1}{r_{1i}^3} - \frac{1}{r_{2i}^3}\right)\left(\frac{\mu}{r_{1i}^3} + \frac{\mu'}{r_{2i}^3}\right)}{\left[\sum\limits_{i=0,f}\left(\frac{\mu}{r_{1i}^3} + \frac{\mu'}{r_{2i}^3}\right)^2\right] + \frac{1}{2}\omega^4} \tag{35}$$

Some comments on the value of α may be made. If a close approach only to the earth or to the moon is made, that is, if either r_{10} or r_{1f} is close to unity while r_{20} and r_{2f} are both large or vice versa, it is readily seen that

$$\alpha \sim -\mu'/(\mu + \mu') \qquad \text{or} \qquad \alpha \sim \mu/(\mu + \mu') \tag{36}$$

corresponding, respectively, to origin at the earth and the moon. If a midcourse portion of the trajectory is to be approximated so that none of the r's is near unity, α will be somewhere between these extreme values, that is, the origin will lie on the line of centers between the earth and the moon. The origin is at the barycenter for $\alpha = 0$.

Integration of the Perturbation Equations

Once the point A has been determined, the nonintegrable terms in the perturbation equations (23) and (24) will be ignored and the equations to be integrated are

$$(d/dt)\,\bar{R}_{AE0}(t) = -\,\bar{\Omega} \times \bar{R}_{AE0}(t) \tag{37}$$

$$\begin{aligned}(d/dt)\,\bar{P}_{AE0}(t) &= -\,\bar{\Omega} \times \bar{P}_{AE0}(t) + \omega^2\,\delta\,\bar{A} \\ &= -\,\bar{\Omega} \times \left(\bar{P}_{AE0}(t) + \delta\,\bar{\Omega} \times \bar{A}\right)\end{aligned}$$

where use has been made of the relation

$$\bar{\Omega} \times (\bar{\Omega} \times \bar{A}) = -\,\omega^2\,\bar{A} \tag{38}$$

The integrals of these equations are, since $\delta\,\bar{\Omega} \times \bar{A}$ is a constant,

$$\begin{aligned}\bar{R}_{AE0}(t) &= M(t)\,\bar{R}_{AE0}(t) \\ \bar{P}_{AE0}(t) &= M(t)\left[\bar{P}_{AE0}(0) + \delta\,\bar{\Omega} \times \bar{A}\right] - \delta\,\bar{\Omega} \times \bar{A}\end{aligned} \tag{39}$$

where the matrix M(t) is a rotation matrix through an angle $-\omega t$ about the $\bar{\Omega}$ direction.

Referring back, now, to Eq. (16), it is seen that, in the rotating system, a solution to the restricted problem valid from the initial time zero to some time t, determined by how long the nonintegrable terms remain negligible, is obtained by substitution of the expressions (39) for $\bar{R}_{AE0}(t)$ and $\bar{P}_{AE0}(t)$ in terms of the two-fixed-center problem. This means that in order to construct the solution of the restricted problem in terms of that of the two-fixed-center problem, it is necessary, for each time t of interest, to compute initial conditions from Eqs. (39) and then obtain the solution, evaluated at the time t, of a two-fixed-center problem with these initial conditions. Thus, if n points on the restricted orbit are desired, n different two-fixed-center problems must be evaluated. This scheme is illustrated in Fig. 5 for the origin at the earth.

One other point should be mentioned. The initial value $\bar{P}_{AE0}(0)$ is to be thought of as given by $\overline{P_{AR0}}$, which in turn is determined by the first Hamilton equation (9) for the restricted problem evaluated at time $t = 0$:

$$\bar{P}_{AR0} = \dot{\bar{R}}_{A0} + \Omega \times \bar{R}_{A0} = \dot{R}_{A0} \tag{40}$$

where $\dot{R}_{A0}$ is the initial velocity in the nonrotating system, since the assumption has been made that the rotating and nonrotating systems have parallel axes at the initial time. Once $\bar{P}_{AE0}(0)$ has been determined $\bar{P}_{AE0}(t)$ is given by Eq. (39) and is to be interpreted as an initial velocity relative to $\bar{A}$ in the rotating system for the two-fixed-center problem, by virtue of the first of the Hamilton equations (14) for this problem. Since, in the rotating system, the earth and moon are fixed, the initial velocity $\bar{P}_{AE0}(t)$ is the same relative to any point in this system. The two-fixed-center solution obtained from this initial velocity $\bar{P}_{AE0}(t)$ and the initial position $\bar{R}_{AE0}(t)$ lead to position $\bar{R}_{AE}(t)$ and velocity $P_{AE}(t)$ for the two-fixed-center problem, which are to be interpreted as position $\bar{R}_{AR}(t)$ and momentum $\bar{P}_{AR}(t)$ for the restricted problem in the rotating system.

Theory for the Inertial System

Derivation and Integration of the Perturbation Equations

A direct approach to an approximation of the solution of the restricted problem by the two-fixed-center problem in an inertial coordinate system can be developed as follows. Recalling the equations of motion for the restricted problem in the inertial system with origin at the barycenter,

$$\ddot{R} = -\mu (R_1/r_1^3) - \mu' (R_2/r_2^3) \tag{1}$$

it is easily shown that the Hamiltonian is

$$H = (1/2)\, P^2 - (\mu/r_1) - (\mu'/r_2) \tag{41}$$

This Hamiltonian has an explicit time dependence since r_1 and r_2 are distances of the vehicle from the earth and moon which are assumed moving in known orbits about the barycenter. The momentum P conjugate to position R relative to the barycenter is just R, the velocity relative to the barycenter. The first Hamilton equation expresses this fact, and the second, together with the first, yields the equations of motion (1).

In this formulation two fixed points are selected for a fixed earth and a fixed moon. The selection of these points is to be made so as to minimize the nonintegrable portion of the perturbation equations. Thus, denoting positions relative to these fixed points by stars, the equations of motion are

$$\ddot{R} = -\mu \frac{R_1^*}{r_1^{*3}} - \mu' \frac{R_2^*}{r_2^{*3}} + \mu\left(\frac{R_1^*}{r_1^{*3}} - \frac{R_1}{r_1^3}\right) + \mu'\left(\frac{R_2^*}{r_2^{*3}} - \frac{R_2}{r_2^3}\right) \tag{42}$$

and the Hamiltonian is

$$H = \frac{1}{2}P^2 - \frac{\mu}{r^*_1} - \frac{\mu'}{r^*_2} + \mu\left(\frac{1}{r^*_1} - \frac{1}{r_1}\right) + \mu'\left(\frac{1}{r^*_2} - \frac{1}{r_2}\right) \tag{43}$$

The Hamiltonian can be expressed as the sum of two terms, the Hamiltonian H_E for the two-fixed-center problem and the perturbation Hamiltonian H_1

$$H_E = \frac{1}{2}P^2 - \frac{\mu}{r^*_1} - \frac{\mu'}{r^*_2}, \quad H_1 = \mu\left(\frac{1}{r^*_1} - \frac{1}{r_1}\right) + \mu'\left(\frac{1}{r^*_2} - \frac{1}{r_2}\right) \tag{44}$$

The perturbation Hamiltonian may be written in the form

$$H_1 = \mu\left(\frac{1}{r^*_{10}} - \frac{1}{r_{10}}\right) + \mu'\left(\frac{1}{r^*_{20}} - \frac{1}{r_{20}}\right) \tag{45}$$

$$- \int\left\{\mu\left[\frac{R^*_1 \cdot \dot{R}^*_1}{r^{*\,3}_1} - \frac{R_1 \cdot \dot{R}_1}{r_1^3}\right] + \mu'\left[\frac{R^*_2 \cdot \dot{R}^*_2}{r^{*\,3}_2} - \frac{R_2 \cdot \dot{R}_2}{r_2^3}\right]\right\} dt$$

Perturbation equations for the initial conditions may now be written as

$$\frac{d}{dt} R_o(t) = \text{grad}_{Po}\, H_1 = 0 - \text{grad}_{Po} \int\{\cdots\} dt \tag{46}$$

$$\frac{d}{dt} P_o(t) = -\text{grad}_{Ro}\, H_1 = \mu\left(\frac{R_{10}^*}{r_{10}^{*3}} - \frac{R_{10}}{r_{10}^3}\right) + \mu'\left(\frac{R_{20}^*}{r_{20}^{*3}} - \frac{R_{20}}{r_{20}^3}\right) +$$

$$\text{grad}_{Ro} \int\{\cdots\} dt$$

If the terms involving the integrals are ignored in the perturbation equations, one obtains

$$R_o(t) = R_o(0) \tag{47}$$

$$P_o(t) = P_o(0) + (t - t_o)\left[\mu\left(\frac{R_{10}^*}{r_{10}^{*3}} - \frac{R_{10}}{r_{10}^{*3}}\right) + \mu'\left(\frac{R_{20}^*}{r_{20}^{*3}} - \frac{R_{20}}{r_{20}^3}\right)\right]$$

since the first of these equations implies also

$$R_{io}^*(t) = R_{io}^*(0) \qquad R_{io}(t) = R_{io}(0) \qquad i = 1, 2 \tag{48}$$

Selection of Fixed Positions for Earth and Moon

It is not easy to see how the fixed positions for the earth and moon should be selected so as to minimize the contribution of the integrals to the perturbation equations (46). Examination of the equations of motion (42), however, suggests that two cases should be considered as follows: (1) motion from earth towards moon: fix earth in its initial and moon in its final position; and (2) motion from moon towards earth: fix earth in its final and moon in its inital position. The initial conditions for the two-fixed-center problem will then be determined by the condition that initial position relative to the barycenter is unmodified and initial velocity relative to the barycenter be determined from Eq. (47), with momentum identified with velocity. The solution $R_E(R_o, P_o(t), t)$ and $P_E(R_o, P_o(t), t)$ for the Euler problem will then be related to that for the restricted problem by

$$R_R(R_o, P_o, t) = R_E(R_o, P_o(t), t) \qquad P_R(R_o, P_o, t) = P_E(R_o, P_o(t), t) \quad (49)$$

where R_R and P_R are to be interpreted as position and velocity relative to the barycenter at time t.

Results of Numerical Comparisons

Two methods of approximating the restricted problem by the two-fixed-center problem have been obtained in the preceding two sections. In addition to these methods, three others, based on the formulation in the rotating system, have been considered. These last three methods are defined as follows:

1) The center of rotation is taken at the center of the moon if the portion of a lunar trajectory to be approximated lies in "moon reference", that is, if all points on this portion are within about 9 earth radii of the moon. For portions of the trajectory outside moon reference, the center of rotation is taken at the earth. The method has not been applied to portions of a lunar trajectory crossing the moon's sphere of influence. Thus, the values of α used for method 1,

$$\begin{aligned} &\alpha = -\mu'/(\mu+\mu') \quad &&\text{earth reference} \\ &\alpha = \mu/(\mu+\mu') \quad &&\text{moon reference} \end{aligned} \qquad (50)$$

are the two extreme values noted in the discussion following Eq. (35) for α. In addition, the parameter δ is taken to be zero.

2) This method uses the value of α determined by Eq. (35). The value of δ is taken to be one.

3) This method also uses the value of α given by Eq. (35), and δ is set equal to zero.

The two methods already derived are identified by

4) The method in the rotating system.

5) The method in the inertial system.

6) Finally, a sixth method was tried in which the effect of the perturbation Hamiltonian in the inertial formulation was neglected. That is, the initial conditions for the two-fixed-center problem are to be just the initial position and velocity relative to the barycenter.

The comparison of the effectiveness of these methods was carried out as follows. First, a typical lunar trajectory was integrated with the effects of moving earth and moon included, but with all perturbations due to sun, other planets, oblateness, etc. eliminated from the program. This trajectory had the character shown in Figs. 2 and 4. The integration was carried out by the Republic Interplanetary Program using the Encke method. In this program the earth is used as origin in earth reference and the moon is the origin in moon reference. Various points on this typical lunar trajectory were taken as initial points and the two-fixed-center approximation was computed at various specified later times. This necessitated the transformation of the initial conditions associated with the various methods (relative to the origin A for the rotating formulations and relative to the barycenter for the inertial formulations) into equivalent initial conditions relative to the earth or moon for portions of the trajectory in earth and moon reference, respectively.

The base lunar trajectory started at time $t = 0$ from about 6590 km from the center of the earth, reached a perisel distance of about 4350 km at 71 hr and reached a perigee distance of 8174 km at 153.9 hr. The entry and exit from moon reference occurred at about 58.7 hr and 84.1 hr, respectively.

Tables 1-4 contain some typical results for the numerical calculations. Tables 1 and 4 are for the earth-reference portions of the trajectory on the first and last legs, respectively. Tables 2 and 3 are for moon reference portions approaching and receding from the moon, respectively. The left hand column contains the initial and final times for the portion of the trajectory to be approximated. These times define, respectively, the initial conditions to be used from the base trajectory and the terminal conditions to be met. The deviations Δx, Δy, and Δz in kilometers for the various methods are entered in columns headed by the corresponding number. These deviations represent the difference in the rectangular coordinates relative to the reference body, the values predicted by the various methods being subtracted from the values given by the base case. The column headed K, which appears in Tables 1 and 4 gives the deviations for the Kepler problem. The last column gives the value of α determined from Eq. (35) for use in methods 2, 3, and 4.

In Table 5 the x, y, and z coordinates of the vehicle relative to the reference body are given for the various times that appear in Tables 1-4. Also given are the distances of the vehicle from the reference body in earth radii. The distance of the earth from the moon is a little less than 60 earth radii.

Some general conclusions on the relative merits of these methods may be drawn. First, it may be noted that methods 1 and 3 are practically the same except for midcourse portions of the trajectory. The reason for this is that except for such portions the value of α is such that the origin is nearly at the earth for earth reference and nearly at the moon for moon reference.

The results of the numerical calculations may be summarized as follows. Method 4 is best for moon reference and for an earth-towards-moon portion of a trajectory starting near the earth. For midcourse portions of the trajectory in earth reference method 1 is, on the whole, slightly superior with methods 2, 3, or 4 nearly as good for certain cases, but not consistently so. Method 1 is also the best fro the earth-reference return to the earth with method 3 nearly identical in most cases. Methods 5 and 6 are decidedly inferior almost everywhere.

The Kepler problem is superior to all of these methods for short to medium range in the neighborhood of the earth and moon. It fails, however, for long range and midcourse portions of the trajectory.

The base trajectory was computed from a digital program using an ephemeris moon, rather than for the restricted problem. Subsequent tests have demonstrated that the deviations tabulated here are typical, and that conclusions on the validity of the theory and the relative merits of the various methods are unaffected by the particular model used for the restricted problem.

It may be noted that the "best" deviations in moon-reference are about 100 times as large as those in earth reference for comparable cases. That this result is not surprising may be seen from the following argument. The vectors A_1 M and Q contain a factor μ for A near the moon and a factor μ' for A near the earth. Thus the non-integrable terms have a factor μ' for the earth-reference cases and a factor μ for the moon reference cases. Thus one would expect the moon-reference deviations to be approximately in the ratio of $\mu/\mu' \sim 80$ to the earth reference deviations.

Conclusions

The results of the numerical comparison made in the previous section show that the formulation in a rotating system is best suited to the approximation of the restricted problem by the two-fixed-center problem. This is not really very surprising because in a rotating system the earth and moon are automatically fixed. This is achieved by introducing terms corresponding to the centrifugal and Coriolis accelerations, which are interpreted as perturbations on the two-fixed-center problem. In the inertial system, on the other hand, fixed positions for the earth and moon had to be selected more or less arbitrarily. As a consequence, the perturbations from the two fixed center problem so selected depends on this selection. Thus, approximations have been introduced before the problem of approximating the effect of the perturbations can even be considered. It would seem, therefore, that a rotating system, in which only the problem of how to treat the perturbations appears, should be the proper choice.

From the numerical results shown in the last section, it is evident that the problem of treating the perturbations is far from an easy one. None of the numerical results obtained can be regarded as satisfactory, or, in fact, as fulfilling the expectations that one might have for the theory. Nevertheless, there are a number of reasons for expecting that further development of the theory might lead to useful and interesting results.

If, for example, one considers the determination of the origin for the rotating system, it is obvious that the method used is fairly crude. The sum of squares of certain vectors appearing in the perturbation equations is minimized. Evidently, if the sum were a weighted sum, different origins would be obtained depending on the weighting factors used. It should, however, be remarked that the present determination yields plausible results, e.g., in the case of motion of an earth or moon satellite, one would certainly expect the rotation of initial conditions implied by Eq. (39) to be about the center of the primary attracting body, or at least about a point very close to its center. A large rotation about a point very far removed from the center would, obviously, drastically distort what should be a stable orbit. Thus, the property that the origin is closer to the earth or moon according as the portion of the restricted problem orbit under consideration is closer to the earth or moon is a reasonable one and shows that the theory is at least qualitatively correct in this respect. For midcourse portions of the trajectory, one cannot use the satellite argument to suggest the proper choice of the origin, though it might be conjectured that the origin should vary continuously with the portion of the trajectory to be approximated.

It is possible to make a few remarks on the parameter δ. Reference to the perturbation Eqs. (23) and (24) shows that if $\delta = 1$ the nonintegrable terms are all integrals from initial to final time, which therefore have zero initial value. It would thus appear that for short range predictions, results for $\delta = 1$, that is, for method 2, would be superior to the others. This result has been observed for some midcourse runs.

It may have been noticed that the perturbation term $\dot{\Omega} \cdot R_A \times P_A$ in the perturbation Hamiltonian H_1 [see Eqs. (15) and (20)] could be treated in the same way as the $R_A \cdot A_1$ term. That is, a factor ϵ could be introduced in the same way as δ. This would change the rotation in the initial conditions, resulting from integration of the perturbation equations, from an angle ωt to an angle $\epsilon \omega t$. To actually introduce the ϵ and obtain a value for it in the same way as for δ would not be easy because the terms in $(1-\epsilon)$, which would appear both inside and outside the integrals, would be far more complex and difficult to treat than the corresponding terms in $(1-\delta)$.

To summarize, then, the various methods so far developed for the rotating system depend on the selection of four parameters α, β, γ (determining the center of rotation A), and δ. At this stage, it appears that some sort of a parameter study using variations from the values of the parameters so far used, and including also, perhaps, variations in the parameter ϵ defined in the last paragraph, might well lead to some useful approximation formulas. There are many ways in which such a study might be carried out, for example, by using weighting factors with the vectors to be minimized, by a systematic variation of the parameters, or by the development of some sort of iteration procedure. From the above discussion, it would appear that β and γ should be close to zero, that ϵ should be close to one, and that α should vary approximately according to Eq. (35). Only for the parameter δ is it difficult to estimate a value except for relatively short range predictions for which one would expect δ to be close to one.*

---

*Some six months after the presentation of this paper at the AIAA Astrodynamics Conference, a possible explanation was discovered for the superiority of the Kepler approximation for portions of the trajectory with a close approach to either the earth or the moon but not to both. In the present theory, the gravitational attractions of the earth and the moon are included in the unperturbed problem and ω, the angular velocity of the moon, is treated as a small parameter. That is, as $\omega \to 0$ with μ and μ' fixed, the problem described by Eq. (6), goes over into the two-fixed-center problem. In the restricted problem, however, $\omega^2 = (\mu + \mu')/\ell^3$, and as $\omega \to 0$ with μ and μ' fixed, either the earth or the moon or both recede to infinity, depending on the location of the origin of the coordinate system used. Thus if the

(Continued on next page)

References

[1] Euler, L., Opera Omnia, Series 2: Opera Mechanica et Astronomica (Societas Scientiarum Naturalium Helveticae, Lausanne, 1957), Vol. 6, pp. 247-293.

[2] Legendre, A. M., Traite des Fonctions Elliptiques et des Integrales Euleriennes, (De Hazard-Courcier, Paris, 1825), Vol. 1, pp. 411-531.

[3] Pines, S. and Payne, M., "Application of the two-fixed-center problem to lunar trajectories", Republic Aviation Corporation Technical Report ARD 837-450 (1961).

[4] Lowy, E. and Payne, M., "Three dimensional closed form solution for negative energy of the two-fixed center problem", Republic Aviation Corporation Technical Report RD-TR-64-3-1 (1964).

[5] Charlier, C. L., Die Mechanik des Himmels (Walter de Gruyter and Co., Berlin 1927), Vol. 1, Chap. 3 and Vol. 2, Chap. 11.

[6] Payne, M. and Pines, S., "The Hamilton-Jacobi formulation of the restricted three-body problem in terms of the two-fixed-center problem", MTP-AERO-62-52 George C. Marshall Space Flight Center, Huntsville, Alabama (1962).

[7] Arenstorf, R. F. and Davidson, M. C. Jr., "Solutions of restricted three-body problem represented by means of two-fixed-center problem", AIAA J. 1, 228-230 (1963).

---

earth is at the origin, the moon recedes to infinity and the restricted problem reduces to the Kepler problem with the earth as the attracting center and similarly for the moon.

Table 1 Earth reference from earth to moon:
$\alpha = -0.012116806$ corresponds to center of rotation at the earth

		1	2	3	4	5	6	K	α
0–1	Δx	0.195	-0.170	0.177	0.008			-0.001	-0.012116746
	Δy	0.1169	-0.0277	0.1189	0.0449			-0.0074	
	Δz	0.0426	-0.0610	0.0506	-0.0045			0.0002	
0–10	Δx	19	-52	19	-16	1033	1025	-6	-0.012116746
	Δy	20	-28	20	-4	310	308	-4.8	
	Δz	8.4	-32	8.3	-1.692	471	467	-0.65	
0–30	Δx		-523	125	-198	3431	3181	-174	-0.012116746
	Δy		-515	282	-115	3020	2817	-109	
	Δz		-377	101	-137	2158	2004	-13	
0–50	Δx		-1425	217	-604			-1036	-0.012116746
	Δy		-1177	1524	176			-596	
	Δz		-917	478	-218			-65	
10–30	Δx	47	-32	65	17	705	669		-0.010659765
	Δy	119	-21	113	46	-590	-447		
	Δz	36	-8.2	358	14	-215	-184		
30–50	Δx	97	702	64	383	900	875		0.093362437
	Δy	393	977	-354	361	-592	-22		
	Δz	98	284	-107	88	-221	-72		

Table 2 Moon reference from earth to moon:
$\alpha = 0.987883194$ corresponds to center of rotation at the moon

		1	2	3	4	5	6	α
59-60	Δx	-4.8	4.2	-3.5	0.38	-1.1	0.77	0.74542049
	Δy	-17	13	-13	0.20	-1.2	0.69	
	Δz	-5.6	4.3	-4.2	0.02	-1.3	0.08	
59-66	Δx	-179	297	-160	67	-219	485	0.95560202
	Δy	-849	842	-811	16	-705	648	
	Δz	-278	267	-268	0.45	-119	63	
59-71	Δx	-213	838	-211	447		13728	0.98774102
	Δy	-2018	1982	-2018	-129		-3695	
	Δz	-715	625	-715	-118		-6390	
66-71	Δx	-13	58	-12	24	-2390	6815	0.98774026
	Δy	-409	368	-409	-24	-2795	-8514	
	Δz	-151	130	-151	-12	-395	-5567	

Table 3 Moon reference moon towards earth:
$\alpha = 0.987993194$ corresponds to center of rotation at the moon

		1	2	3	4	5	6	α
72-73	Δx	-2.6	3.8	-2.5	0.66	-3930	-3930	0.98746055
	Δy	-18	18	-18	0.02	119	119	
	Δz	-5.9	5.7	-5.9	-0.10	227	227	
72-75	Δx	-21	46	-18	14	-14509	-14509	0.98750712
	Δy	-160	151	-158	-3.6	-439	-439	
	Δz	-53	46	-53	-3.3	1463	1463	
72-84	Δx	-12	828	48	424	-53692	-53677	0.98751836
	Δy	-2607	2015	-2579	-295	1277	1277	
	Δz	-883	581	-877	-151	17063	17050	
75-80	Δx	-26	73	-7.0	33	-19552	-19552	0.97489278
	Δy	-457	184	-454	-135	497	496	
	Δz	-153	138	-158	-10	1146	1146	
75-84	Δx	-13	234	62	147	-36241	-36237	0.97552707
	Δy	-1500	1371	-1497	-62	822	822	
	Δz	-523	435	-529	-46	2605	2605	
80-84	Δx	-3.8	30	9.3	20	-15278	-15277	0.84840941
	Δy	-296	446	-231	108	-169	-170	
	Δz	-100	76	-91	-7.7	487	487	

Table 4 Earth reference moon towards earth:
$\alpha = -0.012116806$ corresponds to center of rotation at the earth

		1	2	3	4	5	6	K	α
85-86	Δx	0.453	0.285	0.174		-0.241	-0.235	7.26	0.63122346
	Δy	-0.04	6.46	-283		-5.02	-5.02	-2.00	
	Δz	-0.199	3.81	-4.071		0.138	0.136	-2.52	
85-100	Δx	536	395	704	540	-321	-304	1036	0.53630503
	Δy	-207	1817	-6132	-2157	221	208	-291	
	Δz	-246	530	-1020	244	274	269	-366	
85-153	Δx	2268	1486	2268	1894	-667	-309	2317	-0.012113913
	Δy	-3553	-14204	-3556	-9812	19431	13105	-4320	
	Δz	430	4819	431	2210	-771	-451	-178	
100-120	Δx	167	168	342	255	-102	-55		0.044721505
	Δy	-57	68	-601	-266	158	92		
	Δz	-90	-31	-260	-145	99	82		
100-153	Δx	477	520	477	501				-0.012114226
	Δy	-1027	-4630	-1028	-2808				
	Δz	-26	954	-26	424				
120-153	Δx	-11	90	-11	40	40	159		-0.12114239
	Δy	60	-962	60	-449	1034	1224		
	Δz	-36	169	-36	63	-206	492		

Table 5 Lunar trajectory: position relative to reference body

Time in hours	X, km	Y, km	Z, km	Distance in Earth Radii	Reference Body
0	47	6300	1800	1.0	Earth
1	-19000	-8000	-10000	3.6	Earth
10	-45000	-100000	-46000	18.6	Earth
30	-53000	-210000	-82000	36.8	Earth
50	-51000	-290000	-103000	48.8	Earth
59	50000	22000	482	8.6	Moon
60	46000	20000	187	7.96	Moon
66	24000	7300	-1500	3.97	Moon
71	1300	-3700	-2100	0.70	Moon
72	-5000	-3500	-681	0.96	Moon
73	-10000	-2100	1080	1.62	Moon
75	-19000	1100	4450	3.0	Moon
80	-38000	9100	12000	6.4	Moon
84	-52000	15000	18000	9.0	Moon
85	-57000	-329000	-97000	54.5	Earth
86	-56000	-327000	-95000	54.2	Earth
100	-50000	-300000	-75000	48.9	Earth
120	-36000	-240000	-41000	38.4	Earth
153	2300	-13000	16000	3.3	Earth

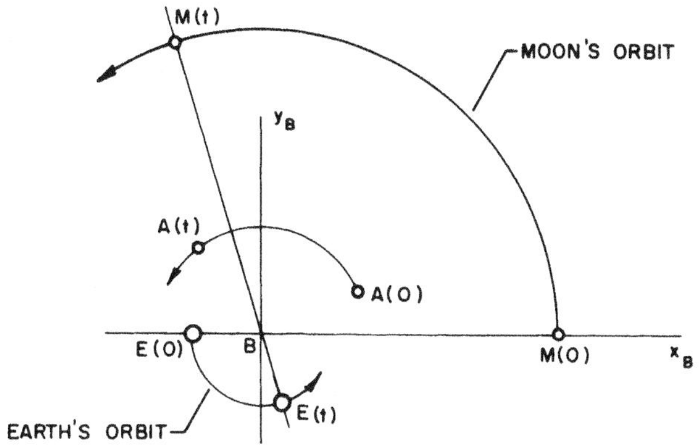

Fig. 1 Motion of the earth, moon and A in an inertial coordinate system with origin at the barycenter.

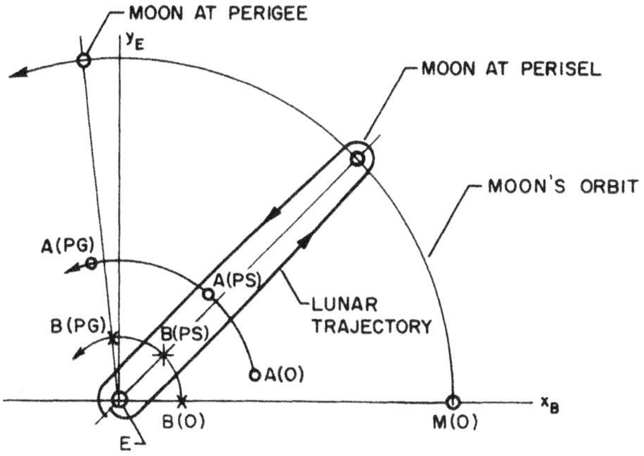

Fig. 2 A typical lunar trajectory in an accelerated coordinate system with the earth at the origin.

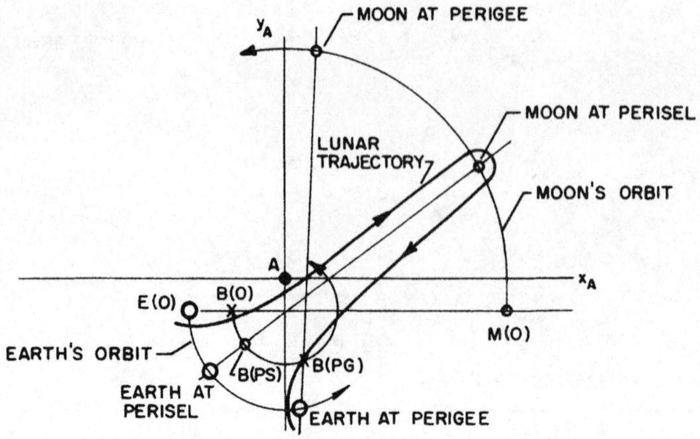

Fig. 3 A typical lunar trajectory in an accelerated coordinate system with origin at a point A sufficiently close to the moon.

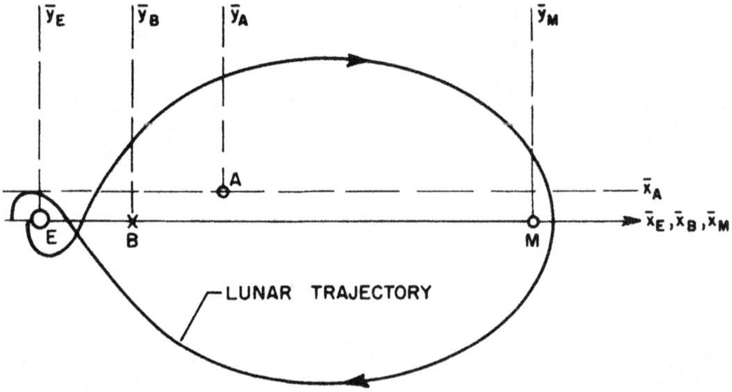

Fig. 4 A typical lunar trajectory in a coordinate system rotating with angular velocity Ω. Origin at E, B, M, or A, as indicated.

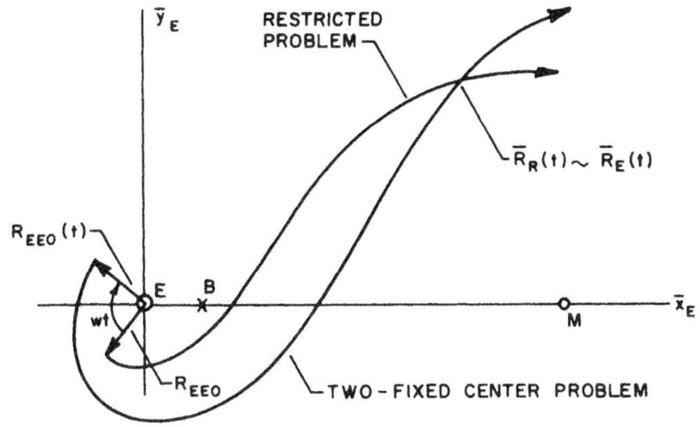

Fig. 5 Modification of initial position to give approximate match of restricted and two-fixed-center problem at time t. Origin is at the earth.

A HIGH-ORDER PERTURBATION THEORY USING RECTANGULAR COORDINATES

William Kizner*

Jet Propulsion Laboratory, California Institute of Technology, Pasadena, California

Abstract

This paper presents basically a method for approximating a system of nonlinear equations by a series of linear equations. The method has two parts. The first is similar to the method of Poisson of developing a solution in terms of a small parameter, the difference being that Poisson used a Taylor series expansion, whereas we use a Runge-Kutta integration. One advantage is that the method can be made of very high order without calculating complicated high-order derivatives. Another advantage is that generally the solution has a smaller error than the Poisson solution of the same order. The second part involves the refinement of the solution to the extent that it has negligible error. Here we show how the Poisson method or ours can make use of any approximate solution to yield an exact solution. This method has been used to solve problems in nonlinear oscillations, and a simple version of it has been applied to the artificial satellite problem. These results will be published elsewhere. Here it will be shown how to derive the differential equations for the osculating elements using rectangular coordinates, and how an analytic solution to these equations may be obtained with the aid of a digital computer.

Presented as Preprint 63-391 at the AIAA Astrodynamics Conference, New Haven, Conn., August 19-21, 1963. This paper presents the results of one phase of research carried out at the Jet Propulsion Laboratory, California Institute of Technology, under Contract No. NAS 7-100, sponsored by NASA.

*Scientist Specialist.

Introduction

The following method is applicable not only to problems in celestial mechanics, but to the solution of any system of nonlinear differential equations where exact solutions are sought. The literature on existing related methods of Poisson and Poincaré is vast. Only a few representative references can be given.[1, 2, 3] First, a modification of the Poisson method of expansion in terms of a small parameter, the Runge-Kutta method, is used to obtain an approximate solution. Next, it is shown how to improve the accuracy of the solution until the prescribed precision is reached.

It is desirable, but not necessary, that slowly varying parameters are available. For the problems of celestial mechanics, these parameters might be any of the ones used in the method of variation of arbitrary constants.[1]

Although it is possible in principle to solve the resulting linear equations for arbitrary initial conditions, it may be impossible to practice. Hence, a method is indicated which utilizes a digital computer in obtaining a solution for the elements in terms of Chebyshev polynomials of the time. These polynomials are rapidly convergent, easily calculated by recurrence relations, and can be truncated to obtain simpler approximate solutions.

Many simple nonlinear equations have been solved using these methods, and the results seem to bear out the theory. In all cases, the solutions by this method were more accurate than solutions by the Poisson method of the same order. Some work has been done on solving problems in celestial mechanics, although quantitative comparisons with existing theories are not yet available.

As observations of artificial satellites and space probes become more accurate, we would like to be able to obtain more information on the gravitational fields and the atmospheres of the planets and natural satellites. Existing artificial satellite theories are of low order and do not treat air drag exactly. The pages of equations of the second-order theory of Kozai[4] illustrate the complexities of these theories even when important perturbative accelerations such as air drag are neglected. With this method, all disturbances ordinarily considered can be exactly accounted for.

Relation of the Poisson Method to Runge-Kutta Integration

For the sake of definiteness, assume we will use the initial osculating elements, q^1, q^2, $\cdots$, q^6, as the reference orbit or the zero-order solution. We order these elements and call the vector thus formed q. The perturbations f(q, t) are functions of q and the time t. Thus, the equations for the variations of constants are given by

$$\dot{q} = f(q, t) \tag{1}$$

Let us introduce the parameter λ which takes on values from 0 to 1, and consider a more general problem where q is not only a function of the initial conditions and time, but also of λ through the equation

$$\dot{q} = \lambda f(q, t) \tag{2}$$

When $\lambda = 0$ we have an exact solution, the zero-order solution. We seek to find the solution when $\lambda = 1$.

For the Poisson method we proceed as follows. We assume that q can be represented, to a reasonable degree of accuracy, by a few terms of a power series in λ; thus

$$q = q_0 + \lambda q_1 + \lambda^2 q_2 + \cdots + \lambda^n q_n \tag{3}$$

By substituting (3) into (2), expanding f(q, t) as a power series in λ, and equating coefficients of like powers of λ, we obtain

$$\begin{aligned} \dot{q}_0 &= 0 \\ \dot{q}_1 &= f(q_0, t) \\ \dot{q}_2^i &= f_j^i(q_0, t)\, q_1^j \\ \dot{q}_3^i &= \frac{1}{2} f_{jk}^i(q_0, t)\, q_1^j q_1^k + f_j^i(q_0, t)\, q_2^j \end{aligned} \tag{4}$$

where

$$f_j^i = \partial f^i/\partial q^j \qquad f_{jk}^i = \partial^2 f^i/\partial q^j \partial q^k$$

the subscripts i, j, k ordinarily take on values from 1 to 6, and the convention is adopted that if an index occurs both as a subscript and a superscript, the product should be summed with respect to this index over the allowable values of the index. The equations may be solved in turn, as each depends only on the previous solutions.

One drawback of Poisson's method is that complicated derivatives $f^i_{jk} \cdots$ arise which are difficult to evaluate. Another is that there is no assurance that the procedure is convergent. Poincaré's theorem[5] can be derived from Cauchy's theorem:

Cauchy's Theorem. Consider the system of equations

$$\dot{q} = F(q, t, \lambda) \tag{2a}$$

If the right-hand member is analytic (expandable in a Taylor series) in q, t, and λ, then the solution of the initial value problem q(0) = C is analytic in the variables C, t, and λ. If This is a local theorem. The theorem was extended by Poincare.

Poincaré's Theorem. Let q(t, λ) denote the solution of (2a) for fixed initial conditions as a function of the time in the interval $0 \leq t \leq T$, and the "small" parameter λ. If F(q, t, λ) can be developed as a Taylor series of the variables q(t, λ) - q(t, 0) and λ, then for sufficiently small λ the solution q(t, λ) is given by a power series in λ:

$$q(t, \lambda) = q_0(t) + \lambda q_1(t) + \lambda^2 q_2(t) + \cdots$$

which is valid for $0 \leq t \leq T$.

Thus under these conditions (3) will hold for sufficiently small λ.

Thus Poincaré's theorem provides some justification of Poisson's method. In contrast, our method appears to converge for any finite time interval with much milder assumptions. For the first part, we need the following theorem about differentiability of solutions[6]: If F(q, t, λ) possesses continuous k^{th}-order derivatives in the variables q, t, and λ, then the solution of the initial value problem q has continuous k^{th}-order derivatives with respect to λ. Also $\dot{q}$ has continuous k^{th}-order derivatives with respect to t and λ.

Thus, the existence of q_λ, defined by

$$q_\lambda(t, \lambda_1) = \lim_{\lambda_2 - \lambda_1 \to 0} \frac{q(t, \lambda_2) - q(t, \lambda_1)}{\lambda_2 - \lambda_1} \tag{5}$$

where

$$q(0, \lambda_1) = q(0, \lambda_2) = q(0)$$

follows, if f(q, t) has continuous first-order derivatives with respect to q and t. In addition, it can be shown that under these conditions

$$\frac{d}{d\lambda}\dot{q}^i = \frac{d}{dt}q_\lambda^i = f^i(q, t) + \lambda f_j^i(q, t)\, q_\lambda^j \qquad i, j = 1, 2, \cdots, 6 \tag{6}$$

Equation (6) is sometimes called the variational equation. The exact solution to the problem is given by

$$q(t, 1) = \int_0^1 q_\lambda(t, \lambda)\, d\lambda + q(t, 0) \tag{7}$$

We use a Runge-Kutta integration[7] to evaluate the integral in (7). Some commonly used Runge-Kutta formulas are (note that the integration interval is one):

$$q(t, 1) \cong q_0 + k_1 \qquad \text{(first order)} \tag{8a}$$

where

$$k_1 = q_\lambda(t, 0) \tag{8b}$$

The first-order solution is the same as for the Poisson method:

$$q(t, 1) \cong q_0 + \frac{1}{2}(k_1 + k_2) \qquad \text{(second order)} \tag{9a}$$

$$k_1 = Q_\lambda(q_0, 0) \tag{9b}$$

$$k_2 = Q_\lambda(q_0 + k_1, 1) \tag{9c}$$

$$q(t, 1) \cong q_0 + \frac{1}{6}(k_1 + 2k_2 + 2k_3 + k_4) \qquad \text{(fourth order)} \tag{10a}$$

$$k_1 = Q_\lambda(q_0, 0) \tag{10b}$$

$$k_2 = Q_\lambda\left(q_0 + \frac{k_1}{2}, \frac{1}{2}\right) \tag{10c}$$

$$k_3 = Q_\lambda\left(q_0 + \frac{k_2}{2}, \frac{1}{2}\right) \tag{10d}$$

$$k_4 = Q_\lambda(q_0 + k_3, 1) \tag{10e}$$

The meaning of the k's as defined symbolically by $Q_\lambda(q, \lambda)$ is that the differential equation for q_λ is solved using the given function q and the second argument for the value of λ. This is illustrated for (9b) and (9c). For (9b)

$$\dot{k}_1 = f(q_0, t) \tag{11}$$

For (9c)

$$\dot{k}_2^i = f^i(q_0 + k_1, t) + f_j^i(q_0 + k_1, t)\, k_2^j \tag{12}$$

Thus, the Runge-Kutta formulas require that the derivative of q with respect to λ be evaluated (in this case, the differential equations must be solved) using previously calculated values of the unknowns and λ, which corresponds to h, the step size. The proof that these formulas are of the order indicated is long-winded but follows the method in Ref. 7.

We will show how to establish (9). The corresponding formulas for the approximation to $q(t, \lambda)$ are

$$\bar{q}(t, \lambda) \cong q_0 + \frac{\lambda}{2}(k_1 + k_2) = q(t, \lambda) \tag{13a}$$

$$k_1 = Q_\lambda(q_0, 0) \tag{13b}$$

$$k_2 = Q_\lambda(q_0 + \lambda k_1, \lambda) \tag{13c}$$

where $\bar{q}(t, \lambda)$ is the approximation for $q(t, \lambda)$. One shows that to second order in λ

$$k_2 = q_\lambda(t, 0) + \lambda q_{\lambda\lambda}(t, 0) \tag{14}$$

From (13b) k_1 is by definition $q_\lambda(t, 0)$. We have from (14) and (13a)

$$\bar{q}(t, \lambda) = q_0 + \lambda q_\lambda(t, 0) + \frac{1}{2}\lambda^2 q_{\lambda\lambda}(t, 0) + 0(\lambda^3) \tag{15}$$

which agrees with the Taylor series expansion to second order in λ. We call methods based on Runge-Kutta formulas Runge-Kutta expansion methods.

Properties of the Solutions

Up to now we have seen that the Runge-Kutta expansion method and Poisson's method are similar in that they both assume that the solution can be approximated by a truncated power series in λ. It is not evident that a solution by one method may be more accurate than a solution by another. There are reasons, some theoretical and some empirical, for expecting that the Runge-Kutta type solution will be more accurate than the Poisson solution of the same order.

Let us illustrate the methods by solving to second order the scalar equation

$$\dot{x} = \epsilon x \qquad x(0) = 1 \tag{16}$$

where ϵ is a constant. We consider the right-hand side of the differential equation as the perturbation. Thus

$$\dot{x} = \lambda \epsilon x \tag{17}$$

when $\lambda = 0$, $\dot{x}_0 = 0$, $x_0 = \text{const} = 1$ to satisfy the initial conditions. The variational equations are

$$(d/dt)x_\lambda = \epsilon x + \lambda \epsilon x_\lambda \tag{18}$$

To find k_1, we have, from (9b) or (11),

$$\dot{k}_1 = \epsilon x_0 + 0\epsilon k_1 = \epsilon \tag{19}$$

$$k_1 = \epsilon t \tag{20}$$

where we have satisfied the initial conditions

$$k_1(0) = 0 \tag{21}$$

For k_2 we have from (9c) or (12)

$$\dot{k}_2 = \epsilon(1 + \epsilon t) + \epsilon k_2 \tag{22}$$

Hence,

$$k_2 = -2 - \epsilon t + 2e^{\epsilon t} \tag{23}$$

From (9a)

$$x = e^{\epsilon t} \tag{24}$$

which is the exact solution. If the Poisson method were used, the second-order solution would be

$$1 + \epsilon t + \left[(\epsilon t)^2/2\right]$$

We now show that the Runge-Kutta expansion type of solution of order two will always solve a system of linear differential equations exactly. Given

$$\dot{q}^i = A^i_j q^j + f^i \qquad i, j = 1, 2, \cdots, n \tag{25}$$

where A and f are continuous functions of the time alone, and n is an integer. We form

$$\dot{q}^i = \lambda\left(A^i_j q^j + f^i\right) \tag{26}$$

$$\dot{q}^i_\lambda = A^i_j q^j + f^i + \lambda A^i_j q^j_\lambda \tag{27}$$

From Eq. (9)

$$\begin{aligned} q^i(0) &= C^i \qquad \dot{k}^i_1 = A^i_j q^j_0 + f^i \\ \dot{k}^i_2 &= A^i_j\left(q^j_0 + k^j_1\right) + f^i + A^i_j k^j_2 \end{aligned} \tag{28}$$

and

$$k_1(0) = k_2(0) = 0$$

If we substitute $y = C + (k_1 + k_2)/2$ in (25), we see that the equation is identically satisified by virtue of (28).

Likewise, it can be shown that the fourth-order formulas (10) will also yield exact solutions to systems of linear equations. Furthermore, these results will hold when the method is generalized in the following section for improving the accuracy of an approximate solution.

Let us now solve the nonlinear equation

$$\dot{x} = x^2 \qquad x(0) = 1 \tag{29}$$

by the Runge-Kutta expansion method and Poisson's method by considering the right-hand side as the perturbation. The exact solution to (29) is

$$x = 1/(1 - t) \tag{30}$$

The nth-order solution by Poisson's method is

$$x_{np} = 1 + t + t^2 + \cdots + t^n \tag{31}$$

For the Runge-Kutta expansion second-order method, the solution is

$$\bar{x}_2 = \frac{1}{4}\left\{3 + t + \exp(t^2 + 2t) + \exp\left[(1 + t)^2\right]\int_1^{1+t} \exp(-u^2)du\right\} \tag{32}$$

In Fig. 1 where these functions are graphed using a semilogarithmic ordinate scale, we see that $\bar{x}_2$ is much more accurate than x_{2p}. In fact, (32) is about equal in accuracy to a seventh-order solution by the Poisson method. If the fourth-order Runge-Kutta formulas are used, the solution does not seem to be expressible in closed form in terms of well-known functions. The accuracy appears to be comparable to a twentieth-order solution by the Poisson method (see Table 1).

The reader may wonder in what sense the solutions based on the Runge-Kutta expansion method are second or fourth order. Let us consider the errors $x - \bar{x}_2$ and $x - \bar{x}_4$, as shown in Table 1.

Table 1 Errors of the second- and fourth-order solution for $\dot{x} = x^2$ by the Runge-Kutta expansion method

t	$x - \bar{x}_2$	$x - \bar{x}_4$
0	0	0
0.1	-0.00017	-0.0000002
0.2	-0.00140	-0.0000146
0.3	-0.00449	-0.0001503
0.4	-0.00862	-0.0008637
0.5	-0.00699	-0.0036525
0.6	0.02726	-0.0126041
0.7	0.19103	-0.0350498
0.8	0.87352	-0.0415062
0.9	4.39599	0.9046431
1.0	∞	∞

From (29) it follows that, for any constant k, $x(t, \lambda) = x(kt, \lambda/k)$, provided that both sides are defined. This enables us to use Table 1 to see how the errors depend on λ. From $x - \bar{x}_2$ at 0.1 and 0.2, we see that

$$\frac{x(0.2, 1) - \bar{x}_2(0.2, 1)}{x\left(0.2, \frac{1}{2}\right) - \bar{x}_2\left(0.2, \frac{1}{2}\right)} = \frac{140}{17} \cong 8 = 2^3 \tag{33}$$

Thus, in this region the error is approximately proportional to the cube of the ratio of λ_s, as it should for a second-order theory. However, for the nominal fourth-order solution, the errors behave more like a fifth-order solution up to about $t = 0.6$.

Another equation that was studied is

$$\ddot{x} + x = ax^2 \tag{34}$$

where a is a constant and the right-hand side is considered as the perturbation. Equation (34) can be solved exactly in terms of elliptic functions. For small $|a|$, the solution to (34) is periodic. Thus, Lindsted's method (Ref. 2, p. 224), which seeks the true fundamental frequency, was used for comparison with the Runge-Kutta expansion method. Although one can obtain solutions to this equation with greater ease by Lindsted's method, solutions by the Runge-Kutta expansion method gave better accuracy, particularly for $|a| \geq 0.04$. For very small $|a|$, both methods gave solutions of comparable accuracy.

A heuristic reason for the success of the Runge-Kutta expansion method is that the derivatives are not all evaluated on the standard trajectory, in contrast to the Poisson method. Thus, the convergence is more like that found with orthogonal functions, rather than with Taylor series.

Improvement of the Solution

We now generalize the method (and also the Poisson method). Suppose we have an approximate solution for q which we call $g_0(t)$, as a function of the initial conditions (which we do not indicate) and the time. Then successive approximations g_1, g_2, $\cdots$, are obtained by making use of the λ factor:

$$\dot{g}_{i+1} = \dot{g}_i(t) + \lambda \left[f(g_{i+1}, t) - \dot{g}_i(t) \right] \tag{35}$$

We proceed as before, setting $\lambda = 0$ and forming the variational equations in accordance with one of the Runge-Kutta formulas such as (8), (9), or (10) and solve.

First, we note that if (8), the first-order process, is used then we have nothing more than a solution by Picard's method of successive approximations which converges to the true solution whenever f is continuous and satisfies a Lipschitz condition with respect to g; g_0 is continuously differentiable and satisfies the initial conditions. Thus we would expect that, in general, the higher-order iterations will also converge, but at a much faster rate.

At present, it has not been determined under what conditions the iterative process converges for methods of order greater than one. There is reason to believe that the process will converge provided that f is continuously differentiable. For such an f, it appears that in a region where the approximation is very good, the higher-order methods will behave like the Picard method, and therefore converge. Succeeding neighborhoods may then be found, and the region of convergence extended indefinitely. There is an analogous theorem in the application of Runge-Kutta methods to the numerical solution of ordinary differential equations (Ref. 7, p. 71) which states roughly that, if a solution by the Euler method converges, then a solution by any of the Runge-Kutta methods converges.

As far as the application of this iterative process to the equations of celestial mechanics goes, the question of convergence of higher-order methods is academic. A solution by the fourth-order Runge-Kutta method should give an excellent approximation for the motion for time periods that are usually of interest. Once a good approximation is obtained, it is known that the Picard method converges rapidly. The Picard method has the further advantage that the functions that have to be evaluated are simplier and can be calculated more accurately.

For other problems for which a good first approximation is unknown, our method of using repeated applications of high-order methods may be practicable.

An alternative method of obtaining better approximations is to use several integration intervals, for instance (9) and (10) might be modified so that h is 1/2. Here, there is no problem in proving convergence as the integration interval is made arbitrarily small. If f has continuous first derivatives the theorem (Ref. 7, p. 71) on the integration of ordinary differential equations can be applied considering the value of t fixed and the integration with respect to λ. If a method of order p is used and there exist continuous derivatives of q with respect to λ of order p+2, then one can apply the Taylor expansion theorem with remainder for the local truncation error to obtain $h^{p+1}\phi(t, \lambda) + 0(h^{p+2})$ where h is the increment in λ. The terms involving h, h^2, $\cdots$, up to h^p vanish, since we assume that the process is of order p. Therefore, for example, if we halve the integration interval and use the classical Runge-Kutta method of order four, the local truncation error will be decreased by 1/32. But, since there will be two intervals, the total error will only be decreased by 1/16. This appears to be a very small increase in accuracy compared to the iterative method.

Chebyshev Polynomials and the Practical Solution of the Resulting Linear Equations

Although the differential equations that are to be solved, such as (8), (9), or (10), are linear, and linear equations are, in general, much more simply solved than nonlinear equations, the solutions are by no means easy to determine. One must first develop the disturbing function in terms of Keplerian elements. This itself is no small task, and may be a source of error. The solution of the resulting equations is also usually accomplished with some approximations. Here we present an extremely simple method in which these sources of error can be eliminated for all practical purposes, but which requires a computing machine and only yields a solution for fixed initial conditions.

The method we are using to solve the equations of celestial mechanics makes use of the initial position and velocity vectors as the parameters.[8,9] The variational equations are derived by differentiating the expressions giving the time derivatives of the parameters with respect to the initial position and velocity vectors. Up to the present time, this was done by numerical differentiation, but future computations will be done using the more accurate but complicated analytical formulas.

The resulting linear differential equations are integrated by a variation of a method due to Lanczos,[10-13] the so-called τ method, which approximates the solution with Chebyshev polynomials. It is far superior to the method of solution in Taylor series in the convergence of the series. Examples where the Taylor series are divergent and the Chebyshev series are rapidly convergent are given in Refs. 10-12. If the linear differential equations are given in normal form, and the vector solution has a time derivative that satisfies the Dirichet conditions (used to establish the convergence of a Fourier series), then it will be evident from the construction of the solution that we will use that this method will yield a convergent solution. A limitation of this procedure is that the solution is only valid in a finite time interval.

Suppose we are given the linear system of equations

$$\mathring{x} = A(t)x + b(t) \tag{36}$$

where x and b are vectors and A is a matrix. We wish to find the solution for $0 \leq t \leq T$. It is convenient to change the independent variable by a linear transformation to the variable τ so that the range is from $0 \leq \tau \leq 1$.

We fit $dx/d\tau = \mathring{x}T$ with Chebyshev polynomials. For the ith component, the derivative can be represented in terms of the shifted Chebyshev polynomials $T^*(\tau)$ as

$$\frac{dx^i}{d\tau} = \frac{1}{2}c_0^i + c_1^i T_1^*(\tau) + c_2^i T_2^*(\tau) + \cdots + c_{n-1}^i T_{n-1}^*(\tau)$$

$$i = 1, 2, \cdots, m \tag{37}$$

where m is the dimension of the vector and n is chosen to be sufficiently large so that the error in the approximation is negligible. For simplicity, assume that n is the same for all components.

The fitting is done on the basis of the evaluation of (36) at n discrete points corresponding to the zeroes of the neglected n^{th} degree shifted Chebyshev polynomial. The zeroes occur at

$$t_i = \frac{T}{2}\left\{1 + \cos\left[\frac{\pi}{2n}(2a_i + 1)\right]\right\} \tag{38}$$

$$a_i = n - i \qquad i = 1, 2, \cdots, n$$

It can be shown that the polynomials are orthogonal at these points. An approximate solution for x is assumed. For want of any, it can be assumed to be a constant. Then the approximate value of $\dot{x}$ is obtained by evaluating the right-hand side of (36) with the trials solution for x. Trial values of the coefficients are found from

$$c_k^i = \frac{2}{n}\sum_{j=1}^{n} \frac{dx^i}{d\tau}(\tau_j)T_k^*(\tau_j) \tag{39}$$

$$i = 1, 2, \cdots, m; \qquad k = 0, 1, \cdots, n-1$$

The values of $T_k^*(\tau_i)$ can be conveniently found with the help of the following identity

$$T_k(y) \equiv T_k^*\left(\frac{1+y}{2}\right) \equiv \cos(k \arccos y) \tag{40}$$

In terms of the variable y, the zeroes are given at

$$y_i = \cos\left[\frac{\pi}{2n}(2a_i + 1)\right] \tag{41}$$

$$a_i = n-1; \qquad i = 1, 2, \cdots, n$$

Consider now

$$T_k(y_i) = \cos\left[\frac{\pi}{2}\frac{k(2a_i + 1)}{n}\right] \tag{42}$$

$$i = 1, 2, \cdots, n$$

The expression

$$k(2a_i + 1)/2$$

can be written in the form of $a + b/n$, where a and b are integers and $b < n$. By this means, the argument of the cosine cosine function of (42) can always be transformed to the first quadrant without any loss of accuracy. Thus the accuracy of the coefficients will normally be determined by the round-off error in calculating the derivatives for (39). If we assume that the errors are unbiased and have a common variance σ^2, then variance of c_k^i can be shown to be $(2/n)\sigma^2$.

Having obtained the trial c_k^i, we integrate the expressions to obtain $x(\tau)$ using

$$\int T_0^*(\tau)d\tau = \frac{1}{2}T_1^*(\tau) + \text{const}$$

$$\int T_1^*(\tau)d\tau = \frac{1}{8}T_2^*(\tau) + \text{const}$$

$$\int T_i^*(\tau)d\tau = \frac{1}{4}\left[\frac{T_{i+1}^*(\tau)}{i+1} - \frac{T_{i-1}^*(\tau)}{i-1}\right] + \text{const} \qquad (43)$$

$$i \geq 2$$

Thus

$$x^i(x) = \frac{1}{4}\left(c_0^i - c_2^i\right) T_1^*(\tau) + \frac{1}{8}\left(c_1^i - c_3^i\right) T_2^*(\tau)$$

$$+ \sum_{j=3}^{n} \frac{1}{4j}\left(c_{j-1}^i - c_{j+1}^i\right) T_j^*(\tau) + \text{const}^i \qquad (44)$$

$$i = 1, 2, \cdots, m$$

The constants in (44) are evaluated by satisfying the initial conditions. Thus

$$\text{const}^i = x^i(0) + \frac{1}{4}\left(c_0^i - c_2^i\right) - \frac{1}{8}\left(c_1^i - c_3^i\right)$$

$$+ \text{coef.}^i \text{ of } T_3^* - \text{coef.}^i \text{ of } T_4^* \cdots \qquad (45)$$

The process is repeated with the new approximation for x. In practice, a good approximation is available from the first-order solution of the nonlinear equation, which requires only one integration. Thus, this procedure will normally converge quickly as well as provide a solution that does not lose accuracy as the time is increased.

In evaluating a Chebyshev series at arbitrary values of τ, the polynomials T_i^* need not be evalusted separately. The following method is extensively used and is rapid and accurate. A version of it can be derived for all orthogonal polynomials based on the recurrence relations. But it is particularly simple for Chebyshev series. Given

$$f(\tau) = \sum_{i=0}^{n} a_i T_i^*(\tau)$$

we form

$$w_{n+2} = w_{n+1} = 0$$

$$w_i = 2(2\tau - 1)w_{i+1} - w_{i+2} + a_i$$

$$i = n, n-1, \cdots 1$$

$$w_0 = (2\tau - 1)w_1 - w_2 + a_0 \tag{46}$$

and $f(\tau) = w_0$. Note that if $2(2\tau-1)$ is calculated at the start then each w_i requires one multiplication and two additions.

It can be shown that, if the solution is many times differentiable, then the Chebyshev polynomial approximation will converge very rapidly after a certain number of terms. The property of Chebyshev series in approximating functions may be used to truncate the series to any desired accuracy. This property is well-known, but it is difficult to state in a mathematically rigorous form. It is roughly that, for functions which are "well-behaved", the Chebyshev polynomial series has maximum convergence among a certain class of orthogonal polynomials, ordinarily used, called the "ultraspherical polynomials." The strong statement made by Lanczos (Ref. 10, p. 453) that the Chebyshev polynomials always have the maximum convergence must be taken within the context of the heuristic discussion. If taken literally, it is incorrect as shown by the example of the approximation of the absolute value function $|x|$, $-1 \leq x \leq 1$ with the error tolerance set at 0.5. The Legendre series provides this accuracy when truncated to the constant term, whereas with the Chebyshev series, it is necessary to include the quadratic term.

It is clear that, for small time intervals, a solution in Chebyshev polynomials is economical when compared with the trigonometric forms that are usually employed. It remains to be seen how they compare for extended time periods. At any rate, one can always break up the desired interval into shorter intervals and have separate solutions for each one.

Acknowledgements

The author wishes to thank Dr. Fumio Yagi, now at Grumman Aircraft Engineering Corporation, for his discussion of this paper, and Mrs. Maxine Linde of Jet Propulsion Laboratory, who did the programming and obtained the numerical results which illustrate the theory.

References

[1] Brouwer, D. and Clemence, G. M., Methods of Celestial Mechanics (Academic Press Inc., New York, 1961), Chap. XI.

[2] Minorsky, N., Nonlinear Oscillations (D. Van Nostrand Co., Inc., Princeton, N. J., 1962), Part II. The introduction to Part II is a short survey of the field.

[3] Moulton, F. R., Differential Equations (Dover Publications, Inc., New York, 1958). Almost all of the book is directly applicable, but Chapters 2, 3, and 4 lay the foundation for expansions in a small parameter.

[4] Kozai, Y., "Second order solution of artificial satellite theory without air drag," Astron. J. 67, 446-461 (1962).

[5] Poincaré, H., Les Méthodes Nouvelles de la Mécanique Céleste T. I. (Dover Publications, Inc., New York, 1957), Chap. II.

[6] Gronwall, T. H., "Note on the derivatives with respect to a parameter of the solutions of a system of differential equations," Ann. Math. 20, 292-296 (1919).

[7] Henrici, P., Discrete Variable Methods in Ordinary Differential Equations (John Wiley and Sons, Inc., New York, 1962), Chaps. 2 and 3.

[8] Pines, S., " Variation of parameters for elliptic and near circular orbits," Astron. J. 66, 5-7 (1961).

[9] Herrick, S., personal communication, Lockheed California Company, a division of Lockheed Aircraft Corporation, 2555 No. Hollywood Way, Burbank, California.

[10] Lanczos, C., Applied Analysis (Prentice-Hall, Inc., Englewood Cliffs, N. J., 1956), Chap. VII.

[11] Tables of Chebyshev Polynomials $S_n(x)$ and $C_n(x)$ AMS 9, Introduction, National Bureau of Standards (1952).

[12] Clenshaw, C. W., "The numerical solution of linear differential equations in Chebyshev Series," Proc. Cambridge Phil. Soc. 53, 134-149 (1957).

[13] Clenshaw, C. W. and Norton, H. J., "The solution of nonlinear ordinary differential equations in Chebyshev series," Computer J. 6, 88-92 (1963).

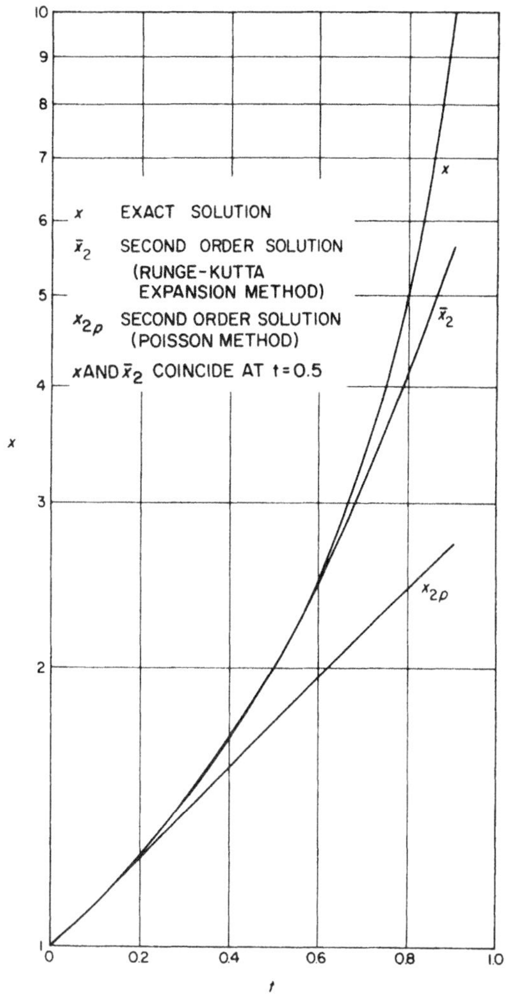

Fig. 1 Comparison of the Poisson second-order solution, Runge-Kutta second-order solution, and the exact solution for $\dot{x} = x^2$.

TRAJECTORIES OF SATELLITES UNDER THE INFLUENCE OF AIR DRAG

Chong-Hung Zee*

Grumman Aircraft Engineering Corporation, Bethpage, N. Y.

Abstract

The decaying trajectory of an artificial satellite, being originally in an elliptical orbit with small eccentricity, is derived by the asymptotic method in nonlinear mechanics. The oscillatory nature of the trajectory is brought out by this analysis. With the actual distribution of atmospheric density, the divergent series in the resulting equations have the characteristics of asymptotic expansion; thus only a few terms are needed to approximate the divergent series, and the decaying trajectory can be easily computed.

Nomenclature

A = reference area of the satellite

B = $\frac{1}{2}(C_D A/M)\, r_o$

C = $g_o/\omega^2 r_o$

C_D = drag coefficient

D = drag force = $\frac{1}{2} C_D \rho A v^2$

e = base of natural logarithms

g_o = gravitational acceleration at $r = r_o$

h = $r^2\dot{\theta}/r_o^2\omega$

Presented as Preprint 63-392 at the AIAA Astrodynamics Conference, New Haven, Conn., August 19-21, 1963.

This paper was prepared while the author was a Project Engineer, Engineering Department, Wright Aeronautical Division, Curtiss-Wright Corporation, Wood-Ridge, N. J.

*Dynamicist, Dynamic Analysis Advanced Development Group.

K_1 = integration constant in Eq. (26)
K_2 = integration constant in Eq. (27)
M = mass of the satellite
r = radial distance from the center of attraction to the satellite
r_o = perigee focal radius of original elliptical orbit
S = defined by Eqs. (9) and (10)
t = time ($t = 0$ at $\theta = 0$, $r = r_o$)
u = r_o/r
v = velocity of the satellite
x = $(\beta r_o/C)\, h^2$
x_o = $\beta r_o/C$
β = reciprocal of scale height
θ = anomaly angle
ρ = atmospheric density at $r = r$
ρ_o = atmospheric density at $r = r_o$
φ = angle between tangent line and the normal of radius r
ψ = defined by Eqs. (9) and (10)
ω = angular velocity at $r = r_o$

Superscripts
$\dot{}$ = first derivative with respect to time "t"
$\ddot{}$ = second derivative with respect to time "t"

Introduction

In a recent paper by Billik,[1] current literature on satellite lifetimes has been well reviewed. From the trajectory point of view, most of the findings are related to the trajectory of an artificial satellite in complete turns around the earth, and the trajectory of the satellite in between the complete turns is not treated in detail, though its oscillatory nature was briefly mentioned by Billik.[1]

In obtaining the lifetimes of satellites, certain model atmosphere must be adopted in the computation. On the contrary, if the trajectory of a satellite is traced out by radar tracking stations on the earth, the atmosphere density can be determined at different altitudes. Studies in this direction have appeared elsewhere.[2,3]

It is generally recognized in nonlinear mechanics that an artificial satellite subject to the perturbation of air drag force will result in oscillatory motion along its mean path.

This oscillatory characteristic has already been partially incorporated in the analysis of variation of satellite position, with uncertainties in the mean atmospheric density,[2] by considering the velocity and the density of air being constant for the period considered. It is the purpose of this paper to derive an integrated equation for the trajectory of an artificial satellite subject to the perturbation of air drag, which will show the oscillations and their damping as the trajectory proceeds.

Analysis

The equations of motion for an artificial satellite under the influence of air drag and the attraction of the earth are (see Fig. 1):

$$\ddot{r} - r\dot{\theta}^2 = -\frac{g_o r_o^2}{r^2} - \frac{D}{M}\sin\varphi \tag{1}$$

$$\frac{1}{r}\frac{d}{dt}(r^2\dot{\theta}) = -\frac{D}{M}\cos\varphi \tag{2}$$

and by definition,

$$v^2 = \dot{r}^2 + r^2\dot{\theta}^2 \qquad \sin\varphi = \dot{r}/v \qquad \cos\varphi = r\dot{\theta}/v$$

Following the usual procedures in eliminating "dt" and letting

$$u = \frac{r_o}{r} \qquad h = \frac{r^2\dot{\theta}}{r_o^2\omega} = \frac{\dot{\theta}}{u^2\omega} \tag{3}$$

Eqs. (1) and (2) become

$$h\frac{d^2u}{d\theta^2} + \frac{dh}{d\theta}\cdot\frac{du}{d\theta} + hu =$$

$$\frac{g_o}{h\omega r_o} - \frac{C_D A}{2M}\cdot\frac{r_o\rho h}{u^2}\cdot\frac{du}{d\theta}\cdot\left[u^2 + \left(\frac{du}{d\theta}\right)^2\right]^{\frac{1}{2}} \tag{4}$$

$$\frac{dh}{d\theta} = -\frac{C_D A}{2M}\cdot\frac{r_o\rho h}{u^2}\left[u^2 + \left(\frac{du}{d\theta}\right)^2\right]^{\frac{1}{2}} \tag{5}$$

Substituting Eq. (5) into Eq. (4) and rearranging yields

$$(d^2u/d\theta^2) + u = g_o/\omega^2 r_o h^2 \qquad (6)$$

which will be solved as a system with Eq. (5).

Letting

$$\frac{C_D A}{2M} r_o = B \qquad \frac{g_o}{\omega^2 r_o} = C$$

Eqs. (5) and (6) can be written as

$$\frac{dh}{d\theta} = -B\frac{\rho h}{u^2}[u^2 + (\frac{du}{d\theta})^2]^{\frac{1}{2}} \qquad (7)$$

$$\frac{d^2u}{d\theta^2} + u = \frac{C}{h^2} \qquad (8)$$

In view of the smallness of ρ, which may be assumed to be zero as the first approximation, Eq. (7) yields h as a constant, and solving Eq. (8) with h being constant results in the following expressions:

$$u = (C/h^2) + S\cos(\theta + \psi) \qquad (9)$$

$$du/d\theta = - S\sin(\theta + \psi) \qquad (10)$$

where S and ψ may be conceived as integration constants.

Now consider S, ψ, and h as functions of θ, and the functional relationship between h and θ is given by Eq. (7). Differentiating Eqs. (9) and (10) with respect to θ once and then with the aid of Eq. (8), the following equations are obtained:

$$\frac{dS}{d\theta} = \frac{2C}{h^3}\cos(\theta + \psi)\frac{dh}{d\theta} \qquad (11)$$

$$\frac{d\psi}{d\theta} = -\frac{2C}{h^3 S}\sin(\theta + \psi)\frac{dh}{d\theta} \qquad (12)$$

Substituting Eqs. (9) and (10) into Eq. (7) yields

$$\frac{dh}{d\theta} = -\frac{B\rho h^3}{C}\left[1 + \frac{Sh^2}{C}\cos(\theta + \psi)\right]^{-2} \cdot \left[1 + 2\frac{Sh^2}{C}\cos(\theta + \psi) + \left(\frac{Sh^2}{C}\right)^2\right]^{\frac{1}{2}} \tag{13}$$

For most artificial satellites the orbit has small eccentricity. If Eq. (9) is conceived as an equation of an ellipse, then the term Sh^2/C represents the eccentricity, which should be very small for the present case. Therefore, by expanding the two brackets in Eq. (13) into series and dropping out the terms containing $(Sh^2/C)^2$ and higher orders, Eq. (13) becomes

$$\frac{dh}{d\theta} = -\frac{B\rho h^3}{C}\left[1 - \frac{Sh^2}{C}\cos(\theta + \psi)\right] \tag{14}$$

As usual, introducing an exponential density

$$\rho = \rho_o e^{-\beta(r-r_o)} = \rho_o e^{\beta r_o} \cdot e^{-\beta r_o(1/u)} \tag{15}$$

and approximating $1/u$ by h^2/C because of the smallness of Sh^2/C, Eq. (15) is simplified as

$$\rho = \rho_o e^{\beta r_o} \cdot e^{-\beta r_o(h^2/C)} \tag{16}$$

Combining Eqs. (14) and (16) and applying the asymptotic method* in nonlinear mechanics[4] gives

$$\frac{dh}{d\theta} = -\frac{B\rho_o}{C} \cdot e^{\beta r_o} \cdot h^3 \cdot e^{-\beta r_o(h^2/C)} \tag{17}$$

or

$$-\frac{\beta\rho_o}{C} e^{\beta r_o} \, d\theta = h^{-3} e^{\beta r_o(h^2/C)} \, dh \tag{18}$$

Set

$$\frac{\beta r_o}{C} h^2 = x \qquad \frac{\beta r_o}{C} = x_o \qquad \text{hence } x_o \geq x \tag{19}$$

*Formerly known as the averaging technique of Kryloff and Bogoluiboff.[5] However, recent extension on K-B method by N. Bogoluiboff and J. Mitropolsky introduces the term "asymptotic method" with further explanation of mathematical foundation behind the K-B method.

and integrating, Eq. (18) results in

$$\theta = \frac{C^2}{2\beta B r_o \rho_o} \left[e^{\beta r_o[(1/C)-1]} \left(1 + \frac{2!}{x_o} + \frac{3!}{x_o^2} + \ldots\right) - \right.$$

$$\left. \frac{1}{h^4} \, e^{\beta r_o[(h^2/C) - 1]} \left(1 + \frac{2!}{x} + \frac{3!}{x^2} + \ldots\right)\right] \quad (20)$$

with the aid of the given initial condition $\theta = 0$, $h = 1$.

The divergent series of x and x_o in the parentheses on the right-hand side of Eq. (20) satisfy the characteristics of an asymptotic expansion if $x \gg 1$.[6] Considering the practical aspects of an artificial satellite, the order of magnitude of all constituents of x can be stated as follows: $\beta r_o \sim O(10^2)$, $C \sim O(10^0)$, and $h^2 \sim O(10^0)$; hence, $x \sim O(10^2) \gg 1$. Thus, in practice, the value of θ in Eq. (20) can be calculated easily by taking only a few terms of the divergent series of x and x_o.

Combining Eqs. (14) and (16) and substituting into Eqs. (11) and (12), the resulting equations are

$$\frac{dS}{d\theta} = -2\,\rho_o B e^{\beta r_o}\left[1 - \frac{Sh^2}{C}\cos(\theta + \psi)\right] \cdot$$

$$\cos(\theta + \psi)\; e^{-\beta r_o(h^2/C)} \quad (21)$$

$$\frac{d\psi}{d\theta} = (2\rho_o B e^{\beta r_o}/S) \cdot \left[1 - \frac{Sh^2}{C}\cos(\theta + \psi)\right] \cdot$$

$$\sin(\theta + \psi)\; e^{-\beta r_o(h^2/C)} \quad (22)$$

Applying the asymptotic method again to Eqs. (21) and (22) gives

$$\frac{dS}{d\theta} = \frac{\rho_o B S h^2}{C}\, e^{\beta r_o}\, e^{-\beta r_o(h^2/C)} \quad (23)$$

$$\frac{d\psi}{d\theta} = 0 \quad (24)$$

Combining Eqs. (17) and (23) yields

$$dh/dS = -(h/S) \tag{25}$$

which has the solution

$$K_1 Sh = 1 \tag{26}$$

when K_1 is an integration constant to be determined. The solution of Eq. (24) is

$$\psi = K_2 \quad \text{(integration constant)} \tag{27}$$

With Eqs. (26) and (27), Eq. (9) becomes

$$u = (C/h^2) + (1/K_1 h)\cos(\theta + K_2) \tag{28}$$

The given initial conditions at perigee as shown in Fig. 1 are $\theta = 0$, $u = 1$, $h = 1$, $du/d\theta = 0$, and $dh/d\theta = -(\rho_o B)/C$ which is obtained by employing Eqs. (15) and (17). Substituting the initial conditions into Eq. (28) and its first derivative, one has

$$(1/K_1)\cos K_2 = 1 - C \tag{29}$$

$$(1/K_1)\sin K_2 = (1 + C)(\rho_o B/C) \tag{30}$$

Combining Eqs. (28), (29) and (30) with

$$\cos(\theta + K_2) = \cos\theta \cos K_2 - \sin\theta \sin K_2 \tag{31}$$

yields

$$u = (C/h^2) + [(1-C)\cos\theta - (1 + C)(\rho_o B/C)\sin\theta] \tag{32}$$

Substituting Eq. (28) into Eq. (3) yields

$$d\theta/dt = h\omega[(C/h^2) + (1/K_1 h)\cos(\theta + K_2)]^2 \tag{33}$$

Since the eccentricity is very small, Eq. (33) can be approximated as

$$d\theta/dt = C^2\omega/h^3 \tag{34}$$

Introducing Eq. (18) into Eq. (34) gives

$$\omega dt = -(1/\rho_o BC)\, e^{-\beta r_o}\, e^{\beta r_o (h^2/C)}\, dh \tag{35}$$

As before, set $(\beta r_o/C)h^2 = x$, $(\beta r_o/C) = x_o$, and integrating with the initial condition $t = 0$, $h = 1$, Eq. (35) results in

$$\omega t = \frac{1}{2\beta B r_o \rho_o} [e^{\beta r_o [(1/C)-1]} \cdot$$

$$(1 + \frac{1}{2} \cdot \frac{1}{x_o} + \frac{1}{2} \cdot \frac{3}{2} \cdot \frac{1}{x_o^2} + \ldots) -$$

$$\frac{1}{h} e^{\beta r_o [(h^2/C)-1]} \cdot$$

$$(1 + \frac{1}{2} \cdot \frac{1}{x} + \frac{1}{2} \cdot \frac{3}{2} \cdot \frac{1}{x^2} + \ldots)] \tag{36}$$

Previously it has been stated that $x \gg 1$ for the present problem; therefore, the divergent series of x and x_o on the right-hand side of Eq. (36) can be approximated by a finite number of terms according to the properties of an asymptotic expansion.[6]

Equations (20, 28, and 36) are required for the calculations of the trajectory of an artificial satellite under the influence of air drag. As is seen from Eq. (28), the term C/h^2 is the average decay path, whereas the term $(1/K_1 h) \cos(\theta + K_2)$ results in oscillations, and the factor $1/K_1 h$ acts as an amplifying factor of the oscillations. Since $u = r_o/r$, this amplification of oscillations in u will result in damping of oscillations in r; thus the oscillation of the trajectory is said to be damped out as the trajectory progresses.

Discussions

In the previous analysis, the smallness of the eccentricity of the trajectory of an artificial satellite in the atmosphere at the beginning of decaying, as well as during the decaying, has been repeatedly mentioned and employed in the process of analysis; therefore, the results derived from the previous analysis are only valid for small eccentricity, say 0.1 or less, which, in fact, is the actual case for most of the artificial satellites launched up to date.[7] In order to compare the present analysis with the existing literature, the following two cases are investigated:

1. Satellite Being Originally in a Circular Orbit

This initial condition leads to the following simplification of the previous analysis. For a circular orbit

$$\omega = (g_o / r_o)^{\frac{1}{2}} \quad \text{and} \quad C = 1$$

hence, Eqs. (29) and (30) result in

$$1/K_1 = 2\rho_o B \qquad K_2 = \pi/2$$

which in turn simplifies Eq. (28) as

$$u = (1/h^2) - (2\rho_o B/h) \sin\theta \tag{37}$$

and Eqs.(20) and (36) become

$$\theta = \frac{M/(C_D A)}{\beta\rho_o r_o^2} \Big[\Big(1 + \frac{2!}{\beta r_o} + \frac{3!}{(\beta r_o)^2} + \ldots\Big) - \frac{1}{h^4} e^{\beta r_o (h^2-1)} \Big(1 + \frac{2!}{\beta r_o h^2} + \frac{3!}{(\beta r_o h^2)^2} + \ldots\Big)\Big] \tag{38}$$

$$t = \frac{M/(C_D A)}{\beta\rho_o g_o^{\frac{1}{2}} r_o^{\frac{3}{2}}} \Big[\Big(1 + \frac{1}{2} \cdot \frac{1}{\beta r_o} + \frac{1}{2} \cdot \frac{3}{2} \cdot \frac{1}{(\beta r_o)^2} + \cdots\Big) - \frac{1}{h} e^{\beta r_o (h^2-1)} \Big(1 + \frac{1}{2} \frac{1}{\beta r_o h^2} + \frac{1}{2} \cdot \frac{3}{2} \cdot \frac{1}{(\beta r_o h^2)^2} + \ldots\Big)\Big] \tag{39}$$

respectively. It is noted that Eq. (38) is exactly the equation (27) in Ref. 1 which is difficult to employ because of the appearance of the exponential integrals as was mentioned by Billik.[1] However, if the practical aspects of the atmospheric model are considered, it is actually not too difficult to deal with because the divergent series has the characteristics of asymptotic expansion.

2. Satellite Being Originally in an Elliptical Orbit with Small Eccentricity (say 0.1 or less)

Recently, Billik[8] has obtained the expression of the change in angular momentum per revolution for a slightly eccentric orbit which may be easily transformed into Eq. (17) in this paper. Again his results are involved in using the exponential integral and tabulated integral. However, in the present paper the corresponding expressions are obtained directly by integrating Eq. (17) and can be easily employed for numerical calculations.

Generally speaking, the question of pure circularity in the decay of an artificial satellite should never happen. If Eq. (28) is considered as the equation of an ellipse, then its eccentricity is h/CK_1. Because C and K_1 are given constants, as the trajectory proceeds, h decreases and can never become zero, it can be seen that the eccentricity of decaying orbit decreases and can never vanish.

Because the distribution of atmospheric density can be best represented by a series of arcs of exponential function,[1] the calculation of the lifetimes of artificial satellites should be obtained by summing up the revolutions over those arcs. If the decay of an artificial satellite is employed to estimate the density distribution of atmosphere, a series of arcs of exponential function should be determined as the satellite decays. For all these computations over arcs of exponential function, it is noted that the initial condition $du/d\theta$ is no more equal to zero; hence, Eq. (30) should be modified as

$$(1/K_1)\sin K_2 = (1 + C)(\rho_o B/C) - (du/d\theta)_o \tag{40}$$

where $(du/d\theta)_o$ is given initial condition. With Eq. (40), Eq. (32) assumes the form

$$u = (C/h^2) + \{(1-C)\cos\theta - [(1 + C)(\rho_o B/C) - (du/d\theta)_o]\sin\theta\}/h \tag{41}$$

This is resulting equation required for trajectory computation.

Conclusion

The integrated equation of the decaying trajectory of an artificial satellite presented in this paper not only shows the mean path, but also the oscillatory nature of the path. Although the term representing mean path is useful for estimating the lifetimes, yet the term showing oscillations should be important when the decaying trajectory is employed in estimating the atmosphere density distribution. From the practical point of view, the locations of ground observation station are geographically limited by nature or political reasons; thus, only certain parts of the trajectory can be observed and located. With these limited local observations, it is believed that the entire decaying trajectory can be traced out by the equations presented in this paper.

References

[1]Billik, B., "Survey of current literature on satellite lifetimes," ARS J. 32, 1641-1650 (1962).

[2]Karranberg, H. K., Levin, E., and Lewis, D. H., "Variation of satellite position with uncertainties in the mean atmospheric density," ARS J. 32, 576-582 (1962).

[3]Lecar, M. and Morrison, R., "A method for determining atmospheric density for short arcs of an almost circular satellite orbit," J. Astronaut. Sci. IX, no. 3, 82-85 (Fall 1962).

[4]Minorsky, N., Nonlinear Oscillations (D. Van Nostrand, Inc., Princeton, N.J., 1962), Chap. 14, pp. 329-355.

[5]Kryloff, N. and Bogoliuboff, N., Introduction to Nonlinear Mechanics (Princeton University Press, Princeton, N.J., 1947).

[6]Whittaker, E. T. and Watson, G. N., A Course of Modern Analysis (Cambridge University Press, N.Y., 1952), 4th ed. (reprinted), pp. 150-152.

[7]Kork, J., "Satellite lifetimes in elliptic orbits," J. Aerospace Sci. 29, 1273-1290 (1962).

[8]Billik, B., "Satellite lifetimes," ARS J. 32, 1926 (1962).

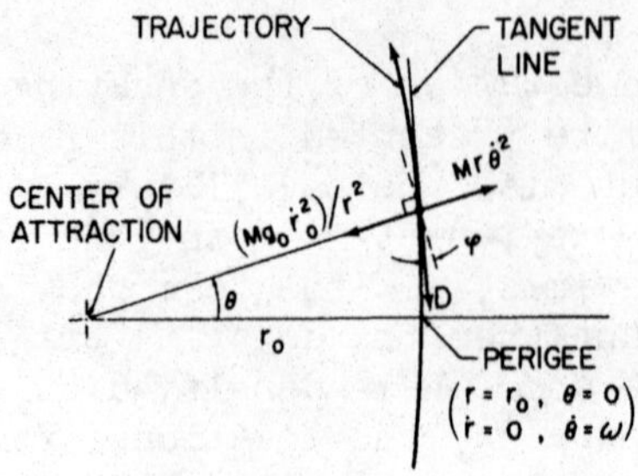

Fig. 1 Notations.

II.

ASTRONOMICAL CONSTANTS AND PHYSICAL PROPERTIES

SYSTEM OF ASTRONOMICAL CONSTANTS

Dirk Brouwer*

Yale University, New Haven, Conn.

Abstract

Recent determinations by new methods of some important constants pertaining to the solar system have made it possible to construct a consistent system of constants which is superior to the system in current use. This paper presents a critical summary of these advances.

Introduction

A gravitational theory or numerical integration of the orbit of a planet furnishes the heliocentric coordinates of the planet as a function of the independent variable of the equations of motion, for which the name ephemeris is currently used.

Astronomical observations made from an observatory on the surface of the earth can yield reliable information on planetary orbits only if adequate allowance is made for the rotation of the observer about the Earth's axis of rotation and of the motion of the center of the Earth about the center of mass of the Earth-moon system, as well as the motion of the earth-moon around the sun. To complicate matters further, the axis of rotation is not fixed in space. As a consequence, the right ascensions and declinations of stars used as reference points in the celestial sphere change from year to year by appreciable amounts called precession. Finally, the velocity of the Earth in its orbit about the sun is about 0.0001 of the velocity of light. This introduces a displacement called

---

Presented at the AIAA Astrodynamics Conference, New Haven, Connecticut, August 19-21, 1963 (not preprinted).

*Director, Yale Observatory.

aberration.

Observations made with a transit circle are referred to the instantaneous equator. In the reduction of such observations, it is necessary to make allowance also for nutation, the periodic variations in the precession. The principal nutation term has a period of 18.6 yr, the period of revolution of the ascending node of the moon's orbit. Its coefficient is called the constant of nutation.

All this refers to the reduction of optical astronomical observations of positions of celestial bodies, until recently the only kind of observations available to study the motions of bodies in the solar system. In the course of the nineteenth century, the procedure used for reducing such observations was greatly improved over the less rigorous earlier methods, but it was only toward the end of the nineteenth century that the need for uniformity was stressed. The issue as presented especially by Simon Newcomb was not merely that of achieving agreement concerning a set of constants for use in ephemeris calculations and reduction of observations; equally important was that the constants should form a consistent set.

The conference of directors of national ephemeris publications and other interested astronomers, held in Paris in 1896 for the purpose of considering this matter, succeeded remarkably well in achieving this goal. Yet a few minor inconsistencies marred its recommendations. The constants adopted were soon introduced in all national ephemeris volumes and have been used to the present time. The resulting uniformity has been of great benefit to positional astronomy and celestial mechanics. No good reasons would exist for making a change, were it not for the fact that new methods and new problems pose more stringent requirements with regard to the system of astronomical constants. The time is favorable for a change just because with new methods, some of the basic constants have recently been determined with accuracies far beyond those attainable with the older methods of positional astronomy.

Constants Related to the Earth

The determination of the size and shape of the Earth and of the Earth's gravitational field on the surface of the Earth is the domain of the geodesist.

A significant advance in this area in recent years is that observations of artificial satellites have yielded accurate values of the coefficients J_2, J_4, etc. in the expression for the Earth's gravitational potential. In order to simplify matters, it will be assumed that a representation with even zonal harmonics is adequate. This is equivalent to assuming that the Earth's gravitational field is rotationally symmetrical about the axis of rotation and also symmetrical about the equatorial plane. The condition that the mean surface of the oceans is an equipotential surface leads to a relationship among the coefficients J and coefficients that describe the geometrical character of this surface. The first-order relation is

$$2 f = 3 J_2 + m$$

in which f is the oblateness of figure, and m is the ratio between the centrifugal acceleration and the acceleration of gravity at the equator. A necessary refinement is that of introducing contributions of the second order on the assumption that J_4, J_6 ... are of $O(J_2^2)$.

A further important advance is that triangulation nets north and south of the equator have been connected and used in more rigorous and more comprehensive solutions for the Earth's equatorial radius than were possible before high-speed computers were available. The fact that these solutions could be made with the value of the flattening obtained from the coefficients J further strengthened the solutions, which yield, for the Earth's equatorial radius, values between 6,378,150 and 6,378,170 with mean error of the order of 25 m. The oblateness used in these solutions was f = 1/298.3, which corresponds closely to the value obtained from the coefficients J derived from satellite observations.

The Distance Earth-Moon and the Geocentric Gravitational Constant GE

Direct determinations by radar echoes at the U.S. Naval Research Laboratory since 1957 have yielded, for the mean center to center distance Earth-moon, the value

$$a_{☾} = 3.844004 \times 10^{8} \text{ m}$$
$$\pm 20$$

The mean error is almost entirely due to the uncertainty in the radius of the moon in the direction of the Earth. [Throughout this article, mean errors (=standard errors) will be used to indicate uncertainties.]

The mean distance $a_{☾}$ is defined without ambiguity by the lunar theory which provides the relation

$$n_{☾}^{2} \; a_{☾}^{3} = 0.99728189 \;\; GE \; (1 + \mu) \qquad (1)$$

in which $n_{☾}$ is the moon's sidereal mean motion in radians per second, μ the mass-ratio moon/Earth, and GE the geocentric gravitational constant. If $a_{☾}$ is expressed in meters, the dimension of GE is meters^3 seconds^{-2}.

The geocentric gravitational constant GE may be obtained from artificial satellite observations. Since the solution depends on triangulation involving positions of observing sites on the earth's surface, its value is strongly correlated with the value of the earth's equatorial radius adopted in the solution.

Finally, the acceleration of gravity at the Earth's equator is related to GE by the formula

$$g_0 \; a_e^2 = GE_1 (1 + (3/2) \; J_2 - (15/8) \; J_4 + \ldots -m) \quad (2)$$

the distinction between E and E_1 being that the latter excludes the mass of the Earth's atmosphere. The ratio

$$E/E_1 = 1.00000086$$

is adequately known, but the factor depending on the coefficients J in the Earth's gravitational potential in uncertain by about 3×10^{-6} owing to uncertainties in the higher J's. A similar factor arises in applications to artificial satellites, but the numerical coefficient of J_{2k} is diminished by the factor $(a_e/a)^{2k}$, a being the semimajor axis of the satellite orbit.

On the other hand, g_o may be obtained from gravity determinations on the surface of the Earth. It involves the difficult problem of achieving a reduction of such measurements to an absolute standard.

It is evident that g_o, $a_{☾}$, and GE are interrelated, so that only one of these three constants can be used as a primary constant. Considering the various aspects, it would appear that the use of GE as a primary constant has merit. However, it is encouraging that the determinations of these three constants from totally independent observational data are accordant, if in the relations among them are substituted values of a_e and μ that follow from recent determinations.

Solar Parallax

Professional astronomers have been accustomed to expressing the scale of the solar system in terrestrial units with the aid of the solar parallax, defined by the relation

$$\sin \pi_{\odot} = a_e / A$$

in which A is the astronomical unit expressed in the same unit of length as the Earth's equatorial radius. The astronomical unit is nearly, but not quite, equal to the semimajor axis of the Earth's orbit. It is defined precisely through the Gaussian constant, which has been fixed by international agreement.

The reason for using the solar parallax, expressed in seconds of arc, is that it appears in this form in the reduction of observations from geocentric to topocentric position. This is also the basis of the direct method of measuring the solar parallax from observations of suitable bodies in the solar system.

Notwithstanding the expenditure of tremendous efforts on the geometric determination of the solar parallax, the results obtained by this method are disappointing. The strongest determination was that from observations of the asteroid Eros during the winter of 1930-1931, when it approached the earth to within 1/5 a.u. (about 28×10^6 km). The large body of observations was masterfully discussed by Jones[1], who arrived at the final result

$$\pi_{\odot} = 8''.7904 \pm 0''.0015 \text{ (mean error)}$$

Among the several indirect methods available, the most confidence-inspiring one is the dynamical method that depends on evaluating the mass-ration (Earth+moon)/sun from the perturbations by the Earth-moon system on the asteroid Eros. It requires the calculation of a precise orbit of Eros over a period of an appreciable number of revolutions. The correction to the mass of the Earth-moon system used in the calculation of the orbit is one of a large number of unknowns, which include corrections to the elements of the orbit of Eros and of the orbit of the center of mass of the Earth-moon system about the sun. Such a solution was carried out by Rabe[2] with the outcome

$$S/(E + M) = 328452 \pm 67$$

S and M representing the masses of the sun and the moon, respectively.

The relation between the solar parallax and the mass-ratio sun/(Earth + moon) may be obtained by combining the equation that defines the astronomical unit

$$A^3 = 2.5226941 \times 10^{13}\ GS$$

with Eq. (1), which may be written

$$a_{☾}^3 = 1.407\ 6669 \times 10^{11}\ G(E+M)$$

giving

$$A/a_{☾} = 5.6379545 \sqrt[3]{S/(E+M)}$$

The numerical factor in the right-hand member depends on observed data through the mean motions of the sun and the moon only, and is correct to the last place given.

By making use of

$$a_{☾}/a_e = \frac{3.844002 \times 10^8 (1 \pm 0.50 \times 10^{-5})}{6.378166 \times 10^6 (1 \pm 0.40 \times 10^{-5})}$$

$$= 60.26814 \pm 0.00039$$

it follows that

$$A/a_e = (339.7890 \pm 0.0022)\sqrt[3]{S/(E+M)}$$

If $\pi_{\odot}$ is expressed in seconds of arc, this may be written

$$\pi_{\odot}\sqrt[3]{S/(E+M)} = 607.0379 \pm 0.0039$$

or

$$\pi_{\odot}^3\, S/(E+M) = 2.236904 \times 10^8 \pm 43$$

With this formula, the solar parallax obtained from Rabe's solution is

$$\pi_{\odot} = 8''.79816 \pm 0''.00060$$

This relation is, in several respects, more advantageous than the relation that may be obtained by using Eq. (2), which may be written

$$\pi_{\odot}^3\, [S/(E+M)] = 2.26442 \times 10^8 (1+\mu)^{-1}$$

The numerical factor involves the use of g_o; also, the mass-ratio $\mu = M/E$ appears in the formula.

In very recent years, the astronomical unit expressed in meters has been obtained directly by measuring the distance Earth-Venus by radar echoes. The most comprehensive discussion of these results, published by Pettengill, et al.[3], deals in detail with observations made near the 1961 inferior conjunction of Venus at the Lincoln Laboratory. It is shown, however, that the Lincoln Laboratory results are in excellent agreement with results obtained by the same method at the Jet Propulsion Laboratory and elsewhere.

The observations yield directly the distance Earth-Venus in light seconds. The ephemeris gives the distance in astronomical units; consequently, the astronomical unit in light seconds is evaluated. For the velocity of light the value

$$c = 2.997925 \times 10^8 \text{ m-sec}^{-1}$$
$$\pm\ 3$$

has become generally accepted. The mean of the results obtained so far may be represented by

$$A = 1.495985 \times 10^{11}\text{m}$$
$$\pm\ 3$$
$$= 499.0068 \text{ light sec}$$
$$\pm\ 10$$

A comparison of the three determinations is given in Table 1 .

Table 1 Comparison of three principal determinations of the solar parallax [a]

$\pi_\odot$	S/(M+E)	A/a_e	Astronomical unit 10^{11}m	light sec
Eros (geometric)				
8".7904	329323	23464.8	1.496624	499s.220
± 15	±170	± 4.0	±256	± 85
Eros (dynamical)				
8.79816	328452	23444.1	1.495303	498.779
± 60	± 67	± 1.6	±102	± 34
Venus (radar echo)				
8.794148	328901.6	23454.78	1.495985	499.0068
± 18	± 2.0	± .05	± 3	± 10

[a] Computed with $a_e = 6.378166 \times 10^6$m

The principal cause for concern is the discordance between the dynamical and the radar echo determinations. The geometric determination differs by 2½ times its mean error from the radar echo determination. The possibility of systematic errors of this magnitude in the measurement of the position of Eros relative to comparison stars cannot be excluded.

In the dynamical method, systematic errors of this nature are absent. Moreover, discussions of earlier Eros observations by Noteboom and by Witt had yielded results in good agreement with the value of the solar parallax obtained in Rabe's solution.

The first indication that Rabe's solution may need a modification was by Eckstein[4], who pointed out that the corrections to the elements of the Earth's orbit obtained by Rabe differ appreciably from the corrections to the same elements obtained by Duncombe[5] in his discussion of the observations of Venus during the past two centuries. By substituting Duncombe's results for these unknowns into Rabe's equations, and also assuming that Newcomb's masses of Mercury, Venus, and Mars need no corrections, Eckstein showed that the solution of Rabe's observation equations yielded for the solar parallax 8".79726, removing approximately one-fourth the discrepancy between the dynamical and the radar echo method.

The subject was further pursued by Marsden[6] who confirmed Eckstein's result by using for the corrections to the elements of the Earth's orbit means obtained from the discussions of observations of Venus, the sun, and Mercury. For the masses of Mercury, Venus, and Mars, he also used mean values obtained from various determinations. Instead of substituting into Rabe's normal equations, Marsden used the equations obtained as additional observation equations, introduced with appropriate weights, and then made a new solution of the combined observation equations. The solar parallax obtained from this solution was 8".7971, and the sum of the squares of the residuals of Rabe's equations was increased from 7.6 to 8.6.

Marsden next combined in a single solution the normal equations for Eros used by Rabe with the more recent normal equations for Venus used by Duncombe, and the equations furnished by the radar echo observations of Venus. In the course of this work, he noted that the correction to the mass of Mars varied appreciably from one solution to the next. By putting this mass correction in the right-hand member and assigning to it a value that leaves a satisfactory set of residuals, he obtained a solution that gives, for the sum of the squares of the residuals of Rabe's equations, 13.7. The value of the astronomical unit expressed in meters is that obtained from the Venus radar echo determination. The essential difference from Rabe's solution is that the mass ratio sun/Mars is reduced from Newcomb's value 3,093,500 to 3,021,700.

A correction of this magnitude to the mass of Mars does not seem too improbable, but two crucial tests will be the following: a) whether this value of the mass of Mars will be confirmed by observation, and b) whether this revised mass of Mars is compatible with observations of Eros outside the 20-yr interval of time covered by the observations used in Rabe's solution. If either of these tests should fail, a puzzling contradiction would remain between the results of the dynamical determination and the radar echo determination of the solar parallax.

Mass of the Moon

Let $r_☾$ be the distance of the moon from the Earth; then the distance of the Earth's center from the center of mass of the Earth-moon system is $\mu r_☾/(1+\mu)$, in which μ is the mass-ratio moon/Earth. If the direction of the moon from the Earth is perpendicular to the direction of an observed body in the solar system at a distance Δ, then the displacement relative to a barycentric ephemeris is $\mu r_☾/(1+\mu)\Delta$, expressed in radians. With the moon at its mean distance $a_☾$, the observed body at one astronomical unit from the Earth, the displacement is the constant of the lunar inequality,

$$L = [\mu/(1+\mu)]\ (a_☾/A)$$
$$\sim 6''.440$$

A close approach of Eros is evidently equally favorable for the determination of the lunar inequality and for determining the solar parallax. Two different treatments of the observations of Eros during the close approach to the Earth in 1930-1931 gave, in Ref. 7, $L = 6''.4378 \pm 0''.0018$

(mean error), and $\mu^{-1} = 81.327 \pm 0.023$, and in Ref. 8, $L = 6''.4429 \pm 0''.0021$ (mean error), and $\mu^{-1} = 81.262 \pm 0.027$ (mean error). The values $\mu^{-1} = E/M$ have been computed with values of $a_{☾}$, A used previously.

An essential limitation on the accuracy attainable is caused by the fact that Eros describes a large arc in the sky during the period of a month that contains the date of closest approach to the earth. Hence the displacements from which the lunar inequality must be obtained involve the use of reference stars in widely different parts of the sky. Consequently, differences in systematic errors in star positions contribute to the uncertainty of the result.

An impressive increase in accuracy in the evaluation of the mass-ratio moon/Earth was achieved by the analysis of two-way Doppler observations during the flight of the spacecraft Mariner II, launched on August 27, 1962. From observations during a period of four months, the result

$$\mu^{-1} = 81.3015 \pm 0.0050 \text{ (mean error)}$$

was obtained, as reported to the Paris Symposium on Astronomical Constants by T.W. Hamilton of the Jet Propulsion Laboratory.

The power of this method may be judged by considering that the mean radius of the orbit described by the center of the Earth is 4671 km, so that the average velocity in the orbit is 12.43 m-sec^{-1}. This is measured by the Doppler shift over four periods with a mean error of one part in 16,000, or about 1/5 the mean error in the evaluation of L by the astrometric method during a close approach of Eros.

Precession and Nutation

The precessional constant as defined by Newcomb may be expressed by the formula

$$P = \left\{ \alpha + \beta \left[\mu / (1 + \mu) \right] \right\} \left[(C - A) / C \right]$$

C and A being the moments of inertia of the Earth about the axis of rotation and about an axis in the equatorial plane, respectively. The lunisolar precession is

$$p_o = P \cos \varepsilon$$

ε being the obliquity of the ecliptic.

For the constant of nutation the formula is

$$N = \gamma [\mu/(1+\mu)] [(C-A)/C] \cos \varepsilon$$

These formulas are valid on the assumption that the Earth is perfectly rigid. The coefficients α, β, γ depend on the elements of the orbits of the Earth and the moon, and are known to at least seven significant figures.

It is evident that, if the formulas for a rigid Earth were applicable, the ratio p_o/N would depend on μ only, so that the values of p_o and N obtained from observations would yield the mass-ratio moon/Earth. The observed values, $p_o = 5039''.4$ and $N = 9''.210$, give $\mu^{-1} = 81.76$. On the other hand $p_o = 5039''.4$, $\mu^{-1} = 81.30$ yields $N = 9''.226$, which contradicts the observed value for the constant of nutation by at least five times its mean error.

This contradiction was noted by Jackson[9] and subsequently accounted for by Jeffreys[10] as a consequence of the fluidity of the core of the Earth. As far as is known, the precessional constant is not so affected, and the formula for a rigid Earth may be used.

With the value of P derived from observation and $\mu = 1/81.30$, the formula gives $(C-A)/C = 0.00327318$.

De Sitter[11] used this formula, with $\mu = 1/81.53$, to obtain $(C-A)/C = 0.00327942$. He next assumed that Clairaut's theory of hydrostatic equilibrium is applicable to the Earth. By introducing two additional parameters, which depend on the distribution of mass in the Earth's interior, he derived from Clairaut's theory the ratio

$$q = 0.50043$$

which appears in the relationship

$$J = 3/2\, J_2 = q[(C-A)/C]$$

yielding

$$J_2 = 0.00109408$$

for which he derived

$$f = 1/296.75$$

With the revised mass-ratio moon/Earth, the result would have been

$$J_2 = 0.00109200$$

which is considerably in excess of the observed value

$$J_2 = 0.00108248$$

according to Kozai's recent determination.

The conclusion is that Clairaut's theory cannot be used to obtain a valid comparison with the observed value of J_2. It is of interest to note that Jeffreys (1936) remarked that the observed large-scale anomalies in surface gravity indicate that the value of J_2 derived from the precessional constant by the application of Clairaut's theory to the actual Earth may be in error by as much as twenty times the apparent uncertainty.

At the present time, the need for using the precessional constant for obtaining J_2 no longer exists. However, the value of (C-A)/C that can be computed from P is an important parameter pertaining to the Earth.

Comparison with Earlier Data

Table 2 Comparison of DeSitter's constants with current values

	DeSitter (1934)		1963	|Diff.|/mean error DeS.
a_e	6378387 ± 50	m	6378166 ± 25	4.4
g_0	9.78053 ± 3	$m\text{-}sec^{-2}$	9.78030 ± 2	7.7
J_2	0.00109408 ± 100		0.00108248 10	11.6
$\pi_\odot$	8''.8030 ± 25		8''.79415 ± 2	3.5
c	299774 ± 5	$km\text{-}sec^{-1}$	299792.5 ± .3	3.7
μ^{-1}	81.53 ± 6		81.300 ± 5	3.8

The first column of data in Table 2 lists the values of six constants used by DeSitter in 1934 when he made an effort to construct a consistent system. The mean errors correspond to the probable errors used by DeSitter.

The next column shows the values of the same constants that the author used early in 1963. For a_e and g_0, he relied on papers by Fisher[12,13]. The value $\mu^{-1} = 81.30$ had been used by the author for the past two years as a convenient rounded number obtained from astrometric determinations of the lunar inequality. It was good fortune that the value was so close to the Doppler determination from Mariner II data. The other constants have been discussed in this article.

The final column gives the ratio between the difference with DeSitter's value and DeSitter's mean error for each constant. The exceptionally large ratio 11.6 for J_2 has been explained as caused by use of a theory that is now known to be not wholly applicable. For the other constants, the average difference is 4.6 times the mean error.

It is a rather common experience in dealing with the interpretation of astronomical observations that the uncertainty of a result is underestimated. In general, three principal causes conspire to produce this condition of affairs. The first is that a solution can make adequate allowance for internal errors only. External effects that produce systematic errors remain unknown. A second cause is the failure or incompleteness of the theory used to represent the relationship between the unknowns and the observed quantities. The third cause is that, in most discussions of observational data, not all the observational evidence available is introduced. An example is the modification by Eckstein of Rabe's discussion of the Eros observations.

It will be interesting to see how the uncertainties currently assigned to these constants will stand up in the light of newer determinations that may be expected in the next 10 or 20 years.

References

1 Jones, H.S., "The solar parallax and the mass of the moon from observations of Eros at the opposition of 1931", Mem. Roy. Astron. Soc. 66, Part II (1941).

2 Rabe, E., "Derivation of fundamental astronomical constants from the observations of Eros during 1926-1945", Astron. J. 55, 112 (1950).

3 Pettengill, G.H., Briscoe, H.W., Evans, J.V., Gehrels, E., Hyde, G.M., Kraft, L.G., Price, R., and Smith, W.B., "A Radar Investigation of Venus", Astron. J. 67, 181 (1962).

4 Eckstein, M.C., "An investigation of the discrepancy of values of the astronomical unit", Astron. J. 68, 231-234 (1963).

5 Duncombe, R.L., "Motion of Venus 1750-1940", Astron. Papers Am. Ephemeris 16, Part I (1958).

6 Marsden, B.G., "An attempt to reconcile the dynamical and radar determination of the astronomical unit", (to be published in Proceedings of the Symposium on Astronomical Constants, 1964).

7 Jeffreys, H., "On the lunar equation", Monthly Notices Roy. Astron. Soc. 102, 194 (1942).

8 Delano, E., "The lunar equation from observations of Eros, 1930-1931", Astron. J. 55, 129 (1950).

9 Jackson, J., " A new determination of the constant of nutation", Monthly Notices Roy. Astron. Soc. 90, 733 (1930).

10 Jeffreys, H.,"On the figures of the Earth and the moon", Monthly Notices Roy. Astron. Soc. Geophys. Suppl. 4, 1 (1937).

11 deSitter, W., "On the system of astronomical constants", Bull. Astron. Inst. Neth. 8,213 (1938), revised and edited by D. Brouwer .

12 Fischer, I.,"An astrogeodetic world datum from gloidal heights based on the flattening f = 1/298.3" J. Geophys.

Res. 65, 2067 (1960).

13 Fischer, I. "Parallax of the moon in terms of a world geodetic system", Astron. J. 67, 373-378 (1962).

THE EVALUATION OF CERTAIN ASTRONOMICAL CONSTANTS FROM THE RADIO TRACKING OF MARINER II

John D. Anderson,* George W. Null,+ and Catherine T. Thornton‡

Jet Propulsion Laboratory,
California Institute of Technology, Pasadena, California

Abstract

The mathematical techniques used to obtain astronomical constant information from the Mariner II data are described from the viewpoint of the parameter estimation procedure, the calculation of the residuals, and the variational coefficients for orbital parameters and physical constants. Numerical illustrations of the reduction are included for 1) data taken during the portion of the flight when the spacecraft was dominated by the sun, 2) data taken during the encounter with Venus, and 3) data taken over the entire mission. Provisional values for the masses of the moon and Venus are given.

I. Introduction

The orbit of the Mariner II spacecraft, which was launched from the Atlantic Missile Range, Cape Canaveral (now Cape Kennedy), Florida, on August 27, 1962, is unique in that it was dominated first by the Earth, then the sun, and finally the planet Venus. On December 14, 1962, the spacecraft came within about

---

Presented as Preprint 63-424 at the AIAA Astrodynamics Conference, New Haven, Conn., August 19-21, 1963. This paper presents the results of one phase of research carried out at the Jet Propulsion Laboratory, California Institute of Technology, under Contract No. NAS 7-100, sponsored by NASA. The construction of a program to reduce the large quantity of Mariner II data naturally involved the efforts of many people. In particular, the reduction would have been impossible without the computer program prepared for the IBM 7090 computer by R. Henry Hudson, Michael R. Warner, and Melba W. Nead of the Jet Propulsion Laboratory. Also, the encouragement and useful criticism provided throughout the development by T. W. Hamilton and P. R. Peabody of the Jet Propulsion Laboratory were invaluable.

*Senior Research Engineer, Systems Analysis.
+Research Engineer, Systems Analysis.
‡Research Engineer, Systems Analysis.

35,000 km of Venus, and its flight path was deflected by about 40° because of its close encounter. In addition, precise two-way doppler data were obtained throughout the 129-day period from launch to January 3, 1963. Fortunately, prior to the launch of Mariner II, an atomic reference had been installed at the Goldstone, California, tracking station of the NASA/JPL Deep Space Instrumentation Facility (DSIF), and this piece of equipment allowed the transmitter frequency to be held to better than one part in 10^{11} over a period of an hour. Equivalently, the doppler shift in the received signal was measured to an accuracy of about 0.03 cps. In terms of velocity units, the corresponding accuracy of the range rate between the probe and station was on the order of 1/2 cm/sec at a received frequency of 960 Mc and a cycle count time of 60 sec.

A combination of the forementioned factors allows the determination of certain constants of the solar system to an accuracy that has been unobtainable in the past. In particular, at least an order of magnitude improvement in the mass of Venus appears possible because of the close approach to the planet. The same order of improvement should be obtained for the mass of the moon because the periodic component in the data which results from the motion of the Earth about the center of mass or barycenter of the Earth-moon system is appreciable. The astronomical unit (AU) can also be obtained from the data, and its accuracy should eventually be comparable to that of the recent radar determinations from the tracking of Venus.

The purpose of the present paper is to describe the computational methods used in the postflight reduction of the Mariner II data. The basic computer program is the Jet Propulsion Laboratory (JPL) Lunar and Planetary Orbit Determination Program which will be described in detail in a forthcoming JPL technical report. Therefore, all of the formulas used to evaluate the quantities required in the solution for the physical constants are not given here. Hopefully, however, the philosophy of the reduction of the data is clarified, and enough of the computational techniques are given so that the detailed formulas required in the computer program follow naturally from the discussion.

In Sec. II the form of the weighted least-squares estimator is given, and the expressions that evaluate the error on the estimated parameters are listed. A formula for the computation of the two-way doppler observable is given in Sec. III. To form residuals, this computed value of doppler shift is simply subtracted from the observed value at the time in question. In Sec. IV the method of finding the coefficients in the equations of condition by a numerical integration of a set of variational equations is described, and in Sec. V four numerical examples of the reduction of a set of Mariner II data are given. The first two examples treat data taken before the probe was primarily influenced by Venus, whereas the third and fourth consider only the data obtained during the encounter with Venus. An evaluation of the results of the determinations is also included in Sec. V. Finally, in Sec. VI the ultimate scientific value of the Mariner II data is discussed.

II. Statistical Formulas

The orbital parameters and physical constants are estimated by the method of least-squares.[1] Thus, if the vector $\vec{x}$ represents the parameters that are subject to estimation and $\hat{\vec{z}}$ is a vector representing the data, then the least-squares estimate $\vec{x}^*$ of $\vec{x}$ is obtained by minimizing the quadratic form $Q = \delta\hat{\vec{z}}^T \delta\hat{\vec{z}}$. The residual vector $\delta\hat{\vec{z}}$ is given by $\delta\hat{\vec{z}} = \hat{\vec{z}} - \vec{z}(\vec{x})$. Actually, the rigorous sum of squares of the residuals is modified by introducing a diagonal weighting matrix W so that a weight can be assigned to each data point. Also, a priori information on the parameter vector $\vec{x}$ is included by treating the a priori values $\tilde{\vec{x}}$ of the parameters as additional observations. In this case, the weighting matrix is written as the inverse of the covariance matrix $\tilde{\Gamma}_x$ on the a priori parameter vector. Thus, the quadratic form Q that is minimized to provide an estimate of $\vec{x}$ is $Q = \delta\hat{\vec{z}}^T W \delta\hat{\vec{z}} + \delta\tilde{\vec{x}}^T \tilde{\Gamma}_x^{-1} \delta\tilde{\vec{x}}$, where $\delta\tilde{\vec{x}} = \tilde{\vec{x}} - \vec{x}$, and the corresponding linearized estimation formula is

$$\left(A^T W A + \tilde{\Gamma}_x^{-1}\right) \delta\vec{x}^* = A^T W \delta\hat{\vec{z}} + \tilde{\Gamma}_x^{-1} \delta\tilde{\vec{x}} \tag{1}$$

The vector $\delta\vec{x}^*$ represents the corrections to the nominal values $\vec{x}$ of the parameters (i.e., $\vec{x}^* = \vec{x} + \delta\vec{x}^*$), whereas the matrix A relates variations in the parameters to variations in the data through the equations of condition $\delta\vec{z} = A\delta\vec{x}$. It is not necessary that the nominal values of the parameters agree with the a priori values, and, therefore, the estimation formula (1) can be successively applied until a convergence is attained. The criterion for convergence is the convergence of the function Q, which is typically stabilized within the limits of computational accuracy after three or four iterations. Of course, the number of iterations depends on the departure of the initial values of the parameters from the estimate $\vec{x}^*$, which minimizes Q.

In addition to obtaining the estimate of the parameters by formula (1), the error on this estimate is also required. For this purpose, two error expressions are available. The first expression assumes that the weighting matrix W is the inverse of the covariance matrix on the data $\hat{\vec{z}}$ and that all parameters that sensibly affect the data are included in the solution. Then the covariance matrix Γ_x on the estimated parameters is given by

$$\Gamma_x = \left(A^T W A + \tilde{\Gamma}_x^{-1} \right)^{-1} \tag{2}$$

A more general expression admits the possibility that a number of sensible parameters $\vec{y}$ are excluded from the solution. In other words, variational coefficients B are introduced such that $\delta \vec{z} = A \delta \vec{x} + B \delta \vec{y}$, and if the covariance matrix on $\vec{y}$ is Γ_y and $\vec{x}$ and $\vec{y}$ are statistically independent, then the more general parameter covariance matrix Γ'_x is given by

$$\Gamma'_x = \Gamma_x + \Gamma_x A^T W B \Gamma_y B^T W A \Gamma_x \tag{3}$$

In either case the square roots of the diagonal elements of Γ_x or Γ'_x are taken to form a standard deviation on each parameter. Also, Γ_x or Γ'_x is normalized so that all of the diagonal elements are unity. The resulting matrix is thus the correlation matrix on the parameters, and the degree to which various parameters arecorrelated by a particular data configuration $\hat{\vec{z}}$ is known.

The usefulness of the resulting errors on the parameter estimate is strictly dependent on the reality of the W and $\tilde{\Gamma}_x$ matrices, which are constructed from a knowledge of the errors on the data and a priori parameters. Thus, $\tilde{\Gamma}_x$ reflects a confidence in parameter determinations prior to the introduction of the data $\hat{\vec{z}}$. It is customary to express results in terms of probable errors, and thus the construction of $\tilde{\Gamma}_x$ follows naturally from the published values of these errors. It follows that if W is also expressed in terms of probable errors, then the errors on the estimate of the parameters are in terms of probable errors. This is true only if the ratio of the standard deviation to the probable error is the same for the data, the a priori parameters, and the estimated parameters. All uncertainties in this paper are assumed given in terms of probable error.

It should be noted that if the a priori information is not used, then for uniform weighting --an option that makes sense only for the same kind of data --the estimate of the parameters is unaffected by assumptions on the accuracy of the data. Also, the errors on the estimated parameters are proportional to the data errors and can be scaled up and down at will. However, if $\tilde{\Gamma}_x^{-1}$ is not null, then the relative importance of data and a priori parameter errors affect both the parameter estimate and the error expressions. Actually, many options exist in the JPL orbit determination program to aid in the selection of meaningful W and $\tilde{\Gamma}_x$ matrices. These are not described here, although the numerical examples presented in Sec. V include a specification of the assumptions on W and $\tilde{\Gamma}_x$. After such assumptions are made, the numerical evaluation of $\vec{x}^*$, Γ_x, or

Γ'_x and the associated correlation matrix is accomplished by the foregoing formulas in a straightforward manner. The description of the computation of the doppler observable $\vec{z}(\vec{x})$ is given in the next section, and the evaluation of the matrix A is described in Sec. IV. The matrix B can always be extracted from a larger A matrix that includes the variational coefficients for all sensible parameters.

III. Doppler Residuals

The doppler residuals are computed by the equation

$$\delta\hat{\vec{z}} = \hat{\vec{z}} - \vec{z}(\vec{x}) \tag{4}$$

where $\hat{\vec{z}}$ is an (n x 1) vector of observed values, and $\vec{z}$ is an (n x 1) vector of the computed values of $\hat{\vec{z}}$.

By doppler data is meant coherent, two-day doppler, which is defined in the following way: An electromagnetic signal is transmitted from a point P_1 at time t_1 and reaches a point P_2 at t_2. At P_2 the received signal is multiplied by a constant k and instantaneously retransmitted. This second signal is then received at a point P_3 at time t_3. For this experiment P_1 and P_3 are stations on Earth, and P_2 is the position of the spacecraft at t_2.

Let ν_T be the frequency transmitted at t_1 and ν_R the frequency received at t_3, and define

$$\phi(t_3) = \nu_R/k\nu_T = \nu_R/b \tag{5}$$

The radar equipment is designed so that a quantity $f(t) = a + b[1 - \phi(t)]$ is continuously received over an interval of time T_c. The constant a is an arbitrary additive frequency used to keep $f > 0$ at all times. Let the total number of cycles counted in this time interval be denoted by N. Then N/T_c is the quantity-designated two-way doppler with a count time T_c, which is measured in seconds so that N/T_c has the dimension cycles per second. The vector $\hat{\vec{z}}$, Eq. (4), is composed of n measurements of N/T_c. Each measurement is tagged with the real time T', the midpoint of the count time associated with the measurement.

The components of the vector $\vec{z}$ are computed as follows: For a given count time T_c and a datum time T',

$$\frac{N}{T_c} \equiv \frac{1}{T_c} \int_{T' - (T_c/2)}^{T' + (T_c/2)} f(t)\,dt \tag{6}$$

The function f(t) is computed as follows:

$$f(t) = a + b\left[1 - \phi(t)\right] \tag{7}$$

where a and b are given constants and

$$\phi(t) = \frac{\nu_R}{k\nu_T} = 1 - \frac{\dot{\rho}_1 + \dot{\rho}_2}{c} + \frac{1}{c^2}\left[\dot{\rho}_2^2 + \dot{\rho}_1\dot{\rho}_2 + H\right] \tag{8}$$

$$H = \vec{r}(t_2) \cdot \left[\frac{\dot{\rho}_2}{\rho_2}\dot{\vec{R}}_i(t_3) - \frac{\dot{\rho}_1}{\rho_1}\dot{\vec{R}}_q(t_1)\right] \tag{9}$$

where

c = speed of light

$\vec{r}(t_2) \equiv$ geocentric position vector of the spacecraft at t_2

$\dot{\vec{R}}_q(t_1) \equiv$ geocentric velocity vector of the transmitting station q at t_1

$\dot{\vec{R}}_i(t_3) \equiv$ geocentric velocity vector of the receiving station i at t_3

$$\vec{\rho}_1 \equiv \vec{r}(t_2) - \vec{R}_q(t_1) \qquad \dot{\vec{\rho}}_1 \equiv \dot{\vec{r}}(t_2) - \dot{\vec{R}}_q(t_1)$$

$$\vec{\rho}_2 \equiv \vec{r}(t_2) - \vec{R}_i(t_3) \qquad \dot{\vec{\rho}}_2 \equiv \dot{\vec{r}}(t_2) - \dot{\vec{R}}_i(t_3)$$

$$\rho_j \equiv \left|\vec{\rho}_j\right| \quad (j=1,2) \qquad \dot{\rho}_j \equiv \frac{\vec{\rho}_j \cdot \dot{\vec{\rho}}_j}{\rho_j} \quad (j=1,2)$$

The time t_3 is the independent variable t; t_2 and t_1 are found from t_3 by an iterative technique. This technique is essentially a light-time correction to the data.

The computation of $\phi(t)$ involves $\vec{r}(t_2)$ and $\dot{\vec{r}}(t_2)$, the geocentric position and velocity vectors of the spacecraft at t_2. These vectors are obtained in the coordinates of date from the trajectory program.[4] Corrections to the data because of atmospheric refraction are applied to the computed values of the range rates $\dot{\rho}_1$ and $\dot{\rho}_2$.

IV. Equations of Condition

The method of computing the variational coefficients in the equations of condition $\delta\vec{z} = A\delta x$ is accomplished by solving a system of differential equations in the coefficients that relate trajectory variations at time t to variations at some epoch t_0. In other words, designate the position and velocity of the spacecraft at time t by the vector $\vec{X}$. This vector is obtained as the solution to the spacecraft equations of motion, which can be written in the form $\dot{\vec{X}} = \dot{\vec{X}}(\vec{X}, \vec{Y}, t)$, where $\vec{Y}$ is a vector of physical constants. The variation in $\dot{\vec{X}}$ can be obtained by differentiating these equations of motion with respect to each position and velocity component and each physical constant. Time variations are ignored. Thus, formulas are available for the elements of the matrices Φ and Θ defined by $\delta\dot{\vec{X}} = \Phi\delta\vec{X} + \Theta\delta\vec{Y}$. On the other hand, the solution to the equations of motion gives $\vec{X}$ in terms of the initial conditions $\vec{X}_0$ at time t_0 and the physical constants so that $\vec{X} = \vec{X}(\vec{X}_0, \vec{Y}, t)$. Therefore, the variation in $\vec{X}$ can be written $\delta\vec{X} = U\delta\vec{X}_0 + V\delta\vec{Y}$, and by differentiating this variation with respect to time, one obtains $\delta\dot{\vec{X}} = \dot{U}\delta\vec{X}_0 + \dot{V}\delta\vec{Y}$. For the two expressions for $\delta\dot{\vec{X}}$ to be equal for arbitrary variations $\delta\vec{X}_0$ and δY, it is necessary that the following differential equations be satisfied:

$$\dot{U} = \Phi U \tag{10}$$

$$\dot{V} = \Phi V + \Theta \tag{11}$$

The numerical solution to the equations of motion, the computation of the Φ matrix, and the numerical solution to Eq. (10) are all described by Holdridge.[4] Initially, U is the unit matrix $U(t_0) = I$, and V is null so that $V(t_0) = 0$. Once U(t) has been obtained, the solution to Eq. (11) is computed in terms of U by numerical quadrature integration of the following formula:

$$V(t) = U(t)\int_0^t U^{-1}(\tau)\Theta(\tau)d\tau \tag{12}$$

It is easily shown by direct substitution that this is a solution to Eq. (11). The computation of the matrix Θ is accomplished by evaluating formulas that represent partial derivatives of the equations of motion of the spacecraft with respect to physical constants.

It still remains to obtain the variation in the observables $\vec{z}$. Partition the A matrix into two parts corresponding to the variation of $\vec{z}$ with respect to the orbital initial conditions $\vec{X}_0$ and the physical constants $\vec{Y}$, respectively. That is, let $\delta\vec{z} = A_1\delta\vec{X}_0 + A_2\delta\vec{Y}$ so that $A = (A_1 \mid A_2)$. Now, by Sec. III the doppler observable can be written in terms of the position and velocity $\vec{X}$ of the spacecraft and also in terms of physical constants $\vec{Y}$ that enter explicitly in the formula for $\vec{z}$. Thus, in general, $\vec{z} = \vec{z}(\vec{X}, \vec{Y}, t)$, although any explicit dependence on the time is unimportant in what follows because of the neglect of time variations. The variation in $\vec{z}$ is, therefore, given by $\delta\vec{z} = G\delta\vec{X} + H\delta\vec{Y}$, where the matrices G and H are obtained by differentiating the expression for the doppler data with respect to $\vec{X}$ and $\vec{Y}$. Again by equating two expressions for the same variation $\delta\vec{z}$, the formulas for A_1 and A_2 are

$A_1 = GU$ and $A_2 = GV + H$, and the required matrix A assumes the following form:

$$A = (GU \mid GV + H) \tag{13}$$

In practice, the vector $\vec{Y}$ contains all of the physical constants that enter either into the trajectory computation or the data computation. Thus, if a constant enters only in the data computation, then the row of the V matrix which corresponds to that constant is zero. An example of this sort of constant is the velocity of light or the geocentric coordinates of the station. Similarly, a constant such as the mass of Venus, which does not enter explicity in the doppler formula, will have a corresponding zero row in the H matrix.

Plots of the partial derivatives computed by the foregoing technique are given in Figs. 1 and 2. Figure 1 shows the partial derivatives of the doppler observable during the heliocentric portion of the flight for an initial epoch eight days after launch (just after the midcourse maneuver on September 5, 1962). Figure 2 shows partial derivatives during the Venus-centered portion of the flight for an epoch on December 7, 1962, which was seven days before closest approach to Venus. Closest approach occurred at 20^h universal time (UT) on December 14, 1962. These partial derivatives have been scaled so that each curve represents the doppler perturbation in cycles per second caused by the indicated perturbation in the parameter. The conversion between doppler and range rate is given by 1 cps = 0.156 m/sec. The definitions of the plotted parameters are

GM_m = universal gravitational constant multiplied by the mass of the moon

ϵ = a constant relating to solar pressure. The solar pressure force on the spacecraft is of the form $F = K(1 + \epsilon)/r^2$, where r is the distance from the sun, K is a constant, and F is directed radially away from the sun.

A_E = astronomical unit

M_V = mass of Venus in solar mass units

V. Numerical Examples

The first example illustrates a least-squares solution for data taken before the Venus encounter period. Only data from the Goldstone, California, DSIF tracking station are used, and, in addition, 90% of the data is neglected by selecting every tenth doppler measurement as data. As a result, 518 measurements are used in this example, with the first measurement occurring on September 5, 1962, and the last on December 1,

1962. The count time for the data is 50 sec over most of the heliocentric portion of the flight. On September 23, 1962, there are three data points with count times of 600 sec and one with a count time of 420 sec. From September 29 to November 25, 1962, there are 49 points with a count time based on 600 sec. Occasionally this is reduced because of the counter missing a cycle before it can count for 600 sec.

The parameters selected for adjustment by the least-squares method are the selenocentric gravitational constant GM_m (the universal gravitational constant times the mass of the moon), the astronomical unit A_E, the geocentric radii R and longitudes λ of the Goldstone transmitter (station 3) and receiver (station 2), and the six geocentric equatorial coordinates of date (X_0, Y_0, Z_0, $\dot{X}_0$, $\dot{Y}_0$, $\dot{Z}_0$) for the spacecraft position and velocity at the epoch t_0. The station latitude ϕ is not included in the solution because a separation of the radius and latitude is impossible with the Mariner II data. In other words, the data essentially measure the rotational velocity of the station about the Earth's polar axis, or, equivalently, the quantity $R \cos \phi$ is obtained from a combination of the data with the known angular velocity of the Earth's rotation. An additional parameter ϵ is added to the solution to account for a heliocentric inverse square perturbative force arising from solar radiation pressure. In other words, the area-to-mass ratio of the spacecraft is relatively large (about 2.16×10^{-11} km^2/g or 10^9 times the same ratio for the Earth), and the solar radiation pressure imparts a perturbative acceleration on the order of 10^{-10} km/sec^2 at one AU. For comparison, the total acceleration is about 6×10^4 times this number. The parameter ϵ represents a percentage change in the nominal value of the acceleration caused by radiation pressure.

The input values of the parameters are the following:

t_0 = September 5, 1962, $0^h23^m32\overset{s}{.}000$ (UT)

$X_0 = -1.4243704 \times 10^6$ km

$Y_0 = -1.9397521 \times 10^6$ km

$Z_0 = -0.10293568 \times 10^6$ km

$\dot{X}_0 = -1.7445233$ km/sec

$\dot{Y}_0 = -2.4233963$ km/sec

$\dot{Z}_0 = -0.10989823$ km/sec

$\epsilon = 0.383$

$GM_m = 4900.7589$ km^3/sec^2

$A_E = 1.4959900 \times 10^8$ km

$R_2 = 6372.0327$ km

$\phi_2 = 35\overset{\circ}{.}20598$

$\lambda_2 = 243\overset{\circ}{.}15175$

$R_3 = 6372.0165$ km

$\phi_3 = 35\overset{\circ}{.}11654$

$\lambda_3 = 243\overset{\circ}{.}19539$

c = 299792.5 km/sec (speed of light)

In this example and those that follow, any constant that is not stated explicitly is given the value specified by Clarke.[2] With the foregoing values of the parameters, the weighted sum of squares of the residuals is $\partial \hat{z}^T W \partial \hat{z} = 649.73$. A uniform weighting is assumed in the first two examples given here, and each data point is assigned a probable error of 0.0569 cps or, equivalently, 0.009 m/sec in range rate. If this figure is multiplied by the square root of 10 to allow for the selection of every tenth measurement, then the equivalent error on each measurement, if used without a selection process, is 0.03 m/sec. This is about a factor of 6 times the estimate of the actual random error on the data, but an allowance is thus provided for possible systematic error sources.

For the few points where the count time is larger than 50 sec, the assumed data error is decreased by $\sqrt{60/T_c}$ to allow for the fact that more information is included in data counted over T_c sec ($T_c > 60$) than there is in a data point counted over 50 sec. The factor 60 is used under the square root instead of 50 because the data counted over 50 sec are sampled every 60 sec. The idea behind this sort of weighting is that the systematic errors in the data dominate the random errors, which behave like the inverse of the count time. However, if systematic errors dominate, then the error on a single data point behaves like the inverse of the square root of the sample time. In other words, the statistics on the solution for the parameters are independent of the sample rate.

Another exception to a uniform weighting is for measurements taken at a low elevation angle where the refraction correction is unreliable. These data are effectively weighted so that they do not influence the solution for the parameters. Figs. 3-5 show residual plots for Goldstone for data taken, respectively, near the beginning, middle, and end of the heliocentric portion of the flight. The ordinate is the doppler residual in cycles per second. These plots would seem to indicate that the errors assigned to the data are not overly optimistic in the sense of being too small. For comparison, data taken from an early pass over the Johannesburg DSIF station (station 5) are shown in Fig. 6. These data are not used in the solution, but the residuals can be computed from the fit to the Goldstone data. The relatively large scatter in the points is attributed to the lack of an atomic reference for the transmitter frequency.

An application of the formulas of the preceding sections without the introduction of a priori information on the parameters ($\tilde{\Gamma}_x = 0$) yields the following corrections and probable errors:

$\delta X_0 = -160 \pm 108$ km $\delta GM_m = 2.016 \pm 0.20\ km^3/sec^2$

$\delta Y_0 = -113 \pm 141$ km $\delta A_E = 16892 \pm 4640$ km

$\delta Z_0 = 2216 \pm 251$ km $\delta R_2 = -170.6 \pm 18.2$ m

$\delta \dot{X}_0 = -0.0856 \pm 0.04$ m/sec $\delta \lambda_2 = -0^\circ.00177 \pm 0^\circ.00043$

$\delta \dot{Y}_0 = 0.108 \pm 0.03$ m/sec $\delta R_3 = -168.6 \pm 17.3$ m

$\delta \dot{Z}_0 = -0.582 \pm 0.111$ m/sec $\delta \lambda_3 = -0^\circ.00171 \pm 0^\circ.00044$

$\delta \epsilon = 0.530 \pm 0.12$

The new weighted sum of squares of the residuals is now reduced to 67.43. A large variation in the AU occurs because a perturbative force has been neglected in these examples. This force results because of a nonstandard operation of the attitude control system, which releases cold nitrogen gas from a number of jets. As a result, a low-thrust perturbative acceleration of about 3×10^{-11} km/sec^2 (about one-third the magnitude of the solar radiation pressure) is imparted to the spacecraft throughout the duration of the mission, and because this effect is neglected, other parameters absorb the low-thrust perturbation to some degree. Clearly, some of the effect is thrown into the solar radiation coefficient ϵ. However, the correlation matrix given in Table 1 indicates a fairly strong correlation (-0.75) between the coefficient ϵ and the AU, and, therefore, the AU would be expected to absorb some of the low-thrust perturbation as well.

The correlation of A_E and ϵ arises because a variation in the AU must also result in a variation in the mass of the sun if the well-known periods of the Earth and planets are to remain constant. Therefore, the inverse square heliocentric distance effect of the AU on the trajectory looks very much like the inverse square effect of radiation pressure so long as the spacecraft is dominated by the sun (see Fig. 1). As a result, the solution just given provides practically no information on astronomical constants, with the notable exception of the mass of the moon. In this case the correlations with other parameters are all small, and a variation in the trajectory has little effect on GM_m. Here the doppler data are measuring the monthly motion of the Earth above the center of mass or barycenter of the Earth-moon system (see GM_m curve in Fig. 1), and variations in the trajectory have practically no effect on this measurement.

As a second numerical example the same input values of the parameters are used as before, but the AU is dropped as a parameter for adjustment. Because the AU can no longer absorb some of the low-thrust perturbation, the expected result of this example is a fairly significant difference from the previous solution for ϵ and those parameters that are highly correlated with the radiation pressure coefficient. The variations given in the listing to follow generally support such a conclusion. However, the important point is that the relatively independent mass of the moon changes only by an amount equal to its probable error, whereas other highly correlated parameters change by as much as four times their estimated error. Therefore, the solution for GM_m is probably not significantly degraded by the low-thrust effect. The sum of squares of the residuals after convergence of Q is 79.86 for this second example, and, thus, by a comparison with 67.43 for the first example, the fit to the data has not been degraded appreciably.

$\delta X_0 = -15.5 \pm 101$ km $\qquad \delta\epsilon = 0.859 \pm 0.086$ m/sec

$\delta Y_0 = 115 \pm 127$ km $\qquad \delta GM_m = 1.803 \pm 0.19$ km^3/sec^2

$\delta Z_0 = 1982 \pm 241$ km $\qquad \delta R_2 = -157.3 \pm 17.9$ m

$\delta \dot{X}_0 = -0.188 \pm 0.028$ m/sec $\qquad \delta\lambda_2 = -0^\circ\!.00277 \pm 0^\circ\!.00034$

$\delta \dot{Y}_0 = 0.182 \pm 0.023$ m/sec $\qquad \delta R_3 = -156.7 \pm 17.1$ m

$\delta \dot{Z}_0 = -0.689 \pm 0.107$ m/sec $\qquad \delta\lambda_3 = -0^\circ\!.00276 \pm 0^\circ\!.00034$

From the first example the estimated value of GM_m is

$$GM_m = 4902.775 \pm 0.20 \text{ km}^3/\text{sec}^2$$

The corresponding value for the mass of Earth GM_E is from Clarke[2]:

$$GM_E = 3.986032(\pm 0.000030) \times 10^5 \text{ km}^3/\text{sec}^2$$

and the mass ratio of the Earth to the moon is

$$GM_E/GM_m = \mu = 81.30155 \pm 0.0034$$

For a third example, a fit to all the Goldstone doppler data during the encounter period of the flight from December 7 to 20, 1962, is given. The count time T_c for the encounter data is normally 600 sec, and therefore, because the sample time is equal to the count time, the data sampling rate is the same as for the foregoing two heliocentric examples. Thus, the weights for the data are the same as before, and again measurements at a low elevation angle are discarded by applying small weights. In addition, a measurement occasionally occurs with a count time shorter than 600 sec, and the corresponding error on such a measurement is increased by a factor of $\sqrt{600/T_c}$ or, what amounts to the same thing, the standard weight is multiplied by the factor $T_c/600$. The total number of measurements used in this encounter example is 610.

The mass M_V of Venus in units of the solar mass is added to the parameter list in this example, but the mass of the moon is dropped, and its value from the first example is used in the trajectory computation. Also, a value of the AU as determined by radar is adopted and an a priori probable error of ±500 km is assigned to it. The least-squares solution is then employed with the ±500-km error inserted in the a priori error matrix $\tilde{\Gamma}_x$. The initial value of the AU is 149, 598, 850 km. For comparison, four independent radar determinations give the following:

Muhleman, et al.[6] (Goldstone)	A_E = 149, 598, 845 ±250 km
Pettengill, et al.[7] (Millstone)	A_E = 149, 597, 850 ±400 km
Thomson, et al.[9] (Jodrell Bank)	A_E = 149, 599, 800 ±5000 km
Kotelnikov[5]	A_E = 149, 599, 500 ±800 km

The initial values of the parameters are

t_0 = December 7, 1962, 0^h(UT)	$\dot{Y}_0 = -10.825346$ km/sec
$X_0 = -3.6494745 \times 10^7$ km	$\dot{Z}_0 = -5.4161879$ km/sec
$Y_0 = -2.9910919 \times 10^7$ km	$\epsilon = 0.383$
$Z_0 = -0.94548403 \times 10^7$ km	$M_V = 2.4471118 \times 10^{-6}$
$\dot{X}_0 = -7.3401189$ km/sec	$A_E = 1.4959885 \times 10^8$ km

The station coordinates are taken from the solution of the first example. The initial sum of squares of the residuals is 336310, and the corrections to the parameters are the following:

$\delta X_0 = -34.5 \pm 181$ km	$\delta M_V = 6.680(\pm 0.591) \times 10^{-10}$
$\delta Y_0 = 33.3 \pm 170$ km	$\delta A_E = 0.0 \pm 498$ km
$\delta Z_0 = 265 \pm 74$ km	$\delta R_2 = -23.8 \pm 16.5$ m
$\delta \dot{X}_0 = 0.115 \pm 0.149$ m/sec	$\delta \lambda_2 = -0^\circ.0001 \pm 0^\circ.00038$
$\delta \dot{Y}_0 = -0.064 \pm 0.215$ m/sec	$\delta R_3 = -23.8 \pm 16.5$ m
$\delta \dot{Z}_0 = -0.369 \pm 0.055$ m/sec	$\delta \lambda_3 = -0^\circ.0001 \pm 0^\circ.00038$
$\delta \epsilon = 0.538 \pm 0.226$	

The new sum of squares of the residuals is 27.66. The correlation matrix for this solution is shown in Table 2. Again the low-thrust perturbation is present, but now, because of the relatively short interval over which data are taken, the radiation pressure can absorb practically all of a radial perturbing force caused by the attitude control system. The spacecraft remains attitude-stabilized over the entire encounter period, so that the solar panels remain pointed at the sun, and the radiation pressure effect is in fact radial. Telemetry information indicates that a large part of the low-thrust perturbation is radial also, but there is a tangential component as well. Therefore, the foregoing solution is not entirely satisfactory.

There is also a loss in the accuracy of the trajectory calculation during a period centered at the time of closest approach to the planet. This arises because the trajectory is computed in geocentric coordinates throughout the mission. Thus, when the probe is in the vicinity of Venus, the direct term in the perturbative acceleration produced by Venus dominates the geocentric equations of motion; this direct term is computed by subtracting the geocentric position of Venus from the geocentric position of the probe. However, during encounter, Venus is about 60,000,000 km from the Earth and because the computer carries only eight places in the trajectory calculations, the difference in these two positions can be computed to only the nearest kilometer. On the other hand, the probe at closest approach is about 35,000 km from Venus and the error in the Venus-centered position occurs in the fifth place instead of the eighth, as it would for a Venus-centered integration. Of course, this is the maximum effect at the time of closest approach.

Correlations of the mass of Venus with the position of the probe are fairly high, although the correlation with the radiation-pressure coefficient is insignificant. Therefore, the exclusion of the low-thrust forces probably does not affect the mass solution appreciably. However, the inaccuracy in the calculation of the trajectory near Venus is a serious matter, and the value of the mass determined with a Venus-centered integration could easily fall outside the probable error as computed from the normal equations associated with this solution.

The mass ratio of the sun to Venus from the foregoing encounter solution is

$$\frac{M_S}{M_V} = 408,533.5 \pm 30$$

The probable error that results from the solution is arbitrarily multiplied by a factor of 3 to allow for systematic errors caused by the inaccurate trajectory calculation and possible nonradial low-thrust forces, if significant.

This provisional value of the mass as determined from Mariner II is consistent with recent determinations by Clemence[3] and Rabe[8]:

Clemence $M_S/M_V = 409,300 \pm 1400$

Rabe $M_S/M_V = 408,645 \pm 208$

It is interesting to perform a least-squares reduction of the encounter data with no assumed a priori information on the AU. The conditions are exactly as in the foregoing solution, except that the matrix $\tilde{\Gamma}_x$ is null. Also, the initial value of the AU is 149, 599, 000 km. The corrections to the initial values with the corresponding probable errors are

$\delta X_0 = 539.5 \pm 1616\,\text{km}$ $\quad \delta M_v = 8.933(\pm 5.997) \times 10^{-10}$

$\delta Y_0 = 438.75 \pm 1145\,\text{km}$ $\quad \delta A_E = -2054 \pm 5342\,\text{km}$

$\delta Z_0 = 517.75 \pm 691\,\text{km}$ $\quad \delta R_2 = -23.6 \pm 16.5\,\text{m}$

$\delta \dot{X}_0 = 0.0525 \pm 0.236\,\text{m/sec}$ $\quad \delta\lambda_2 = -0{:}00009 \pm 0{:}00038$

$\delta \dot{Y}_0 = 0.0467 \pm 0.378\,\text{m/sec}$ $\quad \delta R_3 = -23.6 \pm 16.5\,\text{m}$

$\delta \dot{Z}_0 = -0.4908 \pm 0.318\,\text{m/sec}$ $\quad \delta\lambda_3 = -0{:}00009 \pm 0{:}00038$

$\delta\epsilon = 0.294 \pm 0.741$

The correlation matrix associated with this solution is given in Table 3. A curious situation arises in that the sum of squares of the residuals is reduced to only 46.43. The introduction of another free parameter, the AU, provides a worse fit to the data than the preceding solution, which absorbed all the correction in other parameters, notably the mass of Venus. This phenomenon is attributed to numerical problems with a near-degenerate set of normal equations.

The determination of both the AU and the mass of Venus from the encounter data as presented here is not in agreement with previous reductions of the same data. Until recently, numerical errors in the computation of the equations of condition for the encounter data resulted in a fictitious decoupling of these parameters. The correlations as given in Table 3 are representative of the actual situation and indicate the necessity of adding the heliocentric data to the solution if independent values of A_E and M_V are to be obtained.

VI. Final Reduction of the Data

The full scientific value of the Mariner II tracking data will not be realized until the heliocentric and encounter data are combined in one least-squares reduction. This is impossible at the present time because 1) the low-thrust forces are neglected, 2) the calculation of the trajectory is inaccurate in the vicinity of Venus, and 3) the effects of uncertainties in the ephemerides of the Earth and Venus are unknown. It is unacceptable to be satisfied with the results of Sec. V without a detailed investigation of these three sources of error. Sometime in 1965 it is expected that the final reduction will be accomplished, with the inclusion of a physically reasonable low-thrust model, a Venus-centered integration of the equations of motion during encounter, and an inclusion of orbital elements of the Earth and Venus as additional free parameters in the solution.

Also, the possibility of uncoupling the AU and the mass of Venus implies an independent determination of both of these constants from the Mariner II data only. This in itself demands that a final reduction be forthcoming.

In order to obtain some idea of the strength of a combined solution in determining the masses of the moon and Venus and the AU, a fourth numerical example is given. Data within the interval from September 5 to December 30, 1962, are fitted by weighted least-squares as in the previous examples. The probable errors and the correlation matrix associated with the solution give some idea of the potential scientific value of the data. It should be remembered, however, that the results are conditioned by the model used to obtain them. Thus, any of the three error sources mentioned in the foregoing could increase the probable errors computed with the present model.

The count time associated with the combined data is primarily 60 sec, although some data occur with a 50-sec count time, particularly in the early portion of the mission, before September 29, 1962. However, the interval between samples is always 600 sec, and therefore the weighting used is the same as for the previous numerical examples; the probable error on each point is assumed to be 0.057 cps. The total number of points is 1342. The epoch for the six initial conditions is again September 5, 1962, $0^h 23^m 32^s\!.000$(UT), and the input values of the parameters are exactly the same as in the heliocentric solutions of the previous section. The mass of Venus is also added, with an initial value of 2.4471118×10^{-6}, as in the encounter solutions.

The probable errors from the matrix Γ_x in the normal equations are listed in the following tabulation:

Parameter	Probable error
X_0	±25 km
Y_0	±7 km
Z_0	±18 km
$\dot{X}_0$	±0.003 m/sec
$\dot{Y}_0$	±0.001 m/sec

Parameter	Probable error
$\dot{Z}_0$	± 0.002 m/sec
ϵ	± 0.013
GM_m	± 0.18 km^3/sec^2
M_V	$\pm 0.135 \times 10^{-10}$
A_E	± 55 km
R_2	± 11.6 m
λ_2	$\pm 0\overset{\circ}{.}00022$
R_3	± 11.1 m
λ_3	$\pm 0\overset{\circ}{.}00021$

The associated correlation matrix is given in Table 4.

As expected, the addition of the encounter data does not appreciably improve the determination of the mass of the moon. However, the AU and the mass of Venus are definitely uncoupled. As a result, the probable error on the mass ratio M_S/M_V is about ± 2.3 and the AU is determined to ± 55 km. Again, it should be emphasized that these errors will most likely be larger when a corrected mathematical model is used to reduce the data.

The corrections to the parameters indicate the erroneous nature of the combined solution given here. The corrections are listed for reference as follows:

$\delta X_0 = -613$ km	$\delta GM_m = 2.791$
$\delta Y_0 = 60$ km	$\delta M_V = 3.494 \times 10^{-10}$
$\delta Z_0 = 1324$ km	$\delta A_E = 1690$ km
$\delta \dot{X}_0 = 0.015$ m/sec	$\delta R_2 = -177.1$ m
$\delta \dot{Y}_0 = 0.056$ m/sec	$\delta \lambda_2 = -0\overset{\circ}{.}00320$
$\delta \dot{Z}_0 = -0.353$	$\delta R_3 = -179.3$ m
$\delta \epsilon = 0.644$	$\delta \lambda_3 = -0\overset{\circ}{.}00330$

References

[1] Brouwer, D. and Clemence, G. M., Methods of Celestial Mechanics (Academic Press, N. Y., 1961), Chap. VIII.

[2] Clarke, V. C., Jr., "Constants and related data used in trajectory calculations at the Jet Propulsion Laboratory," Jet Propulsion Lab. TR 32-273 (May 1, 1962).

[3] Clemence, G. M., "The motion of Mercury, 1765-1937," Astronomical Papers Prepared for the Use of the American Ephemeris and Nautical Almanac, Vol. 11, Part 1, U. S. Naval Observatory (Superintendent of Documents, U. S. Govt. Printing Office, Washington, D. C., 1943).

[4] Holdridge, D. B., "Space trajectories program for the IBM 7090 computer," Jet Propulsion Lab. TR 32-223 (March 2, 1962), pp. 4-6 and 80-85.

[5] Kotelnikov, U. A., "Radar contact with Venus," paper presented at the 12th International Astronautical Congress, Washington, D. C. (October 1961).

[6] Muhleman, D. O., Holdridge, D. B., and Block, N., "The astronomical unit determined by radar reflections from Venus," Astron. J. 67, 191-203 (May 1962).

[7] Pettengill, G. H., Briscoe, H. W., Evans, J. V., Gehrels, E., Hyde, G. M., Kraft, L. G., Price, R., and Smith, W. B., "A radar investigation of Venus," Astron. J. 67, 181-190 (May 1962).

[8] Rabe, E., "Derivation of fundamental astronomical constants from the observations of Eros during 1926-1945," Astron. J. 55, 112-125 (May 1950); also "Additional note on the solar parallax from Eros," Astron. J. 59, 409-411 (November 1954).

[9] Thomson, J. H., Taylor, G. N., Ponsonby, J. E. B., and Roger, R. S., "A new determination of the solar parallax by means of radar echoes from Venus," Nature 190, 519 (1961).

Table 1. Correlation matrix of estimated parameters (heliocentric data)

X_0	Y_0	Z_0	$\dot{X}_0$
0.99999999 00	0.96520441 00	-0.81608672 00	-0.85505868 00
	0.99999999 00	-0.82121404 00	-0.89405078 00
		0.10000000 01	0.56217142 00
			0.10000000 01

$\dot{Y}_0$	$\dot{Z}_0$	ϵ
0.84143814 00	-0.90310363 00	0.70740512 00
0.89278047 00	-0.93536808 00	0.80383706 00
-0.52519772 00	0.62124607 00	-0.41863089 00
-0.99388224 00	0.86011139 00	-0.96346607 00
0.10000000 01	-0.88869873 00	0.97215246 00
	0.10000000 01	-0.80063977 00
		0.10000000 01

GM_m	A_E	R_2
-0.20143232 00	-0.36452017 00	0.69345473 00
-0.19435844 00	-0.43841335 00	0.70738159 00
0.20803739 00	0.25862856 00	-0.70537353 00
0.22959995 00	0.70996388 00	-0.54139221 00
-0.20667874 00	-0.66715355 00	0.53461257 00
0.90344363-01	0.26610538 00	-0.62413774 00
-0.18169434 00	-0.75017612 00	0.44942844 00
0.10000000 01	0.28883922 00	-0.13767263 00
	0.99999999 00	-0.20239079 00
		0.99999997 00

λ_2	R_3	λ_3
-0.47995839 00	0.70654134 00	-0.47320807 00
-0.62711255 00	0.71906652 00	-0.62209836 00
0.30677602 00	-0.70420323 00	0.28952056 00
0.75731903 00	-0.55255681 00	0.76235378 00
-0.76258093 00	0.54822414 00	-0.76839046 00
0.61741365 00	-0.64746739 00	0.60928615 00
-0.84248237 00	0.46047395 00	-0.84936448 00
0.14715431 00	-0.11787525 00	0.16066238 00
0.63193730 00	-0.18919596 00	0.65323931 00
-0.31985184 00	0.48365797 00	-0.30900698 00
0.99999999 00	-0.32536693 00	0.91680225 00
	0.10000000 01	-0.31293735 00
		0.10000000 01

Table 2. Correlation matrix of estimated parameters (encounter data with radar value of astronomical unit)

X_0	Y_0	Z_0	$\dot{X}_0$
0.99999999 00	0.87135188-01	0.97709008 00	-0.64543258 00
	0.99999999 00	0.20048585 00	0.70433611 00
		0.99999999 00	-0.53852265 00
			0.99999999 00

$\dot{Y}_0$	$\dot{Z}_0$	ϵ
0.66159760 00	-0.87809492 00	-0.76401981 00
-0.68929597 00	0.26703228 00	0.55535539 00
0.56025328 00	-0.87711782 00	-0.66783237 00
-0.99933945 00	0.82325760 00	0.97031111 00
0.99999998 00	-0.84048032 00	-0.97456546 00
	0.99999999 00	0.87081263 00
		0.99999999 00

M_V	A_E	R_2
0.72212500 00	-0.83068052 00	-0.24360008-01
0.69440477 00	-0.62682203 00	0.33784237-01
0.80494571 00	-0.87840870 00	0.48896376-03
0.21040822-01	0.11090040 00	0.47002149-01
0.24835319-02	-0.13236153 00	-0.42257276-01
-0.46509647 00	0.54169167 00	-0.35746852-02
-0.14008742 00	0.28690804 00	0.54727647-01
0.10000000 01	-0.95412323 00	0.37043528-01
	0.10000000 01	-0.21548057-02
		0.99999999 00

λ_2	R_1	λ_1
0.12506240-01	-0.24324718-01	0.12555988-01
-0.23272287-01	0.33744321-01	-0.23318626-01
-0.16642510-01	0.52443880-03	-0.16585466-01
-0.21827929-01	0.46949303-01	-0.21896638-01
0.31746378-01	-0.42196190-01	0.31817856-01
0.39886074-01	-0.36227709-02	0.39799020-01
-0.48000510-01	0.54603650-01	-0.48065545-01
-0.56525975-01	0.37058219-01	-0.56509231-01
-0.82880517-03	-0.21622593-02	-0.84321566-03
0.10701837 00	0.33540558 00	0.10267242 00
0.99999999 00	0.11051900 00	0.87433770 00
	0.10000000 01	0.10617150 00
		0.10000000 01

Table 3. Correlation matrix of estimated parameters (encounter data)

X_0	Y_0	Z_0	$\dot{X}_0$
0.10000000 01	0.98429213 00	0.99971603 00	-0.82010586 00
	0.10000000 01	0.98657806 00	-0.70623228 00
		0.10000000 01	-0.81131441 00
			0.10000000 01

$\dot{Y}_0$	$\dot{Z}_0$	ϵ
0.86309836 00	-0.99586318 00	-0.97269596 00
0.76038331 00	-0.96768193 00	-0.91738338 00
0.85555240 00	-0.99561592 00	-0.96891939 00
-0.99666737 00	0.85661373 00	0.92753089 00
0.10000000 01	-0.89548566 00	-0.95421597 00
	0.10000000 01	0.98397955 00
		0.10000000 01

M_V	A_E	R_2
0.99689218 00	-0.99808291 00	0.33024180-01
0.99429653 00	-0.99332922 00	0.40554557-01
0.99795235 00	-0.99872138 00	0.35803872-01
-0.77503727 00	0.78311934 00	0.13484853-02
0.82304973 00	-0.83020063 00	0.59762225-02
-0.98834047 00	0.98962636 00	-0.36036583-01
-0.95242362 00	0.95679223 00	-0.17655036-01
0.99999999 00	-0.99957282 00	0.39395602-01
	0.10000000 01	-0.35998082-01
		0.10000000 01

λ_2	R_3	λ_3
0.33020682-02	0.33116870-01	0.34642521-02
-0.15435612-02	0.40636969-01	-0.13945233-02
0.15199166-03	0.35896437-01	0.31477319-03
-0.15165302-01	0.12456749-02	-0.15331252-01
0.19450652-01	0.60844832-02	0.19621187-01
0.49221383-02	-0.36132797-01	0.47519448-02
-0.16405383-01	-0.17777804-01	-0.16575314-01
-0.36040185-02	0.39485912-01	-0.34455041-02
-0.19917025-02	-0.36087703-01	-0.21499905-02
0.10701897 00	0.33626564 00	0.10268149 00
0.10000000 01	0.11051713 00	0.87433808 00
	0.10000000 01	0.10617814 00
		0.10000000 01

Table 4. Correlation matrix of estimated parameters (all data)

X_0	Y_0	Z_0	$\dot{X}_0$
0.10000000 01	0.72780269 00	-0.23335174 00	-0.92219384 00
	0.10000000 01	-0.40996823-01	-0.63361675 00
		0.99999997 00	0.43859638 00
			0.99999999 00

$\dot{Y}_0$	$\dot{Z}_0$	ϵ	GM_m
0.85482378 00	-0.26340828 00	-0.91507193 00	-0.54060934-01
0.70668631 00	-0.10663809 00	-0.75653090 00	-0.19759797 00
-0.36094973 00	0.99200970 00	-0.96708035-01	-0.32982747 00
-0.72632764 00	0.47818597 00	0.70010974 00	0.23176035-01
0.99999999 00	-0.34837122 00	-0.83167356 00	-0.25531074 00
	0.10000000 01	-0.77626910-01	-0.38391856 00
		0.99999999 00	0.19440309 00
			0.10000000 01

M_V	A_E	R_2
0.23572644 00	-0.94725787 00	-0.33949891-01
0.26192750 00	-0.71079446 00	-0.51922631-01
-0.67237449-01	0.11676041 00	-0.14919415-01
-0.21460117 00	0.76431842 00	0.47345684-01
0.31275864 00	-0.90538336 00	0.86875536-03
-0.63609141-01	0.12982825 00	-0.50575017-02
-0.22798802 00	0.96241290 00	0.24543235-01
-0.21606586 00	0.84540892-01	0.56303384-02
0.10000000 01	-0.25973417 00	0.19085203-01
	0.99999999 00	0.12047822-01
		0.99999999 00

λ_2	R_3	λ_3
0.42779268 00	-0.30949278-01	0.42198741 00
0.98904198-01	-0.56228849-01	0.27907746-01
-0.40476323-01	-0.34321404-01	-0.83343981-01
-0.19448978 00	0.42987743-01	-0.20540782 00
0.55926981 00	0.29643447-02	0.54038560 00
0.77632490-02	-0.24769904-01	-0.29403631-01
-0.52446706 00	0.26911641-01	-0.49543539 00
0.80728181-02	0.28206999-01	0.36693079-01
0.12620668 00	0.13933549-01	0.11521332 00
-0.61352473 00	0.10944943-01	-0.59445646 00
0.44177124-01	-0.31430317 00	0.43355151-01
0.99999999 00	0.49472272-01	0.65175586 00
	0.10000000 01	0.52254804-01
		0.99999999 00

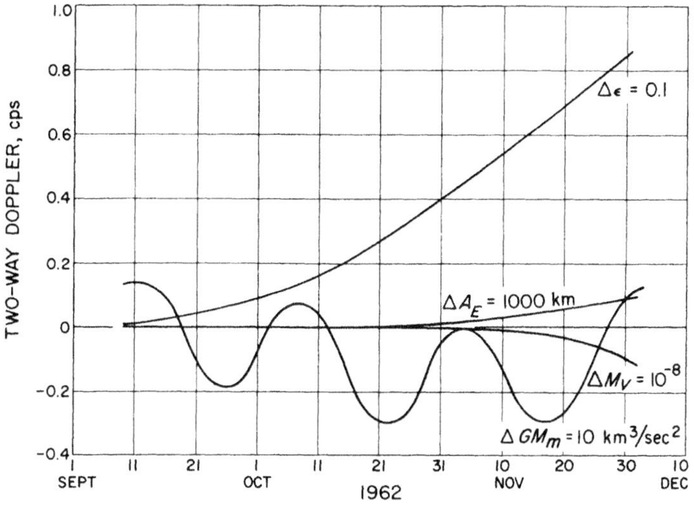

Fig. 1 Sensitivity of two-way doppler to variations in astronomical constants during heliocentric portion of flight.

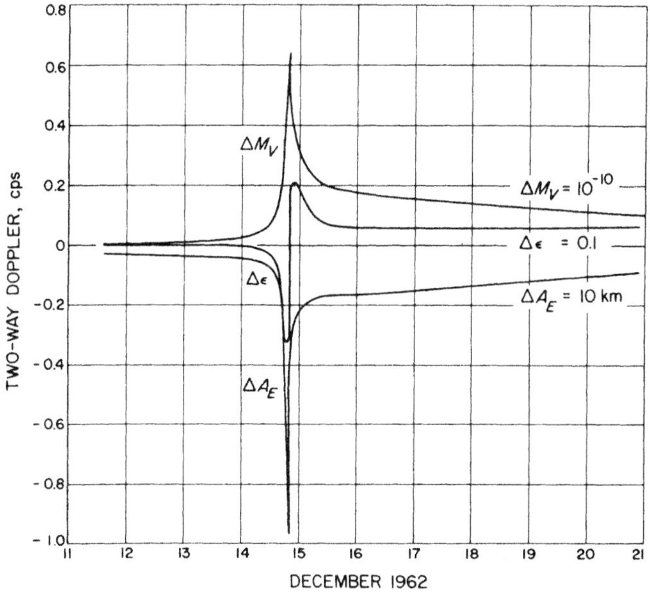

Fig. 2 Sensitivity of two-way doppler to variations in astronomical constants during encounter portion of flight.

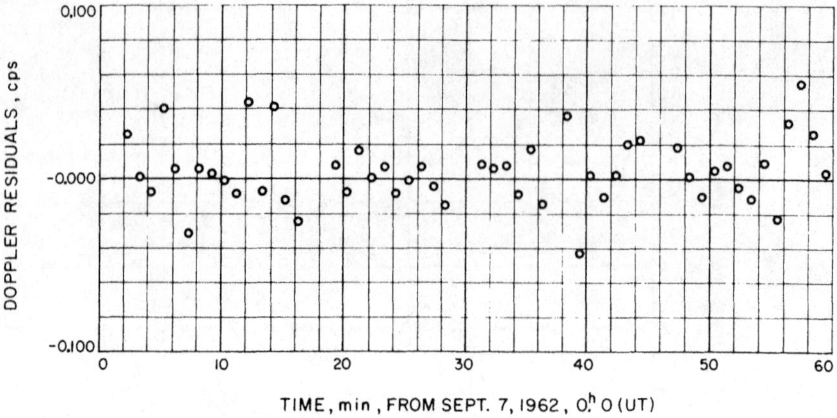

Fig. 3 Early Goldstone residuals.

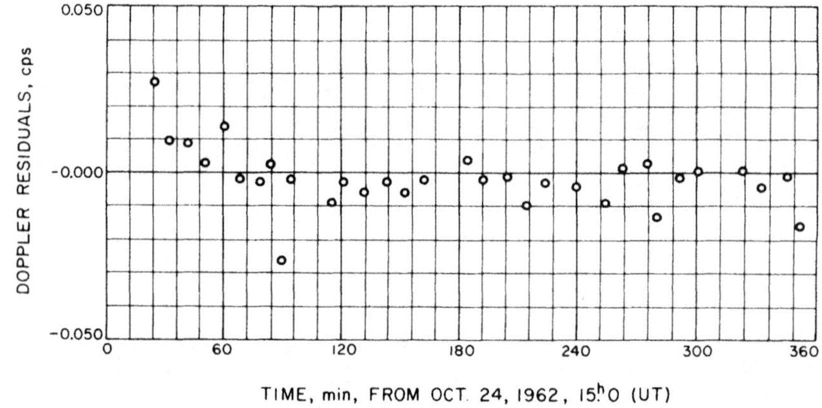

Fig. 4 Goldstone residuals near midpoint of mission.

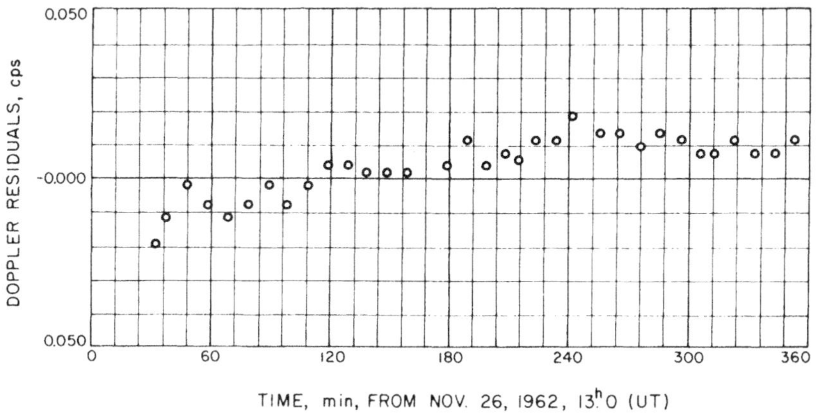

Fig. 5 Late Goldstone residuals.

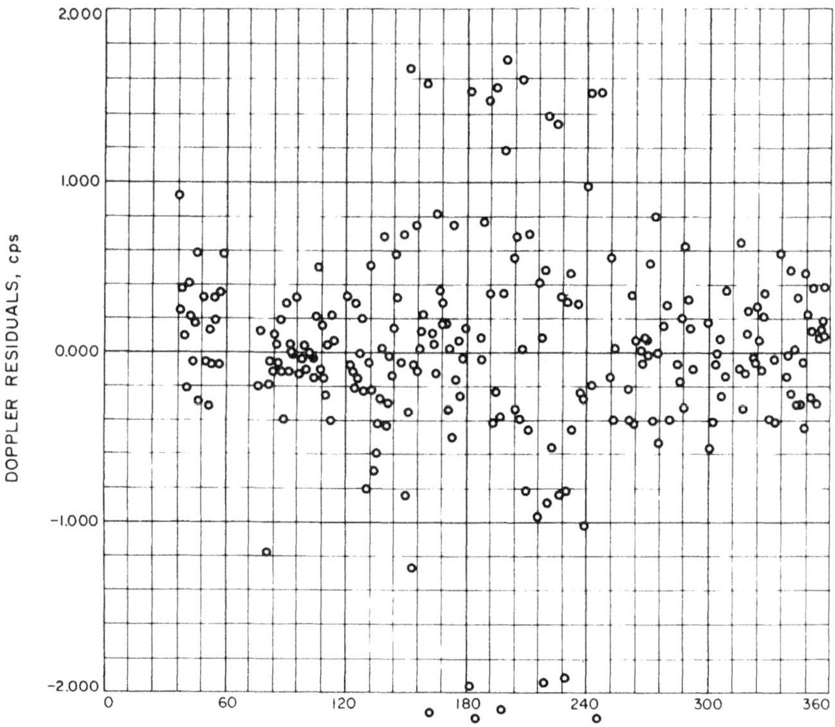

Fig. 6 Early Johannesburg residuals.

ATMOSPHERIC DENSITY DETERMINATION FROM SATELLITE OBSERVATIONS

John P. Rossoni,* P. Sconzo,† and R. J. Greenfield‡

IBM Space Systems Research, Cambridge, Mass.

and

K. S. Champion§

Air Force Cambridge Research Laboratories, Cambridge, Mass.

Abstract

A funded contract with the Air Force Cambridge Research Laboratories supported three years of research leading to the development and computer applications of methods for the derivation of atmospheric density at perigee heights from an analysis of satellite observations. The computer programs represent the closest approach to an almost automatic calculation of upper atmospheric density, patterned after theories developed by L. Jacchia of the Smithsonian Astrophysical Observatory. These programs provide the scientific community with an objective tool of investigation, free of subjective interpretation of observational data. The computer program developed includes: 1) differential correction of orbital parameters, based on optical and/or radar observations; 2) representation of satellite motion over extended time periods by least-squares fit of orbital parameters, using empirical expressions; 3) determination of rate-of-change of anomalistic period (secular accelerations) from local behavior of mean anomaly residuals; and 4) derivation of density values from secular accelerations, based on numerical integration of a rotating, nonspherical model atmosphere, exhibiting the diurnal solar bulge. Density results have been obtained from five satellites, including 1959 Alpha 2 (Vanguard II Rocket Body), 1962 Eta 1 and the recent Explorer XVII, 1963 IOTA.

Presented as Preprint 64-31 at the AIAA Aerospace Sciences Meeting, New York, N. Y., January 20-22, 1964.

*Manager.

†Chief Scientist.

‡Staff Engineer.

§Director, Atmospheric Physics Branch.

Introduction

The determination of upper atmosphere density is carried out by means of artificial satellites, either directly by special measuring instruments installed in the vehicle itself, as in the EXPLORER XVII density gauge, or indirectly, by a careful reduction of the observations made over a long period of time (months or years). The latter method is possible because the motion of the satellite in its orbit is slowly affected by atmospheric drag. The variation of selected orbital elements, if properly computed, gives an indication that can be used for the determination of the density causing such a variation.

The project, sponsored by the Air Force Cambridge Research Laboratories, supported a three-year research effort to develop methods and computer programs to derive atmospheric density values at perigee heights from an analysis of satellite observations. The methods, analytical formulations, and special techniques are due to Jacchia.[1-3] Modifications to his approach were necessary in attempting an automatic procedure. It is believed that the resulting computer programs represent the first attempt at an automatic calculation of density, where as many other techniques, including Jacchia's original one,[4] use human intervention between computer runs. These interventions give a subjective interpretation of partial results, and lead to final values of density which differ from one investigator to the other. The differences have particular meaning when the density values are used for the development of accurate atmospheric models. It is then necessary to adopt results from one source only, to avoid a general, average model.

This paper describes the procedure developed to obtain detailed quantitative information about the perturbing effects of atmospheric drag by appropriate analysis of satellite motion, and, from this analysis, determine density values.

Orbit analysis

Differential Orbit Correction

Improved mean orbital elements are obtained by a technique for differential orbit correction and generation of ephemeris tables (DOCET) based on Brouwer's[5] adaptation to artificial satellites of the Delaunay[6]-von Zeipel[7] lunar theory. This theory takes into account, with great accuracy, the effects of the earth's oblateness on the motion of a satellite.

Observations of a particular satellite are arranged in ascending chronological order and subdivided into groups of two to six days. The length of time covered by each group depends on the number of observations available. The differential orbit correction is performed on a set of initial orbital elements at the same epoch within each interval of time, using all available observations, properly weighted. The types of observations most commonly used are: 1) optical measurements of right ascension and declination; 2) optical measurements of azimuth and elevation, and 3) radar measurements of range, azimuth, and elevation.

The assumed precision of the optical observations corresponds to field reduced Baker-Nunn Camera plates (about 2-3 min of arc in both angles and about 0.01 sec in time). This assumption makes it possible to neglect extraterrestrial gravitational perturbations as well as the small short periodic perturbations due to the ellipticity of the equator and to gravitational anomalies (sectorial and tesseral harmonics of the geopotential), all of which would be concealed by the inaccuracies of the observations. These effects, neglected here, would be taken into account in high-precision work based on extremely accurate observations (photoreduced Baker-Nunn plates, for instance).

The differential orbit correction produces a set of refined orbital elements for each group of observations processed. Since the groups are not overlapping, each set of elements is independent and characterizes the behavior of the orbit at the epoch chosen within each group. At that epoch, the standard error (sigma of an observation whose weight is unity) and the mean-square error for each element are also computed, as measures of the goodness of the orbit correction. Large values of sigma indicate the presence of bad observations; large values of mean-square errors are caused mainly by an unfortunate distribution of the observations along the orbit, for instance, when they are grouped together along a short arc.

We obtain as a result of the differential orbit correction several sets of orbital elements and the corresponding standard errors for each epoch. These sets of elements represent the orbit at intervals of two to six days apart.

Empirical Expressions for the Orbital Elements

A continuous representation of the variation of the orbit with time can be attempted by fitting each orbital element with empirical expressions. These must represent the effects of the secular and long periodic gravitational perturbations as well as the effects of air drag. The functions used to fit the elements are chosen so as to follow the classical formulation for the gravitational perturbations. Let

$$\epsilon_j(t-t_o) = P_j^{(n)}(t-t_o) + A_j \begin{matrix}\sin\\ \cos\end{matrix}\; 2\,[\omega_o + \dot{\omega}_o(t-t_o)]$$

$$+ B_j \begin{matrix}\cos\\ \sin\end{matrix}\; [\omega_o + \dot{\omega}_o(t-t_o)]$$

$$(j = 1 \rightarrow 5)$$

be the approximating function of time for the classical orbital element ϵ_j in the order: mean anomaly, argument of perigee, right ascension of node, inclination, and eccentricity; an explicit expression for the semimajor axis is not included, since this element may be determined from the time derivative of the mean anomaly.

$P_j^{(n)}$ is a polynomial of nth degree, and ω_o and $\dot{\omega}_o$ are the values of the argument of perigee and its time derivative at the origin of time, $t = t_o$.

The numerical coefficients A_j, B_j, and those of the polynomial $P_j^{(n)}$ are determined by a general least-squares procedure, based on the sets of orbital elements computed at consecutive epochs which cover a time interval of two to four months.

As a result of this step, each orbital element is expressed in analytical form as a function of time. The least-squares fit gives a reasonably good representation of the variation of the orbital elements. Residuals between this representation and the original elements are computed at each epoch. The size of the residuals is expected to be of the same order of magnitude as the mean-square error associated with each orbital element at that epoch.

Once the goodness of the least-squares fit has been established, the functions expressing the orbital elements are evaluated at each time of the observations and residuals are formed between observed and computed quantities. A regular trend in these residuals indicates an effect that has not been accounted for in the fitting procedure. This trend exhibits small regular oscillations due to short-periodic perturbations from air drag, satellite tumbling, etc., and, in particular, it reveals irregular fluctuations due to the variation of the air density. These fluctuations are best studied by examining the residuals in the mean anomaly, which is the most sensitive of the orbital elements to the drag perturbations.

Residuals in Mean Anomaly

The mean anomaly M fixes the position of the satellite in its orbit. The mean anomaly, like the eccentric anomaly, is not an observable quantity. When we speak, therefore, about mean anomaly residuals (ΔM), it is with the tacit understanding that a pseudo-observed mean anomaly has been defined.

Definitions of mean anomaly residuals have been developed by Kulikov and Batrokov[8] and later by Izsak.[9] These authors consider rotated residuals, obtained from the usual residuals $\cos \delta \; d\alpha$ and $d\delta$ by means of projections along two unit vectors, the first of which is perpendicular to the orbit plane and the second "as close as possible" to that of the velocity vector $\dot{\underline{r}}$. The equation of condition arising from the second projection is the only equation that has a nonzero coefficient for ΔM. Under the assumption that the remaining orbital elements are known exactly, this equation may be used to define ΔM.

For the purpose of this outline, we can say that the mean anomly residual is a measure of the lag of the satellite in its orbital path, caused by the drag effect.

If these residuals are kept small enough by a carefully performed fitting procedure of the mean anomaly, then a trend is clearly exhibited, and it reveals the small variation of atmospheric density.

The procedure computes now a mean anomaly residual for each observation taken during the interval of time encompassed by the fitting function. The residuals may then be plotted as a function of time. The resulting graph would display a cluster of points scattered along a smooth curve, the form of which must be inferred. Figure 1 clearly shows the trend of the mean anomaly residuals, plotted from observations on Satellite 1959 ALPHA 2 for the period March 13, 1959 to April 22, 1959 (36640 to 36680 modified Julian days).

The original procedure followed by Jacchia[4] was to draw this curve manually, relying on experience. He then read off the ordinates of the curve at equally spaced intervals of time and applied a numerical differentiation formula, using a desk calculator, to obtain the first and second derivative of the mean anomaly residuals needed for the next step of the computation.

We have successfully modified this subjective approach into an automatic procedure that computes the rate of change of the satellite period from a double differentiation of the analytic representation of the mean anomaly and of its residual curve.

Rate of Change in the Anomalistic Period

The anomalistic period of a satellite is defined by

$$P = \frac{1}{n}$$

where n is the time derivative of the mean anomaly $\epsilon_1 = M$, expressing the mean motion in revolutions per unit of time

$$n = \frac{dM}{dt}$$

We may thus compute the rate of change of P from the relation

$$\dot{P} = -\left(\frac{d^2M}{dt^2}\right) \bigg/ \left(\frac{dM}{dt}\right)^2$$

$\dot{P}$ is a small quantity that decreases with increasing height of the satellite above the earth's surface. Typical values range from 10^{-5} to 10^{-8} rev/day for satellites with heights of between 200 and 700 km. The effect of the variation of dM/dt is minor compared to the effect of changes in d^2M/dt^2; an accurate determination of $\dot{P}$ demands a precise knowledge of d^2M/dt^2.

As we have already said, the representation of the mean anomaly is obtained by an analytic function $\epsilon_1(t - t_0)$, the coefficients of which are determined empirically. We have also computed the residuals in mean anomally ΔM. Writing $\overline{M}$ (t) for the analytical approximation, we may now express the mean anomaly as

$$M(t) = \overline{M}(t) + \Delta M(t)$$

and deduce from this

$$\frac{d^2M}{dt^2} = \frac{d^2\overline{M}(t)}{dt^2} + \frac{d^2\Delta M(t)}{dt^2}$$

The second derivative of the mean anomaly has thus been expressed as the sum of two parts. The first part may be computed in closed form from the expression for $\epsilon_1 (t - t_o) = M(t)$. The second part is evaluated from the trend of the mean anomaly residuals by means of a new procedure called the "barycenter points" technique. This procedure has been devised especially to automate the computation of the smooth curve fit of the mean anomaly residuals and their second derivative. It replaces the manual curve tracing and numerical differentiation of Jacchia's technique.

The procedure may be outlined as follows. A sequence of unweighted geometric barycenter point B_p are determined as the center of mass of successive groups of (t_i , ΔM_i) - points having t_i as abscissa and ΔM_i as ordinate. The groups cover a time interval of 2d days with an overlap of d days between intervals. With d = 1, for example, we obtain 1 barycenter/day. The local representation of the smooth curve is then obtained by means of a low-order polynomial fitted, in a least-squares sense, to a certain number of barycenter points. To determine now the second derivative of the trend of the mean anomaly residuals at a chosen epoch t_p , we select the barycenter B_p closest in time to t_p and take an equal number of barycenters on either side of B_p. The second derivative of the least-squares fit polynomial through these points is used as the value of the second derivative of $\Delta M(t)$ at t_p. The order of the polynomial and the number of barycenters used in an individual fit are parameters that must be chosen beforehand on the basis of experience and experimentation. Representative values used in the test case were seven points and a third-degree polynomial.

The second derivative computed from the local least-squares fit to the barycenter points is added, finally, to the contribution from the analytic part, to yield the final value for d^2M/dt^2 at time t_p. Figure 2 clearly shows the computed barycenter points (indicated by crosses) among the scatter of (t_i, ΔM - points). This figure represents the first portion of Fig. 1 in a larger scale and shows how the barycenter points closely follow the trend of the mean anomaly residuals.

Several additional details must be mentioned. Although the abscissas t_i may be assumed to be free of gross error, the same cannot be said for the ordinates ΔM_i. To prevent unusually large residuals from distorting the barycenter points, the following rejection logic for bad observations is applied. Before computing the barycenter corresponding to a particular 2d-day interval, we fit all residuals in the interval by means of a linear function of time y(t), which indicates the rough trend of the ordinates ΔM_i during the interval. Each ΔM_i in turn is compared with the corresponding ordinate $y(t_i)$ and is retained only if

$$|\Delta M_i - y(t_i)| < L$$

where L is a preassigned tolerance and is assumed to be twice the maximum error expected in the observations. The barycenter computation is based, then, only on the points that have passed the previous test. Furthermore, no barycenter is computed if fewer than three residuals are available.

A possible source of systematic error in the determination of d^2M/dt^2 is the presence of nonlinear terms in the representation of the argument of perigee ω. The correction C_ω to d^2M/dt^2 arising from the acceleration of perigee may easily be computed from the well-known equation of center. We obtain

$$C_\omega = (1 - 2e + \frac{3}{2} e^2) \frac{d^2\omega}{dt^2}$$

By defining the corrected value of d^2M_c/dt^2 as

$$\frac{d^2M_c}{dt^2} = \frac{d^2M}{dt^2} + C_\omega$$

we obtain, finally, the corrected value for the rate of change of P:

$$\dot{P}_c = -\left(\frac{d^2M_c}{dt^2}\right) \Big/ \left(\frac{dM}{dt}\right)^2$$

Density Determination

Determination of Atmospheric Density of Perigee Height

The basic relation used in deducing atmospheric density from the rate of change in anomalistic period is Sterne's equation[10]:

$$\dot{P} = -\frac{3}{2} C_D \frac{A}{m} \rho_\pi a \int_0^{2\pi} \frac{\rho}{\rho_\pi} \frac{(1 + e \cos E)^{3/2}}{(1 - e \cos E)^{1/2}} \cdot (1 - d \frac{1 - e \cos E}{1 + e \cos E})^2 \, dE$$

where P is the rate of change of the anomalistic period; ρ_π the atmospheric density at the perigee height; C_D is the aerodynamic drag coefficient; A/m is the area-to mass ratio; ρ is the atmospheric density at the height corresponding to a given value of the independent variable E, the accentric anomaly; and $d = P \omega_s (1 - e^2)^{1/2} \cos i$, where ω_s is the angular velocity of atmospheric rotation. The other symbols have their conventional meaning.

In the atmospheric model used in Jacchia's formulation[3] the expression for ρ is given by

$$\rho (z) = \rho_o(z) F_{20} [1 + \varphi(z) \cos^6 \frac{\psi'}{2}]$$

where z is the height of the satellite above the geoid, $\varphi(z) = 4.6 + \tan^{-1} [0.005(z-600)]$, ψ' is the geocentric angular distance from the satellite to the center of the atmospheric bulge, which is assumed to have rotational symmetry, and $\rho_o(z)$ is the minimum nighttime density.

The form of the functions $\varphi(z)$ and $\rho(z)$ has been deduced experimentally for the range 200 km $\leq z \leq$ 700 km.[11,12] The factor F_{20}, representing the 20-cm solar flux, is irrelevant because it is eliminated in the ratio ρ/ρ_π.

Furthermore, a lag angle $\lambda = 30°$ was assumed for the position of the bulge with respect to the subsolar point. Finally, ρ was directly computed from Sterne's equation, using numerical integration over the interval $(0, 2\pi)$.

Computer Programs

There is little point in presenting a description of the computer program developed for this analysis. It may be of interest, however, to know that they have been organized into independent processors under the control of a monitor program. The three main processors perform the following functions: 1) differential orbit correction, accepting as input observations and initial orbital elements of a satellite producing improved orbital parameters at closely spaced epochs; 2) ephemeris tables, which use a general least-squares fitting procedure to express the elements as functions of time, and use these expressions, together with the mean anomaly residuals, to obtain values of the rate of change of the anomalistic period P at regular intervals; and 3) atmospheric density integration calculation, which accepts the output of 2 and the characteristics of the satellite, and generates density values at the perigee height of the satellite.

Results on Satellite 1963 IOTA (Explorer XVII)

Values of atmospheric density at perigee heights were obtained from observations of Satellite 1963 IOTA. Figure 3 gives the parameters associated with the position of the perigee for the period from April 11, 1963 to May 17, 1963. In Fig. 4, the density values are plotted against time. Also plotted are the daily geomagnetic index Ap and the 10.7-cm and 20-cm solar flux.

The relevant values of the solar flux were then inserted into the atmospheric models of Jacchia and Slowey[13,14] and into a revised version[16] of Harris and Priester's[15] model. A comparison of the derived densities and the densities predicted by the models are shown in Fig. 5. The perigee height of the satellite is nearly constant, but a substantial increase in density is noticeable when the perigee moves into the solar bulge.

Densities obtained with on-board ionization gauges at 16:00 hours local time during the interval April 10 to 20 agree with drag densities obtained at this time within the experimental accuracy of the gauge itself, which is of about 20%.

Results on Satellite 1962 ETA 1

Density values were also derived from observations of Satellite 1962 ETA 1. These observations cover the period from May 16, 1962 to May 12, 1963. Fig. 6 gives the perigee height and the angle Ψ'. This satellite has an almost polar orbit; thus the perigee height varies considerably, being much greater over the pole than over the equator.

Density values at perigee have been plotted in Fig. 7. The main variation shown in this graph is due to the changing perigee height. To remove this variation, the density values at perigee were reduced to density values at a standard height of 250 km. This was accomplished by multiplying the determined value of the density at perigee by a ratio; this ratio is given by the density at 250 kms., obtained from Jacchia's model, divided by the density at the perigee as given by the same model.

The computed density values at a standard height of 250 km are plotted in Fig. 8. At this height the solar bulge seems to be present, and quite broad.

Conclusions

The results obtained from processing observational data for two satellites are in fair agreement with those obtained by independent techniques. In particular, the density values for Satellite 1963 IOTA were compared with those obtained by Jacchia and found to be within a 10% agreement.

The results indicate that the method described previously is sensitive enough to reveal time variations of the atmosphere such as the "27-day" fluctuations, that is, oscillations of the atmospheric density which are in phase with the decimetric solar flux. However, transient variations of the type occurring during magnetic storms may go undetected because of their short duration. It is possible to detect also these short-lived variations by an appropriate choice of selected parameters.

The technique outlined in this paper is providing a tool for the computation of density, which should relieve a researcher from the most tedious manual tasks encountered in this process. It is hoped that criticism of this procedure will lead to further improvements to be incorporated in the programs.

The ultimate goal would be to provide the scientific community with the best possible automatic procedure to compute atmospheric densities from satellite observations.

References

[1] Jacchia, L. G., "The determination of atmospheric drag on artificial satellites," Dynamique des Satellites, edited by M. Roy (Springer-Verlag, Berlin, 1963), p. 136.

[2] Jacchia, L. G., "Variations in the earth's upper atmosphere as revealed by satellite drag," Rev. Mod. Phys. 35, 973-991 (1963).

[3] Jacchia, L. G. and Slowey, J., "Accurate drag determinations for eight artificial satellites; atmospheric densities and temperatures," Smithsonian Contrib. Astrophys. 8, no. 1 (1963).

[4] Smithsonian Astrophysical Observatory, Special Rept. Series 62, Cambridge, Mass. (1962).

[5] Brouwer, D., "Solution of the problem of artificial satellite theory without drag," Astron. J. 64, 378 (1959).

[6] Delaunay, C. E., Theorie du Mouvement de la Lune (Paris, 1864-67), 2 vols.

[7] Von Zeipel, H., Ark Astro. Math Fys. II, no. 1 (1916).

[8] Kulikov, D.K. and Batrokov, Y.V., "Method for improving orbits," ARS J. 30, 865 (1960).

[9] Izsak, I G., "Differential orbit improvement with use of rotated residuals," Smithsonian Astrophysical Observatory, Special Rept. 73 (1961).

[10] Sterne, T.E., "Effect of the rotation of a planetary atmosphere upon the orbit of a close satellite," ARS J. 29, 777 (1959).

[11] Jacchia, L.G., "A variable atmospheric density model from satellite accelerations," J. Geophys. Res. 65, 9 (1960).

[12] Jacchia, L.G. and Slowey, J., "Preliminary analysis of the atmospheric drag of the twelve-foot balloon satellite (1961 DELTA 1)," Smithsonian Astrophysical Observatory, Special Rept. 84 (1962).

[13] Jacchia, L.G. and Slowey, J., "Accurate drag determination for eight artificial satellites," Smithsonian Astrophysical Observatory, Special Rept. 100 (1962).

[14] Jacchia, L.G. and Slowey, J., "Atmospheric heating in the auroral zones," Smithsonian Astrophysical Observatory, Special Rept. 136 (1963).

[15] Harris, I. and Priester, W., "Theoretical models for the solar cycle variation of the upper atmosphere," NASA TN D-1444 (1962).

[16] Harris, I. and Priester, W., "Relation between theoretical and observational models of the upper atmosphere," J. Geophys. Res. 68, 5891 - 5894 (1963).

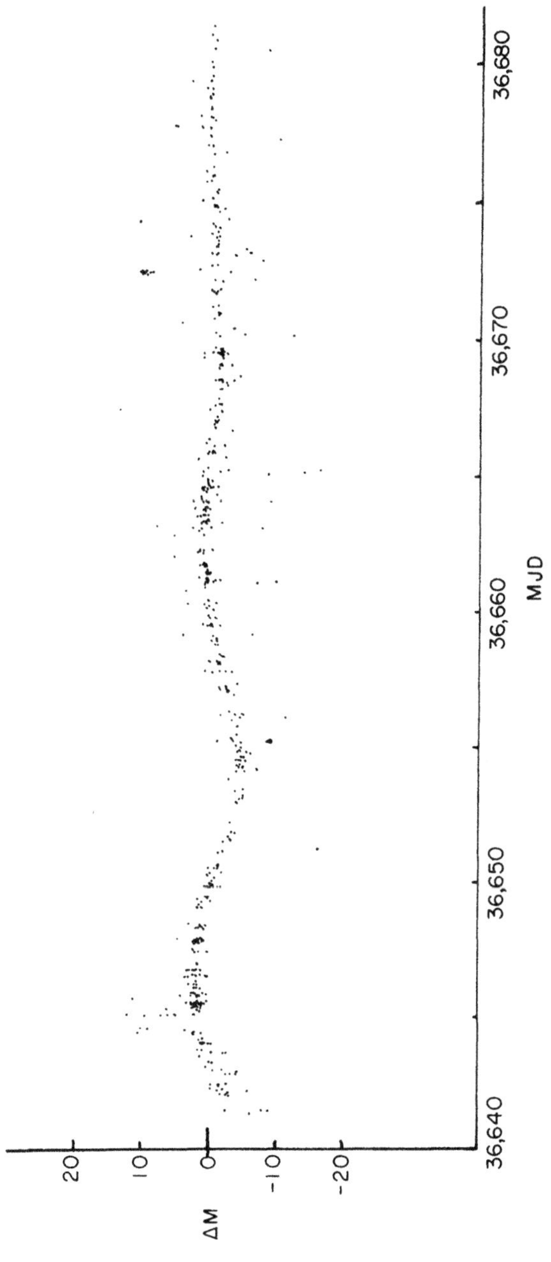

Fig. 1 Mean anomaly residuals ΔM for the period March 13 to April 22, 1959 (Satellite 1959 Alpha 2).

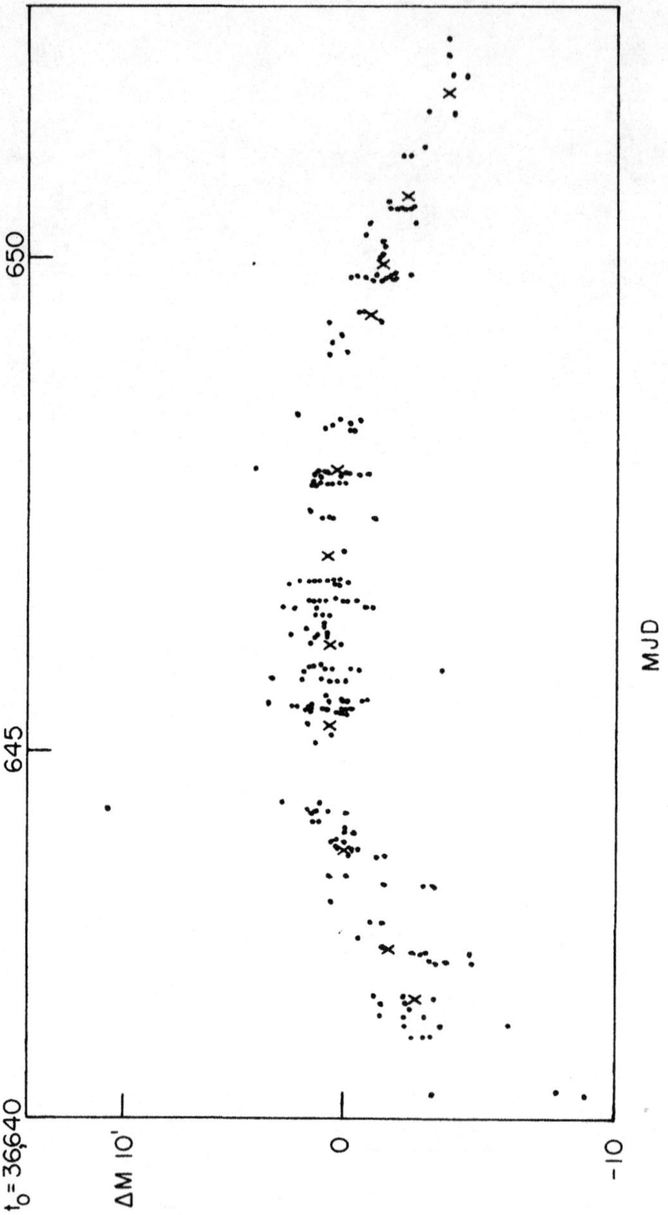

Fig. 2 Barycenter points (crosses) computed from mean anomaly residuals (dots).

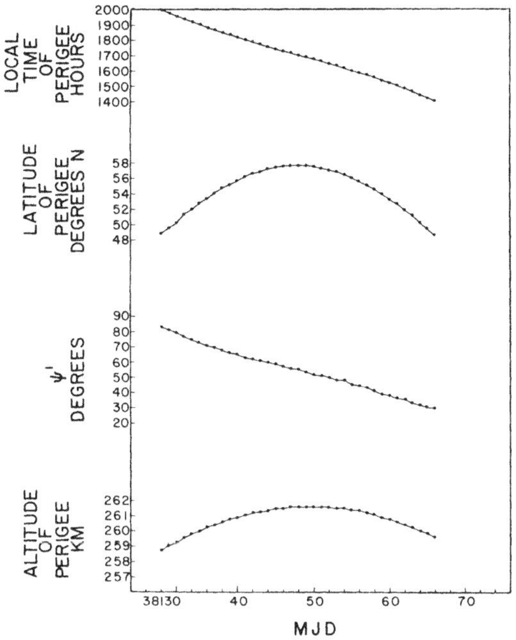

Fig. 3 Parameters describing the perigee position (Satellite 1963 IOTA).

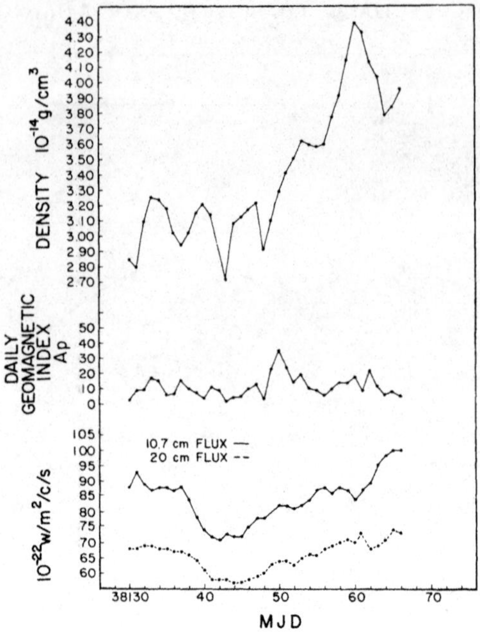

Fig. 4 Perigee density, Ap index, and solar flux values (Satellite 1963 IOTA).

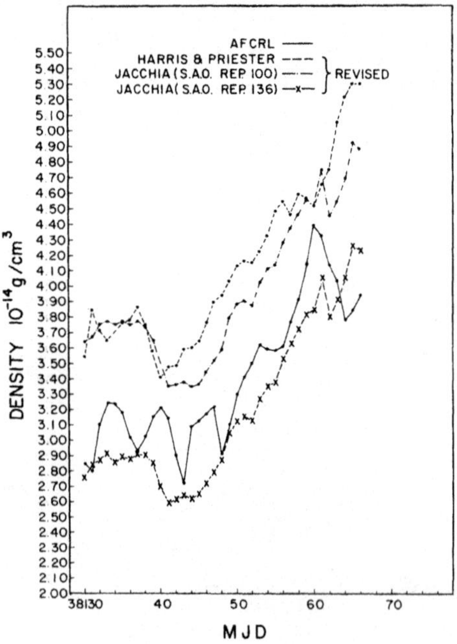

Fig. 5 Comparison of derived densities with atmospheric model predictions (Satellite 1963 IOTA).

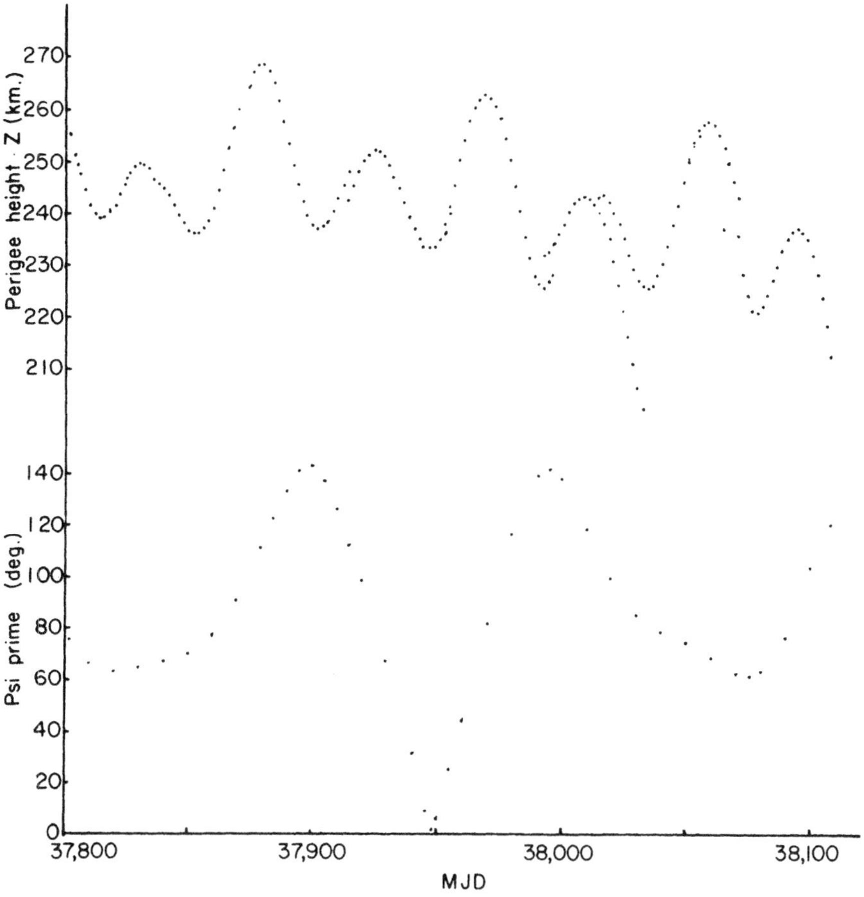

Fig. 6 Perigee height and angle Ψ' (Satellite 1962 ETA 1).

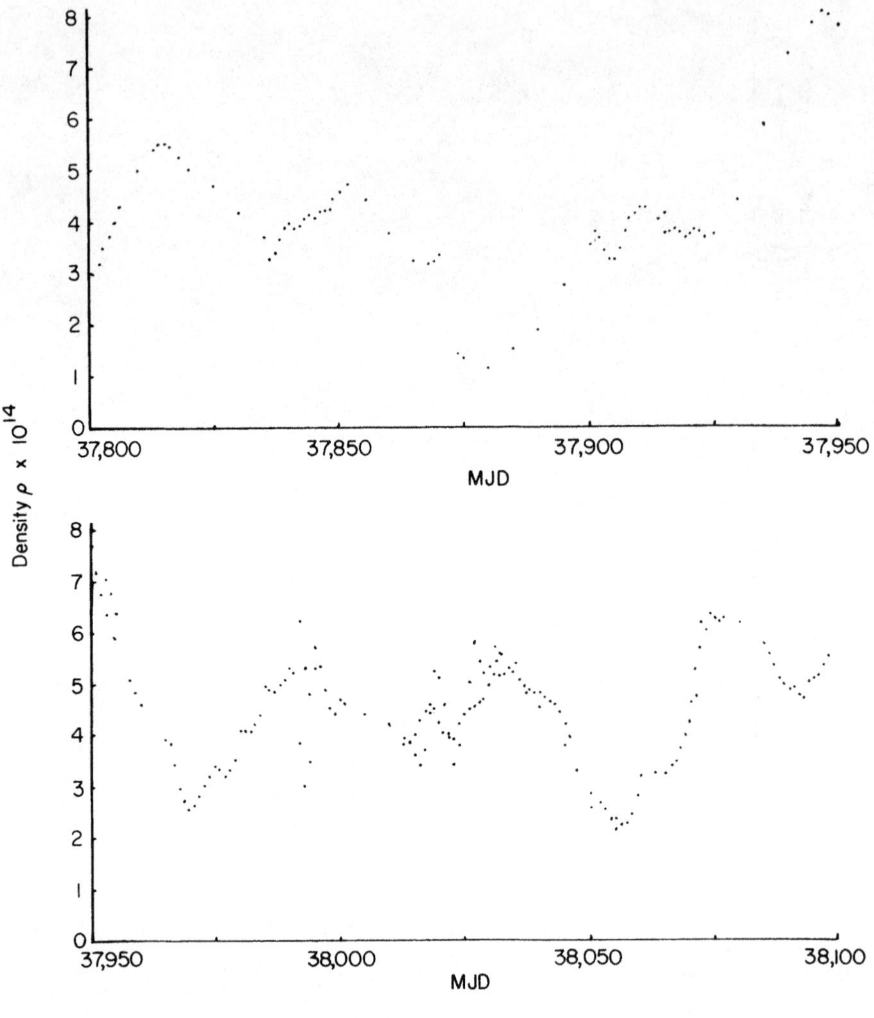

Fig. 7 Density at perigee height (Satellite 1962 ETA 1).

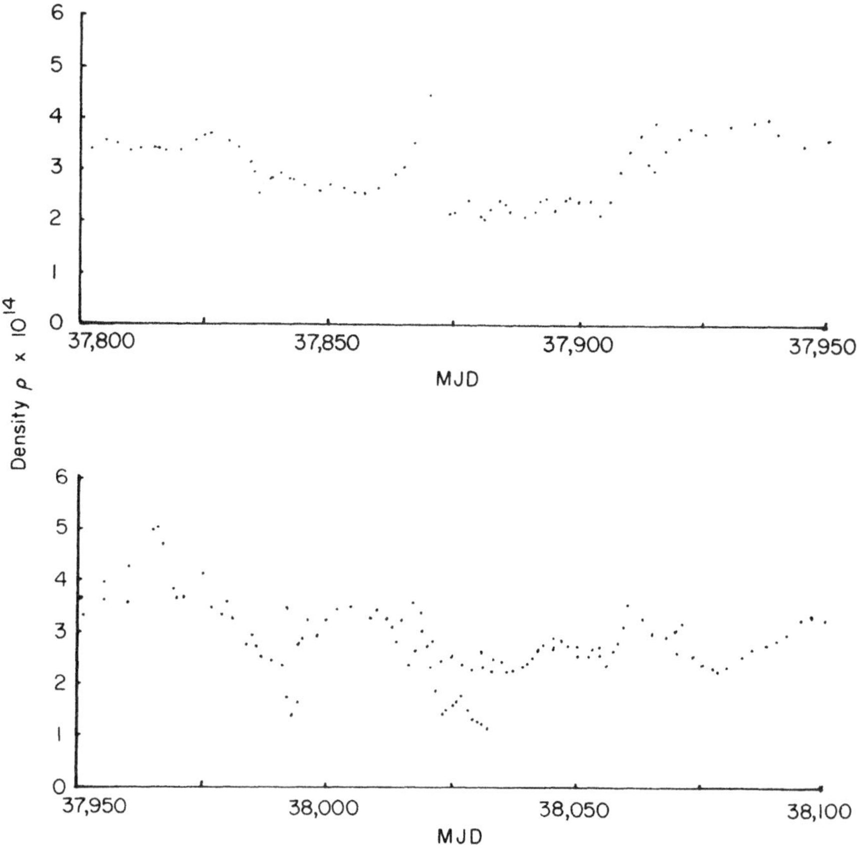

Fig. 8 Density at standard height of 250 km (Satellite 1962 ETA 1).

III.

ORBIT SELECTION

Artificial Satellite Orbits

PROBABILISTIC EVALUATION OF SATELLITE MISSIONS INVOLVING GROUND COVERAGE

Barry Boehm*

The RAND Corporation, Santa Monica, California

Abstract

A relation $P(\alpha,s,h,i)$ is developed for the probability of occurrence of the following event: that a point Q on earth at latitude α comes within a distance s of the ground track of a satellite in orbit, at an altitude h and an inclination i, during a single satellite orbital revolution, under the assumption that the point Q is randomly located in longitude with respect to the ascending node of the orbit. Some methods of using the relation to evaluate coverage capabilities of satellite systems are presented; applications of the methods are discussed, and numerical results given for some particular applications. Limitations due to analytical simplifications are discussed.

I. The Coverage Problem

A question which arises frequently in dealing with satellites is the following: "How often can a desired operation involving the satellite and a point, or set of points, on earth be performed?" It often happens in such satellite studies that there exists a critical side range s away from the satellite's ground track such that the operation can be carried out during a particular period if and only if the point on earth comes within the distance s of the satellite's ground track during that period. In such a case, the question simplifies to: "How often can one expect a satellite's ground track to come within a certain distance s from a point or set of points on earth?" This paper presents a method for determining the answer to this question.

Presented as Preprint 63-396 at the AIAA Astrodynamics Conference, New Haven, Conn., Aug. 19-21, 1963. This paper is based on Ref. 6. This research is sponsored by the U.S. Air Force under Project RAND--contract No. AF 49(638)-700 monitored by the Directorate of Development Planning, Deputy Chief of Staff, Research and Development, Hq. USAF. Views or conclusions contained herein should not be interpreted as representing the official opinion or policy of the United States Air Force.

*Mathematician, Computer Sciences Department.

Examples of satellite operations involving a critical side range are fairly common. In problems of satellite interception or rendezvous, the critical side range depends upon the satellite's orbital altitude and the performance characteristics of the booster vehicle. Similarly, in problems involving the ejection of objects from orbit to points on earth, the critical side range is a function of the satellite's altitude and the propulsion capability of the ejection rocket. Observation and communication satellites also possess a critical side range, which can be determined by the range and viewing angle limitations of the equipment involved.

Coverage Relationships

As a first step, a relation is developed for the fraction of a parallel of latitude which can be covered by a satellite with a given side-range capability during a single revolution in its orbit. The satellite is assumed to be traveling in a circular orbit at an inclination i with respect to a spherical, atmosphere-free earth; its altitude is thus a constant denoted by h (see Fig. 1). The satellite is assumed to have a critical side range s; it will, as it moves along its orbit, generate a swath of points on earth of half-width s within which successful operations involving the satellite can be carried out. The fraction of a parallel of latitude covered by the satellite can then be obtained by determining the longitudinal extent of the intersection of the swath with the parallel of latitude.

Relationships in Inertial Space

One first observes that the boundaries of the swath form planes parallel to the orbit plane in inertial space. One may then look at the problem edge-on, as in Fig. 2, defining a right-handed rectangular inertial coordinate system with the z axis pointing toward the north pole and the x axis intersecting the equator at the ascending node (pointing up out of the paper in Fig. 2). In this coordinate system, the relations

$$\begin{aligned} z &= \sin \alpha \\ y &= \cos \alpha \sin \lambda_I \end{aligned} \qquad (1)$$

hold, where α is the latitude of a point on the earth's surface, and λ_I is its inertial longitude, measured eastward from the ascending node of the orbit. The equations for the boundaries of the swath are

$$z \cos i - y \sin i = \pm \sin s \tag{2}$$

where s is expressed in earth radius units.

Substitution for z and y produces relations for the longitude of the intersection of the boundaries of the swath with the parallel of latitude α :

$$\sin \lambda_{I1} = \frac{\sin \alpha \cos i + \sin s}{\sin i \cos \alpha}$$

$$\sin \lambda_{I2} = \frac{\sin \alpha \cos i - \sin s}{\sin i \cos \alpha} \tag{3}$$

Each relation is (3) specifies two values of λ_I, one the supplement of the other, resulting from the symmetry about the y axis and representing intersections on the upward and downward legs of the orbit. Taking account of the possible non-intersections, one obtains the following relations:

$$\lambda_{I1} = \begin{Bmatrix} \sin^{-1}\left(\dfrac{\sin \alpha \cos i + \sin s}{\cos \alpha \sin i}\right), & |\alpha| \leq |i| + s \\ \pi/2 & , |\alpha| \geq |i| + s \end{Bmatrix} \tag{4}$$

$$\lambda_{I2} = \begin{Bmatrix} \sin^{-1}\left(\dfrac{\sin \alpha \cos i - \sin s}{\cos \alpha \sin i}\right), & |\alpha| \leq |i| - s \\ \pi/2 & , |\alpha| \geq |i| - s \end{Bmatrix} \tag{5}$$

$$\Delta\lambda_I = 2\left(\lambda_{I2} - \lambda_{I1}\right) \tag{6}$$

Each value of λ_I is taken to be the principal value:

$$-\frac{\pi}{2} < \lambda_{I1},\ \lambda_{I2} \leq \frac{\pi}{2}$$

Effect of Earth Rotation

To find the range of longitude $\Delta\lambda$ covered by the satellite over a rotating earth, the assumption of a circular orbit at altitude h is first employed to determine the satellite's rotation rate ω

$$\omega = \sqrt{\mu_E}\ (h+R_E)^{-3/2} \tag{7}$$

where R_E is the radius of the spherical earth and μ_E the earth's gravitational constant. Then, if $\Delta\theta$ is the distance that the satellite travels while covering points at latitude α, the distance the earth travels in the corresponding time is $(\omega_E/\omega)\Delta\theta$, where ω_E is the rate of the earth's rotation, and thus

$$\Delta\lambda = \Delta\lambda_I - (\omega_E/\omega)\Delta\theta \tag{8}$$

The quantity $\Delta\theta$ is determined with the aid of Fig. 2. The end points θ_1 and θ_2 of that part of the satellite's path in which it covers points at latitude α are obtained from their projections onto the (y,z) plane:

$$\begin{aligned} \sin\theta_j = \bar{B}\cdot\bar{A}_j &= \cos\alpha \sin\lambda_{I_j} \cos i \\ &- \sin\alpha \sin i \qquad (j = 1,2) \end{aligned} \tag{9}$$

with $\Delta\theta$ then given simply as

$$\Delta\theta = 2\left(\theta_2 - \theta_1\right) \tag{10}$$

II. Coverage Probabilities

The amount of coverage $\Delta\lambda$ given to a parallel of latitude α by a satellite in a circular orbit at altitude h and inclination i, with a side-range capability s, is given by Eq. (8), with $\Delta\theta$ defined in (10). If one may then assume that the longitude of the ascending node of the satellite's orbit has an equal likelihood of being in any location with respect to the longitude of a point Q at latitude α, it is possible to invoke the principle of geometric probability to state that the probability of covering the point Q, within a single orbital revolution of the satellite, is

$$P(\alpha, h, i, s) = \Delta\lambda/2\pi \tag{11}$$

Average Value Formulation

It is also possible to consider the function P as the average number of conjunctions N_c per revolution of a satellite during one revolution in its orbit under the assumption that its orbital period is incommensurate with the period of the earth's revolution. This conclusion follows by a theorem on almost periodic functions[1] that the integral multiples of a real number $N_1 \neq 0$ taken modulo another real number $N_2 \neq 0$ incommensurate with N_1 are uniformly dense in the interval $[0, N_2]$. The advantage of this formulation is that N_c per revolution is additive; e.g., the average number of conjunctions during two revolutions is twice the average number during one revolution. This property will be used later to assess the effects of orbital altitude upon coverage. The assumption of incommensurability is a reasonable one, since the set of orbital periods commensurate with that of the earth's rotation is (although dense on the real line), of Lebesgue measure, zero there.

III. Numerical Results

The average number of coverages per revolution is essentially independent of orbital altitude throughout the region of altitudes less than two earth radii. This is borne out by Fig. 3, which shows the average number of coverages per revolution (N_c per revolution made by a satellite with an orbital inclination of 60° and a side range of 500 nautmiles for two choices of orbital altitude, 200 nautmiles and 3440 nautmiles (one earth radius). Another coverage characteristic shown by Fig. 3 is that maximum coverage is obtained for points at a latitude equal to the orbital inclination minus the side range:

$$\alpha_{max} = i - s \tag{12}$$

Since a side range of 500 nautmiles is equal to about 8°, it is seen that points at a latitude of about 52° (north or south, due to symmetry about the equator) will have on the average more conjunctions than points at other latitudes. With a satellite at an altitude of 200 nautmiles, for example, a point at latitude 52° will have on the average about 33 conjunctions in 100 satellite orbital revolutions.

Figures 4-9 show how the average number of conjunctions per revolution varies with orbital inclination, side range, and earth latitude for the single orbital altitude $h = 200$ nautmiles. In each case, the values would be slightly lower for satellites at higher altitudes. Coverage of points at a number of representative latitudes is shown as a function of satellite orbital inclination for a side range of 500 nautmiles in Fig. 4. Here again it is noted that for each point at a given latitude, there exists a choice of orbital inclination for which coverage of the point is maximized. As might be expected, a relationship similar to (12) holds:

$$i_{max} = \alpha + s \tag{13}$$

Figure 5 shows how well a satellite at a number of representative orbital inclinations would be able to cover points on the earth at different latitudes, again for a side range of 500 nautmiles. Interesting special cases are the equatorial orbit, which covers points at latitudes $\alpha \leq s$ during every orbital revolution and points at latitudes $\alpha > s$ not at all; and the polar orbit, which covers points at latitudes $\alpha \geq 90^\circ - s$ during every revolution and is able to cover points at all latitudes at least some of the time. (In fact, any satellite with an orbital inclination $i \geq 90^\circ - s$ will be able to cover points at all latitudes at least some of the time.)

Figures 6-8 show how coverage varies with different values of side range capability. Figure 6 displays the $\alpha = 40^\circ$ curve of Fig. 4, with its side range of 500 nautmiles, along with the curves resulting from other values of side range. Figures 7 and 8 show the effects of assuming different side range capabilities for the $i = 60^\circ$ and $i = 90^\circ$ curves of Fig. 5.

A latitude of particular interest for rendezvous or intercept problems is that of Cape Kennedy, about 28°.5. Figure 9 shows the average number of conjunctions per revolution of a satellite and Cape Kennedy as a function of the available out-of-plane capability for various values of orbital inclination. Thus, if one has the capability of reaching satellites in orbital planes 200 nautmiles away from the launch point, one will be able to reach a satellite in a 30° orbit an average of 17 times/100 rev, or a satellite in a polar orbit an average of 4 times/100 rev.

Of course, the orbital altitude of the satellite will make a great difference in the amount of time it takes to complete a given number of orbital revolutions. To assess the effect of variations in altitude, a number of graphs of N_c per day, the average number of conjunctions per day, are given in Figures 10-13. Figures 10 and 11 show the average number of conjunctions per day of a satellite at inclination 40° with a point on the earth at latitude 40° as a function of orbital altitude h and various values of out-of-plane capability s. Figure 12 shows, for points at the latitude of Cape Kennedy and s = 250 nautmiles, the average number of conjunctions per day as a function of orbital inclination for various values of orbital altitude. Such curves can be expanded to show the average number of conjunctions in a time period T as a function of T. This is done in Fig. 13 for the h = 200-naut-mile curve of Fig. 12, using various orbital inclinations. Again, the caveat should be issued that the values presented here are asymptotic in nature and do not necessarily predict the behavior of a single satellite over a short period of time.

Multisatellite Coverage Probabilities

As a simple example of the use of the coverage probabilities, suppose that one has a number N of communication satellites at an altitude of 300 nautmiles, an inclination of 50°, and with an out-of-plane operating capability of 500 nautmiles, and a ground communications station at latitude 40°. Assuming that the satellites are randomly oriented with respect to the station, one wishes to determine the probability P(N) that at least one of the N satellites is in position to communicate with the station within a time equal to one orbital period (95.6 min). Figure 6 shows the result P(1) for one satellite to be 0.25. For more than one satellite, the probability that at least one satellite can communicate is equal to unity minus the probability that none of the satellites can communicate, i.e.,

$$P(N) = 1 - [1 - P(1)]^{N} \tag{14}$$

The function P(N) for this case is shown as Fig. 14; for instance, if one has ten satellites, the probability that at least one will be in position to communicate within a time equal to one orbital period is 0.944.

IV. A General Mission-Evaluation Technique

The ability of the function P(α, h, i, s) to impart a feeling for the relative coverage capabilities of various satellite configurations is diminished by the fact that it is a function of four variables, a rather difficult thing to visualize. The previous figures have presented various two-dimensional sections of this function in order to achieve some degree of intelligibility, but this has been done at the expense of holding two or so of the variables at constant values, giving an overspecialized picture for most mission considerations.

If, instead of holding these extra variables constant, one were to take suitably defined averages over their range of variation, a reduction in dimension could be accomplished which would make the data not only easier to visualize, but also more meaningful with respect to the mission being considered. This is the idea behind the general mission-evaluation technique described below.

Mathematical Formulation

Suppose that one has a method of evaluating the relative importance of a satellite configuration as a function of some subset of the variables α, h, i, s . Let the subspace of the four-dimensional Euclidean space R_4 (α, h, i, s) spanned by this subset be denoted as X; denote the quotient space $R_4 \div X$ (the subspace spanned by the remaining variables) as Y. The function $P(\alpha, h, i, s)$ can then be considered as a function $P(x, y)$, where x represents a point in X , y a point in Y .

If one then expresses the relative importance of points x_o in X by means of a distribution function $V(x_o)$, representing the probability that points x such that $x \leq x_o$ (the inequality sign is assumed to refer to all components of x) are of interest to the mission, one may consider the function $P(x, y)$ as a random variable with X as sample space and $V(x)$ as its associated distribution function. The expectation of this random variable will be a function defined on Y , representing the expected number of coverages per revolution of points of interest to the mission:

$$E(y) = \int_X P(x, y) \; d\,V(x) \tag{15}$$

The integral defined in this way is a Stieltjes integral whose existence is guaranteed by the fact that $P(x, y) = P(\alpha, h, i, s)$ is continuous in all its variables, and $V(x)$ is monotone in x .

In order to compute the function $E(y)$, one uses the density function $v(x)$, associated with $V(x)$ by the relation

$$V(x) = \int_{-\infty}^{x} \cdots \int v(z) \; dz \tag{16}$$

The function $v(x)$ may be continuous in x , in which case

$$E(y) = \int_X P(x, y) \; v(x) \; dx \tag{17}$$

or it may be a convex combination of unit delta-functions, taking on nonzero values only at the points $x_1, x_2, \ldots x_n$, and satisfying

$$v(x_j) > 0 , \quad j = 1,2,\ldots,n$$
$$v(x_1) + v(x_2) + \ldots + v(x_n) = 1 \tag{18}$$

In this case

$$E(y) = \sum_{j=1}^{n} P(x_j,y) \; v(x_j) \tag{19}$$

Since all integrations have been carried out on a digital computer, the first case reduces, in practice, to the second.

Application I: Ground System Coverage

Consider the problem of placing a satellite with an altitude $h = 300$ nautmiles and an out-of-plane operating capability $s = 500$ nautmiles at an orbital inclination i, which provides maximal coverage of a certain area on the earth, say the continental U.S.A. In this problem, the relative importance of a satellite configuration can be characterized in terms of ground point latitude α : latitudes covered by large stretches of the continental U.S.A. are more important to the mission than latitudes covered by small stretches, and latitudes outside the range of the continental U.S.A. are of no importance to the mission. Thus, the relevant subspace X is the space of latitudes α ; the corresponding space Y consists of points of the form (h, i, s), and the expectation relation is

$$E(h, i, s) = \int_{-\infty}^{\infty} P(\alpha, h, i, s) \; d\,V(\alpha) \tag{20}$$

The density function $v(\alpha)$ associated with $V(\alpha)$ is a constant k multiplied by the extent of land in the continental U.S.A. at latitude α, as obtained from a map. This $v(\alpha)$ is shown in Fig. 15; it is a piecewise continuous function with k, chosen so that

$$\int_{-\infty}^{\infty} v(\alpha) \; d\alpha = 1 \tag{21}$$

A number of numerical integrations of Eq. (20) have been performed, using $h = 300$ nautmiles, $s = 500$ nautmiles, and different values of inclination, with the results given in Fig. 16. It is seen that the maximum coverage is obtained at an inclination of about 50°, and that at this inclination the expected fraction of the continental U.S.A. covered by the satellite during a single orbital revolution is about 0.26.

Application II: A Rendezvous-Intercept Problem

In this problem, it is desired to place the launch site of a rendezvous or intercept vehicle at that latitude at which its operational effectiveness is maximized, over the range of altitudes and side ranges $500 \le h \le 3000$, $500 \le s \le 3000$. The effectiveness of a launch site-satellite combination as a function of altitude h and side-range capability s is taken to be proportional to the vehicle mass ratio required to effect an intersection of the trajectory of the launched vehicle with the orbit of the satellite (i.e., effectiveness is proportional to the amount of payload which can be placed into orbit). To simplify the analysis, an approximate solution for the mass ratio incorporating minimum-energy trajectories with respect to a spherical, atmosphere-free earth is used; the resulting density function is

$$v(h, s) = k_1 \exp\left[-270\left(1+0.1\,\frac{s}{h}\right)\sqrt{\frac{h}{r_o(r_o+h)}}\right] \qquad 500 \le h,\ s \le 3000 \tag{22}$$

$v(h, s) = 0$, elsewhere; where r_o is the radius of the earth in nautical miles, and k_1 is determined by the requirement

$$\int_{500}^{3000}\int_{500}^{3000} v(h, s)\ dh\, ds = 1 \tag{23}$$

Three examples are considered, representing different specifications of the orbital inclinations of interest to the mission. In examples A and B, the single inclinations $i = 35^\circ$ and $i = 60^\circ$, respectively, are of sole interest; in example C, a probability distribution of inclinations of interest is considered, with the probability mass divided discretely as follows:

i(deg)	0	15	30	45	60	75	90
$v_1(i)$	0.1	0.05	0.1	0.15	0.25	0.1	0.25

In examples A and B, the relevant subspaces X and Y consist of points of the form $x = (\alpha, i)$ and $y = (h, s)$; the expectation relation is

$$E(\alpha, i) = \int_{500}^{3000} \int_{500}^{3000} P(\alpha,h,i,s)\ v(h,s)\ dh\ ds \qquad (24)$$

In example C, the subspace X consists of points of the form $x = (h, i, s)$, and Y is the subspace of latitudes α ; here the expectation relation is

$$E(\alpha) = \sum_i v_1(i) \int_{500}^{3000} \int_{500}^{3000} P(\alpha,h,i,s)\ v(h,s)\ dh\ ds \qquad (25)$$

These equations have been numerically integrated for a number of values of latitude α ; the results are given in Fig. 17. Effective coverage of satellites at an inclination of 35° is maximized by near-equatorial launch sites; coverage capability, with respect to the mission defined, is strong for launch sites with latitudes up to about 30°; for higher latitudes, coverage begins to degrade considerably--satellites at an inclination of 60° are covered reasonably well by launch sites at any latitude; best coverage is attained at latitudes of about 50°. Curve B can be considered as a distillation of the curves in Fig. 7, averaged with respect to the mission of interest not only with respect to side range but also with respect to orbital altitude. For the mixture of inclinations considered in example C, it is seen that launch sites at extremely low, or high latitudes provide somewhat better coverage than those at the middle latitudes, but the differences in coverage capability are not large.

General Remarks

For the purpose of clearer exposition, the examples presented here have considered somewhat simplified definitions of performance criteria; the method can be, and has been, used for more detailed mission analyses. It must, of course, be kept in mind that relative coverage capability is not the only criterion by which satellite system parameters are selected. However, the results of applying the present technique provide a quantitative measure of coverage which can be balanced with other mission considerations.

V. Discussion of Assumptions and Limitations

Equal Likelihood

Unless a satellite is able to apply corrective thrusts to maintain itself in a particular relation with respect to a point on the earth, the assumption of equal likelihood is a reasonable one. Indeed, the nodal longitude exhibits a number of secular, periodic, and sporadic variations because of such effects as earth rotation and asphericity, atmospheric drag, radiation pressure, and the gravitational influences of other bodies; many of these effects and variations are imprecisely determined. Some care must be exercised in the interpretation of these results, however; they are asymptotic in nature and do not necessarily predict the behavior of a single satellite over a short period of time.

An exceptional case in which the equal likelihood assumption appears least reasonable is that of the earth-period or 24-hr satellite, for which $\omega = \omega_E$. However, this case is also exceptional from the standpoint of the basic coverage equation (8), which reduces to the geometric relation $\Delta\lambda = \Delta\lambda_I - \Delta\theta$, equal to zero for an equatorial orbit and nearly zero for low inclinations.

Eccentric Orbits

Another limitation on the analysis is the assumption of a circular orbit, which allowed the use of a constant satellite angular velocity ω. Since a satellite in an orbit of nonzero eccentricity travels fastest at perigee and slowest at apogee, the quantities of interest in judging deviations from constant angular velocity are the ratio of angular velocity at perigee (ω_P) and apogee (ω_A) with respect to the angular velocity (ω_C) of a satellite in a circular orbit with the same semimajor axis. The vis viva integral yields the following relationships:

$$\frac{\omega_P}{\omega_C} = \frac{1}{1-e}\sqrt{\frac{1+e}{1-e}}$$
$$\frac{\omega_A}{\omega_C} = \frac{1}{1+e}\sqrt{\frac{1-e}{1+e}} \qquad (26)$$

Some representative values are given in Table 1.

Table 1 Variation of angular velocity with eccentricity

	e=0	e=0.1	e=0.2	e=0.4	e=0.6	e=0.8
ω_P/ω_C	1.00	1.22	1.45	2.54	5.00	15.0
ω_A/ω_C	1.00	0.85	0.67	0.47	0.31	0.185

The assumption of a constant rotation rate ω was used in Eq. (8) in order to simplify calculation of the correction term due to earth rotation. For a satellite at an altitude of about 200 nautmiles, the correction term is small, and eccentricities up to the order of 0.4 will not produce appreciable differences in the coverage probabilities. For a satellite at an altitude of about one earth radius, the correction term is larger, and the results will begin to differ appreciably for eccentricities on the order of 0.1 or 0.2.

Earth Oblateness and Atmosphere

The main effects of the earth's asphericity upon a satellite's orbit are small secular variations of the argument of perigee and longitude of the ascending node. Neither of these effects produces more than slight modifications of the coverage probabilities; thus the assumption of a spherical earth can be made with no significant loss of generality. Atmospheric perturbations of satellite orbits with altitudes greater than 100 nautmiles are not of the size or nature to affect the assumptions upon which the coverage calculations are based. Figure 18 shows the region of applicability of the analysis in terms of orbital eccentricity e and perigee altitude h_P.

References

[1]Besicovitch, A. S., Almost Periodic Functions (Dover Publications, Inc., New York, 1954).

[2]Bennett, F. V., Coleman, T. L., and Houbolt, J. C., "Determination of the required number of randomly spaced communication satellites," NASA TN D-619 (January 1961).

[3]Gundel, B. H., "Satellite coverages," Astronaut. Sci. Rev. 2, 19-23 (October/December 1960).

[4]Krause, H., "The motion of a satellite station around the earth in an elliptical orbit inclined to the earth's equator," The RAND Corp. Trans. T-52 (October 21, 1955).

[5]Doob, J. L., Stochastic Processes (John Wiley and Sons, Inc., New York, 1953).

[6]Boehm, B., "Satellite coverage probabilities," The RAND Corp. Memo. RM-3297-PR (October 1962).

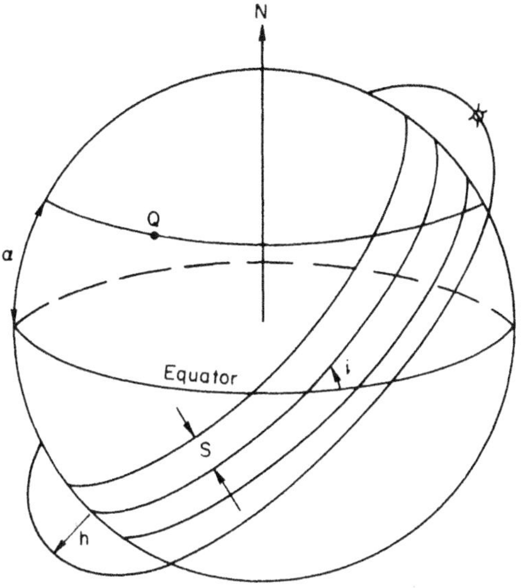

Fig. 1 The coverage problem

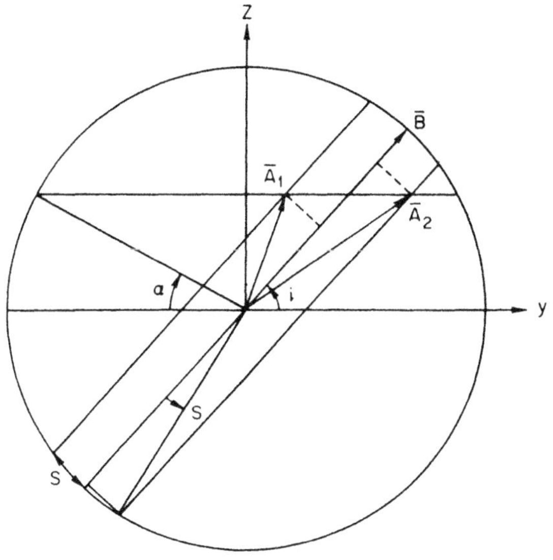

Fig. 2 Coverage relationships in the plane

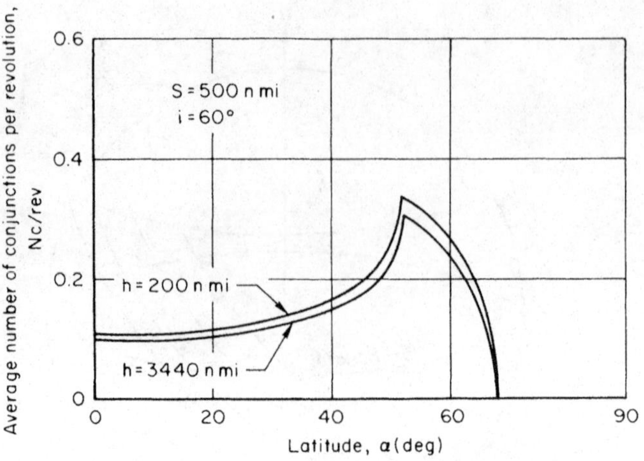

Fig. 3 Average number of conjunctions per revolution, (N_c per revolution) latitude, two altitudes

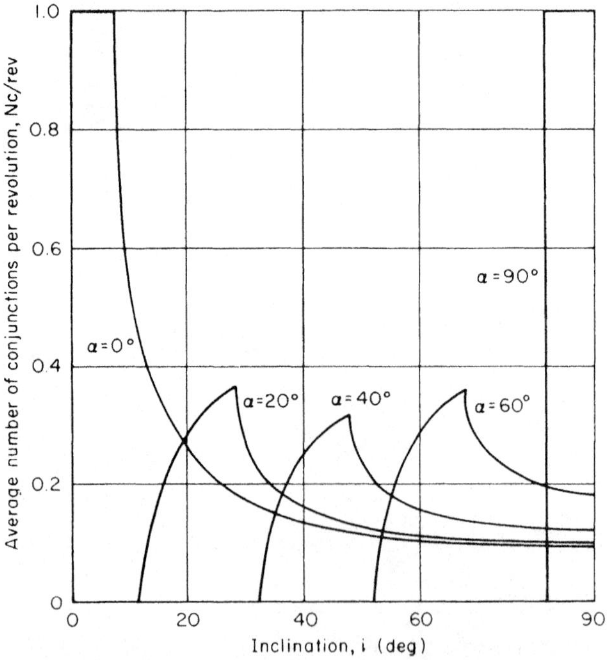

Fig. 4 N_c per revolution inclination, various latitudes

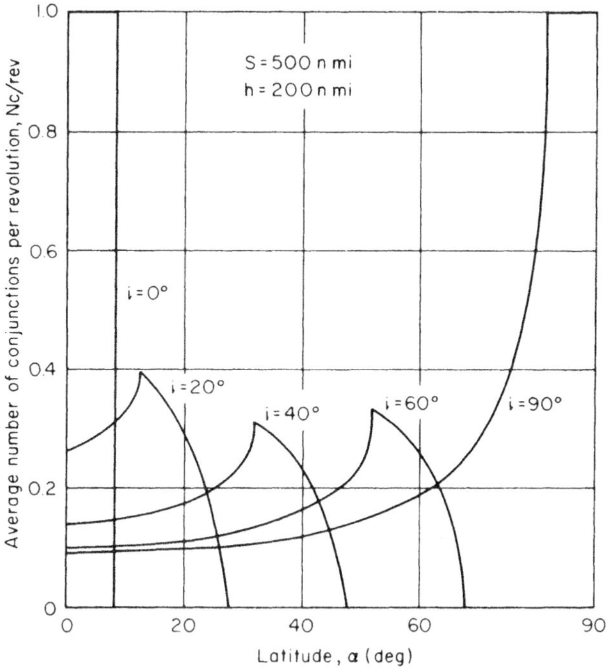

Fig. 5 N_c per revolution latitude, various inclinations

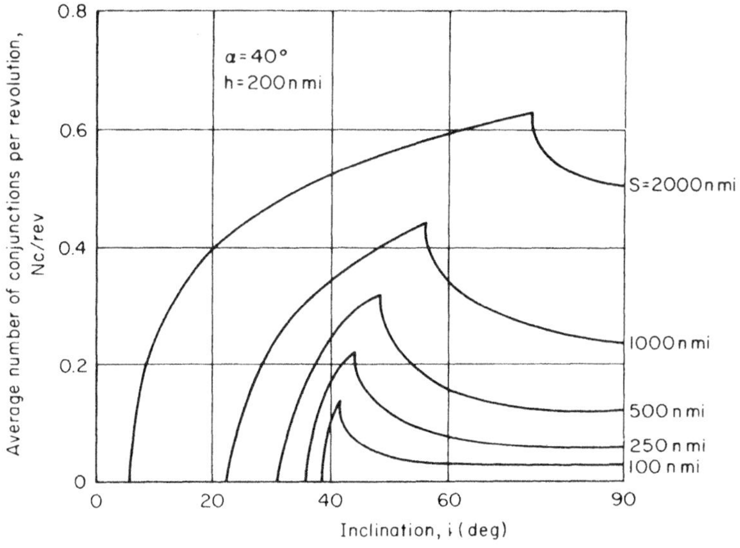

Fig. 6 N_c per revolution inclination, various side ranges

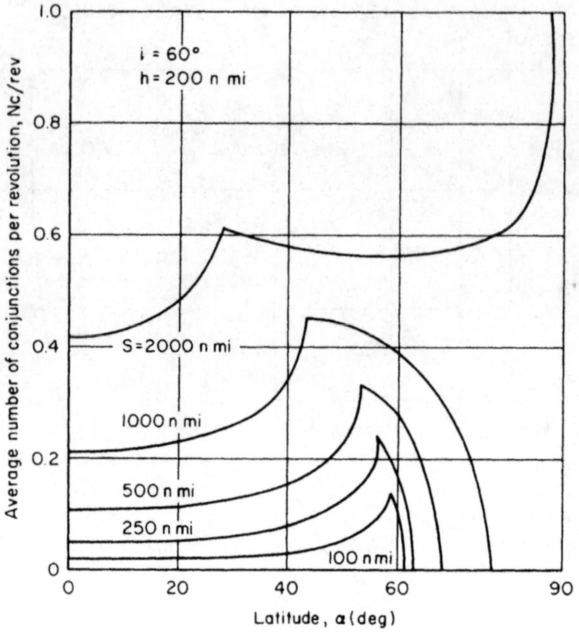

Fig. 7 N_c per revolution latitude, various side ranges (60° orbit)

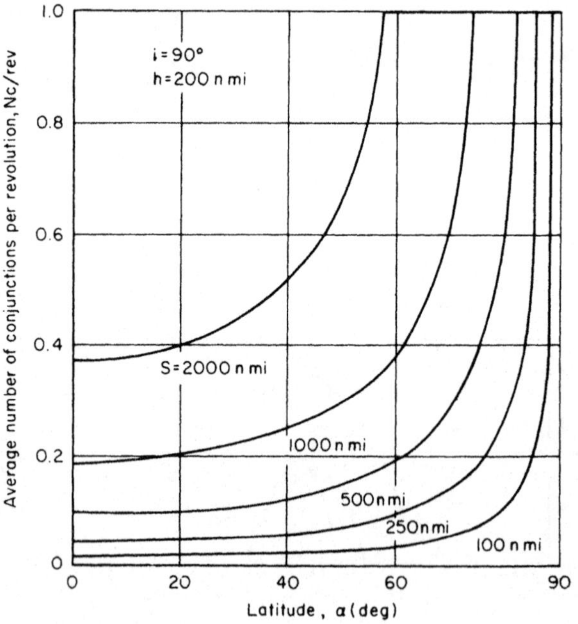

Fig. 8 N_c per revolution latitude, various side ranges (90° orbits)

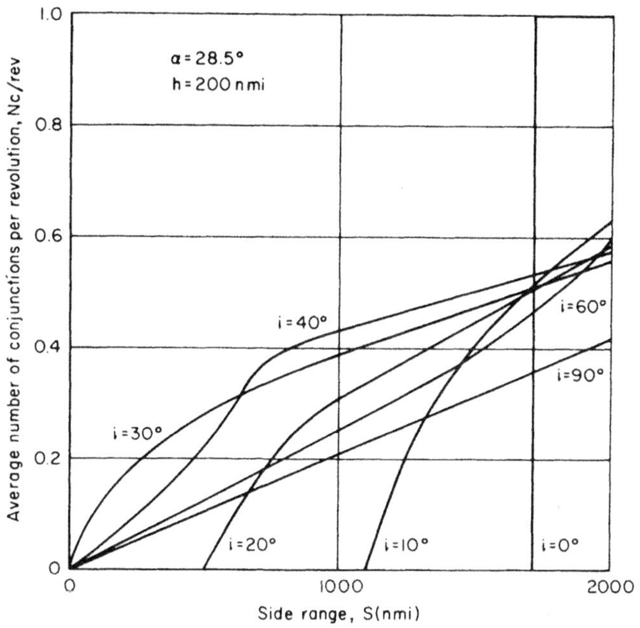

Fig. 9 N_c per revolution side range, various inclinations

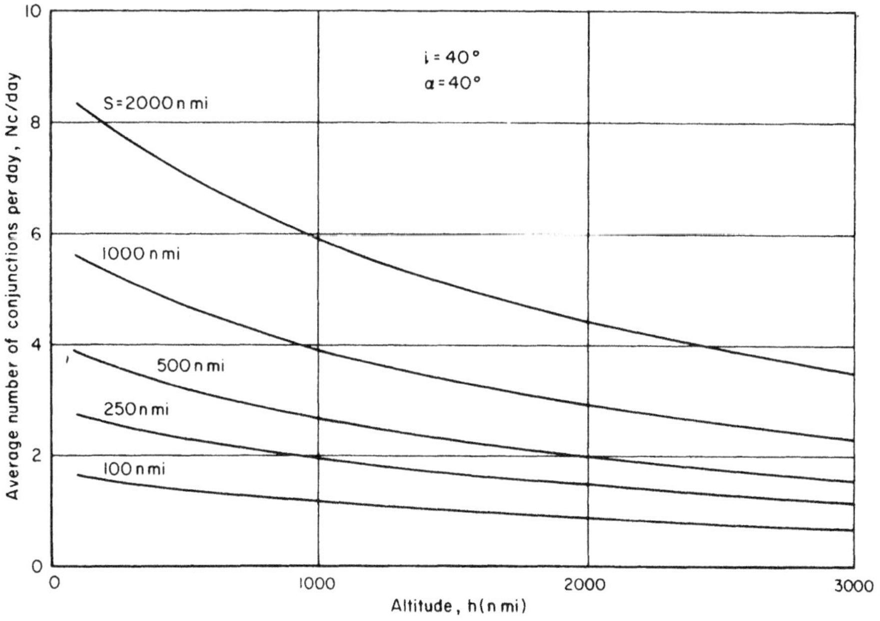

Fig. 10 Average number of conjunctions per day, (N_c per day) altitude, various side ranges

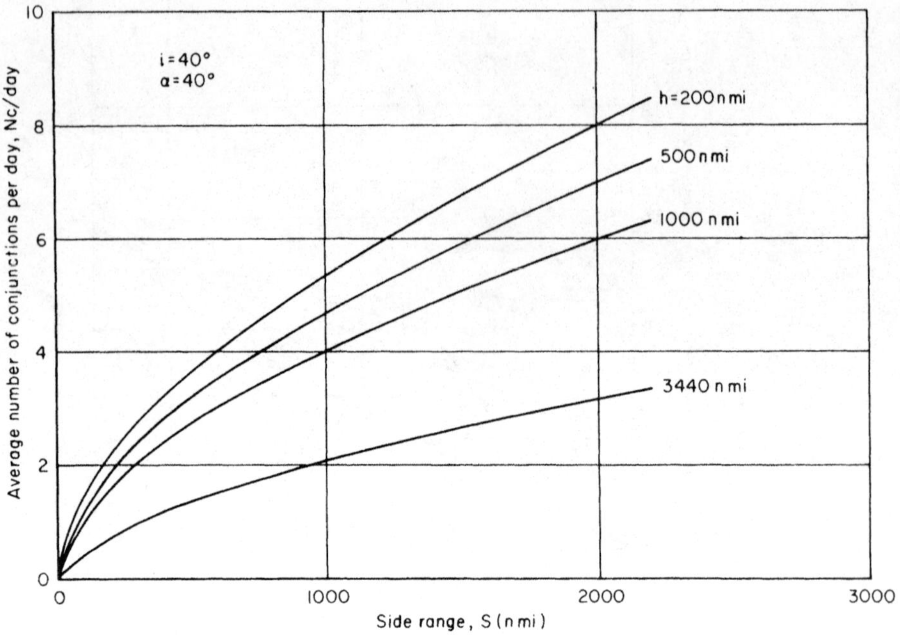

Fig. 11 N_c per day side range, various altitudes

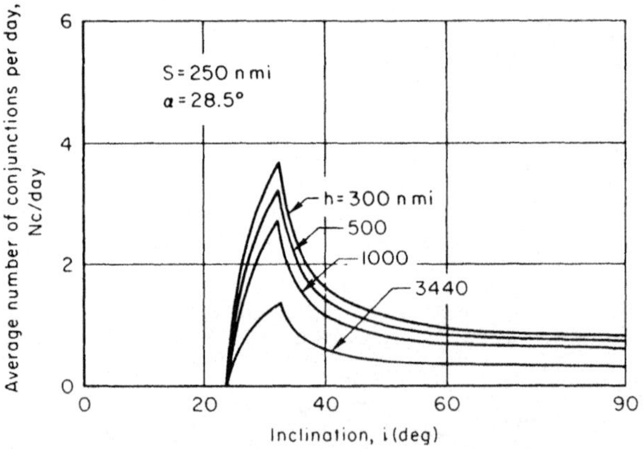

Fig. 12 Average number of conjunctions per day (N_c per day) inclination, various altitudes

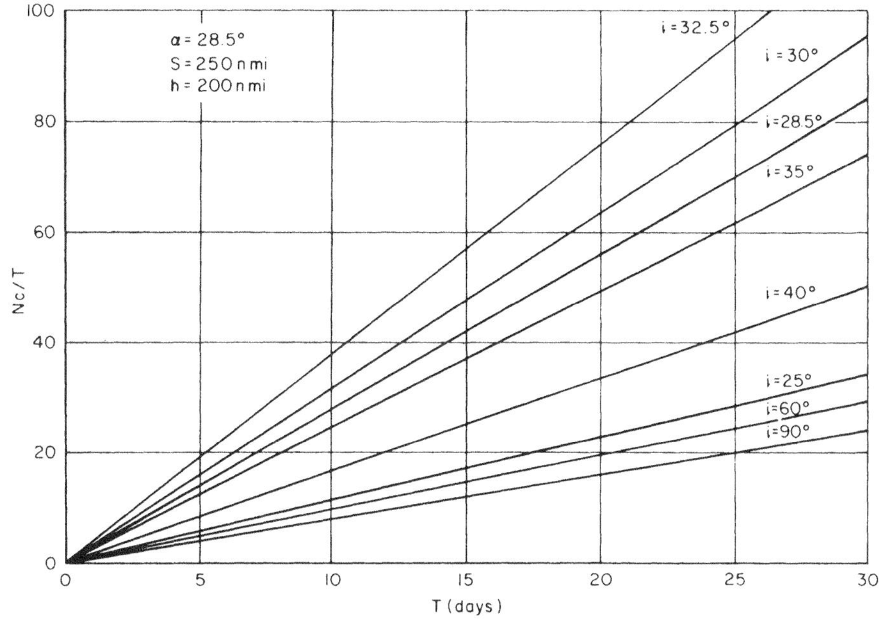

Fig. 13 Average number of conjunctions within a given period, T

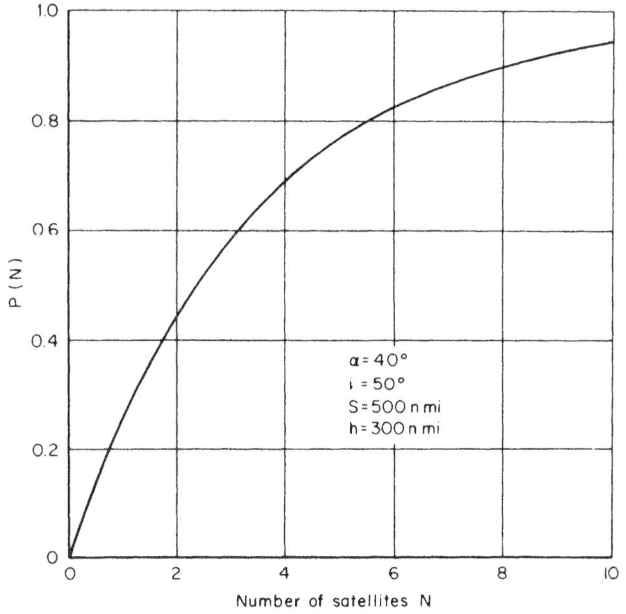

Fig. 14 Probability of communication with at least one satellite within one orbital revolution

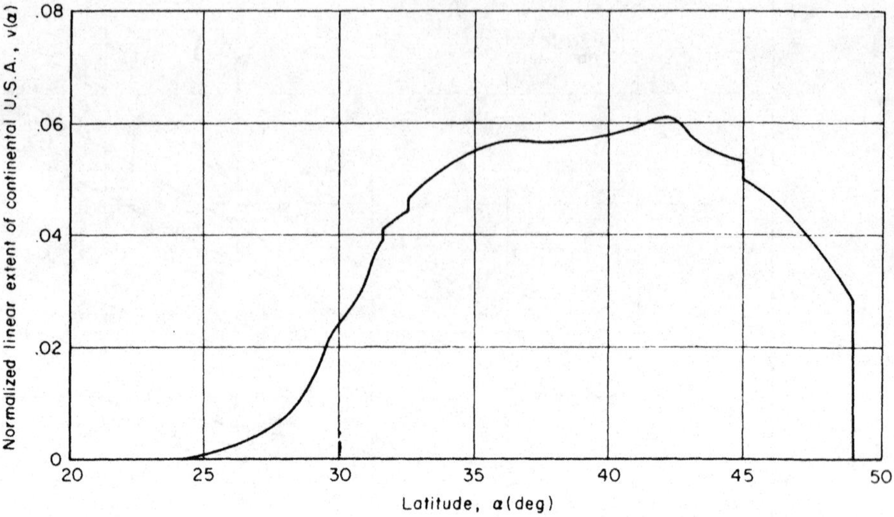

Fig. 15 Normalized linear extent of continental U.S.A. latitude

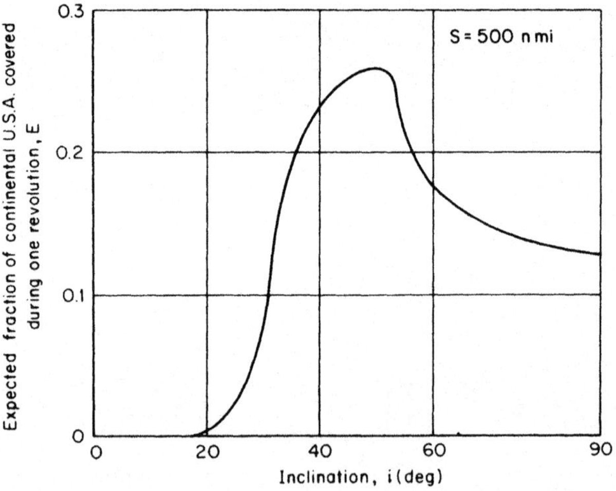

Fig. 16 Expected fraction of continental U.S.A. covered during one revolution inclination

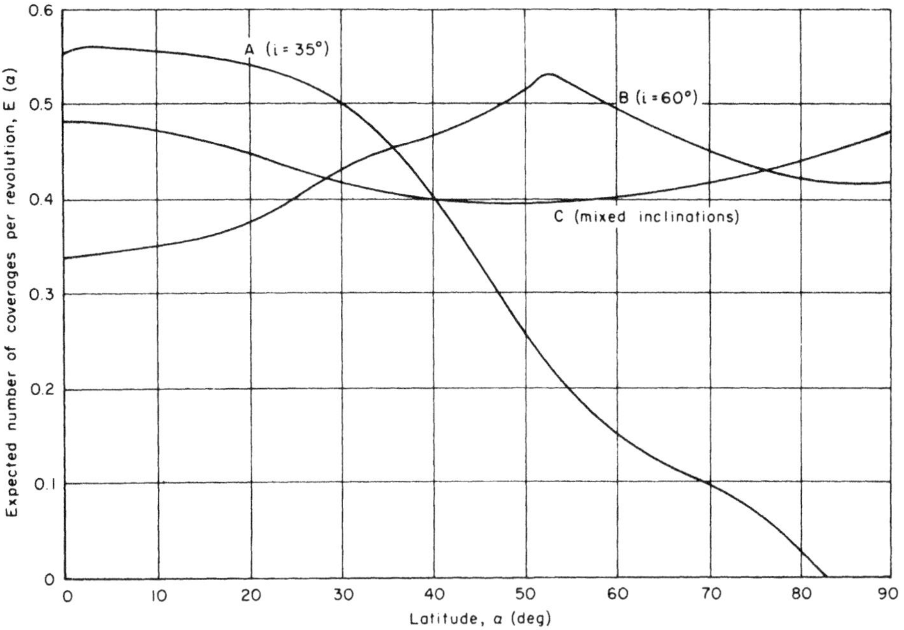

Fig. 17 Effective coverage versus latitude, rendezvous-intercept mission

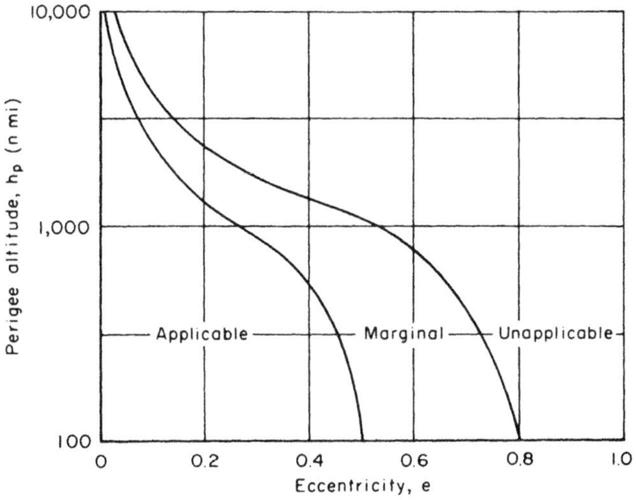

Fig. 18 Region of applicability of coverage probabilities

ORBITAL ASPECTS OF NONSYNCHRONOUS COMMUNICATION SATELLITE SYSTEMS

Hans K. Karrenberg* and R. David Lüders†

Aerospace Corporation, El Segundo, Calif.

Abstract

Effects of the following system parameters upon the communication probability (i.e., the long-term average probability of being able to communicate between various pairs of ground terminals by means of satellite relays arranged in certain nonsynchronous orbital systems) were examined: the number and inclination of the several orbit planes used, the total number of satellites in the system and their apportionment among the orbit planes, and the mode of distribution (random or evenly spaced) of the satellites within the planes. The ground station pairs chosen for consideration form a worldwide communication network with a large variety of geographic separations and orientations. Several attractive 5000-naut-mile-alt orbital systems of evenly spaced and randomly distributed satellites were found and compared. These include: 1) systems of equally separated polar orbits; 2) systems of equally separated orbits commonly inclined at an optimum angle; and 3) mixed-inclination, equatorial-polar systems with the satellites optimally apportioned among the orbits. The communication fraction was determined as a function of total number of satellites for each of these systems. Owing to the adverse geographic location or separation of some station pairs, their communication fractions were found to be much worse than values for the majority of the links considered. The enhancement of communications on

Presented as Preprint 63-397 at the AIAA Astrodynamics Conference, New Haven, Conn., August 19-21, 1963.

This paper was prepared under U. S. Air Force Contract No. AF 04(695)-169.

*Manager, Orbit Designation Section, Astrodynamics Department.

†Member of the Technical Staff, Astrodynamics Department.

such difficult links by means of an intermediate ground relay station, used in conjunction with two satellite relays, was investigated. Studies were also made to determine the probable durations of interrupted and unavailable service and the relative distribution of communication "holes."

Nomenclature

CR = circular communication region for a ground station

$d_j(\psi)$ = arc of intersection of the jth orbit plane with the RMC for an orientation angle ψ (see Fig. 3)

h_j = altitude of the satellites in the jth orbit plane

i_j = inclination of the jth orbit plane to the equator

L = longitudinal separation of ground stations 1 and 2; $0^\circ \leq L < 180^\circ$

MWTC = mean-waiting-time-to-communicate; by definition, 50% of the communication holes are no longer than the MWTC

n_1 = number of orbit planes in the system

n_{2_j} = number of satellites in the jth orbit plane

N = total number of satellites in the orbital system

$$N = \sum_{j=1}^{n_1} n_{2_j}$$

opt = subscript denoting an optimum value

P_D = down probability; probability of not being able to communicate between a pair of ground stations due to the lack of a mutually visible satellite relay

$\overline{P}(t)$ = probability of having to wait no longer than time t to begin (or resume) communication via satellite relay

R = radius of the (spherical) earth; 3440 naut miles

RMC = region of mutual communication for a pair of ground stations (defined in Fig. 1)

t = waiting time for a communication hole to end

T = orbital period of each of the satellites in a common altitude system

θ_j = true anomaly at which a satellite in the jth orbit plane rises (or sets) relative to a ground station

λ_1, λ_2 = latitudes of ground stations 1 and 2

σ = minimum station elevation angle permitting communication between a satellite and a ground station (see Fig. 1a)

φ_j = great circle arc measure of the region within which a satellite in the jth orbit can communicate with a ground station (see Figs. 1a and 2)

ψ = orientation angle of an orbital system relative to the rotating earth (see Fig. 3)

ω = angular velocity of each of the satellites in a common altitude system

Ω_j = longitude of the ascending node of the jth orbit plane measured eastward from the ascending node of the first orbit plane; Ω_1 is zero by definition

I. Introduction

The subject of communication between pairs of widely separated ground terminals by means of satellite relays has received considerable attention in the recent technical literature. In particular, high-altitude orbital systems containing satellites which move in synchronism with the rotational period of the earth have been examined at length, and the advantages of such systems have been clearly indicated. However, the inherent feature of synchronous systems is that they employ oriented, active repeaters and consequently require both attitude control and station-keeping. In view of the current state-of-the-art in boost vehicles, attitude control devices, and over-all system reliability, it has been argued that synchronous communication satellite systems are not as yet technically feasible.

Systems of nonsynchronous satellites at medium altitudes (about 5000 naut miles), on the other hand, are currently practicable. Such systems employing more but simpler satellites could provide an immediate global communication capability, and furnish a proving ground for the development of more advanced hardware and techniques.

This paper introduces an analytical technique that is useful in determining the probability that communication is possible between various pairs of ground stations by means of satellite relays arranged in certain systems of nonsynchronous orbits. Although the generality of the technique introduced permits it to be applied to systems consisting of any number of satellites in any number of circular orbits with free choice of their altitudes, orbit inclinations, and nodal locations, the application of the technique in the present work is confined to systems in which the satellites are in common-altitude circular orbits and are divided among a few distinct orbit planes. The use of only a few planes permits the utilization of multiple-satellite injection. The technique is demonstrated in the examination of several promising types of orbital systems to determine which ones provide the best coverage of a typical worldwide network of communication links.

II. Study Comparison

Various authors[1-7] have analyzed some of the orbital aspects of nonsynchronous communication satellite systems. In order to compare the scope, emphasis, and fundamental differences of the studies, a tabular summary of the contents of those references and this paper is presented in Table 1. The contents of the studies are divided into five categories that tend to indicate the scope of the analyses. In each case, the results are a measure of communicability between pairs of ground stations.

The first category simply states the orbital altitudes that were considered, or, if altitude was a study parameter, the range of altitudes that were considered. Since all of the studies examine circular orbits about a spherical earth, there is no ambiguity in describing the orbits by their altitudes. However, in addition to circular orbits, Ref. 4 presents a brief discussion of elliptic orbits. For these orbits, the apogee altitudes range from about 4000 to about 38,000 naut miles. Note that in five of the nine studies (Ref. 6 presented two studies), a parametric examination of the altitude was made. The second category indicates the values of the minimum station elevation angle which were considered. Communications between a satellite and a ground station are assumed to be impossible if the elevation angle of the satellite at the ground station is less than the specified minimum. The altitude and minimum station elevation angle can be combined to define a lenticular region within which a satellite is mutually visible to two ground stations. Any satellite in that region can serve as a communication relay between the pair of ground stations. Reference 2 and parts of Ref. 6 differ from the other studies in that they assume that the region of mutual communication is irregular in shape, rather than lenticular, and may be approximated by a circle.

The third category describes the number and kinds of satellite distributions considered. Note that only Refs. 4 and 7 and this paper consider a distribution other than one for which "the satellites are deployed singly in randomly spaced circular orbits." The fourth category indicates either the orbit inclinations that were considered or the range if inclination was a study parameter. These last two categories are study areas in which this paper makes significant advances beyond the earlier studies. Reference 4 presents results for a system of satellites which are randomly distributed in a single circular polar orbit. This paper and Ref. 7 generalize this concept to include systems of satellites which are randomly distributed in positioned circular orbits. This paper and Ref. 7 analyze orbital systems of mixed inclination; in addition, this paper determines the optimum apportionment of satellites among the orbit planes. Systems of commonly inclined orbit planes are studied, and the optimum inclinations are determined. In every case, the number of orbit planes is optimized. Reference 4 presents results for a system of satellites which are spaced in circular polar orbits. This paper generalizes this concept to include systems of satellites which are equally spaced in positioned circular orbits whose inclinations are identical (eight values between 0° and 90° were considered) or mixed. Studies are made of the number of satellites required for 100% communicability and of the variation in communicability with the number of satellites per plane if less than 100% communicability is acceptable. Since each orbit plane contains a number of satellites (which are either randomly distributed or equally spaced), the utility of multiple-satellite launchings and their effect on orbit inclination were discussed.

The fifth category presents the links or pairs of ground stations between which communication is desired. The ability to provide communications between two stations via a satellite at a specified altitude is a function of the following parameters: 1) the great circle distance between the stations; 2) the azimuthal orientation of the stations; and 3) the latitude of the center of the region of mutual communication. Only Refs. 6 and 7 and this paper examine enough station pairs to allow sufficient variation in the preceding parameters and their combinations to permit comparison. In addition, this paper presents an analytical treatment of double hop relays, the probabilities of interrupted and unavailable service and the relative distribution of time intervals during which communication is impossible. Similar work was done in Ref. 7, which presents computer results concerning double hop relays and communication downtimes.

In brief summary, the contents of this paper which are unique or different in scope from comparable papers by other authors fall into three categories. They are 1) satellite distributions: equally spaced satellites in positioned circular orbits; 2) orbit inclination: optimum satellite apportionment in mixed inclination systems and optimum inclinations for commonly inclined systems; and 3) global communication network: the advantages of the double hop mode, the probabilities of interrupted or unavailable service and the relative distribution of downtimes.

III. Geometry of Mutual Visibility

In order that communication via satellite relay be possible between a pair of ground stations, it is necessary that there be at least one satellite with which both stations can simultaneously communicate. The necessary and sufficient geometric condition for line-of-sight communication between a satellite and a ground station is illustrated in Fig. 1a. The satellite must be located within the cone whose apex is at the ground station S and whose elements make an angle of $(90^\circ - \sigma)$ with the station radius vector $\overline{OS}$, where σ is the minimum station elevation angle that permits a signal to be transmitted to or received from a satellite by a ground station.‡

The requirement that a satellite lie within the station cone, combined with the condition that all satellites in the

‡Sinden and Mammel, in Ref. 2, do not assume that the minimum acceptable elevation angles at the transmitter and receiver are equal; instead, the criterion for communication which they use is that the signal-to-noise ratio at the receiving station must exceed some prescribed value. However, later in their analysis, they make an approximation that is avoided here to the region of mutual communication.

same orbit plane are in a circular orbit of altitude h, implies that only those satellites located within the shaded segment of the orbit sphere (of altitude h) can communicate with S. The communication region (CR) for station S is the radial projection of the shaded segment of the orbit sphere onto the surface of the earth. A satellite in the orbit plane in question can communicate with station S if and only if its subpoint falls within CR. The great circle arc radius φ of the communication region is related to the altitude h of the orbit and to σ, the minimum elevation angle at which communication is possible, as follows:

$$\phi = \left[\cos^{-1}\left(\frac{R \cos \sigma}{h + R}\right) - \sigma\right]$$

Figure 2 shows φ as a function of satellite altitude h for several typical values of σ.

Figure 1b illustrates the communication geometry for a pair of ground stations whose latitudes are λ_1 and λ_2 and whose longitudes differ by L. The communication region for the first station is the area CR_1 within the small circle of radius φ centered at point 1. Similarly, CR_2 is bounded by a small circle of the same size centered at point 2. Only a satellite whose subpoint falls within the shaded lenticular area common to CR_1 and CR_2 can serve as a communication relay between stations 1 and 2. Thus, the problem reduces to an analysis of intersections of the subsatellite traces with the region of mutual communication (RMC).

IV. Analytical Technique

The problem of determining the fraction of time during which at least one satellite is mutually "visible" to a pair of ground stations can be solved in either of two ways. The direct approach to the problem makes use of a computer simulation program that generates the ephemeris for the system of satellites starting from the prescribed initial configuration. At each time step in this process, the program determines which satellites are mutually visible to the given set of station pairs. The output from such an ephemeris program enables one to ascertain the precise nature of the communication coverage provided by a particular orbital system; for example, the extent and distribution of communication downtimes (i.e., times during which communication is impossible due to the absence of all the satellites from the RMC), the minimum handover interval from a "setting" satellite to another in the RMC, and the usage levels of a given satellite by more than one station pair can all be readily determined.

The second method of determining the communication fraction is by means of a probability program that calculates the long-term average probability of being able to communicate between a pair of ground stations by means of the satellites in a given orbital system. This latter technique was used to generate the results presented in this paper. The basic steps in the calculation procedure are as follows:

1) For an arbitrary rotational orientation of the earth within the system of orbits, determine the arcs of intersection of the orbit planes with the RMC.

2) Using that set of arc lengths, calculate the probability for each orbit plane at that instant that no satellite is located in the RMC.

3) The combined probability of there being no satellite in any of the orbital arcs intersected by the RMC is the product of the separate probabilities found in step 2.

4) Steps 1-3 are repeated for all orientations of the RMC relative to the system of orbit planes; i.e., for all positions of the earth realized in a day.

5) An average "down probability" P_D (of not being able to communicate via satellite relay) is then determined by averaging the products found previously.

The calculations just outlined can be expressed mathematically as follows:

$$P_D = \frac{1}{2\pi} \int_0^{2\pi} \prod_{j=1}^{n_1} p_j(\psi)\, d\psi$$

where P_D is the long-term average down probability sought; n_1 is the number of orbit planes in which the satellites are arranged; ψ is an angle which specifies the rotational orientation of the earth relative to the system of orbit planes (see Fig. 3); and $p_j(\psi)$ is the probability that none of the satellites in the jth orbit, for the orientation angle ψ, lie within the RMC.

The quantity $p_j(\psi)$ depends upon the longitude of the ascending node of the jth orbit with respect to the ascending node of the first (reference) orbit, the altitude and inclination of the jth orbit plane, the number and mode of distribution of the satellites in the jth orbit, and the coordinates of the pair of ground stations involved. Figure 3 illustrates the geometry of the intersection of the jth orbit plane with the region of mutual communication on the earth's surface. Let ψ denote the longitudinal position of the ascending node of the first orbit plane measured eastward from a reference meridian through the first (the westernmore) of the two ground stations. Then $d_j(\psi)$ denotes the length of the jth suborbital arc intercepted by the RMC when the orbital system orientation angle is ψ. The principal steps in the procedure for determining $d_j(\psi)$ are as follows:

1) Solve a pair of quadratic equations (see Appendix A) for the angular position of the (at most four) intersections of the jth suborbital trace with the two communication station circles; $d_j(\psi)$ is zero unless each pair of roots is real and distinct.

2) Order the four roots to determine whether the roots "interweave" (i.e., whether the regions of visibility to the two separate stations overlap); $d_j(\psi)$ is zero unless such an intermingling of the roots occurs.

3) Finally, $d_j(\psi)$ is the difference of the innermost pair of roots.

The quantity $p_j(\psi)$ is related to $d_j(\psi)$ and the manner in which the n_{2_j} satellites are distributed in the jth orbit. One of the kinds of satellite distributions which can be treated with the probability program is the random mode. If the satellites are distributed randomly in the jth orbit whose intersection with the RMC at "time" ψ is $d_j(\psi)$, the probability that a given satellite is not mutually visible to the station pair is

$$\left[1 - \frac{d_j(\psi)}{2\pi}\right]$$

Then the probability that none of the n_{2_j} satellites in the jth orbit lies within the RMC is

$$p_j(\psi) = \left[1 - \frac{d_j(\psi)}{2\pi}\right]^{n_{2_j}}$$

Another kind of satellite distribution which can be handled by the program is the evenly spaced mode. If the n_{2_j} satellites in the jth orbit are spaced at regular intervals of $2\pi/n_{2_j}$ around the orbit, then the probability that no satellite in the jth orbit is located within the RMC is given by

$$p_j(\psi) = \left[1 - n_{2_j} d_j(\psi)/2\pi\right] \quad \text{if} \quad d_j(\psi) \leq 2\pi/n_{2_j}$$

or

$$p_j(\psi) = 0 \qquad\qquad \text{if} \quad d_j(\psi) > 2\pi/n_{2_j}$$

In the latter case, it is a certainty that at least one satellite in the jth orbit is mutually visible at time ψ.

The assumption is made in the analytical technique presented previously that the angular position of any satellite in its orbit plane is independent of ψ, the rotational orientation of the earth relative to the orbit plane. It is this assumption that is valid for a system of satellites whose trace repeats (i.e., closes on itself) for the first time after a large number of satellite revolutions that limits the applicability of the analytical technique to nonsynchronous orbital systems.

It is important to note that, because of the assumption of "random epoch" just mentioned, any enhancement of communication coverage that might be effected by interorbital phasing (i.e., synchronism of the satellites in one orbit plane with those in another) cannot be determined with the technique introduced here.

The virtue of the probability program lies in its ability to determine rapidly an important performance characteristic for a communication satellite system, the long-term average down probability. With this tool, one can examine a large variety of orbital systems to determine which merit further, more intensive study.

V. Communication Links

The system of communication links chosen for study is indicated in Table 2 in which the latitude and longitude of each station and the longitude difference of each station pair are tabulated. The communication links, displayed graphically on the world map in Fig. 4, form a worldwide communication network.

Each ground station chosen in the system studied has at least one of the following qualifications: 1) it is a large population center, 2) it is a center of political importance such as a national capitol, and 3) although the station itself is unimportant when judged by either of the preceding criteria, it may serve as a vital intermediate ground relay point between two or more key terminals.

Some stations that qualify for inclusion in the communication network on as many as two counts were excluded because of their proximity to, and hence their approximate identity to other stations which were included. For example, though each is a capital city of considerable size, neither Paris nor Washington, D.C. was included in the system of ground stations selected, since London and New York, respectively, were. Three stations that were included in the system primarily because of their utility as intermediate ground relays between widely separated terminals are Point Barrow, Honolulu, and Dakar. The double hop mode of communication used with such ground relays will be discussed in a later section.

The station pairs that were chosen provide a large variety of regions of mutual communication: from polar to equatorial in location, from north-south to east-west in azimuthal orientation, and of various sizes depending upon station separation. By incorporating such variety in the system it is hoped that, even though a particular station pair of interest may not appear, a station pair whose RMC is of similar size and orientation will have been considered.

VI. System Parameters and Their Effects on Communicability

Although the generality of the analytical technique previously described permits its application to systems consisting of any number of satellites in any number of circular orbits with free choice of satellite altitudes, orbit

§Actually, the communication stations would be placed at suitably isolated sites near the cities indicated.

inclinations, and nodal locations, the application of the technique in the present work is restricted to systems in which the satellites are in common altitude circular orbits for which $\varphi = 61^{\circ}$ (see Fig. 2). One combination of altitude and minimum station elevation angle is h = 5000 naut miles and $\sigma = 5^{\circ}$. The results of the analyses which follow are stated for this pair of values but are equally applicable to all h, σ pairs whose $\varphi = 61^{\circ}$.

Effects of the following system parameters upon communicability will be examined in subsequent sections: the number and inclination of the several orbit planes used, the total number of satellites in the system and their apportionment among the orbit planes, and the mode of satellite distribution (random or evenly spaced) within the planes.

A. Number of Planes

Figure 5 shows the variation of communication down probability P_D with the number of equally separated polar orbit planes n_1 among which 24 satellites are evenly divided and in which they are randomly distributed. For simplicity, only eight representative communication links are shown on the figure. Note that for all the links considered, the reduction in P_D is substantial as n_1 increases from one to two. For high-latitude links such as London-Moscow and Peiping-Tokyo there is no appreciable further reduction in P_D as n_1 increases. However, for low-latitude links such as Los Angeles-Caracas and Dakar-Capetown, the improvement in P_D continues up to $n_1 = 4$. Assuming that multiple-satellite launchings are employed, the total cost of establishing an orbital system will be proportional to the number of distinct orbit planes rather than to the total number of satellites which would be the case for individually launched satellites. Since there is negligible reduction in P_D beyond $n_1 = 4$ for any of the links, there is not sufficient reason on the basis of P_D alone to justify the use of more than four polar orbit planes.

Although inclinations other than polar were not examined, the general trend exhibited by polar orbits is also expected for systems of commonly inclined nonpolar orbits. As attractive satellite systems are sought in subsequent sections, the optimum values of n_1 will be reported.

B. Number of Randomly Distributed Satellites in a Polar System

Figures 6 show the dependence of P_D upon N, the total number of satellites equally divided among and randomly distributed within four equally separated polar planes. Curves for all 25 communication links are presented. The fact that the curves are straight lines on the semi-logarithmic grid is in agreement with the proof given in Ref. 2 that P_D is an exponential function of N. Thus, for example, on the Rio de Janeiro-Dakar link, the addition of 20 satellites (five more in each of the four polar planes) reduces P_D by a factor of 10.

The station pairs for which the communication fractions are the best and worst are London-Moscow and Dakar-Capetown, respectively. The communication down probabilities for these extreme cases differ by a large ratio (more than 10^3 for $N = 32$, for example) that increases with N.

C. Number of Evenly Spaced Satellites in a Polar System

Consider a system of polar orbits at least one of whose planes always intersects the region of mutual communication for a station pair. Then it is possible for a finite number of satellites, evenly spaced in those orbit planes, to provide uninterrupted coverage of that station pair. The minimum number of evenly spaced satellites required to make P_D zero depends upon the particular link and the number of polar orbit planes in the system.

To provide uninterrupted coverage of each of the 25 station pairs considered here it was found that: 17 satellites would be required in each of two orthogonal polar orbits; 7 satellites would be required in each of three equally separated polar orbits; and 7 satellites would be required in each of four equally separated polar orbits. In each case, the critical link is Dakar-Capetown. Evidently, the polar orbital system which requires the fewest evenly spaced satellites is the three-plane system of 21.

Figure 7 shows the variation of P_D with the number of evenly spaced satellites in the system of three polar planes. The curves for evenly spaced satellites, unlike those in Fig. 6 for randomly distributed ones, are highly nonlinear and, in fact, have vertical asymptotes. This feature of each curve bears out the statement that uninterrupted communication ($P_D = 0$) can be assured on any of the links with a finite number of evenly spaced satellites. For example, the vertical asymptote of the London-Moscow link indicates that only 3.34 (or 4) satellites are required in each of the three polar orbits for a total of 12 satellites. The most difficult link is Dakar-Capetown which requires 7 satellites in each polar orbit plane. Although curves for only 7 of the 25 communication links appear in Fig. 7, curves for the remaining 18 links would fall between those shown for the extreme cases.

It is interesting to compare P_D's for equal numbers of randomly distributed and evenly spaced satellites. Figures 6 showed that, for a system of 20 randomly distributed satellites divided equally among four polar planes, the down probabilities would range from about 0.2% for London-Moscow to 18% for Dakar-Capetown. However, Fig. 7 indicated that as few as 21 satellites equally divided among and evenly spaced within three polar planes would make P_D zero for all 25 of the links.

D. Systems of Evenly Spaced Satellites in Inclined Orbits

Table 3 shows the number of evenly spaced satellites required in each of three equally separated, commonly inclined orbit planes to assure uninterrupted communication coverage (i.e., to make P_D zero) for each of the 25 station pairs. Orbit inclinations from 0° to 90° are considered. The vertical asymptotes of Fig. 7 are listed in the 90° column. Note that for every link but two, the orbit inclination that requires the smallest number of satellites for "saturation" is either equatorial or polar.

The items parenthesized in the table are the maximum numbers of evenly spaced satellites required in each of the inclined orbit planes to make P_D zero for all 25 communication links. Infinite values appear in the table for those station pairs whose regions of mutual communication are not always intersected by the system of orbit planes for then, regardless of the number of satellites used, uninterrupted communication cannot be possible. Evidently the best common inclination for a three-plane system is polar since the minimum value (7.00) of all the parenthesized maxima occurs in the $i = 90^\circ$ column (for Dakar-Capetown as was indicated in Fig. 7). Note that the omission of a critical link from the communication network can effect a drastic change in the optimum orbital inclination. For example, if the Point Barrow-London link were dropped from the network, the preferred orbital inclination would be approximately 40°.

Another kind of orbital system appears promising as a means of obtaining complete communication coverage with the fewest satellites. It is a mixed-inclination system in which the satellites are divided among the equatorial orbit plane and one or more equally separated polar orbit planes. Table 4 indicates the number of evenly spaced satellites in various orbital systems which would be required to make P_D zero. Evidently the regions of mutual visibility for station pairs 12, 23, 24 and 25 are not intersected by the equator. Consequently, those links must be covered by satellites in one or more polar planes. A mixed system of one equatorial and one polar plane must contain 8 satellites in the equatorial plane (because of link 21) and 8 satellites in the polar plane (because of links 1 and 13) for a total of 16 satellites.

Similarly, the minimum number of evenly spaced satellites required in a one-equatorial, two-polar system is 18 with 8 satellites in the equatorial orbit and 5 in each of the two orthogonal polar orbits.

Finally, the minimum "one-and-three" equatorial-polar system must employ 8 satellites in the equatorial plane and 5 in each of the three polar planes for a total of 23.

Thus, for the network of links considered, the equatorial-polar system that requires a minimum number of evenly spaced satellites is one comprising 16 satellites, 8 in the equatorial plane and 8 in one polar plane. Note that this is substantially fewer satellites than the 21 required in the three-polar system considered in Fig. 7. Note also that the optimum number of planes in this case is two.

In the search for the minimum system of randomly distributed satellites, consideration will be given to the three types of systems just examined for evenly spaced satellites; these are 1) several polar orbits; 2) commonly inclined orbits; and 3) mixed-inclination systems comprising one equatorial orbit and one or more polar orbits.

E. Apportionment of Randomly Distributed Satellites in a Mixed-Inclination System

Figure 8 shows the dependence of P_D upon the apportionment of 60 randomly distributed satellites among one equatorial and three polar orbits. The two critical links (i.e., the two links of the total 25 which form the envelope of maximum P_D as a function of the number of equatorial satellites) are New York-Dakar and Tokyo-Point Barrow. P_D for the former link decreases as the number of equatorial satellites increases, but P_D for the latter, a high-latitude link, increases as the number of equatorial satellites increases. Since communication is assumed equally vital over all the links, the optimum apportionment of satellites in a mixed inclination system is the one that minimizes the largest value of P_D over all the station pairs for a given total number of satellites N. For N = 60, the optimum apportionment corresponds to the value indicated by the intersection of the P_D curves for New York-Dakar and Tokyo-Point Barrow. This value is approximately 26 equatorial satellites. However, the remaining 34 satellites cannot be equally divided among the three polar planes. If N were equal to 59, then 26 equatorial and 33 (3 x 11) polar satellites would undoubtedly form an optimally apportioned system. But for N = 60 with three polar planes, the number of equatorial satellites must be divisible by three. The values which are operationally practical are circled in Fig. 8. The best (i.e., lowest) P_D of these is the system composed of 24 equatorial satellites and 12 satellites in each of the three polar planes.

Other equatorial-polar systems of 60 satellites (e.g., one equatorial, two polar) were examined and optimized but were found inferior to the one equatorial, three polar system discussed previously. Thus, the optimum number of planes in this case is four. Furthermore, it was found that for a total of 30 satellites the best equatorial-polar system is again "one-and-three," and that the optimum apportionment of satellites is 12 in the equatorial and 6 in each of the three polar planes. Thus, as was found for 60 satellites, 40% of the satellites should be placed in the equatorial plane and 20% in each of the three polar planes.

Figures 9 show the variation in P_D with N for an optimally apportioned (i.e., 40, 60%) one-equatorial, three-polar system. Curves are presented for all 25 communication links.

As indicated previously, all values of N cannot, for practical reasons, be apportioned in the 40, 60% manner. In such cases, the apportionment should be made as nearly 40, 60% (or better, 26/60 = 43, 57%) as possible. P_D values for these slight variations in apportionment will be very nearly those indicated in Figs. 9.

The optimally apportioned equatorial-polar system of Figs. 9 may be compared with the four polar plane system of Figs. 6. For all values of N, the P_D's for New York-Dakar, the worst link of the equatorial-polar system, are much lower than the P_D's for Dakar-Capetown, the worst link of the polar system. For example, for N = 60, P_D = 0.000408 for New York-Dakar, whereas for Dakar-Capetown, $P_D = 0.00616$. From another viewpoint, the number of satellites in the equatorial-polar system required to reduce the New York-Dakar P_D to 1% is approximately 35, whereas the number of satellites required in the polar system to reduce the Dakar-Capetown P_D to 1% is approximately 54.

F. Systems of Randomly Distributed Satellites in Commonly Inclined Orbit Planes

Satellite systems in which all the orbit planes have a common inclination were previously considered for evenly spaced satellite distributions. Similar systems, but with satellites randomly distributed in each plane, will now be analyzed.

Figure 10 indicates the selection of an orbit inclination for a system of 60 satellites, 20 in each of three planes. An optimum inclination is one which minimizes the maximum value of P_D for all 25 communication links. As the figure indicates, Point Barrow-London and Dakar-Capetown form the envelope of maximum P_D's as a function of inclination. The minimum P_D occurs at their intersection (circled) and corresponds to an inclination of 48.7^o. A similar investigation for $N = 30$ yielded a slightly different optimum inclination, 50.1^o.

These data were then used in Figs. 11 to show the variation in P_D with N for systems of three orbit planes commonly inclined at an optimum value. Curves for all 25 communication links are presented. Although optimum inclinations were determined for only two values of N (30 and 60), these values were so nearly the same that only slight variations in P_D from those presented in Figs. 11 are anticipated for other values of N if an inclination of 49° or 50° is assumed.

The results of Figs. 11 may be compared with the previously discussed polar and equatorial-polar systems. As was intended, the two worst links for the commonly inclined system are Dakar-Capetown and Point Barrow-London. Their common P_D at $N = 60$ is 0.00702. Comparable values for the polar and equatorial-polar systems are 0.00616 and 0.000408, respectively. The number of satellites in the commonly inclined system required to reduce the worst P_D to 1% is approximately 55 compared with 54 and 35 for the polar and equatorial-polar systems, respectively. Thus, the polar and commonly inclined systems are comparable in quality, whereas the equatorial-polar system is by far superior. However, it must be remembered that the selection of an operational system depends on many considerations, such as launch site locations and booster payload capabilities, whose influences are not considered in this paper.

The most dramatic comparison of all is that between the best system of evenly spaced satellites and the best system of randomly distributed satellites. The best evenly spaced system requires only 16 satellites for 100% coverage of all 25 links, whereas the best randomly distributed system requires 35 satellites for a 99% probability of communication across the worst of the 25 links. Of course, the evenly spaced satellites require a reliable station-keeping capability, which the randomly distributed ones do not.

VII. Double Hop Mode

Section V, which described the system of communication links, indicated that the primary purpose for including three ground stations, Honolulu, Dakar and Point Barrow, was to capitalize on their ability to serve as intermediate ground station relays. These intermediate relays are required for widely separated and/or poorly oriented communication links such as Tokyo-Los Angeles, New York-Capetown and Rio de Janeiro-London. Direct communication via a satellite from Tokyo to Los Angeles or vice versa, is much more difficult than double hop mode communication; i.e., Tokyo to a satellite to Honolulu (or Point Barrow) to a satellite to Los Angeles. Similarly New York-Capetown and Rio de Janeiro-London communications are greatly enhanced if Dakar is used as an intermediate relay.

Consider the previously described optimum system of evenly spaced satellites. Sixteen satellites, 8 in a polar plane and 8 in the equatorial plane, were required to assure uninterrupted communications between all 25 links. In this determination, the double hop mode was utilized to provide communications for the three difficult links just mentioned. If the double hop mode had not been used, the optimum system would have consisted of 57 satellites, 24 in a polar plane required by Tokyo-Los Angeles and 33 in the equatorial plane required by New York-Capetown.

A substantial difference in performance is also demonstrated if systems of randomly distributed satellites are considered. In this case, the optimum system that was determined consists of an equatorial orbit plane which contains approximately 40% of the total number of satellites and three polar orbit planes, each of which contains approximately 20% of the satellites. Using this system, the variations of P_D with N for the Tokyo-Los Angeles, Tokyo-Honolulu and Honolulu-Los Angeles links are separately presented in Fig. 12. The down probabilities associated with the joint Tokyo-Honolulu-Los Angeles link may be described as follows. The joint P_D cannot be less than the larger individual P_D (Tokyo-Honolulu, in this case). Furthermore, the joint P_D cannot be greater than the sum of the individual P_D's. These two statements define upper and lower bounds on the joint P_D. Consider, as an example, the P_D's associated with a total of 30 satellites, 12 in the equatorial plane and 6 in each of the three polar planes. From Fig. 12, $P_{D_{\text{Tok-Hono}}} = 0.0178$ and $P_{D_{\text{Hono-LA}}} = 0.0032$. The joint P_D cannot be less than 0.0178 and cannot be greater than 0.0210. This latter value may be of more relative importance for it guarantees that, on a long-term average basis, joint communications between Tokyo and Los Angeles will be possible (1 - 0.021 = 0.979) or 97.9% of time. This may be described as an "up probability." The other bound indicates that joint communications may be possible as much as, but no more than 98.22% of the time.

The bounds on joint P_D are the curves in Fig. 12 labeled Tokyo-Honolulu and Tokyo-Honolulu-Los Angeles. The two curves are quite close together and converge with increasing N.

If the performances of the two separate links are independent, then the single best estimate of their joint performance is the product of their individual up probabilities. This product may be expressed in terms of a down probability as $[P_{D_{\text{Tok-Hono}}} + P_{D_{\text{Hono-LA}}} - (P_{D_{\text{Tok-Hono}}})(P_{D_{\text{Hono-LA}}})]$. For 30 satellites, this P_D is $[0.0178 + 0.0032 - (0.0178)(0.0032)]$ or 0.020943. This value lies between the bounds as it must.

If the performances of the two separate links are correlated, then the correlation must be determined and used in the joint P_D evaluation. However, in the cases discussed the bounding curves are so close together that, for most purposes, any value of P_D which lies on or between the boundary curves may be chosen for a given total number of satellites. Thus, a knowledge of the independence or correlation of the separate link performances is not required.

The important consequence of this investigation is the striking reduction in P_D which can be obtained by using the double hop mode. The Tokyo-Honolulu-Los Angeles boundary curve P_D is less than the Tokyo-Los Angeles curve by a factor of 3.8 for 30 satellites and 19.4 for 60 satellites. Comparing the curves in another way, 55 satellites are required to reduce the Tokyo-Los Angeles P_D to 1%, whereas only 36 satellites are required to reduce the Tokyo-Honolulu-Los Angeles P_D to 1%.

The ability of a different intermediate ground relay station, Point Barrow, to enhance communications between Tokyo and Los Angeles, was also investigated. The Tokyo-Point Barrow-Los Angeles boundary curve is also presented in Fig. 12. It is seen that Point Barrow is not as effective as Honolulu as a relay in improving Tokyo-Los Angeles communications.

Similar results for Rio de Janeiro-London and New York-Capetown communications using Dakar as an intermediate relay are presented in Fig. 13. The Rio de Janeiro-Dakar-London boundary curve P_D is less than the Rio de Janeiro-London curve by a factor of 8.6 for 30 satellites and 138 for 60 satellites. The New York-Dakar-Capetown boundary curve P_D is less than the New York-Capetown P_D by a factor of 20.6 for 30 satellites and 714 for 60 satellites.

VIII. Duration and Frequency of Communication Holes

It was indicated earlier that an inherent feature of any system of randomly distributed satellites is its inability to provide uninterrupted coverage of a pair of ground stations; that is, P_D, however small, cannot be zero. It is important then to appreciate not only the degree of coverage a system provides (measured by P_D), but also the distribution of downtimes during which communication via satellite relay is impossible due to the absence of all satellites from the region of mutual communication. As yet, no correlation has been established between the P_D of a satellite system and the duration and frequency of downtimes or holes in the coverage.

A technique is now introduced which makes it possible to determine the duration probability of communication holes that occur for certain orbital systems of randomly distributed satellites, and the technique is used to analyze an orbital system of interest.

Questions concerning communication hole duration will arise on two occasions: 1) communication between a pair of ground stations is attempted at some time and found to be impossible because there is no mutually visible satellite; or 2) communication in progress is interrupted when the only satellite mutually visible to the pair of stations sets with respect to either of them. The two instances are treated separately, but the question in either case takes the form: What is the probability that the communication hole will persist no longer than some time t?

A. Case 1: A Discovered Communication Hole

The probability of having to wait no longer than time t to begin communication after the discovery of a hole is denoted by $\overline{P}(t)$. In Appendix B, the following expression is derived for the waiting probability:

$$\overline{P}(t) = \frac{1}{2\pi} \int_0^{2\pi} \left\{ 1 - \prod_{j=1}^{n_1} \left[1 - P_j(\psi)\right]^{n_{2_j}} \right\} d\psi$$

where

$$P_j(\psi) = \frac{t/T}{\left[1 - d_j(\psi)/2\pi\right]} \qquad \text{if} \quad d_j(\psi) > 0$$

or

$$P_j(\psi) = 0 \qquad \text{if} \quad d_j(\psi) = 0$$

One of the most attractive systems of randomly distributed satellites found earlier involved 30 satellites in a one-equatorial, three-polar plane system with 12 satellites in the equatorial plane and 6 in each of the three-polar planes. The preceding equations were applied to that system to determine the probabilities associated with the discovery of communication holes of various durations for the links that received the worst and the best coverage; these were New York-Dakar (P_D = 0.0186) and Peiping-Tokyo (P_D = 0.000271), respectively. Due to the regularity of the system geometry, it was necessary to integrate ψ only over the range 0° to 60°. Several sets of $d_j(\psi)$'s, which were obtained graphically for the RMC's of the two communication links, were used to determine $P(\psi)$ and hence $\overline{P}(t)$. The results are shown by the solid curves in Fig. 14 for the New York-Dakar and the Peiping-Tokyo links.

Note that a discovered hole for the former, more difficult link entails a mean-waiting-time-to-communicate (MWTC) of 6.4 min; i.e., $\overline{P}$ (6.4 min) = 0.50. The other link, whose P_D is smaller by a factor of about 70, still has a MWTC of 5.6 min.

B. Case 2: The Hole Following a Communication Interruption

The probability of having to wait no longer than time t to resume communication after an interruption is shown in Appendix B to be given by:

$$\overline{P}(t) = \frac{1}{2\pi}\int_0^{2\pi}\left\{1 - [1 - P_k(\psi)]^{\left(n_{2_j} - 1\right)} \times \prod_{j=1}^{n_1} [1 - P_j(\psi)]^{n_{2_j}}\right\} d\psi$$

where

$$j = 1, 2, \ldots, n_1 \quad \text{but} \quad j \neq k \quad ;$$

$$\left.\begin{aligned} P_j(\psi) &= \frac{t/T}{[1 - d_j(\psi)/2\pi]} \quad \text{if} \quad d_j(\psi) > 0 \\ &\text{or} \\ P_j(\psi) &= 0 \quad \text{if} \quad d_j(\psi) = 0 \end{aligned}\right\} j = 1, 2, \ldots, n_1$$

and

$$d_k(\psi) = \max d_j(\psi)$$

The expression in the product under the integral sign is the same as that in case 1 except for the reduction of the exponent of the smallest $[1 - P_j(\psi)]$ term from n_{2_j} to $(n_{2_j} - 1)$.

The expression for $\overline{P}(t)$ given previously was used to obtain the dashed curves in Fig. 14 for the New York-Dakar and Peiping-Tokyo links. Note that the MWTC's for these links after an interruption in communication are 6.6 and 5.8 min, respectively. These MWTC's are just 0.2 min longer than the times for a discovered hole.

The information presented in Fig. 14 may be expressed in another way; differentiation of the curves yields density distributions of communication downtimes. Fig. 15 shows the density distribution of holes following the discovery that communication is impossible over the New York-Dakar (solid curve) and the Peiping-Tokyo (dashed curve) links. One can see that for the New York-Dakar link, 9% of the communication holes are between 1 and 2 min in duration and that holes less than 1 min in duration are five times as frequent as 14- to 15-min holes. Also 77.1% of the holes are less than 15 min in duration.

The fact that the curve for the Peiping-Tokyo link is everywhere above the one for New York-Dakar indicates that shorter communication holes are more frequent for the former easier link. Note that 87.6% of the communication holes are of 15 min duration or less.

IX. Summary

An analytical technique was introduced and employed to determine the communication fractions provided by a variety of nonsynchronous orbital systems for a worldwide network of links. The effects upon communicability of the following orbit system parameters were studied: the number and inclination of the orbit planes, the number of satellites used, and their apportionment among and mode of distribution in the orbit planes.

It was found that, for a system of randomly distributed satellites, the communication down probability decreases exponentially (but never reaches zero) with an increasing number of satellites. However, some orbital systems that involve a finite number of evenly spaced satellites can provide uninterrupted communication coverage of the entire network of station pairs.

Of the orbital systems examined, the two which proved most attractive are:

1) A system of 16 satellites with 8 evenly spaced in an equatorial orbit and 8 evenly spaced in a polar orbit. Such a system provides uninterrupted coverage of all 25 communication links.

2) A system of 30 satellites with 12 randomly distributed in the equatorial plane and 6 distributed randomly in each of 3 equally separated polar orbit planes. The communication

fraction for the worst link (New York-Dakar) is 98.14%.

The improvement in communicability, effected by the use of intermediate ground relay points between widely separated ground stations, was studied. It was found that, for the one-equatorial, three-polar system of 30 satellites, the insertion of Dakar and Honolulu as ground relays made the double hop communication fractions between such distant stations as New York and Capetown and Tokyo and Los Angeles comparable to the single hop communication fractions for links such as New York-Dakar and Tokyo-Honolulu.

A method was devised and applied to estimate the probable duration of communication delays of two kinds: discovered holes and interruptions in communication in progress. From these results, it is possible to infer the density distribution of communication holes and, hence, to obtain an additional indication of the acceptability of an orbital system for communication purposes.

Appendix A: Determination of $d_j(\psi)$

Figure 3 illustrates the geometry of the intersection of the jth orbit plane with CR_1, the communication circle for ground station 1. Let the angle θ_j (measured from the ascending node) denote the intersections of the jth orbit plane with CR_1, i.e., those points at which a satellite in the jth orbit rises or sets relative to the first station. The following expression, derived from the application of spherical trigonometry to the geometry in Fig. 3, relates θ_j to the latitude of the ground station λ_1, the orientation angle ψ, the longitudinal separation of the first and the jth ascending nodal points Ω_j, the inclination of the jth orbit i_j, and the coverage circle radius φ_j:

$$\cos \theta_j[\cos \lambda_1 \cos (\psi + \Omega_j)]$$

$$+ \sin \theta_j[\sin \lambda_1 \sin i_j - \cos \lambda_1 \cos i_j \sin (\psi + \Omega_j)] - \cos \phi_j = 0$$

The preceding equation can be written as a quadratic in $\sin \theta_j$, and hence will yield at most two roots, the true anomalies of a satellite at rise and set.

Similarly, the following expression was found which relates θ_j for CR_2 to the latitude of the second ground station λ_2, the orbital parameters for the jth plane, and L the longitudinal separation of the pair of communication stations:

$$\cos\theta_j[\cos\lambda_2\cos(L-\psi-\Omega_j)]$$

$$+\sin\theta_j[\sin\lambda_2\sin i_j+\cos\lambda_2\cos i_j\sin(L-\psi-\Omega_j)]-\cos\phi_j=0$$

The second equation also yields at most a pair of roots. (Note that the second equation reduces to the first if L is set equal to zero.)

The next step in the calculation of the communication down probability is to determine the arc $d_j(\psi)$ of the jth orbit plane within which a satellite is visible to both ground stations. As is indicated in the discussion in the text, $d_j(\psi)$ is nonzero if and only if: 1) the two pairs of roots of the preceding equations are real and distinct; and 2) the regions of visibility of the satellite to the two separate stations overlap; i.e., there is an intermingling of the two pairs of roots. Then $d_j(\psi)$ is the difference of the innermost two intermingled roots.

Appendix B: Probable Communication Hole Duration

Case 1: A Discovered Communication Hole

Let there be n_{2_j} satellites distributed randomly in the jth plane where $j = 1, 2, \cdots, n_1$. At time t_o (when the orientation angle of the orbital system is ψ) the jth orbit plane intersects the RMC for a station pair in arc $d_j(\psi) \geq 0$ (see Fig. 3).

Since communication is not possible at time t_o, all the satellites in the jth orbit plane lie in the arc of length $[2\pi - d_j(\psi)]$. Hence, the probability that one of the satellites in the jth orbit plane will enter the RMC during time t is just the probability that the satellite will be found at time t_o in the arc ωt adjacent to $d_j(\psi)$; ω is the common angular velocity of all satellites in the system. Therefore, the previous probability is

$$P_j(\psi) = \frac{\omega t}{2\pi - d_j(\psi)} = \frac{t/T}{[1 - d_j(\psi)/2\pi]} \quad \text{if} \quad d_j(\psi) > 0$$

or

$$P_j(\psi) = 0 \qquad \text{if} \quad d_j(\psi) = 0$$

where the relation between angular velocity and orbit period T was employed.

Clearly then the probability that none of the n_{2_j} satellites in the jth orbit plane will enter the RMC during time t is

$$[1 - P_j(\psi)]^{n_{2_j}}$$

Furthermore, the probability that none of the satellites in any of the n_1 orbit planes will enter the RMC during time t is given by the product

$$\prod_{j=1}^{n_1} [1 - P_j(\psi)]^{n_{2_j}}$$

Then the probability $P(\psi)$ that a hole discovered at time t_o (and orientation angle ψ) will persist no longer than time t is just the complement of that product:

$$P(\psi) = \left\{1 - \prod_{j=1}^{n_1} [1 - P_j(\psi)]^{n_{2_j}}\right\}$$

The quantity $P(\psi)$ must next be averaged over all orientation angles realized as the earth rotates within the system of orbit planes to obtain the average probability $\overline{P}(t)$ of being able to begin communicating within time t of the discovery of the hole. Unless the system geometry is repetitive, the averaging process must be carried out over the entire range on ψ.

Symbolically,

$$\overline{P}(t) = \frac{1}{2\pi}\int_o^{2\pi} P(\psi)\, d\psi = \frac{1}{2\pi}\int_o^{2\pi} \left\{1 - \prod_{j=1}^{n_1} [1 - P_j(\psi)]^{n_{2_j}}\right\} d\psi$$

In the preceding derivation, it was tacitly assumed that the set of $d_j(\psi)$ was constant during the time interval t. Actually, of course, the lengths of the arcs change with the passage of time owing to the rotation of the earth, but these changes are slight if t is small as compared with the orbital period T which is the case for the examples discussed in the text.

Case 2: The Hole Following a Communication Interruption

An interruption in communication occurs when the only satellite mutually visible to a pair of ground stations sets with respect to either of them. The situation at the instant communication is interrupted is identical to that at the discovery of a hole (case 1) except that one of the satellites has just left the arc of intersection of its orbit plane with the RMC and, therefore, will not re-enter the RMC during the (short) time interval t. Consequently, the expression for $P(\psi)$ derived in case 1 must be modified to account for the unavailability of the just-setting satellite.

The conservative assumption is now made that the only mutually visible satellite before communication interruption is in the kth orbit plane whose arc of intersection with the RMC is the largest. Since only $(n_{2_k} - 1)$ of the satellites in the kth orbit have any chance of entering the RMC in time t, while n_{2_j} satellites ($j = 1, 2, \cdots, n_1$ but $j \neq k$) in the other $(n_1 - 1)$ orbits have a chance, the expression for $P(\psi)$ must take the form

$$P(\psi) = \left\{ 1 - [1 - P_k(\psi)]^{\left(n_{2_j} - 1\right)} \prod_{j=1}^{n_1} [1 - P_j(\psi)]^{n_{2_j}} \right\}$$

where

$$j = 1, 2, \ldots, n_1 \quad \text{but} \quad j \neq k$$

$$\left.\begin{aligned} P_j(\psi) &= \frac{t/T}{[1 - d_j(\psi)/2\pi]} && \text{if} \quad d_j(\psi) > 0 \\ &\text{or} \\ P_j(\psi) &= 0 && \text{if} \quad d_j(\psi) = 0 \end{aligned}\right\} j = 1, 2, \ldots, n_1$$

and

$$d_k(\psi) = \max[\, d_j(\psi)]$$

As in case 1, $P(\psi)$ must be averaged over all ψ to obtain $\overline{P}(t)$. Again the validity of the method depends upon t being small compared with the orbital period T.

References

[1] Pierce, J.R. and Cutler, C.C., "Interplanetary communications," Advances in Space Science (Academic Press Inc., New York, 1959), Vol. I, pp. 55-109.

[2] Sinden, F.W. and Mammel, W.L., "Geometric aspects of satellite communication," Inst. Radio Engrs. Trans. Space Electron. Telemetry SET-6 (September-December 1960), pp. 146-157.

[3] Bennett, F.V., Coleman, T.L., and Houbolt, J.C., "Determination of the required number of randomly spaced communication satellites," NASA TN D-619 (January 1961).

[4] Mueller, G.E., Hebenstreit, W.B., and Spangler, E.R., "Communication satellites - how high?," Astronautics 6 (July 1961), pp. 42-45 and 82-89.

[5] Pierce, J.R., "Communication satellites," Scientific American 205 (October 1961), pp. 90-102.

[6] Bennett, F.V., "Further developments on the required number of randomly spaced communication and navigation satellites," NASA TN D-1020 (February 1962).

[7] Rinehart, J.D. and Robbins, M.F., "Characteristics of the service provided by communications satellites in uncontrolled orbits," Bell System Tech. J. (September 1962), pp. 1621-1670.

Table 1 Study comparison

Ref. 1	Ref. 2	Ref. 3	Ref. 4	Ref. 5	Ref. 6		Ref. 7	This paper
Altitude is a parameter (1000-6000 naut miles).	Altitude, minimum station elevation angle and station separation distances are involved in a parameter which takes on various values. The altitudes are intended to be low (under 2500 miles).	Altitude is a parameter (1000-5000 naut miles).	Altitude is a parameter (1000-20,000 naut miles).	Altitude is a parameter (3000-9000 naut miles).	One altitude (5000 naut miles).	Four altitudes (1000,2000, 3000 and 5000 naut miles).	Four altitudes (2500, 4000, 6000 and 8000 naut miles).	One altitude (5000 naut miles).
One value of the minimum station elevation angle ($\sigma = 7.5^{\circ}$).		Three values of σ (0°, 5°, 10°).	One value of σ (7.5°).	One value of σ (7°).	One value of σ (5°).	Three values of σ (0°, 5°, 10°).	One value of σ (7.5°).	One value of σ (5°).
One type of satellite distribution: the satellites are deployed singly in randomly spaced circular orbits.	One type of satellite distribution: the satellites are deployed singly in randomly spaced circular orbits.	One type of satellite distribution: the satellites are deployed singly in randomly spaced circular orbits.	Two types of satellite distributions: 1) the satellites are randomly distributed in a circular polar orbit; 2) satellites are spaced in circular polar orbits. Elliptic orbits are discussed.	One type of satellite distribution: the satellites are deployed singly in randomly spaced circular orbits.	One type of satellite distribution: the satellites are deployed singly in randomly spaced circular orbits.	One type of satellite distribution: the satellites are deployed singly in randomly spaced circular orbits.	Two types of satellite distribution: 1) the satellites are deployed singly in randomly spaced circular orbits; 2) the satellites are randomly distributed in positioned circular orbits. Multiple satellite launchings are considered.	Two types of satellite distribution: 1) satellites are randomly distributed in positioned circular orbits; 2) satellites are equally spaced in positioned circular orbits. In every case the number of orbits is optimized. Multiple satellite launchings are considered.
Three orbit inclinations are considered ($i=0^{\circ}$, 60° and 90°).	Inclination is optimized for various regions of mututal communication.	Inclination is a parameter (0°-90°) optimum values are indicated.	Two orbit inclinations are considered ($i=0^{\circ}$ and 90°).	Two orbit inclinations are considered ($i=0^{\circ}$ and 90°).	Inclination is a parameter (0°-90°).	Inclination is a parameter (0-90°).	Six orbit inclinations are considered ($i=0^{\circ}$, 28°, 35°, 50°, 70°, and 90°). One mixed inclination ($i=28^{\circ}$ and 90°) system is considered.	Ten orbit inclinations are considered (0°, 10°, 20°, ..., 90°). Optimum inclinations for commonly inclined systems are found. Various equatorial-polar mixed systems are considered, and the apportionment of satellites among the orbit planes is optimized.
Two communication links (pairs of stations) are considered: Maine-Europe, Hawaii-Nevada.	Station pairs whose regions of mutual communication are centered at latitudes 0°, 15°, 30°, 45°, 60°, 75° and 90° are considered.	One station pair is considered: Newfoundland-Ireland.	Four station pairs New York-Paris, Portland-New York, Tokyo-Portland, Portland-Washington.	Six station pairs: Miami-Rio, Honolulu-Sydney, Maine-Brest, Seattle-Honolulu, Seattle-Tokyo, Brest-New Delhi.	A global network of 70 station pairs.	A network of 86 station pairs with 4 station separation distances (30°, 45°, 60°, 75°), 5 link directions (links inclined at 0°, 30°, 45°, 60°, 90°) and various station latitudes.	A global network of 24 station pairs. Computer results pertaining to double hop relays and to the frequency, duration and distribution of downtimes are presented. The effects of multiple-satellite visibility, mutual communication region overlap and orbit plane launching are presented.	A global network of 25 station pairs. Analytical studies pertaining to double hop relays, the probability of interrupted or unavailable service and the relative distributions of downtimes are presented.

Table 2 Network of 25 communication links

	Station pair	Latitude, deg	Longitude, deg	Longitude difference, deg
1.	New York-London	41N/52N	74W/0	74
2.	New York-Dakar	41N/15N	74W/17W	57
3.	New York-Caracas	41N/10N	74W/67W	7
4.	Los Angeles-Caracas	34N/10N	118W/67W	51
5.	Caracas-Rio de Janeiro	10N/23S	67W/43W	24
6.	Rio de Janeiro-Dakar	23S/15N	43W/17W	26
7.	Dakar-Capetown	15N/34S	17W/19E	36
8.	Dakar-Cairo	15N/30N	17W/31E	48
9.	Dakar-London	15N/52N	17W/0	17
10.	London-Cairo	52N/30N	0/31E	31
11.	London-Moscow	52N/56N	0/38E	38
12.	Point Barrow-London	71N/52N	157W/0	157
13.	Moscow-Peiping	56N/40N	38E/116E	78
14.	Moscow-Bombay	56N/19N	38E/73E	35
15.	Cairo-Bombay	30N/19N	31E/73E	42
16.	Bombay-Manila	19N/15N	73E/121E	48
17.	Bombay-Peiping	19N/40N	73E/116E	43
18.	Peiping-Tokyo	40N/36N	116E/140E	24
19.	Manila-Tokyo	15N/36N	121E/140E	19
20.	Manila-Melbourne	15N/38S	121E/145E	24
21.	Tokyo-Honolulu	36N/21N	140E/158W	62
22.	Honolulu-Los Angeles	21N/34N	158W/118W	40
23.	Tokyo-Point Barrow	36N/71N	140E/157W	63
24.	Point Barrow-Los Angeles	71N/34N	157W/118W	39
25.	Point Barrow-New York	71N/41N	157W/74W	83

Table 3 Number of evenly spaced satellites required in each orbit plane to assure uninterrupted communication[a]

Station pair	0°[b]	10°	20°	30°	40°	50°	60°	70°	80°	90°
1. New York-London	25.57	12.27	8.19	6.28	5.34	5.01	5.03	5.26	4.66	4.10
2. New York-Dakar	6.80	6.06	5.55	5.30	5.36	5.69	6.54	(7.67)	5.92	5.36
3. New York-Caracas	3.60	3.45	3.40	3.45	3.55	3.73	4.39	5.01	4.74	4.44
4. Los Angeles-Caracas	5.65	5.28	5.06	5.03	5.28	5.89	(7.36)	6.88	5.54	5.16
5. Caracas-Rio de Janeiro	3.80	3.73	3.74	3.97	4.56	4.80	4.92	5.01	5.09	4.99
6. Rio de Janeiro-Dakar	3.91	3.83	3.85	4.15	4.80	5.06	5.19	5.38	5.52	5.49
7. Dakar-Capetown	4.61	4.45	4.47	5.06	6.49	7.31	7.32	7.44	(7.35)	(7.00)
8. Dakar-Cairo	5.31	5.00	4.78	4.75	4.93	5.43	6.55	5.81	4.86	4.58
9. Dakar-London	4.73	4.20	3.96	3.82	3.85	3.98	4.26	5.49	5.17	4.72
10. London-Cairo	5.71	4.82	4.32	4.04	3.90	3.87	4.00	4.30	4.17	3.90
11. London-Moscow	12.02	6.89	5.11	4.28	3.83	3.64	3.62	3.71	3.46	3.34
12. Point Barrow-London	(∞)	(∞)	(∞)	(∞)	(∞)	(8.60)	5.86	5.30	4.42	4.25
13. Moscow-Peiping	137.00	17.08	9.53	6.82	5.64	5.20	5.12	5.28	4.88	4.20
14. Moscow-Bombay	6.66	5.41	4.83	4.49	4.32	4.36	4.62	5.24	5.36	4.78
15. Cairo-Bombay	4.92	4.64	4.43	4.36	4.28	4.83	5.60	5.03	4.30	4.11
16. Bombay-Manila	5.07	4.87	4.73	4.85	5.21	5.66	5.56	5.72	5.25	4.46
17. Bombay-Peiping	5.38	4.92	4.59	4.43	4.47	4.69	5.21	5.25	4.63	4.37
18. Peiping-Tokyo	4.50	4.11	3.82	3.61	3.50	3.55	3.71	3.82	3.52	3.36
19. Manila-Tokyo	3.83	3.66	3.56	3.57	3.65	3.85	4.27	4.47	4.21	3.97
20. Manila-Melbourne	4.10	3.93	4.10	4.45	5.57	6.76	7.22	7.29	7.18	6.86
21. Tokyo-Honolulu	7.21	6.42	5.82	5.47	5.45	5.77	6.56	8.49	6.07	5.07
22. Honolulu-Los Angeles	4.94	4.61	4.36	4.25	4.30	4.56	5.12	4.73	4.27	4.06
23. Tokyo-Point Barrow	(∞)	(∞)	16.98	7.16	5.39	4.74	4.60	4.65	4.89	4.50
24. Point Barrow-Los Angeles	(∞)	(∞)	8.85	5.51	4.54	4.30	4.25	4.32	4.58	4.39
25. Point Barrow-New York	(∞)	(∞)	(∞)	10.12	6.33	5.14	4.82	4.78	4.85	4.25

[a]Note: The satellites are assumed to be equally divided among and evenly spaced within three equally separated, commonly inclined orbit planes.

[b]Since an equatorial system comprises only one orbit plane, numbers in the $i=0^{\circ}$ column should be compared with three times the numbers in any other column.

Table 4 Number of evenly spaced satellites required to assure uninterrupted communication

Station pair	In one equatorial orbit plane	In one polar orbit plane	In each of two polar orbits	In each of three polar orbits
1. New York-London	25.57	7.38	4.94	4.10
2. New York-Dakar	6.80	∞	10.23	5.36
3. New York-Caracas	3.60	∞	5.50	4.44
4. Los Angeles-Caracas	5.65	∞	13.14	5.16
5. Caracas-Rio de Janeiro	3.80	∞	7.99	4.99
6. Rio de Janeiro-Dakar	3.91	∞	9.41	5.49
7. Dakar-Capetown	4.61	∞	16.12	7.00
8. Dakar-Cairo	5.31	∞	10.00	4.58
9. Dakar-London	4.73	∞	5.62	4.72
10. London-Cairo	5.71	12.71	4.52	3.90
11. London-Moscow	12.02	4.40	3.62	3.34
12. Point Barrow-London	∞	5.49	4.74	4.25
13. Moscow-Peiping	137.00	7.15	5.00	4.20
14. Moscow-Bombay	6.66	∞	5.92	4.78
15. Cairo-Bombay	4.92	∞	6.54	4.11
16. Bombay-Manila	5.07	∞	∞	4.46
17. Bombay-Peiping	5.38	∞	6.37	4.37
18. Peiping-Tokyo	4.50	7.12	3.80	3.36
19. Manila-Tokyo	3.83	∞	4.94	3.97
20. Manila-Melbourne	4.10	∞	12.61	6.86
21. Tokyo-Honolulu	7.21	∞	10.35	5.07
22. Honolulu-Los Angeles	4.94	∞	5.79	4.06
23. Tokyo-Point Barrow	∞	5.83	4.96	4.50
24. Point Barrow-Los Angeles	∞	6.02	4.74	4.39
25. Point Barrow-New York	∞	5.64	4.70	4.25

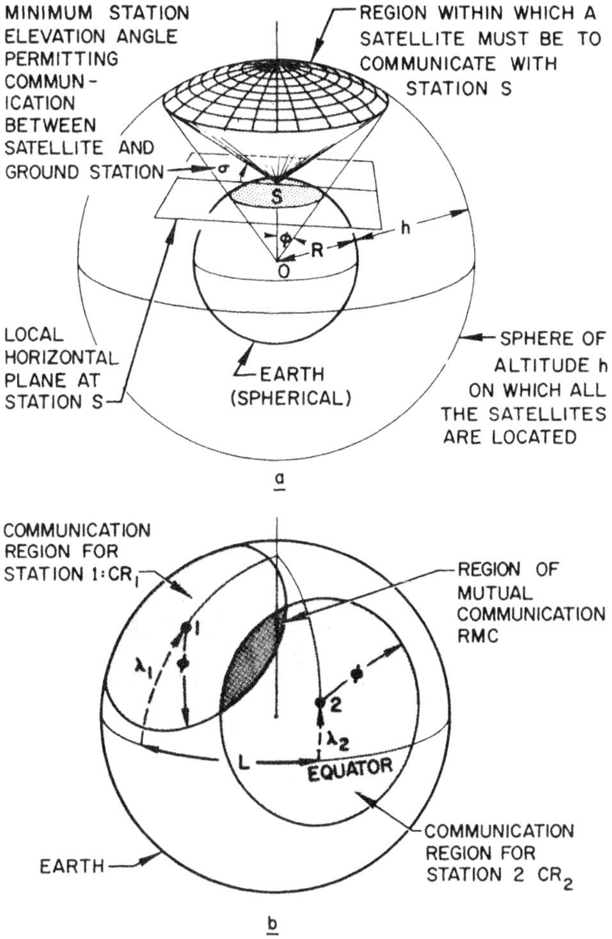

Fig. 1 Geometry of mutual satellite visibility.

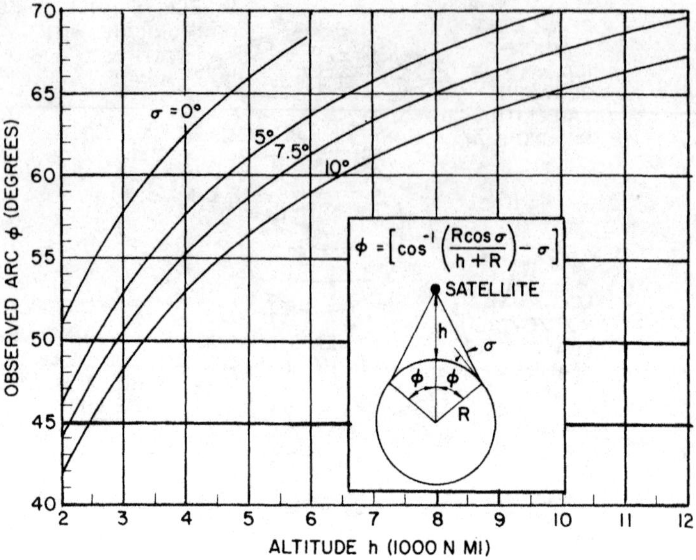

Fig. 2 Observed arc φ vs altitude.

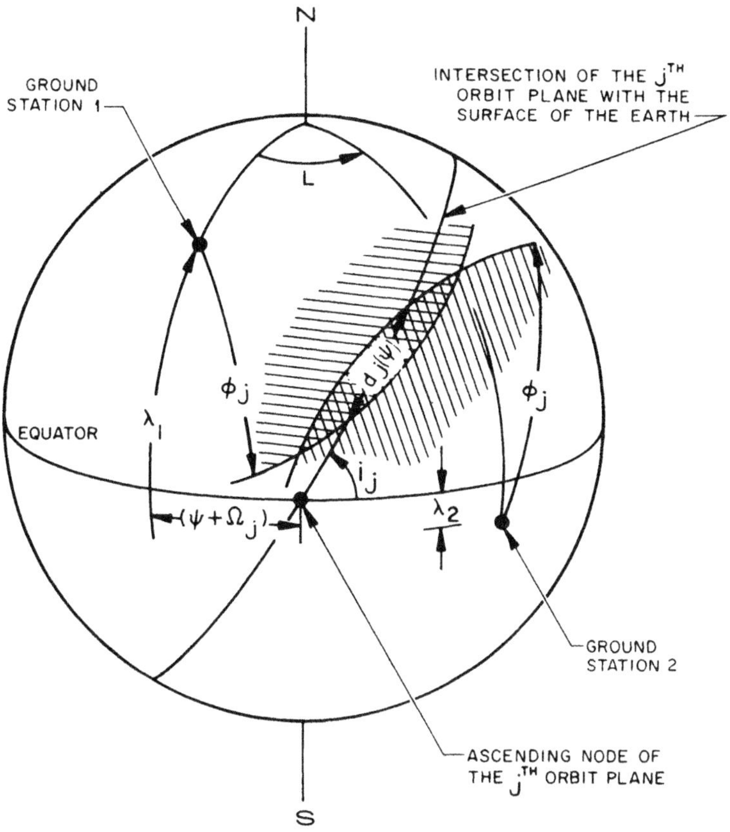

Fig. 3 Orbital arc $d_j(\psi)$ within which a satellite is mutually visible to a pair of stations.

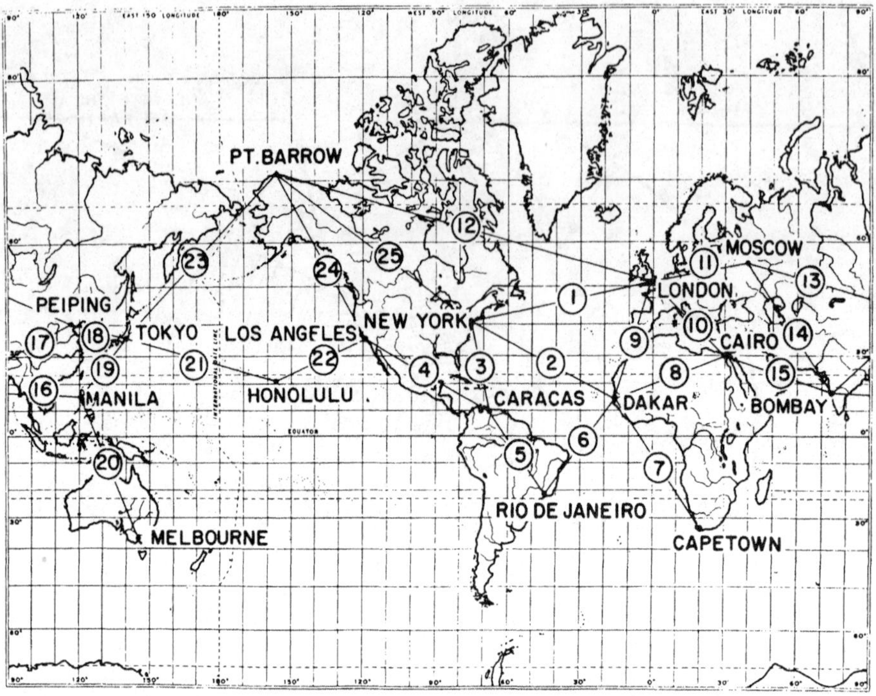

Fig. 4 Worldwide communication network of 25 links.

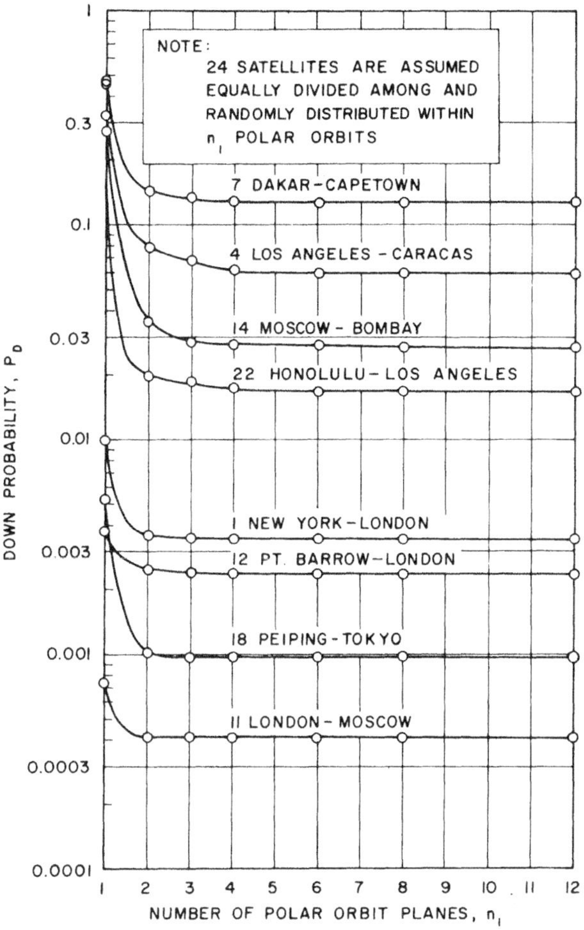

Fig. 5 Communication down probability vs number of equally separated polar orbit planes.

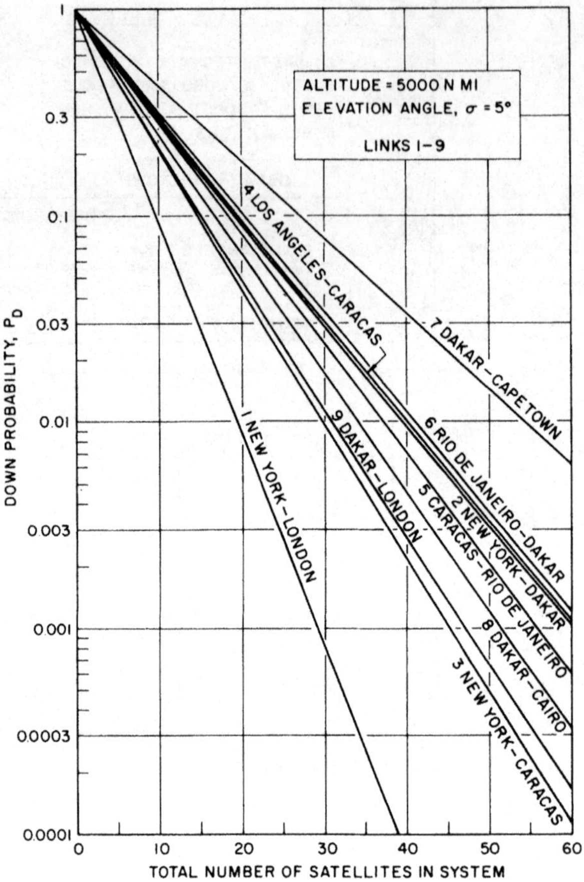

Fig. 6a P_D vs the total number of randomly distributed satellites equally divided among four-polar planes.

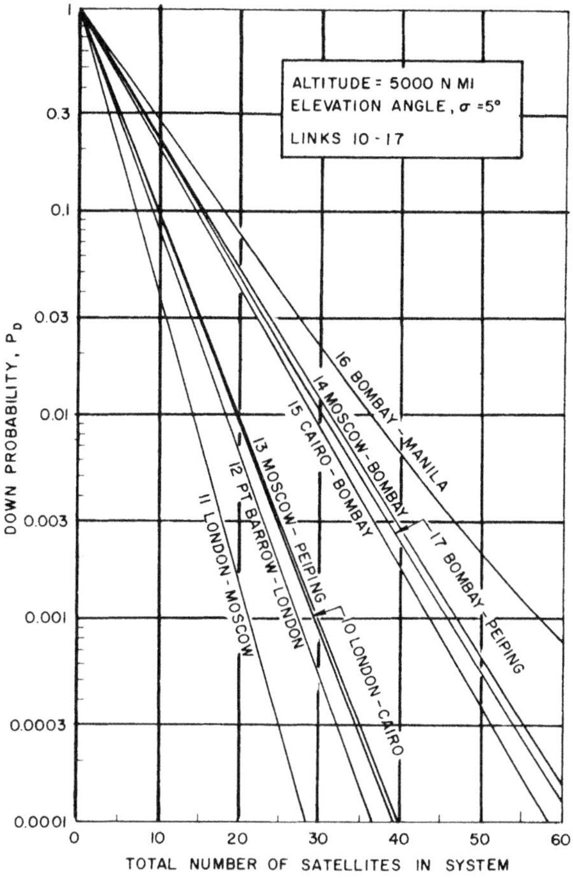

Fig. 6b P_D vs the total number of randomly distributed satellites equally divided among four-polar planes.

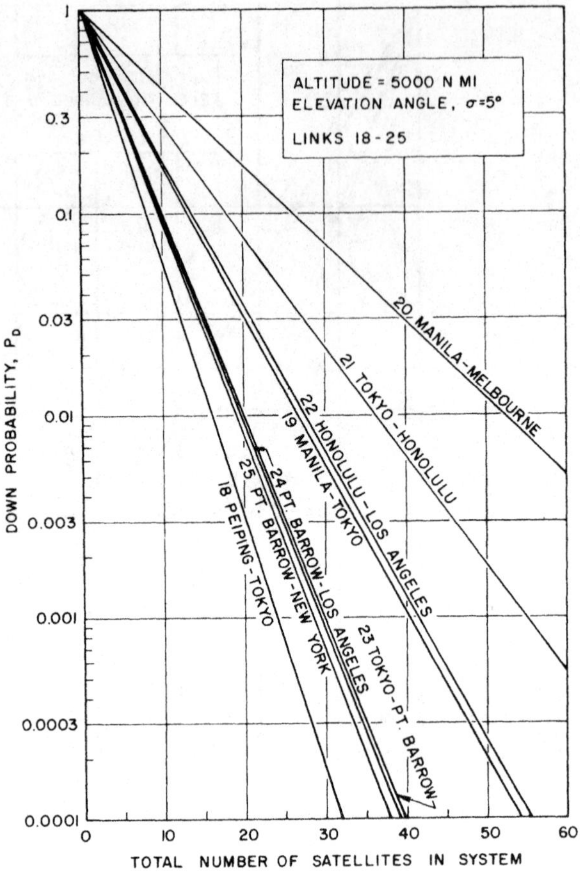

Fig. 6c P_D vs the total number of randomly distributed satellites equally divided among four-polar planes.

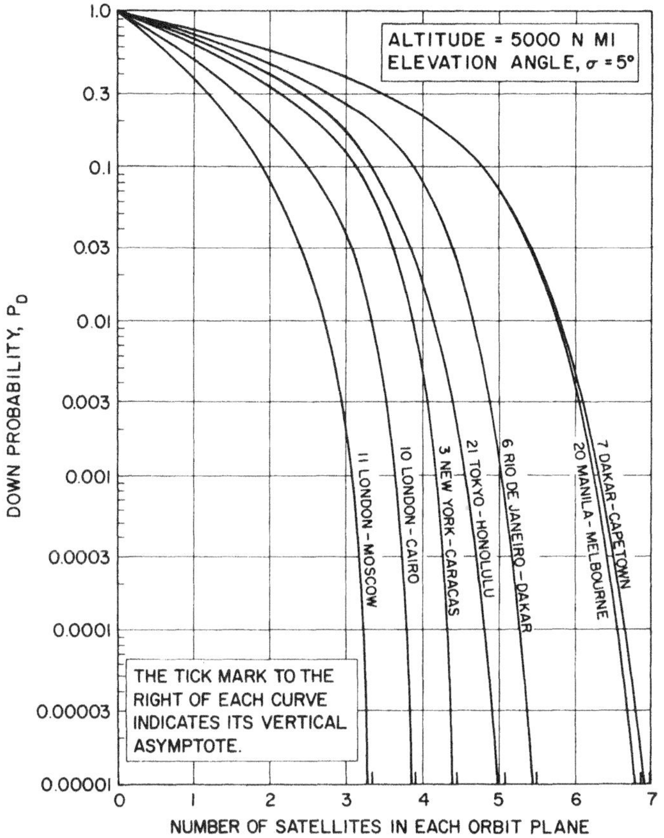

Fig. 7 P_D vs the number of evenly spaced satellites in each of three polar planes.

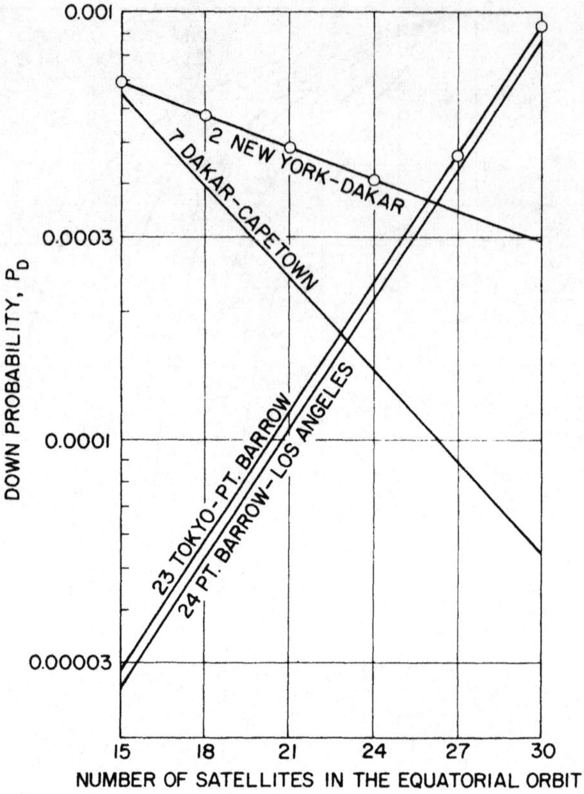

Fig. 8 P_D vs the apportionment of 60 randomly distributed satellites in a one-equatorial, three-polar system.

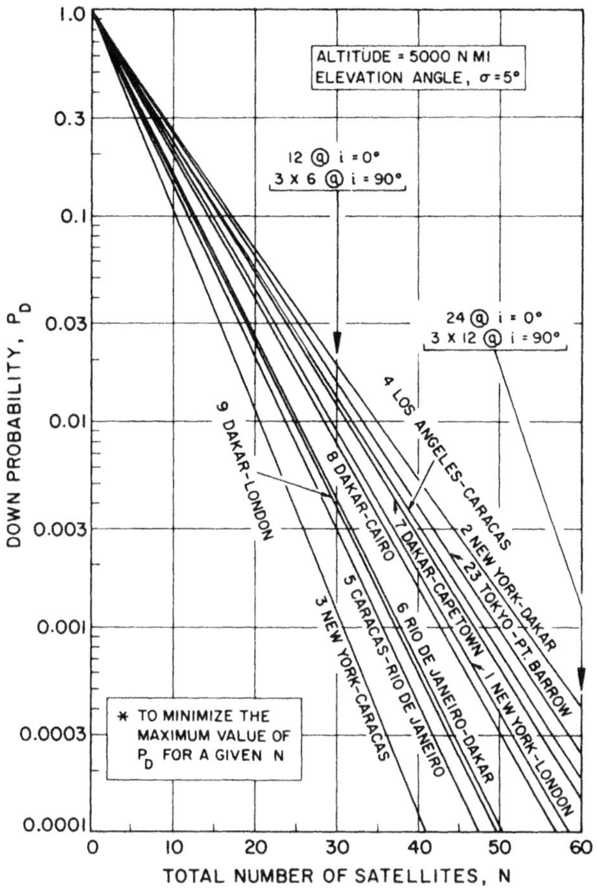

Fig. 9a P_D vs the total number of randomly distributed satellites optimally apportioned in a one-equatorial, three-polar system.

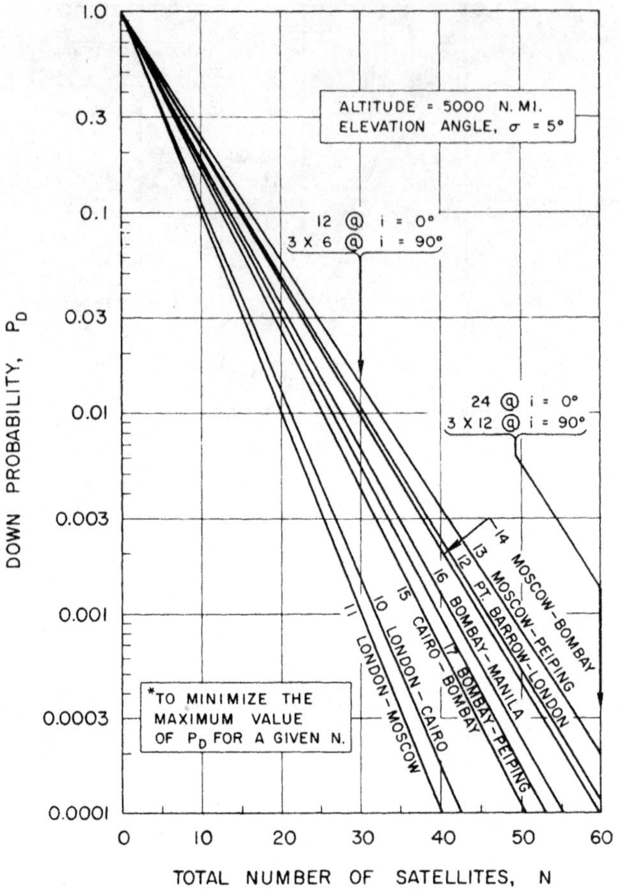

Fig. 9b P_D vs the total number of randomly distributed satellites optimally apportioned in a one-equatorial, three-polar system.

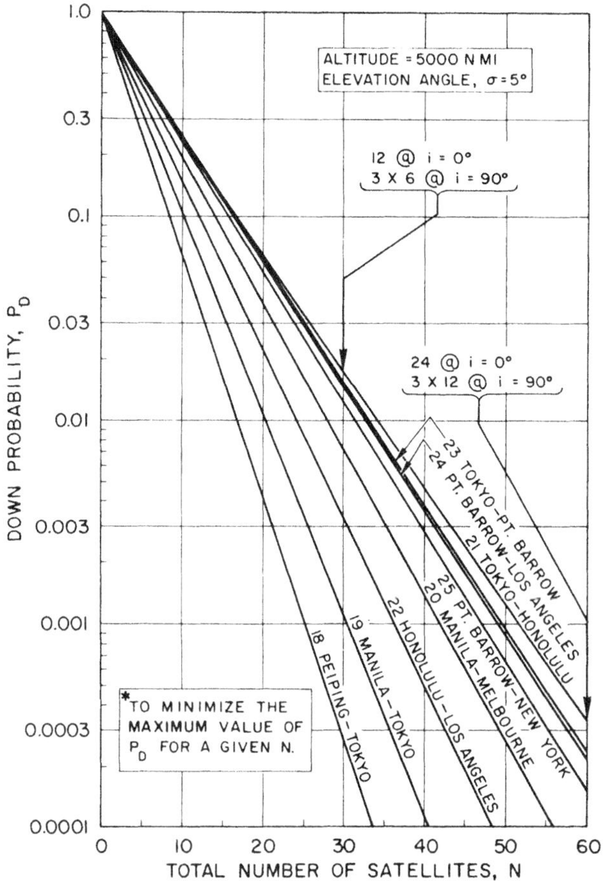

Fig. 9c P_D vs the total number of randomly distributed satellites optimally apportioned in a one-equatorial, three-polar system.

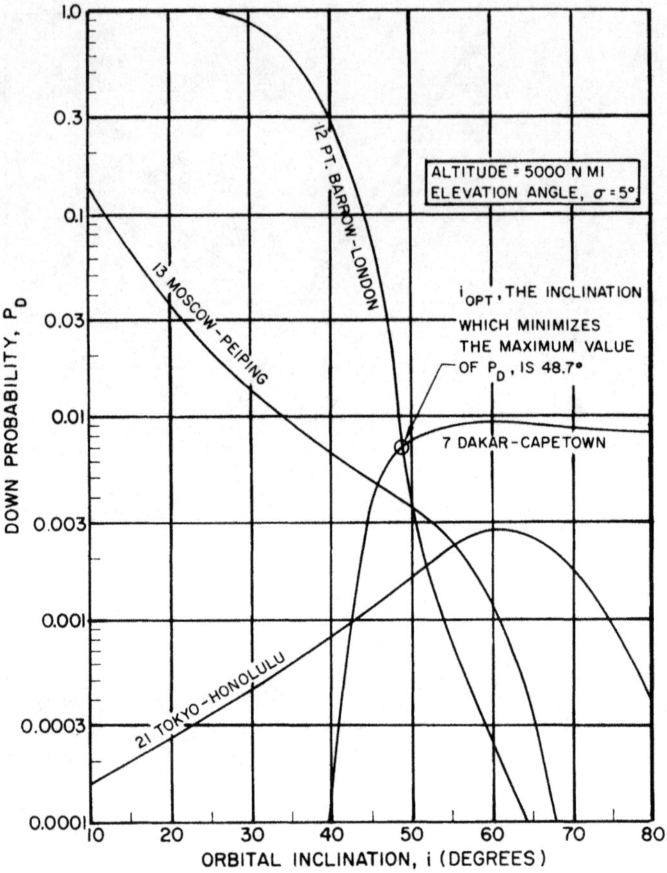

Fig. 10 P_D vs i for 60 randomly distributed satellites equally divided among three commonly inclined planes.

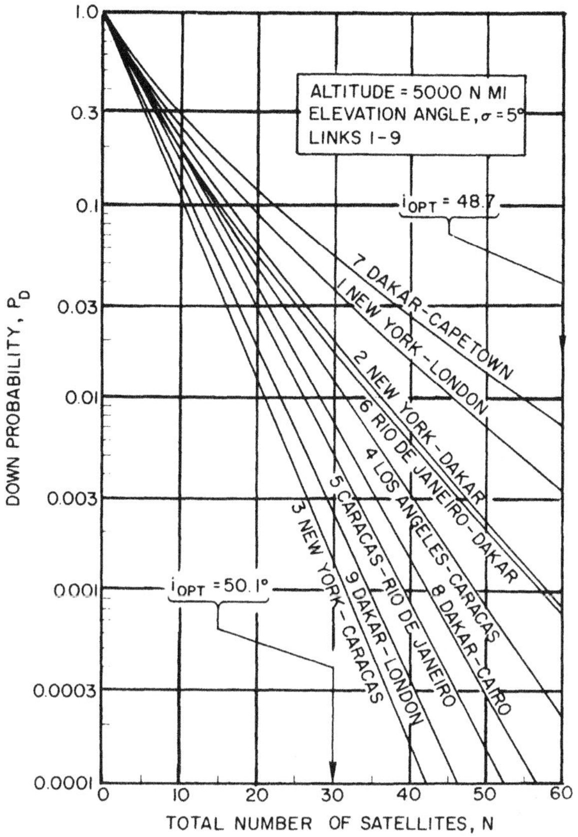

Fig. 11a P_D vs N for a commonly and optimally inclined three-plane system of randomly distributed satellites.

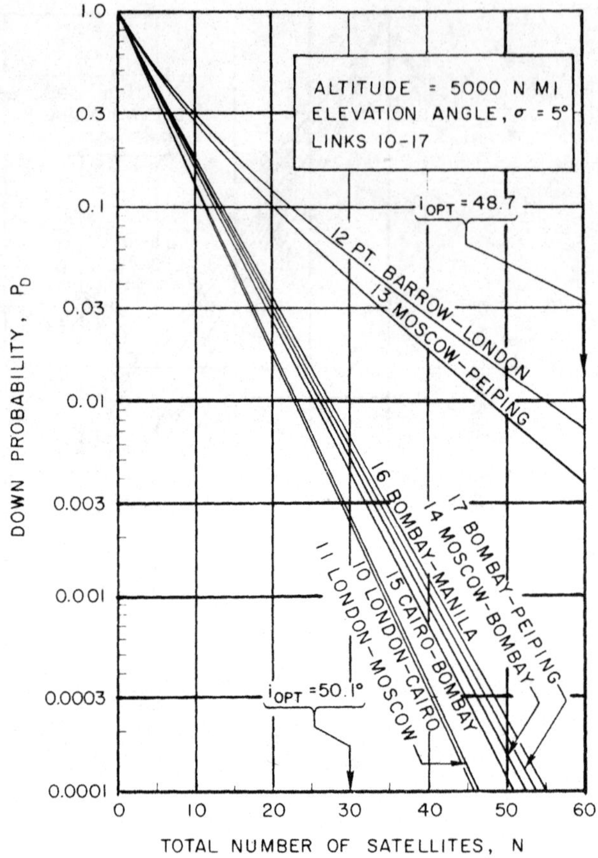

Fig. 11b P_D vs N for a commonly and optimally inclined three-plane system of randomly distributed satellites.

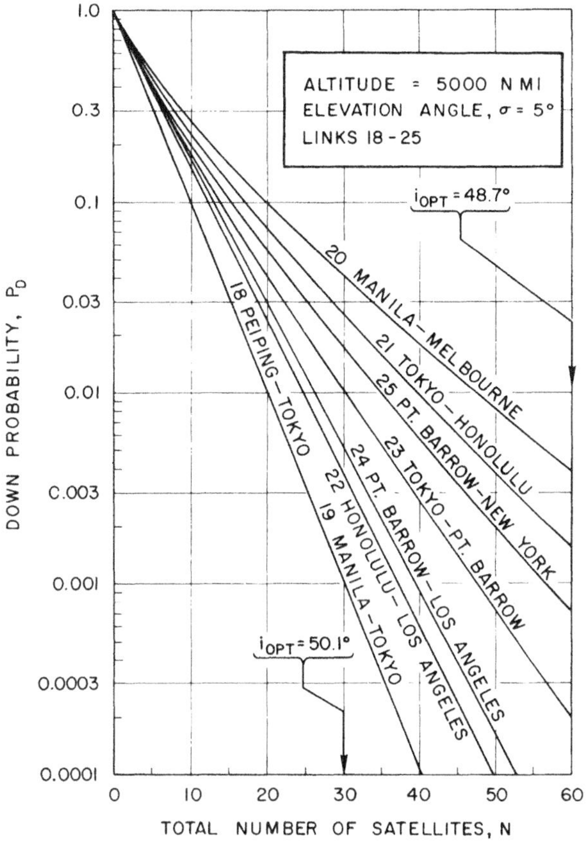

Fig. 11c P_D vs N for a commonly and optimally inclined three-plane system of randomly distributed satellites.

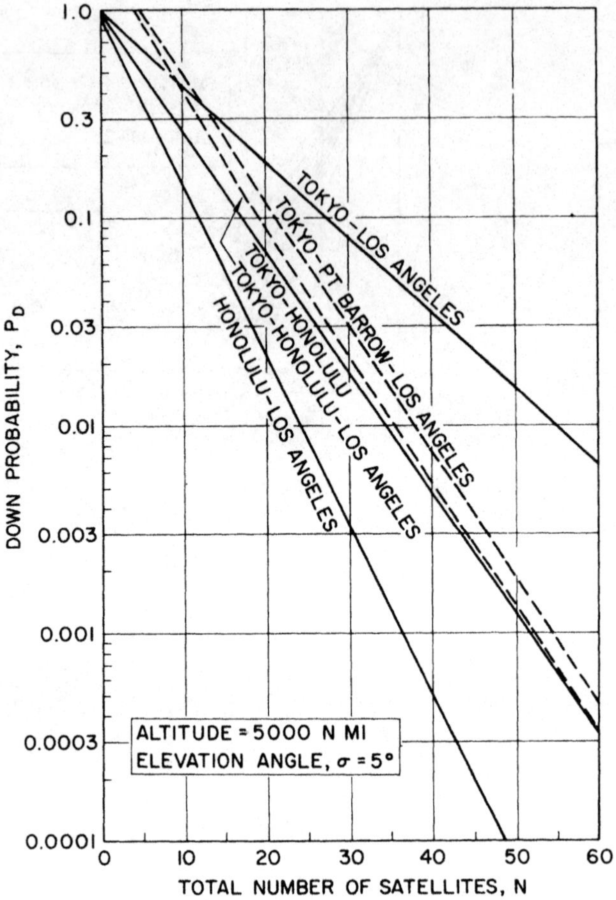

Fig. 12 The use of Point Barrow and Honolulu as intermediate ground relays.

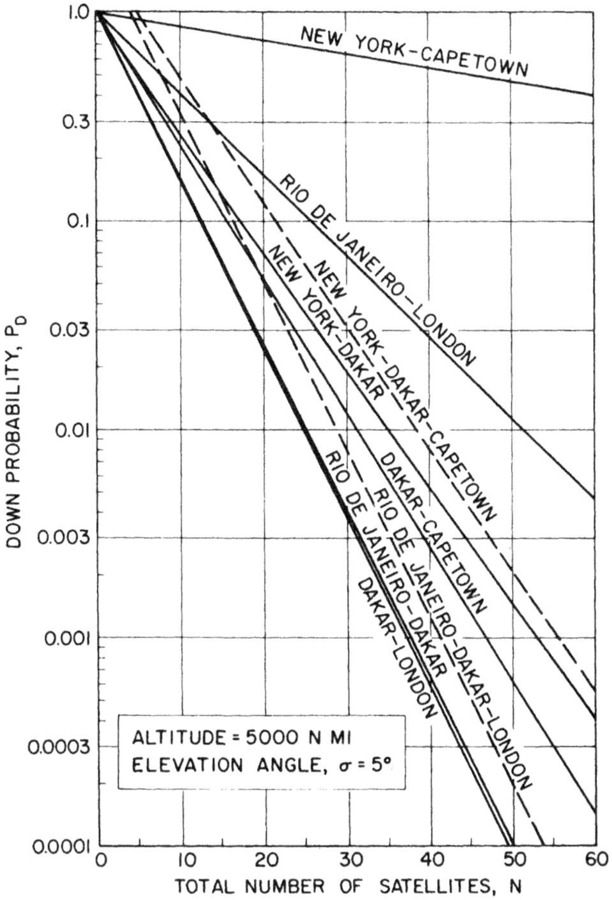

Fig. 13 The use of Dakar as an intermediate ground relay.

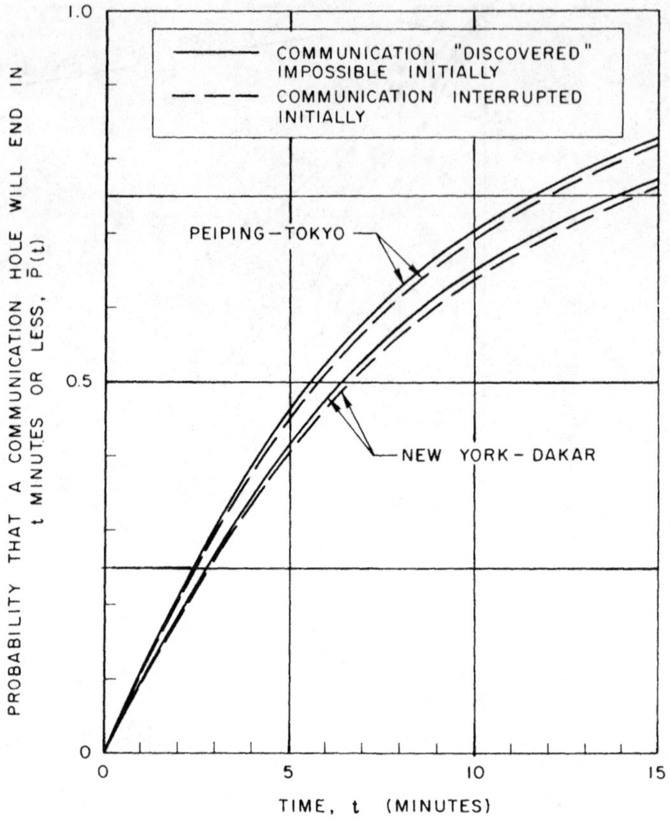

Fig. 14 Communication hole duration probability for a one-equatorial, three-polar system of 30 randomly distributed satellites.

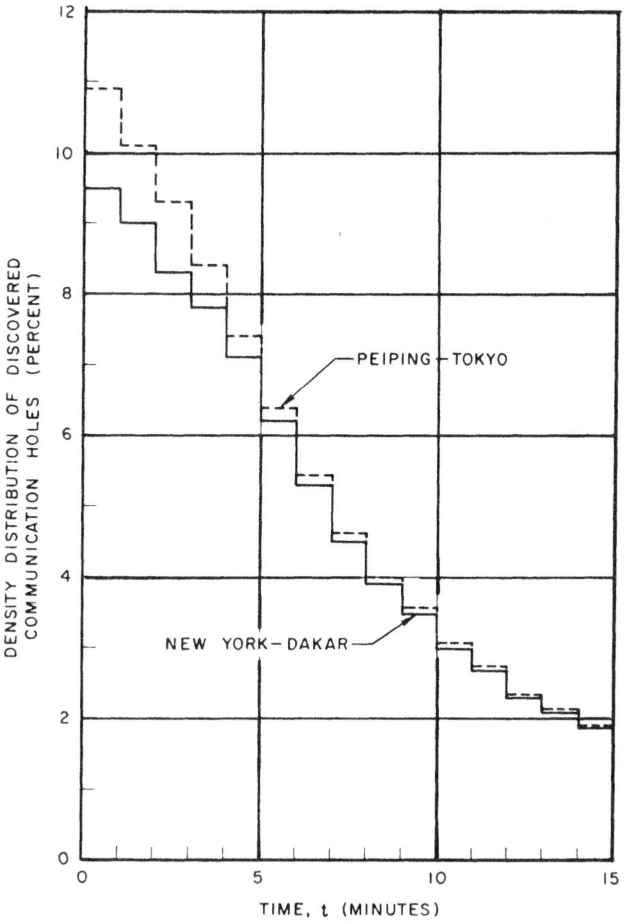

Fig. 15 Density distribution of discovered communication holes for the one-equatorial, three-polar system of 30 randomly distributed satellites.

III.

ORBIT SELECTION

Guidance in the Earth-Moon Space

GUIDANCE OF UNMANNED LUNAR AND INTERPLANETARY SPACECRAFT

Carl G. Pfeiffer*

Jet Propulsion Laboratory, California Institute of Technology, Pasadena, Calif.

Abstract

This paper describes mathematical techniques which have been developed for accomplishing Earth-based guidance of unmanned lunar and interplanetary spacecraft. The orbit determination and guidance correction aspects of the problem are developed from linear perturbation techniques, well known in exterior ballistics problems. An expression for the effect of neglected second order terms is developed. The treatment of correlated noise on the orbit determination data is discussed. Policies for determining when to perform impulsive guidance corrections to the orbit are described, and a technique for optimal guidance of powered flight trajectories is outlined.

1. Introduction

A lunar or interplanetary mission would normally consist of a boost portion of flight, during which the launch vehicle injects the spacecraft into some desired near-Earth orbit, and a postinjection portion of flight, during which the spacecraft is separated from the launch vehicle and coasts or thrusts toward the target. The objective of the spacecraft guidance system is to predict the target error due to injection errors and postinjection disturbances, and to take corrective action by commanding impulsive and/or continuous thrust acceleration of the spacecraft (Fig. 1). The purpose of this paper is to describe some of the mathematical techniques employed to accomplish the postinjection guidance of an unmanned lunar or interplanetary spacecraft, where there

Presented as Preprint 63-399 at the AIAA Astrodynamics Conference, New Haven, Conn., August 19 - 21, 1963. This paper presents the results of one phase of research carried out at JPL under Contract No. NAS 7-100, sponsored by NASA.

*Group Supervisor, Systems Analysis Section.

is the capability of performing computations and transmitting commands from Earth.[1] The mission objectives might be: a lunar or planetary impact; flyby; soft-land; or orbit, with trajectory characteristics as described in Ref. 2.

The guidance problem, as approached here, is an application of final value control theory. It has two basic parts:

1. Orbit Determination--to process in real time whatever measurements (observations) are available for an estimate of the parameters which define the state of the spacecraft trajectory, and to compute the statistics of the error in this estimate.
2. Guidance Policy--based on the information obtained in 1. above, to choose times t_i to accomplish the impulsive velocity corrections to the orbit, and to determine what these corrections should be, or, in the case of continuously applied thrust acceleration, to develop a steering and propulsion program as a function of time along the trajectory.

The techniques for accomplishing these tasks could be thought of as an extension of methods well known in celestial mechanics, but there are certain novel features about the guidance problem which affect the development of the analysis. Examples are:

1. The data types available for determining the orbit are new to celestial mechanics, consisting of such measurements as highly accurate Earth-based two-way doppler data, and spacecraft-based angle data with rather complex noise characteristics.
2. The orbit can be physically altered by the application of impulsive or continuously applied velocity corrections, with complicated constraints placed on the application of the corrections.
3. A very precise final orbit is required, with target errors relative to Venus and Mars on the order of 100 km.
4. The characteristics of the orbits are unusual.[2] For example, an interplanetary spacecraft trajectory might be a hyperbola near the Earth, with eccentricity (e)

equal to 1.25; an ellipse around the Sun, with e = 0.25; and a hyperbola near the target, with e = 5.0. Earth-moon transfer trajectories are usually highly eccentric ellipses, with e = 0.98.

5. Certain new kinds of trajectory error sources must be considered. For example, in the case of continuously applied thrust acceleration it is necessary to consider the state of the attitude control and propulsion subsystems, for errors in these devices can have the same effect on the trajectory as an uncertainty in the value of the astronomical unit or an unknown solar wind.

The analytical techniques described below are suited to these conditions.

The notation employed is as follows: the independent variable is the time t; T is a fixed final time, which may be thought of as target encounter; other capital letters are matrices; I is the identity matrix; the bar (-) indicates a column vector; the superscript(') indicates a transpose; the subscript (s) refers to the standard trajectory; δ refers to a variation taken at a fixed time; and $E[\cdots]$ indicates the statistical expectation of the bracketed quantity. When appropriate, (t) will be omitted when referring to time varying quantities.

2. The Linear Perturbation Equations

The development of the guidance problem is based upon well-known differential correction techniques which have long been employed in exterior ballistics celestial mechanics problems.[3] This approach assumes that a standard (reference) trajectory is available and that only the first variation of the coordinate deviations from the standard need be considered when constructing the orbit determination and guidance policy equations. The linear perturbation equations employed to describe the coordinate deviations will be developed here. The effect of the neglected second order terms is discussed in Part 3.

Let the equations of motion of the spacecraft be

$$\frac{d}{dt}(\bar{x}) = \bar{f}(\bar{x}, t) + \bar{g}(\bar{y}, t) + \bar{h}(\bar{z}, t) \tag{1}$$

where the x_i are the position and velocity coordinates, the y_i are the m control variables which are available for regulating the trajectory acceleration (such as the attitude angles and magnitude of the thrust acceleration), the z_i are the k (constant) error-source variables which physically affect the trajectory accelerations (such as the magnitude of a solar wind or an error in the applied rocket thrust), t is time, and the components of the vector functions f, $\bar{g}$, and $\bar{h}$ are piecewise differentiable with respect to the x_i, y_i, and z_i, respectively. The function $\bar{h}(\bar{z}, t)$ will be assumed linear in the z_i. A more general formulation would not assume that the effects of $\bar{x}$, $\bar{y}$, and $\bar{z}$ are separable, and that the $\bar{h}(\bar{z},t)$ is nonlinear, but the simpler form of Eq. (1) is sufficient for the purpose of this discussion. The quantity $\bar{q}(t) = [\bar{x}(t), \bar{z}]$ can be considered the "state" of the system.

Suppose that the initial condition x_0 is given, $\bar{y}(t)$ is specified, and $\bar{z}(t)$ is assumed known, and let the resulting trajectory be defined as the standard trajectory. In a neighborhood of this path the variation of the vector $\bar{x}(t)$ obeys the differential equation

$$\begin{aligned}\frac{d}{dt}(\delta\bar{x}) = {} & F\delta\bar{x} + G\delta\bar{y} + H\delta\bar{z} \\ & + \text{(second order terms of the type } \delta x_i \delta x_j) \\ & + \text{(second order terms of the type } \delta y_i \delta y_j) \\ & + \text{higher order terms}\end{aligned} \tag{2}$$

where

$$F = \left[\frac{\partial \bar{f}}{\partial \bar{x}}\right], \qquad G = \left[\frac{\partial \bar{f}}{\partial \bar{y}}\right], \qquad H = \left[\frac{\partial \bar{f}}{\partial \bar{z}}\right]$$

are evaluated along the standard trajectory. Define the state-transition matrix U (τ, t) by (the adjoint equations):

$$\frac{d}{dt} U + UF = 0 \tag{3}$$

where the time τ is considered fixed ($\tau \geq t$), and

$$U(\tau, \tau) = I \tag{4}$$

The motivation for this definition follows from the relationship

$$\frac{d}{dt}\left[U(\tau, t)\delta x(t)\right] = G\delta\overline{y} + H\delta\overline{z} + \text{higher order terms} \tag{5}$$

for, by integrating Eq. (5), it follows that

$$\delta\overline{x}(\tau) = U(\tau, t)\delta\overline{x}(t) + \int_t^\tau U(\tau, s)\left[G(s)\delta\overline{y}(s) + H(s)\delta\overline{z}\right]ds + \text{higher order terms} \tag{6}$$

The control $\delta\overline{y}(t)$ could be applied as an impulse (delta function) at time t_i, as discussed in Part 5.

When the higher order terms are ignored Eq. (6) is called the fundamental equation of linear guidance analysis, providing the basis for applying the powerful tools of linear theory to the guidance problem. It is employed in developing the orbit determination equations (Part 4), in computing the required impulsive corrections to the orbit (Part 5) and in specifying the optimality conditions for powered flight guidance (Part 6). The elements of $U(\tau, t)$ are called "differential coefficients."

The variation of a differentiable "boundary function" $\beta_i(\overline{x}, t)$ at some fixed final time T is

$$\delta\beta_i = [\text{grad } \beta_i]'\delta\overline{x}(T) = \left[\frac{\partial\beta_i}{\partial x_1(T)}, \ldots, \frac{\partial\beta_i}{\partial x_n(T)}\right]\delta\overline{x}(T) \tag{7}$$

Thus, from Eq. (6)

$$\delta\beta_i = \overline{\lambda}_i'(t)\,\delta\overline{x}(t) + \int_t^T \left[\overline{\eta}_i'(\tau)\delta\overline{y}(\tau) + \overline{\rho}_i'(\tau)\,\delta\overline{z}\right]d\tau \\ + \text{higher order terms} \tag{8}$$

where

$$\overline{\lambda}_i'(t) = [\text{grad } \beta_i]'\, U(T, t) \tag{9}$$

$$\overline{\eta}_i'(t) = \overline{\lambda}_i'(t)G(t) \tag{10}$$

$$\overline{\rho}_i'(t) = \overline{\lambda}_i'(t)H(t) \tag{11}$$

The elements of $\overline{\lambda}_i(t)$ are called "differential corrections" for the boundary function β_i, and the elements of $\overline{\eta}_i(t)$ and $\overline{\rho}_i(t)$ are called "influence functions" or "Green's functions" for β_i. When the higher order terms are ignored Eq. (8) is said to describe the "first variation" of β_i, while if only the second order terms are considered (discussed below) the resulting expression becomes the "second variation" of β_i. It is the task of the guidance system to estimate the quantities $\delta\overline{x}(t)$ and $\delta\overline{z}$ in order to choose the control $\delta\overline{y}(t)$ which will achieve some desired combination of the boundary function variations $\delta\beta_i$. A typical requirement is that the position error be nulled at the fixed final time T (at the target), in which case $\delta\beta_i$ for $i = 1, 2, 3$ are the position coordinates.

3. The Second Variation of the Boundary Functions

The application of linear perturbation theory in the operational situation is accomplished by choosing the pre-flight reference trajectory as the standard, and evaluating the differential coefficients along this path. If the orbit determination later indicates that the coordinate deviations are outside the range of linearity, a new standard trajectory is defined based on the best estimate of the orbit parameters. This process is iterated as many times as necessary to cause the effect of higher order terms in the guidance problem to be arbitrarily small. At each step of the process, however, linear theory is assumed to hold for the purpose of estimating the orbital elements and computing the guidance corrections. A technique for analytically determining the effect of the neglected second order terms is presented here. To simplify the exposition it will be assumed that there is only one control function $y(t)$ to consider, and that the boundary functions β_i are linear functions of the final state $\overline{x}(T)$.†

Let twice‡ the contribution of the second order terms in Eq. (8) be denoted as $\delta^2\beta_i$, which is called the second variation of β_i. Thus,

$$\delta^2\beta_i = \int_0^T \delta\overline{x}'(t)\, C_i(t)\, \delta x(t)\, dt + \int_0^T \xi_i(t)\delta y^2(t)\, dt \tag{12}$$

where $C_i(t)$ is an n by n matrix with jk^{th} element

$$(c_i)_{jk} = \sum_{a=1}^{n} \lambda_{ia} \left(\frac{\partial^2 f_a}{\partial x_j \partial x_k} \right) \tag{13}$$

and

$$\xi_i = \sum_{a=1}^{n} \lambda_{ia} \left(\frac{\partial^2 f_a}{\partial y^2} \right) \tag{14}$$

Define the modified state transition matrix

$$\widetilde{U}(\tau, t) = \begin{cases} U(\tau, t) & \text{if } \tau \geq t \\ 0 & \text{if } \tau < t \end{cases} \tag{15}$$

and substitute this quantity into the integral term of Eq. (6), so that the upper limit of that integral becomes T instead of τ. Employing this modified expression, Eq. (12) can be written (to second order) as

---

†Assuming $\beta_i(\overline{x})$ to be linear is no loss of generality, for the state variables x_j can always be redefined so as to make this true.

‡It is conventional to delete the factor 1/2 when discussing the second variation.

$$\delta^2\beta_i = \delta\bar{x}'(0)\left[\int_0^T U'(\tau,0)C_i(\tau)U(\tau,0)d\tau\right]\delta\bar{x}(0)$$

$$+ 2\delta\bar{x}'(0)\left[\int_0^T\int_0^T U'(\tau,0)C_i(\tau)\tilde{U}(\tau,t)G(t)\delta y(t)dtd\tau\right]$$

$$+ \left[\int_0^T\int_0^T\int_0^T G'(s)\tilde{U}'(\tau,s)C_i(\tau)\tilde{U}(\tau,t)G(t)\delta y(t)\delta y(s)dtd\tau ds\right]$$

$$+ \int_0^T \xi_i(\tau)\delta y^2(\tau)d\tau \qquad (16)$$

If no control y(t) is applied in the interval $[0, T]$ (a coast period), $\delta^2\beta_i$ reduces to only the first term on the right hand side of Eq. (16), which is a quadratic form in terms of the initial conditions disturbances.

Numerical studies have thus far indicated that the second order terms can be ignored during coasting periods without significantly affecting the overall guidance system performance,[4] but this question has yet to be thoroughly explored. Probably the most important application of Eq. (16) is in the area of continuous control of powered flight trajectories, briefly discussed in Part 6.

4. Orbit Determination

The orbit determination process consists of predicting the uncorrected final values of the boundary functions β_i (see Part 2) when given noise-contaminated tracking data. This is accomplished by estimating the initial condition $\bar{q}(0)$, or, equivalently, $\bar{q}(t)$, and determining the $\delta\beta_i$ from Eq. (8). The $\bar{q}$ is the state of the system as defined in Part 2. The "initial" time $t = 0$ can be taken to be any point along the trajectory, such as injection, or some later time where an impulsive correction is accomplished (which introduces new uncertainty into the system), or where a powered flight phase begins (see Part 6). The well-known orbit determination equations are reviewed below.

Suppose that noise-contaminated observation vectors $\bar{\phi}(t)$, which are related to the instantaneous spacecraft position and velocity, are being sampled at times $t_0, t_1, \cdots, t_n$. From Eq. (6) the first variation of the observations is found to be

$$\delta\overline{\phi}_i = \overline{\phi}(t_i) - \overline{\phi}(t_i)_{\text{standard}} = A(t_i)\,\delta\overline{q}_0 + \overline{n}(t_i) \tag{17}$$

where

$$\delta\overline{q}_0' = \left[\delta\overline{x}'(0), \delta\overline{z}'\right] \tag{18}$$

$$A(t_i) = \left[\frac{\partial\overline{\phi}(t_i)}{\partial\overline{x}(t_i)}\right]\left[U(t_i, 0) \;\Big|\; \left\{\int_0^{t_i} U(t_i, s)H(s)ds\right\}\right] \tag{19}$$

and $\overline{n}(t_i)$ is random noise on the data. (The notation $|$ indicates a partitioning of the matrix.) Suppose all the observations at all the times t_i are collected to construct

$$\delta\overline{\phi}' = \left[\delta\overline{\phi}'(0), \delta\overline{\phi}'(t_1), \cdots, \delta\overline{\phi}'(t_n)\right] \tag{20}$$

which can be written in the form

$$\delta\overline{\phi} = A\delta\overline{q}_0 + \overline{n} \tag{21}$$

where

$$A = \left[A(0) \;|\; A(t_1) \;|\; \cdots \;|\; A(t_n)\right] \tag{22}$$

$$\overline{n}' = \left[\overline{n}'(0) \;|\; \overline{n}'(t_1) \;|\; \cdots \;|\; \overline{n}'(t_n)\right] \tag{23}$$

Let $\overline{q}_0$ and $\overline{n}$ be considered as random vectors over the ensemble of all missions, having zero means and known a priori covariance matrices Λ_0 and Γ, respectively. The Λ_0 is obtained from an error analysis of the injection guidance system if $t = 0$ corresponds to injection, and includes the effect of postinjection orbit determination information and uncertainty arising from errors in the postinjection guidance corrections if $t = 0$ is some later time. The Γ is found from studying the characteristics of the measuring devices, where Γ is diagonal for "white" noise. Seeking a linear unbiased, minimum variance estimate, the Gauss-Markoff theorem is applied[5] to obtain (after some manipulation)

$$\delta\bar{q}_0^* = \left[\Lambda_0^{-1} + A'\Gamma^{-1}A\right]^{-1} A'\Gamma^{-1}\delta\bar{\phi} \qquad (24)$$

where the * indicates the estimated quantity. Equation (24) is the weighted least squares form of the estimate. It follows that the covariance of the error in the estimate is

$$E\left[\left(\delta\bar{q}_0^* - \delta\bar{q}_0\right)\left(\delta\bar{q}_0^* - \delta\bar{q}_0\right)'\right] = \Lambda_0^* = \left[\Lambda_0^{-1} + A'\Gamma^{-1}A\right]^{-1} \qquad (25)$$

If Λ_0 is "large," so that the a priori information matrix Λ_0^{-1} is negligible, then

$$\left[\Lambda_0^{-1} + A'\Gamma^{-1}A\right] \rightarrow \left[A'\Gamma^{-1}A\right] = J \qquad (26)$$

where J is called the "normal matrix." If the normal matrix is nonsingular the quantity $\delta\bar{q}_0$ is said to be "observable." For an application of the weighted least squares estimation technique, consult Ref. 6.

Another approach to orbit determination is to process the data sequentially[5] where each incoming observation vector $\delta\bar{\phi}(t_i)$ is processed individually. Let

$\delta\hat{\bar{q}}_i$ = best estimate of $\delta\bar{q}(t_i)$ based on all data up to but not including time t_i

$\hat{\Lambda}_i$ = covariance of the error in the estimate $\delta\hat{q}_i$

Given the new data $\delta\bar{\phi}(t_i)$, the revised estimate and covariance matrix become

$$\delta\bar{q}_i^* = \delta\hat{\bar{q}}_i + \hat{\Lambda}_i A_i'\left[A_i\hat{\Lambda}_i A_i' + \Gamma_i\right]^{-1}\left[\delta\bar{\phi}_i - A_i\delta\hat{\bar{q}}_i\right] \qquad (27)$$

$$\Lambda_i^* = \hat{\Lambda}_i - \hat{\Lambda}_i A_i'\left[A_i\hat{\Lambda}_i A_i' + \Gamma_i\right]^{-1} A_i\Lambda_i \qquad (28)$$

The $\delta\hat{\bar{q}}_{i+1}$ and $\hat{\Lambda}_{i+1}$ are then computed by mapping $\delta\bar{q}_i^*$ and Λ_i^* forward with the state transition matrix; the data vector $\delta\bar{\phi}(t_{i+1})$ is processed as before; and the process is iterated to completion. In this way the instantaneous state of the system is continually estimated, which implies that an optimal "filter" has been constructed. This approach, which is mathematically identical to the weighted least squares technique if the noise is "sequentially correlated" (see below), effectively treats the advanced estimate as an observation, and optimally combines it with the new measurement to obtain the minimum variance estimate. The technique was developed by R.E. Kalman in his important analysis of linear systems excited by uncorrelated Gaussian error sources,[7] and applied to the orbit determination problem by S. F. Schmidt. The sequential estimation technique might be attractive for powered flight orbit determination, where the time-varying random behavior of the propulsion system must be described, and the guidance correction continuously determined.

It should be noted that the sequential estimation technique does not always lead to a minimum variance estimate, for in general it is necessary to consider all data prior to and including time t_i in order to estimate $\delta\overline{q}(t_i)$. This is because all the information obtained about $\delta\overline{q}(t_i)$ prior to t_i cannot always be stored in the n by n covariance matrix Λ_i^*. It can be shown[5] that the necessary and sufficient condition to be satisfied if the sequential estimation technique is to yield a minimum variance estimate is that the $\delta\overline{q}(t_i)$ and $\overline{n}(t_i)$ be generated by a "sequentially correlated" stochastic process. Such a process is characterized by

$$R_{ki} = R_{kj} R_{ji} \tag{29}$$

for all $t_i \leq t_j \leq t_k$, where the "normalized correlation matrix" R_{ji} for a stochastic vector $\delta\overline{w}(t)$ is defined by

$$R_{ji} = \left\{ E\left[\delta\overline{w}(t_j)\delta\overline{w}'(t_i)\right] \right\} \left\{ E\left[\delta\overline{w}(t_i)\delta\overline{w}'(t_i)\right] \right\}^{-1} \tag{30}$$

If the $\delta\overline{w}(t_i)$ are Gaussian variables it can be shown that the process is Markoff. It is easily shown that the coordinate deviations $\delta\overline{x}(t_i)$, the constant error sources $\delta\overline{z}$, and uncorrelated noise $\overline{n}(t_i)$ are sequentially correlated variables, the normalized correlation matrices being, respectively, the state transition matrix, the identity matrix, and the matrix delta function. Examples of other sequentially correlated processes, which might be appropriate for describing the time-varying errors in applied thrust acceleration during powered flight phases, are given in Ref. 5. These more general cases have yet to be thoroughly analyzed.

If the time between observations becomes very small it may be convenient to express the sequential estimate in a continuous form. This is accomplished by putting Eq. (27) and (28) into the more general form presented in Ref. 5. Assuming that the elements of these equations are all differentiable functions of time (including the data), they can be expressed as a Taylor series expansion in terms of $(t_j - t_i)$. Letting $t_j \rightarrow t_i$, the difference equations become differential equations. This analysis is developed in Ref. 8.

If the noise vector can be decomposed into components with no cross correlation, it follows that sequential correlation exists if each component is autocorrelated according to

$$E\left[n_k(t_j)n_k(t_i)\right] = \sigma_k^2 \exp -\alpha_k(t_j - t_i) \tag{31}$$

where k denotes the kth component of the noise vector, σ_k^2 is the (constant) variance of that component, and α_k is some positive constant. The case $\alpha_k = 0$ corresponds to a "bias"; the case $\alpha_k = \infty$ corresponds to "white noise"; and the any case in between corresponds to "drift." In an unpublished extension of Ref. 9, Gordon made an examination of the relationship between the parameter α and the uncertainty in the estimate of the target error as a function of tracking time for the case of an approach guidance system using optical (stellar) observations as the data type. An idealized model of the problem is shown in Fig. 2, where the observations are given by

$$\delta\phi(t) = \frac{\delta\beta}{v(T - t)} + n_1(t) + n_2(t) \tag{32}$$

The $\delta\phi(t)$ is the variation in star angle observed from a spacecraft coasting in a straight line toward a target planet; $\delta\beta$ is the unknown target error to be determined; $v(T-t)$ is the distance-to-go to the final time (target encounter); $n_1(t)$ is uncorrelated noise ($\alpha_1 = \infty$); and $n_2(t)$ is noise which has $0 \leq \alpha_2 \leq \infty$. The standard deviations of the noise are (approximately) $\sigma_1 = 0.1\ (10^{-3})$ rad. and $\sigma_2 = 2\ (10^{-3})$ rad. The variance of the error in the estimate $\delta\beta^*$ vs. distance-to-go is plotted in Fig. 3 for various values of α_2 for the case $v = 6.34$ km/sec. This corresponds to a representative Mars mission.

5. Impulsive Guidance Corrections

Suppose that the control $\delta\bar{y}(t)$ consists of the three components of thrust acceleration, lasting for a very short period of time (an impulse), and is applied in such a way as to null the predicted miss (position) components at the target $\delta\bar{m}(T)$. Letting $\delta\beta_i = \delta m_i$ for $i = 1, 2, 3$, from Eq. (8) it follows that

$$0 = \delta\bar{m}(T) = M\delta\bar{x}^*(t) + K\Delta\bar{v}(t) \qquad (33)$$

where

$$M = \left[\frac{\partial\bar{m}(T)}{\partial\bar{x}(t)}\right] = \left[\frac{\partial\bar{m}(T)}{\partial\bar{x}(T)}\right]\left[U(T, t)\right] \text{ (a 3 x 6 matrix)}$$

$$K = \left[\frac{\partial\bar{m}(T)}{\partial\bar{v}(t)}\right]$$ = submatrix of M composed of differential corrections for velocity components (a 3 x 3 matrix)

$\Delta\bar{v}(t)$ = velocity impulse applied at time t

$$= \int_{t-\epsilon}^{t+\epsilon} \delta\bar{y}(t)\,dt \qquad \text{where } \epsilon \text{ is small}$$

$\delta\bar{x}^*(t)$ = estimate of the state vector deviation at time t (from the orbit determination process)

The required $\Delta\bar{v}$, sometimes called "velocity-to-be-gained," is calculated from Eq. (33) to be

$$\Delta\bar{v}(t) = -K^{-1}M\delta\bar{x}^*(t) \qquad (34)$$

assuming the indicated inverse exists. (The case of fractional corrections of the estimated target error will not be considered here.)

The impact parameter coordinate system is usually chosen to describe the target errors at the final time T, where the axis m_1 is aligned along the incoming asymptote of the approach hyperbola (Fig. 4). This system has the advantage that m_2 and m_3 are the only coordinates which affect target error, and variations in m_1 effectively yield time-of-flight errors. This can be verified by imagining the spacecraft approaching a massless target, and visualizing the effect of errors in the m_i. Also convenient for analysis purposes is the "proper" coordinate system, wherein the K matrix is diagonal. This is constructed by applying the orthonormal matrices L_2 and L_1 to Eq. (33) to obtain

$$\delta\hat{\bar{m}}(T) = L_2 \delta\bar{m}(T) = L_2 M \delta\bar{x}^*(t) + \hat{K}\Delta\hat{\bar{v}}(t) \tag{35}$$

where L_2 is the transformation which diagonalizes the real symmetric matrix $[KK']$, and

$$\hat{K} = L_2 K L_1' = \left[L_2 K K' L_2'\right]^{1/2} = \text{diagonal}$$

$$L_1 = \left[L_2 K K' L_2'\right]^{-1/2} \left[L_2 K\right]$$

$$\Delta\hat{\bar{v}}(t) = L_1 \Delta\bar{v}(t)$$

It can be verified that L_1 is orthonormal (notice that L_2 does not equal L_1). In Ref. 10 numerical values of the elements of $\hat{K}$ are presented for a representative set of elliptical transfer trajectories. Figure 5 shows the elements of $\hat{K}$ for a 66-hr lunar trajectory.

In general, there would be a sequence of times at which impulsive corrections might be made, for imperfect orbit determination information and errors in applying the desired impulse at any time t_i implies that further control might be called for. Assuming that the task of the guidance system is to null the predicted target error, that no fractional corrections are allowed, and that the orbit determination data to be taken are specified, the sequence of times for performing the corrections defines the guidance policy. The selection of these times is an example of an optimal stochastic control problem, and it has been attacked from various points of view by various authors. The techniques developed in Ref. 11, 12, and 13 are discussed in Ref. 14, and numerically compared by applying each method to an idealized guidance problem. Briefly, these three schemes for determining the guidance policy are:

1. The variance ratio method.[11] Corrections are performed whenever the rms error in the correction due to orbit determination uncertainty falls to a prespecified percentage of the rms correction one expects to make.
2. The spacing-ratio method.[12] Corrections are applied at the times, $t_1, t_2, \cdots, t_N$, where t_N is specified by the allowable error and $(T-t_i/T-t_{i+1})$ is a fixed ratio which minimizes the average total propellant expended.
3. The minimum error method.[13] Corrections are made whenever the predicted final mean-squared miss resulting from the correction is less than can be obtained by waiting at least until the next correction opportunity. The probability of depleting the correction capability is taken into account. Corrections occur at different times on each trajectory, depending upon the particular perturbation.

In all cases the statistical expectations are taken over the ensemble of all perturbed trajectories. Methods 1 and 2 above lead to a sequence of decision times which is fixed for all trajectories, while method 3 is "adaptive" in the sense that at any time t the decision to correct the orbit or not depends upon the size of the error to be nulled, the amount of correction capability available, and the variance of the error in the estimate.

It is shown in Ref. 14 that these three schemes require approximately the same correction capability (propellant) expenditure and lead to approximately the same final target error uncertainty for the particular example studied. The general problem of defining an optimal guidance policy is yet to be fully understood, however, especially for the case of highly nonstandard mission operations. Further work is called for in this area.

6. Powered Flight Guidance

Powered flight guidance is necessary when thrust acceleration is continuously applied over a long interval relative to the total mission time, such as in the boost vehicle phase of flight, or when electric propulsion is used during interplanetary transfer. The orbit determination aspect of the problem could be treated similarly to the free-fall case if the thrust acceleration error can be described by a set of random constants, or the acceleration error might be treated as a time-varying sequentially correlated process as described above.

The guidance policy aspect of powered flight is actually an easier analysis task than the impulsive case, for the control acceleration is already committed and it only remains to find some rational way to compute small variations in steering angles,and perhaps thrust acceleration level, which will cause desired end conditions to be achieved. Various rudimentary guidance schemes could be constructed, or a more sophisticated "optimal" method could be employed. The elements of such a technique will be summarized here, following the approach of Ref. 15. This problem is further discussed in Ref. 16 and 17.

Suppose that the control $\overline{y}(t)$ is applied on the standard trajectory in such a way that some linear function of the end conditions, $\beta_0(\overline{x})$, is minimized at the final time T, and some other linear functions of the end conditions, $\beta_i(\overline{x})$, are made equal to zero, where $i = 1, 2, \cdots r$. For example, β_0 might be the time-of-flight coordinate discussed in Part 5, and β_1 and β_2 the remaining two positions coordinates. To simplify the exposition to follow, let it be assumed that there is only one control function y(t) to consider. For example, y(t) might be the angle of the thrust vector from some inertial reference (the steering angle). Since the standard trajectory has been optimized, it follows that the influence functions $\eta_i(t)$ discussed in Part 2 must be linearly dependent. This can be shown by introducing the performance index

$$p = \sum_{i=0}^{r} \nu_i \beta_i(\overline{x}) \tag{36}$$

where the ν_i are constant Lagrange multipliers. (This is the Mayer formulation of the problem.) Assuming no initial condition or random errors, Eq. (8) yields

$$\delta p = \int_0^T \eta_p(t)\,\delta y(t)\,dt + \text{higher order terms} \tag{37}$$

If the first variation of p is to be zero it follows that

$$\eta_p(t) = \sum_{i=0}^{r} \nu_i \eta_i(t) = 0 \tag{38}$$

which proves the linear dependence. A further necessary condition (the Legendre condition) is that

$$\xi_p(t) = \sum_{i=0}^{r} \nu_i \xi_i(t) \geq 0 \tag{39}$$

This is heuristically established by imagining a plus and minus impulsive perturbation of the control at time t, which preserves the desired end conditions to first order, and observing that the $\xi_i(t)$ terms dominate the expression for the second variation given by Eq. (16). Equations (38) and (39) also follow directly from the well-known Pontryagin Principle.

Given that a standard trajectory which meets these conditions has somehow been devised, suppose that a small perturbation to the initial conditions, $\delta\overline{x}(0)$, is detected by the orbit determination process, and that there are no random errors after time zero. An approximate neighboring optimal trajectory can be constructed to correspond to the new initial condition by employing the expressions given above for the first and second variations of the $\beta_i(\overline{x})$. Clearly the second variation must be considered, for Eq. (38) shows that a control variation has no first order effect on the performance index. Consistent with the linearity assumptions discussed above, let the terms $(\partial^2 f_i/\partial x_j \partial x_k)$ be deleted in the expression for the second variation, given by Eq. (16) (that is, $C_i = 0$) so that Eq. (8) becomes (approximately)

$$\delta\beta_i = \overline{\lambda}_i'(0)\,\delta\overline{x}(0) + \int_0^T \eta_i(t)\,\delta y(t)\,dt + \int_0^T \xi_i(t)\,\delta y^2(t)\,dt \tag{40}$$

$$+ \text{ higher order terms}$$

Assuming that the β_i have been suitably normalized to make them dimensionally compatible (the normalization is arbitrary, but usually is suggested by the physics of the problem), it is shown in Ref. 15 that it is possible to introduce the transformed boundary function variations

$$\delta\hat{\beta} = L\,\delta\overline{\beta} \tag{41}$$

where L is an r + 1 by r + 1 orthonormal matrix chosen so that $\hat{\beta}_0$ is the "uncontrollable" direction. This means that the function $\delta y(t)$ has no first order effect on $\hat{\beta}_0$, a condition which follows from the optimality of the standard trajectory. The corresponding influence functions have the property

$$\int_0^T \hat{\eta}_i(t)\hat{\eta}_j(t)\,dt = \begin{cases} 0 & \text{if } i \neq j \\ \mu_i & \text{if } i = j \end{cases} \tag{42}$$

Thus, if $\delta\hat{\beta}_0$ is to be minimized and

$$\delta\hat{\beta}_i = \hat{\lambda}_i'(0)\,\delta\overline{x}(0) + \Delta\hat{\beta}_i(\delta y) = \text{specified for } i = 1, 2, \cdots r \tag{43}$$

where the $\Delta\hat{\beta}_i(\delta y)$ are the first order effects of $\delta y(t)$ in the controllable directions, it follows that

$$\delta y(t) = \sum_{i=1}^{r} \left[\xi_p(t)\mu_i^2\right]^{-\frac{1}{2}} \Delta\beta_i(\delta y)\hat{\eta}_i(t) \tag{44}$$

The $\Delta\hat{\beta}_i(\delta y)$ are determined by Eq. (43). The resultant $\delta\hat{\beta}_0$ is

$$\delta\hat{\beta}_0 = \hat{\lambda}_0'(0)\delta\overline{x}(0) + \frac{1}{2}\sum_{i=1}^{r}\left[\mu_i\right]^{-1}\left[\Delta\hat{\beta}_i^2(\delta y)\right] \tag{45}$$

Note that it is necessary to assume that the inequality holds strictly in Eq. (39) and that $\mu_i \neq 0$ in Eq. (42) if the control variation is to be defined by Eq. (44).

The specified $\delta\hat{\beta}_i$ might not be selected to cause the standard end conditions to be attained, for this could lead to excessive performance loss on some perturbed trajectories. The selection of the "aiming point" is an aspect of the guidance policy problem which is highly mission dependent, and will not be discussed here.

7. Problem Areas

It appears now that the mathematical techniques discussed here provide an adequate theoretical basis for accomplishing Earth-based guidance of unmanned spacecraft, but more evaluation and mission simulation is called for. The potential problems seem to lie more in the orbit determination than in the guidance policy area, for an adequate correction philosophy can usually be constructed if the mission objectives are well defined and the spacecraft correction capability is properly specified. The cost associated with a less-than-optimal policy is often expressible in terms of a slightly increased correction capability requirement on the spacecraft. The problems in the orbit determination area tend to be of two types:

1. Uncertainty in the mathematical model, where unknown error sources are not being considered in the supposedly minimum variance estimate of the orbit parameters.
2. Complicated "noise" on the observations, requiring a prohibitive amount of computation in order to optimally weight the data and/or necessitating the consideration of a large number of state variables. Such noise can come from either correlated external distances which distort the vehicle coordinates, (i. e. variation in propulsion system parameters) or from actual noise on the measuring instruments.

The first type of problem is well known in celestial mechanics and is solved by analyzing data from enough missions to be able to construct an adequate mathematical model. Indeed, one important goal of space missions of the near future will be to chart the solar system more precisely so that guidance operations can be more accurately accompublished. The second type of problem is well known to guidance system designers, who are often called upon to invent ingenious approximations to the "optimal filter" equation. The sequential correlation model offers a good starting point for finding data-weighting techniques which will cause the orbit to be determined within mission tolerance.

The ultimate test of any guidance scheme is a complete launch-to-impact mission simulation, with Monte Carlo selection of all random disturbances that affect the measurements and trajectory coordinates. Such a program is under construction at the Jet Propulsion Laboratory and has been used to evaluate some of the techniques discussed in this paper.

References

[1]Noton, A. R. M., E. Cutting, and F. Barnes, Analysis of Radio-Command Midcourse Guidance, Technical Memorandum 32-38, Jet Propulsion Laboratory, Pasadena (June 1959).

[2]Clarke, V. C., Jr., "Design of Lunar and Interplanetary Ascent Trajectories," Technical Report 32-30, Rev. 1, Jet Propulsion Laboratory, Pasadena (March 15, 1962).

[3]Bliss, G. A., Mathematics for Exterior Ballistics (John Wiley and Sons, New York, 1953).

[4]Gordon, H. J., "Nonlinear Analysis of Midcourse Maneuver Requirements," JPL Space Programs Summary No. 37-16, Jet Propulsion Laboratory (August 1962).

[5]Pfeiffer, C. G., "Sequential Estimation of Correlated Stochastic Variables," Technical Report 32-445, Jet Propulsion Laboratory, Pasadena (June 1963).

[6]Hamilton, T. W., L. Sjogren, W. Kirhofer, J. Fearey, and D. L. Cain, The Ranger-4 Flight Path and Its Determination From Tracking Data," Technical Report 32-345, Jet Propulsion Laboratory, Pasadena (September 15, 1962).

[7]Kalman, R. E., "A New Approach to Linear Filtering and Prediction Problems," ASME J. of Basic Engineering (March 1960).

[8]Pfeiffer, C. G., "Continuous Estimation of Sequentially Correlated Random Variables," Technical Report 32-524, Jet Propulsion Laboratory (October 1963).

[9]Gordon, H. J., "Study of an Approach Guidance System," Space Programs Summary No. 37-19, Vol. IV, Jet Propulsion Laboratory, Pasadena (February 28, 1963).

[10]Soong, T. T., C. G. Pfeiffer, R. Hamburg, "Unified Guidance Analysis in Design of Space Trajectories," Technical Report 32-577, Jet Propulsion Laboratory (January 1964).

[11]Battin, R. H., "A Statistical Optimizing Navigation Procedure for Space Flight," ARS J. (November 1960).

[12]Breakwell, J., "The Optimum Spacing of Corrective Thrusts in Interplanetary Navigation," Mathematics in Science and Engineering, Vol. 5 (Academic Press, New York, 1962).

[13]Pfeiffer, C. G., "A Dynamic Programming Analysis of Multiple Guidance Corrections of a Trajectory," Technical Report 32-513, Revision 1, Jet Propulsion Laboratory (to be published).

[14]Curkendall, D. W. and C. G. Pfeiffer, "A Discussion of Guidance Policies for Multiple-Impulse Correction of the Trajectory of a Spacecraft," AIAA Progress in Astronautics and Aeronautics: Guidance and Control - 11, edited by M. Summerfield (Academic Press, New York, 1964), Vol. 13, pp. 667-687.

[15]Pfeiffer, C. G., "On Neighboring Optimal Final Value Control," Jet Propulsion Laboratory Technical Report (to be published).

[16]Breakwell, J. V., J. L. Speyer, and A. E. Bryson, "Optimization and Control of Nonlinear Systems Using the Second Variation," SIAM J. on Control, Series A, Vol. 1, No. 2, Society for Industrial and Applied Mathematics, Philadelphia (1963).

[17]Kelley, H. J., "Guidance Theory and Extremal Fields," Trans. IRE Prof. Group Automatic Control (1962).

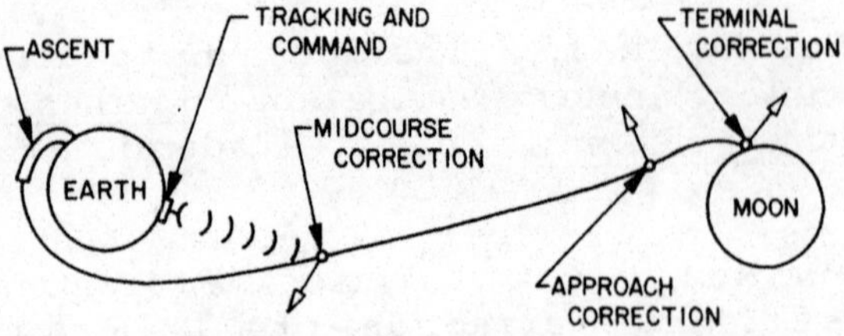

Fig. 1 Cartoon view of postinjection guidance.

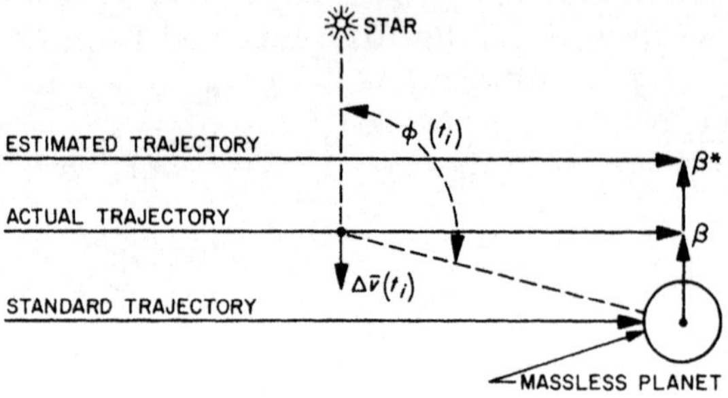

Fig. 2 Idealized model of approach guidance optical measurements.

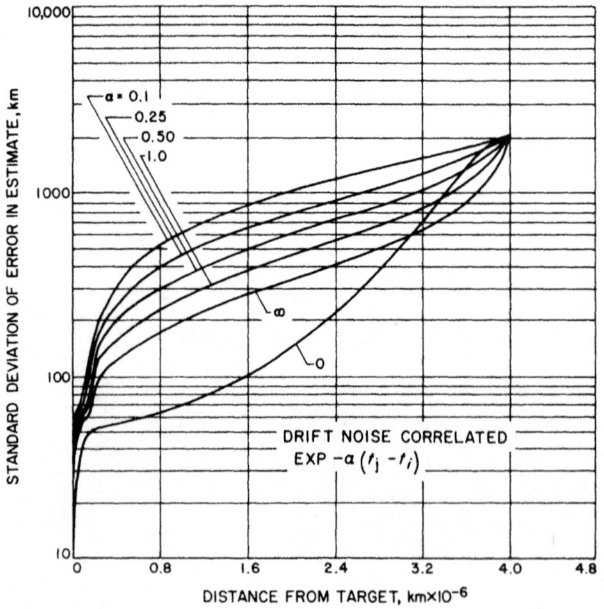

Fig. 3 Effect of drift noise on target error uncertainty.

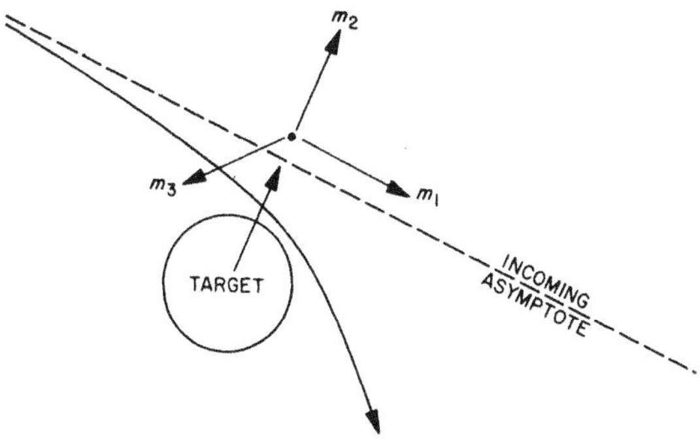

Fig. 4 The impact parameter coordinate system.

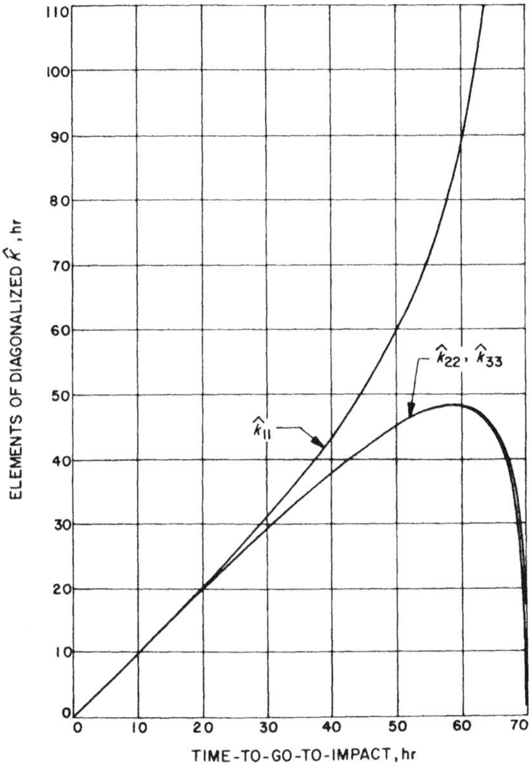

Fig. 5 Velocity differential corrections in proper coordinate system for a 66-hr lunar trajectory.

AN APPROXIMATE METHOD FOR COMPUTING ERROR COEFFICIENT MATRICES FOR LUNAR TRAJECTORIES

John D. McLean*

NASA-Ames Research Center, Moffett Field, Calif.

Abstract

This paper is concerned with the problems involved in the computation of error coefficient matrices for use in midcourse navigation for lunar missions. An investigation is made of the possibility of computing the matrices in closed form using patched conic approximations to the reference trajectory. It is shown that there is an inherent limit to the accuracy which can be obtained by the patched conic approach and that this accuracy is considered insufficient for midcourse guidance. A numerical method is presented for improving the accuracy to an acceptable level, and the velocity penalties associated with using the improved and unimproved matrices for midcourse guidance are compared. A qualitative evaluation of the computer time and storage requirements for the approximate method is also given.

Introduction

Studies conducted at Ames[1,2] on midcourse navigation for a circumlunar mission used linear matrix operations for trajectory determination and guidance. These operations require the knowledge of "error coefficient" or "transition" matrices, which are the matrices of first partial derivatives of the variational parameters at one time on a given reference trajectory with respect to those at an earlier time. The variational parameters may be any set of variables that completely defines deviations from the reference position and velocity.

This paper is concerned with the problems involved in the computation of the transition matrices. The results presented are from a theoretical study conducted in an attempt to produce a simpler, though perhaps less accurate, method of

Presented as Preprint 63-400 at the AIAA Astrodynamics Conference, New Haven, Conn., August 19-21, 1963.

*Research Scientist.

computation than that used previously. This alternate method is based on the patched conic approach. Although the results are inconclusive in many respects, it appears that the method can be used to advantage in some cases.

The work can be divided into four parts as follows: First, it is shown that computation of the matrices in closed form using patched conic approximations is inherently limited in accuracy, and that this accuracy is considered inadequate for lunar missions. Second, a numerical method is presented for improving the accuracy to an acceptable level. Third, either the matrices from the strictly two-body computations or the improved ones can be used for midcourse guidance at the expense of additional corrective velocity. The magnitudes of the velocity penalties for the improved and unimproved matrices are compared. Fourth, a qualitative evaluation of the computer time and storage requirements for the approximate method is presented.

Theory

At this point, a brief outline of the method used for computing the transition matrices in the midcourse guidance studies mentioned earlier will help to illustrate the object of the present study. The computation was carried out using the components of position and velocity as variational parameters; however, the same approach could conceivably be used with other variational parameters. The procedure is as follows: A set of linear differential perturbation equations is obtained by expanding the vehicle's equations of motion about the reference trajectory. The coefficients in the equations are prescribed functions of position on the reference trajectory, hence, of time. The simultaneous integration of six sets of these equations with appropriate initial conditions will yield the desired matrices. The accuracy of this calculation is limited only by the accuracies of the mathematical model and of the numerical operations, but the integration is quite time consuming. In addition, if a high precision numerical integration is used, a large amount of computer storage may be required.

Use of Two-Body Approximation

One approach to the problem of simplifying the computation of the matrices is the use of two-body approximations. That is, if the trajectory is assumed to be a conic over a given time interval, then the state vectors[+] at the initial and final times can be related to each other in closed form. Hence, the transition matrix over this interval can also be found in closed form. If the total trajectory is divided into n of these intervals, then the transition matrix relating perturbations from the reference trajectory at the final time to those at the initial time can be approximated by multiplying the n individual, two-body matrices together.

As n is increased, that is, as the size of the intervals is reduced, the individual conics become better approximations to the actual trajectory. Therefore, if no accuracy is lost in the numerical operations, the matrices are obtained with increasingly greater accuracy.

There is, however, a limit to the accuracy which can be obtained by the reduction of the time intervals over which the reference trajectory is approximated by individual conics. To illustrate this point, consider the linear perturbation equations of motion

$$\dot{\bar{x}} = F\bar{x} \tag{1}$$

where $\bar{x}$ is the vector of perturbations from the reference trajectory and F is a matrix of coefficients that are functions of position on the reference trajectory. It can be shown that the transition matrix Φ also satisfies Eq. (1); that is,

$$\dot{\Phi} = F\Phi \tag{2}$$

The matrix φ is defined as the transition matrix that would be obtained if only a single homogeneous spherical central body were present and F_c, as the value that the matrix F would have in such a case. The matrices ψ and F_p can then be defined by the following equations:

$$\left.\begin{aligned} \Phi &= \varphi + \psi \\ F &= F_c + F_p \end{aligned}\right\} \tag{3}$$

Equation (2) can now be written as

$$\dot{\varphi} + \dot{\psi} = F_c\varphi + F_p\varphi + F_c\psi + F_p\psi \tag{4}$$

but

$$\dot{\varphi} = F_c\varphi \tag{5}$$

so

$$\dot{\psi} = F_p\varphi + F_c\psi + F_p\psi \tag{6}$$

[+]For the present, "state vector" will be defined as any set of six variables from which the vehicle's instantaneous position and velocity can be determined.

Now consider the case illustrated in Fig. 1. The reference trajectory is being approximated by conics over a series of short time intervals for the purpose of computing the transition matrices. Here, t_E is the terminal reference time and t_o is the time of injection. The transition matrix $\Phi(t_n; t_{n-1})$ from t_{n-1} to t_n is written, for convenience, as Φ_n and similarly for φ and ψ.

From Eq. (6),

$$\psi_n = \int_{t_{n-1}}^{t_n} (F_p\varphi_n + F_c\psi_n + F_p\psi_n)dt \tag{7}$$

and ψ_n is the error in approximating the transition matrix Φ_n across the nth interval with the matrix φ_n computed from the two-body equations.

For use later in the paper, it is desired to know the limit approached by ψ_n as n becomes very large, that is, as the time interval approaches zero. The F matrices on any practical trajectory will be nonnull and finite if the proper variables are chosen to describe the trajectory. Therefore, since at time t_{n-1}, φ_n is the unit matrix and ψ_n is the null matrix, the last two terms in the integrand become negligibly small with respect to the first as the time interval is made very small. Thus, according to the classic definition of the integral,

$$\lim_{\Delta t \to o} \psi_n = F_p(t_n)\Delta t \tag{8}$$

where $\Delta t = t_n - t_{n-1}$.

The transition matrix Φ_E, relating deviations in the state vector at time t_E to those at t_o, can be written in the following form:

$$\Phi_E = (\varphi_n + \psi_n)(\varphi_{n-1} + \psi_{n-1}) \ . \ . \ . \ (\varphi_2 + \psi_2)(\varphi_1 + \psi_1) \tag{9}$$

In order to find the relationship between Φ_E and its two-body approximation, it is necessary to expand Eq. (9). Setting n equal to 4 is sufficient to show the form of the expansion. In this case,

$$\begin{aligned}\Phi_E = {} & \varphi_4\varphi_3\varphi_2\varphi_1 + (\varphi_4 + \psi_4)(\varphi_3 + \psi_3)(\varphi_2 + \psi_2)\psi_1 + \\ & (\varphi_4 + \psi_4)(\varphi_3 + \psi_3)\psi_2\varphi_1 + (\varphi_4 + \psi_4)\psi_3\varphi_2\varphi_1 + \psi_4\varphi_3\varphi_2\varphi_1 = \\ & \varphi_4\varphi_3\varphi_2\varphi_1 + \Phi_4\Phi_3\Phi_2\psi_1 + \Phi_4\Phi_3\psi_2\varphi_1 + \\ & \Phi_4\psi_3\varphi_2\varphi_1 + \psi_4\varphi_3\varphi_2\varphi_1\end{aligned} \tag{10}$$

Equation (10) can be generalized to n terms to give

$$\Phi(t_E;t_o) = \varphi(t_E;t_o) + \sum_{i=1}^{n} \Phi(t_E;t_i)\psi_i\varphi(t_{i-1};t_o) \tag{11}$$

The second term on the right represents the error $\psi(t_E;t_o)$ in approximating the transition matrices over the short intervals with two-body equations.

As n is allowed to become very large, Eq. (11) becomes

$$\Phi(t_E;t_o) = \varphi(t_E;t_o) + \lim_{n\to\infty} \sum_{i=1}^{n} \Phi(t_E;t_i)\psi_i\varphi(t_{i-1};t_o) \tag{12}$$

but from Eq. (8),

$$\psi_i = F_p(t_i)\Delta t$$

so that

$$\Phi(t_E;t_o) = \varphi(t_E;t_o) + \lim_{n\to\infty} \sum_{i=1}^{n} \Phi(t_E;t_i)F_p(t_i)\Delta t\varphi(t_{i-1};t_o)$$

or

$$\Phi(t_E;t_o) = \varphi(t_E;t_o) + \int_{t_o}^{t_E} \Phi(t_E;t)F_p(t)\varphi(t;t_o)dt \tag{13}$$

Thus, in the limit, the error ψ_E, in approximating the transition matrix, approaches the integral on the right-hand side of Eq. (13). In order for this integral to yield the null matrix for all values of t_o the integrand must be identically the null matrix. Since the transition matrices are Jacobians of the variational parameters at one time, with respect to those at a previous time, they must be nonsingular. Some of the terms of the matrix F_p are the second partial derivatives of the perturbing potential function and, because of the form of that function, must be nonzero. For these reasons, the integrand in Eq. (13) cannot be identically zero and ψ_E must, in general, have nonzero terms. In other words, the integral in Eq. (13) represents a minimum error in approximating the transition matrices by closed-form computation.

It is of interest to know what effect the reduction of the time intervals has on the matrices from the closed-form computation. From Eq. (5) the closed-form matrix over the nth interval could be computed from

$$\varphi_n = \int_{t_{n-1}}^{t_n} F_c\varphi_n$$

where the terms of the matrix F_c are calculated as functions of the state vector on the approximating conic between t_{n-1} and t_n. As the time interval is made arbitrarily small, φ_n

approaches its initial value, the unit matrix, and F_c approaches the value which would be calculated using the state vector on the reference trajectory at time t_n. From the classic definition of the integral

$$\lim_{n\to\infty} \varphi_n = I + F_c\Delta t \tag{14}$$

where F_c is evaluated on the reference trajectory. If φ_E is written as the product to n matrices of the form in Eq. (14), then it can be expanded to give

$$\varphi_E = I + F_c(t_1)\Delta t + F_c(t_2)\varphi(t_1;t_o)\Delta t + F_c(t_3)\varphi(t_2;t_o)\Delta t + \ldots F_c(t_n)\varphi(t_{n-1};t_o)\Delta t$$

Hence,

$$\lim_{n\to\infty} \varphi_E = \int_{t_o}^{t_E} F_c(t)\varphi(t;t_o)dt \tag{15}$$

where $\varphi(t_o) = I$ and F_c is evaluated on the reference trajectory. The integral in Eq. (15) represents the ultimate accuracy available, except in isolated fortuitous cases, from using closed-form equations to approximate the matrices. Therefore, this integral is useful as a standard of comparison for the closed-form matrices computed using finite time intervals.

Computation of Matrices in Closed Form

For convenience, the two-body matrices were computed in closed form with the components of the Cartesian position and velocity used both as the variational parameters‡ and as the components of the state vector. The procedure is one of linearization, as in the case of the differential equations, which uses a Taylor series expansion in which all derivatives of higher than first order are neglected.

‡It is wise to point out that Pines[3,4] has shown that transition matrices computed over long time arcs become ill-conditioned as a result of secular terms and that this difficulty can be reduced greatly if variational parameters different from the ones used here are chosen. No difficulty because of this phenomenon has been encountered in the Ames lunar studies, presumably because the time arcs involved have never exceeded about half an orbital period. Such ill conditioning need not affect the comparison of results from an approximate method and a more accurate model, but it should be considered if the method is to be incorporated into a navigation system.

The equations relating position and velocity at two points are written according to the method of Laplace[5] and differentiated. The closed-form equations are

$$\left.\begin{aligned} R &= f\bar{R}_o + g\bar{V}_o \\ V &= \dot{f}\bar{R}_o + \dot{g}\bar{V}_o \end{aligned}\right\} \tag{16}$$

where $\bar{R}_o$ and $\bar{V}_o$ are the initial position and velocity vectors, $\bar{R}$ and $\bar{V}$ are their final values, while f, g, $\dot{f}$, and $\dot{g}$ are scalar functions of the initial position and velocity and of the time interval being used.

The first partial derivatives of the above equation are

$$\left.\begin{aligned} \partial\bar{R}/\partial\bar{R}_o &= fI + \bar{R}_o(\nabla_R f)^T + \bar{V}_o(\nabla_R g)^T \\ \partial\bar{R}/\partial\bar{V}_o &= gI + \bar{R}_o(\nabla_V f)^T + \bar{V}_o(\nabla_V g)^T \\ \partial\bar{V}/\partial\bar{R}_o &= \dot{f}I + \bar{R}_o(\nabla_R \dot{f})^T + \bar{V}_o(\nabla_R \dot{g})^T \\ \partial\bar{V}/\partial\bar{V}_o &= \dot{g}I + \bar{R}_o(\nabla_V \dot{f})^T + \bar{V}_o(\nabla_V \dot{g})^T \end{aligned}\right\} \tag{17}$$

The vector partial derivatives in Eqs. (17) are of the form

$$\frac{\partial\bar{R}}{\partial\bar{R}_o} = \begin{bmatrix} \dfrac{\partial x}{\partial x_o} & \dfrac{\partial x}{\partial y_o} & \dfrac{\partial x}{\partial z_o} \\ \dfrac{\partial y}{\partial x_o} & \cdots\cdots & \\ \dfrac{\partial y}{\partial z_o} & \cdots\cdots & \end{bmatrix}$$

The notation ∇_R and ∇_V indicates the gradient with the components of position and velocity, respectively, used as independent variables, and I is the 3×3 unit matrix. The vector partial derivatives are submatrices of the two-body transition matrix that can be written

$$\varphi = \begin{bmatrix} \dfrac{\partial\bar{R}}{\partial\bar{R}_o} & \dfrac{\partial\bar{R}}{\partial\bar{V}_o} \\ \dfrac{\partial\bar{V}}{\partial\bar{R}_o} & \dfrac{\partial\bar{V}}{\partial\bar{V}_o} \end{bmatrix} \tag{18}$$

Improvement of Two-Body Matrices

It will be shown later that, with finite time intervals, the error in the matrices was considered unacceptable even though they compared favorably with the results of Eq. (15). This inaccuracy was the reason for performing the second part of the work, that is, finding a means for improving the approximation. The approach to this problem was to perform an approximate integration of Eq. (7).

If Cartesian coordinates are used, the matrices in Eq. (7) can be partitioned in the following form:

$$\left.\begin{aligned} F_c &= \begin{bmatrix} 0 & I \\ \overline{F}_c & 0 \end{bmatrix} \\ F_p &= \begin{bmatrix} 0 & 0 \\ \overline{F}_p & 0 \end{bmatrix} \\ \varphi &= \begin{bmatrix} \varphi_1 & \varphi_2 \\ \varphi_3 & \varphi_4 \end{bmatrix} = \begin{bmatrix} \dfrac{\partial \overline{R}}{\partial \overline{R}_o} & \dfrac{\partial \overline{R}}{\partial \overline{V}_o} \\ \dfrac{\partial \overline{V}}{\partial \overline{R}_o} & \dfrac{\partial \overline{V}}{\partial \overline{V}_o} \end{bmatrix} \end{aligned}\right\} \tag{19}$$

In this notation the zeros are 3×3 null matrices, and $\overline{F}_c$ and $\overline{F}_p$ are 3×3 matrices of partial derivatives of accelerations due to the spherical central body and of the perturbing accelerations, respectively. The matrix ψ is partitioned in a form corresponding to that of φ.

Substituting Eqs. (19) in Eq. (6) gives

$$\begin{bmatrix} \dot{\psi}_1 & \dot{\psi}_2 \\ \dot{\psi}_3 & \dot{\psi}_4 \end{bmatrix} = \begin{bmatrix} 0 & 0 \\ \overline{F}_p & 0 \end{bmatrix} \begin{bmatrix} \varphi_1 & \varphi_2 \\ \varphi_3 & \varphi_4 \end{bmatrix} + \begin{bmatrix} 0 & I \\ \overline{F}_c & 0 \end{bmatrix} \begin{bmatrix} \psi_1 & \psi_2 \\ \psi_3 & \psi_4 \end{bmatrix} +$$

$$\begin{bmatrix} 0 & 0 \\ \overline{F}_p & 0 \end{bmatrix} \begin{bmatrix} \psi_1 & \psi_2 \\ \psi_3 & \psi_4 \end{bmatrix} \tag{20}$$

Expanding Eq. (20) results in

$$\left.\begin{aligned} \dot{\psi}_1 &= \psi_3 \\ \dot{\psi}_2 &= \psi_4 \\ \dot{\psi}_3 &= \overline{F}_p\varphi_1 + \overline{F}_c\psi_1 + \overline{F}_p\psi_1 \\ \dot{\psi}_4 &= \overline{F}_p\varphi_2 + \overline{F}_c\psi_2 + \overline{F}_p\psi_2 \end{aligned}\right\} \tag{21}$$

so that

$$\left.\begin{aligned} \ddot{\psi}_1 &= \overline{F}_p\varphi_1 + \overline{F}_c\dot{\psi}_1 + \overline{F}_c\psi_1 \\ \ddot{\psi}_2 &= \overline{F}_p\varphi_2 + \overline{F}_c\dot{\psi}_2 + \overline{F}_c\psi_2 \end{aligned}\right\} \tag{22}$$

If ψ_1 and ψ_2 are neglected,

$$\left.\begin{aligned} \ddot{\psi}_1 &\approx F_p\varphi_1 \\ \ddot{\psi}_2 &\approx F_p\varphi_2 \end{aligned}\right\} \tag{23}$$

It is assumed that the accelerations in Eqs. (23) are constant at their average value over the nth interval. Since φ_1 at the beginning of the interval is the unit matrix and φ_2 is the null matrix,

$$\left.\begin{aligned} \psi_1(t_n;t_{n-1}) &= \overline{F}_p(t_{n-1}) + \overline{F}_p(t_n)\varphi_1(t_n)\ \frac{(t_n - t_{n-1})^2}{4} \\ \psi_2(t_n;t_{n-1}) &= \overline{F}_p(t_n)\varphi_2(t_n)\ \frac{(t_n - t_{n-1})^2}{4} \\ \psi_3(t_n;t_{n-1}) &= \overline{F}_p(t_{n-1}) + F_p(t_n)\varphi_1(t_n)\ \frac{(t_n - t_{n-1})}{2} \\ \psi_4(t_n;t_{n-1}) &= \overline{F}_p(t_n)\varphi_2(t_n)\ \frac{(t_n - t_{n-1})}{2} \end{aligned}\right\} \tag{24}$$

Equations (24) represent, in partitioned form, the results of an approximate integration of Eq. (7). The perturbation matrices thus computed can be added to the two-body matrices for corresponding intervals in order to improve the approximation.

Evaluation of Accuracy

The significance of the inherent inaccuracy in the closed-form matrices and the effectiveness of the improvement have been judged on the basis of a numerical example. Likewise, a comparison of the velocity penalties discussed earlier, resulting from use of the improved and unimproved matrices, was carried out for the same numerical example. This section of the paper describes the method of evaluating the errors in prediction and guidance, the error criteria used, and the method of computing the velocity penalty.

Method of evaluation: The analysis of the accuracy of the approximate method is carried out by comparison of the approximate matrices with those of a "correct" solution. The correct solution in this study is assumed to be obtained by the integration of the perturbation differential equations from a mathematical model considered sufficiently accurate for computing the reference trajectory. For determination of the errors in prediction, an error matrix E is found by subtracting the approximate matrix from the correct one. This error matrix is partitioned in the form of Eq. (18).

$$E = \begin{bmatrix} E_1 & E_2 \\ E_3 & E_4 \end{bmatrix} = \begin{bmatrix} \Delta \dfrac{\partial \overline{R}}{\partial \overline{R}_o} & \Delta \dfrac{\partial \overline{R}}{\partial \overline{V}_o} \\ \Delta \dfrac{\partial \overline{V}}{\partial \overline{V}_o} & \Delta \dfrac{\partial \overline{V}}{\partial \overline{V}_o} \end{bmatrix} \tag{25}$$

The matrix on the extreme right indicates which of the E_i corresponds to the error in partial derivatives of position with respect to initial position, position with respect to initial velocity, etc. For example, if an initial position deviation from the reference were multiplied by E_1 the result would be the vector error in predicting the position deviation from the reference at the final time. The E_i will be referred to in the remainder of the paper as the prediction error matrices.

It is also desired to know the error arising from using the approximate matrices for guidance. For this purpose the fixed time of arrival guidance law from previous Ames studies[1,2] will be used. The transition matrix Φ from the time of a velocity correction to the terminal point is partitioned in the form described previously

$$\Phi = \begin{bmatrix} \Phi_1 & \Phi_2 \\ \Phi_3 & \Phi_4 \end{bmatrix}$$

Then the velocity vector $\delta\overline{V}_G$ to be gained is

$$\delta\overline{V}_G = -\Phi_2^{-1}\Phi_1\delta\overline{R} - \delta\overline{V} \tag{26}$$

where $\delta\overline{R}$ and $\delta\overline{V}$ are the position and velocity deviations from the reference trajectory at the time of the correction. If Φ, $\delta\overline{R}$, and $\delta\overline{V}$ are known exactly, the position deviation $\delta\overline{R}_E$ at the end point after the application of the velocity correction will be

$$\delta\overline{R}_E = \Phi_1\delta\overline{R} + \varphi_2\delta\overline{V} - \Phi_2(\Phi_2^{-1}\Phi_1\delta\overline{R} + \delta\overline{V}) = 0 \tag{27}$$

Using the approximate matrices φ_1 and φ_2 in computing the velocity correction gives

$$\delta\bar{R}_E = (\Phi_1 - \Phi_2\varphi_2^{-1}\varphi_1)\delta\bar{R} = B\delta\bar{R} \tag{28}$$

The matrix B will be referred to as the guidance error matrix.

The results to be presented later show that as the terminal point is approached, the guidance errors resulting from use of the approximate matrices become negligibly small. Therefore, a guidance error resulting from an earlier correction can be eliminated at the expense of a subsequent correction nearer the terminal point. It is assumed that this second correction will be made at some time t_2 when the residual error will be negligible. The deviations from the reference at this time will be given by

$$\begin{bmatrix} \delta\bar{R}_2 \\ \delta\bar{V}_2 \end{bmatrix} = \Phi^{-1}(t_E;t_2) \begin{bmatrix} \delta\bar{R}_E \\ \delta\bar{V}_E \end{bmatrix} \tag{29}$$

The terminal position deviation $\delta\bar{R}_E$ is given by Eq. (28). Similarly, the velocity deviation $\delta\bar{V}_E$ can be shown to be

$$\delta\bar{V}_E = (\Phi_3 - \Phi_4\varphi_2^{-1}\varphi_1)\delta\bar{R} = C\delta\bar{R} \tag{30}$$

where $\delta\bar{R}$ and the matrix C are evaluated at the time of the first correction.

It is shown in Ref. 2 that if Φ is partitioned as before, then

$$\Phi^{-1} = \begin{vmatrix} \Phi_4^T & -\Phi_2^T \\ -\Phi_3^T & \Phi_1^T \end{vmatrix} \tag{31}$$

Therefore, with $\delta\bar{R}_E$ and $\delta\bar{V}_E$ substituted from Eqs. (28) and (30), and the subscript 1 added to indicate the time of the first velocity correction,

$$\begin{matrix} \delta\bar{R}_2 \\ \delta\bar{V}_2 \end{matrix} = \begin{bmatrix} \Phi_4^T & -\Phi_2^T \\ -\Phi_3^T & \Phi_1^T \end{bmatrix} \begin{bmatrix} B_1\delta\bar{R}_1 \\ C_1\delta\bar{R}_1 \end{bmatrix} \tag{32}$$

The second velocity correction is given by

$$\delta\bar{V}_{G_2} = -\varphi_2^{-1}\varphi_1\delta\bar{R}_2 - \delta\bar{V}_2 \tag{33}$$

where it is understood that the Φ_i and φ_i in Eqs. (32) and (33) are evaluated at the time of the second correction. Values of $\delta\bar{R}_2$ and $\delta\bar{V}_2$ for substitution into Eq. (33) are obtained from Eq. (32).

$$\delta\overline{V}_{G_2} = (\Phi_3^T - \varphi_2^{-1}\varphi_1\Phi_4^T)B_1\delta\overline{R}_1 + (\varphi_2^{-1}\varphi_1\Phi_2^T - \Phi_1^T)C_1\delta\overline{R}_1 \qquad (34)$$

If it is assumed that the second correction is made at a range where the differences between the correct and approximate matrices are negligible, then Eq. (34) becomes

$$\delta\overline{V}_{G_2} \approx (\Phi_3^T - \Phi_2^{-1}\Phi_1\Phi_4^T)B_1\delta\overline{R}_1 + (\Phi_2^{-1}\Phi_1\Phi_2^T - \Phi_1^T)C_1\delta\overline{R}_1 \qquad (35)$$

From Eq. (31) it can be shown that

$$\Phi_2^{-1}\Phi_1 = \Phi_1^T(\Phi_2^{-1})^T$$

$$\Phi_1\Phi_4^T = I + \Phi_2\Phi_3^T$$

so that

$$\delta\overline{V}_{G_2} \approx -\Phi_2^{-1}B_1\delta\overline{R}_1 = D\delta\overline{R}_1 \qquad (36)$$

The matrix D will be referred to as the velocity penalty matrix.

It is difficult to assess the velocity penalty and the errors in prediction and guidance by considering the individual terms in the matrix, and a simpler criterion is needed. One possibility would be to multiply each matrix by its transpose and compute the eigenvalues of the resulting matrix. The square root of the maximum eigenvalue in each case would then represent the maximum velocity penalty, prediction error, or guidance error resulting from using the approximate matrices with a unit deviation from the reference trajectory. Instead of this maximum value, the error was represented by the norm of the appropriate matrix because of the computational simplicity. Since the norm is equal to the square root of the sum of the eigenvalues, the resulting information is essentially the same.

An idea of the error in prediction relative to the total predicted error can be obtained by expressing the norm of the error matrices as percentages of the norms of the corresponding correct submatrices. The guidance error, because of the use of the approximate matrices, is a function only of the initial position deviation. Therefore, a similar percentage error in guidance can be found from the norm of the guidance error matrix expressed as a percentage of the norm of the correct submatrix of partials of position with respect to position. The velocity penalty is also a function only of the initial position deviation, so that computation of a relative velocity correction penalty should involve only that portion of the initial correction arising from position deviations from the reference. The criterion used for this comparison is the ratio of the norm of the velocity penalty matrix to that of the matrix $\varphi_2^{-1}\varphi_1$, computed at the time of the initial correction.

Digital computer study: The accuracy was evaluated by means of a digital computer study of a sample moon-to-earth trajectory. The computational methods are described here briefly. The trajectory has a flight time of about 3-1/4 days from perilune to return perigee and lies approximately in the moon's orbital plane. The flight time is fairly long and therefore allows the perturbing forces not accounted for in the two-body equations to have a relatively large effect. For comparison purposes, the transition matrices obtained from integrating the differential perturbation equations from a four-body model are taken as correct. This model includes the second harmonic term of the earth's gravitational potential as well as the vehicle, sun, earth, and moon.

The two-body transition matrices were computed, as described in the section on theory, both by integration and in closed form. It was found better in this computation to take a larger sphere of influence for the moon than the generally accepted value given by the 2/5 power of the mass ratios. The reason for this can be seen if the perturbation differential equations instead of the vehicle's equations of motion are used to compute the moon's sphere of influence. As a result of differentiation, the term representing the action of the central body on the perturbing body disappears and the perturbing actions are equal where the accelerations due to the two bodies are equal. In the one-dimensional case, this equality occurs when the ratio of the ranges to the two bodies is equal to the cube root of the mass ratios.

The integration for both the reference trajectory and the perturbation equations uses a Cowell "second-sum" method. A fourth-order Runge-Kutta method is used to start the integration and to change step size during the flight. The equations given in the section on theory were used to compute the conic approximations, and the time intervals over which the conic approximations were used were chosen to coincide with an integral number of intervals from the Cowell integration.

Results and Discussion

The data resulting from the digital computer study will now be presented. These results will, first, indicate why the unimproved two-body approximation is considered inadequate; second, show that the suggested improvement is adequate; third, compare the velocity penalties resulting from the use of improved and unimproved matrices.

Accuracy of Unimproved Matrices

The norms of the prediction error matrices resulting from integrating the two-body perturbation differential equations along the reference trajectory are presented in Fig. 2. The prediction error matrices from Eq. (25) are indicated for the appropriate curves. The prediction submatrix to which the error matrix corresponds is also indicated along with the units used. Note that the increase in prediction error, with range from the earth, is approximately exponential. The norms of the corresponding submatrices of the correct transition matrices increase in a similar fashion with increasing range. For this reason, the percentage errors in prediction exhibit a much less drastic increase.

These percentages for the data just presented are plotted in Fig. 3. The curves for prediction error matrices E_3 and E_4 have been omitted because they are nearly identical with those for matrices E_1 and E_2, respectively. Note that even though the error in prediction increases continually with range from the earth, the percentage error decreases near the moon. This decrease occurs, of course, because the norms of the correct prediction submatrices increase more rapidly in this region than those of the error matrices.

For the closed-form computation of the two-body matrices the time intervals used were sufficiently short to give results close to those given by the integration. A single time interval was satisfactory for ranges less than about 330,000 km, but for the portion of the trajectory between that range and perilune, it was found necessary to rectify more often. Intervals of about 0.1 day were found to give satisfactory results and were used over the entire trajectory.

The percentage errors in prediction for the closed-form computation are presented in Fig. 4. Comparison of these curves with those of Fig. 3 shows little difference in the accuracy of the two methods of computation except in the region of greatest error. The maximum percentage for E_1 from the closed-form computation is about 13.5% as compared to 12.5% for the integration.

The norms of the guidance error matrices for the two methods of computation are plotted together in Fig. 5. The correspondence between the two methods is quite good at ranges beyond 200,000 km. The pronounced differences between the two curves at the shorter ranges occur because the earth's oblateness causes larger relative differences between the reference trajectory and the conic approximations. The errors for both methods are quite small in this region and the difference between the two is not significant when the guidance error is considered on a percentage basis as in Fig. 6.

As in the case of the prediction error, the only noticeable difference between the percentage errors for the two methods of computation occurs near the range of 350,000 km. The data in Fig. 6 indicate that the two-body matrices could probably be used successfully for guidance at ranges of less than 250,000 km. The large errors at longer ranges indicate the need for a correction method such as was outlined earlier in the paper. The next data to be presented will show the improvement in accuracy resulting from the use of the correction method.

Accuracy of the Improved Matrices

The same 0.1 day time intervals used for the closed-form computation just discussed were used to compute the corrected matrices. It was found that including the earth's oblateness terms in the correction computation decreases rather than improves the accuracy, unless the time intervals are greatly reduced near the earth. For this reason the oblateness terms were omitted from this calculation, and the resulting percentage errors in prediction are presented in Fig. 7. In this case all the errors are less than 0.5%. The irregular form of the curves is to be expected because of the assumptions made in computing the corrections. These fluctuations, as well as the average value, could be reduced to the limits of computational accuracy by reduction of the time intervals and inclusion of the earth's oblateness.

The improvement in guidance accuracy resulting from the use of the corrected matrices is illustrated in Fig. 8. For comparison, the norms of the guidance error matrices from the corrected and uncorrected closed-form computations are both presented. Except at short ranges, where oblateness is a significant factor, the guidance error is reduced by at least an order of magnitude at every point. The percentage error in guidance was also calculated using the corrected matrices and was found to reach a maximum value of about 0.2% at a range of 150,000 km.

The Velocity Penalty

The final data to be presented compare the velocity penalties resulting from use of the improved and unimproved matrices. For the purpose of computing the velocity penalty that resulted from using the approximate matrices, it was assumed that the final velocity correction would be made at the shortest range for which data have been presented so far, about 45,000 km. It was further assumed, as discussed in the theory, that at this range the errors in the approximate matrices are negligible. The norms of the velocity penalty matrices computed at the longer ranges from the closed-form transition matrices, both corrected and uncorrected, are presented in Fig. 9. The additional velocity correction required is quite small in both cases for corrections made at ranges of less than 200,000 km. At longer ranges, however, the penalty for using the uncorrected matrices rises rapidly, while that for the corrected matrices remains quite small.

The ratio of the norm of the velocity penalty matrix to that associated with the initial correction is plotted in Fig. 10 for both cases discussed above. Elimination of the errors arising from the use of the approximate matrices at long ranges requires a velocity penalty of as much as seven times the initial correction. Of course, the use of additional corrections at intermediate ranges and the proper choice of a correction schedule would reduce the penalty from that shown here. On the other hand, it seems apparent from these results that the use of the uncorrected two-body matrices will result in a substantial velocity penalty. The corrected matrices result in a maximum velocity penalty of about 6% of the initial correction and this penalty also may be reduced by proper scheduling.

Qualitative Evaluation of Computer Requirements

Finally, it is desired to consider whether the use of the approximate method has any advantages in terms of required computer time and storage. The study presented here was carried out on an IBM 7090 computer and the IBM Fortran compiler. For this reason, any precise comparison of computer storage requirements is impractical, but some general conclusions can be drawn.

First, if a Cowell integration is used for the reference trajectory, it has been found in another Ames study that the integration of the perturbation equations using a fourth-order Runge-Kutta method requires less storage than is needed for the closed-form computation of the two-body matrices. No study has been conducted on the accuracy of this integration, but a few preliminary results indicate about the same accuracies as in the corrected matrices described here. This integration requires quite short step sizes, and for this reason it consumes about four times as much computer time as the approximate method. The need for such short steps indicates that it would be impractical to use this simple Cowell integration of the perturbation equations in conjunction with an Encke integration of the reference trajectory.

In the case of an Encke integration of the reference trajectory, it would be necessary to compute the two-body matrices so that only the additional computer capacity required to account for the perturbing accelerations needs to be considered. The Encke differential equations could be expanded in a Taylor series about the reference trajectory in order to get the perturbation differential equations. These equations could then be integrated to give the corrections to the two-body transition matrices. The correction scheme discussed previously represents a simplification of this procedure, but a comparison of the computer requirements could be arrived at only by programming the two methods. Use of the correction method as described here will require approximately double the storage needed for computing only the two-body matrices.

Conclusions

The conclusions which have been drawn from the results of the study are listed below:

1) The transition matrices computed in closed form using conic approximations to the reference trajectory can be used for midcourse guidance at the expense of a substantial penalty in corrective velocity. While this velocity penalty is large compared with the basic midcourse requirements, it might be acceptable in terms of the over-all requirements of the mission.

2) The accuracy of the two-body matrices can be improved to almost any desired level at the expense of increased computer capacity and time. The computation of these corrections approximately doubles the computer capacity required over that needed for computing the two-body matrices only.

3) If a Cowell integration is used for the reference trajectory, the approximate method appears to be inferior on the basis of computer storage requirements. The approximate method does show a substantial savings in time.

4) Since an Encke integration of the reference trajectory would require computation of the two-body matrices, only the additional computer capacity required for improving these

matrices needs to be considered. The approximate method presented for improving the matrices is actually an approximate method of integrating the perturbed Encke differential equations, and, therefore, it represents some saving in computer capacity and time. Considerably more study would be required to evaluate this savings.

References

[1]Smith, G.L., Schmidt, S.F., and McGee, L.A., "Application of statistical filter theory to the optimal estimation of position and velocity on board a circumlunar vehicle," NASA TR R-135 (1962).

[2]McLean, J.D., Schmidt, S.F., and McGee, L.A., "Optimal filtering and linear prediction applied to a midcourse navigation system for the circumlunar mission," NASA TN D-1208 (1962).

[3]Pines, S., Wolf, H., and Payne, M, "Asymptotically singular differential correction matrices," Republic Aviation Corp., Rept. RAC 708-453(328) (April 1961).

[4]Pines, S. and Sistino, A., "The use of variational parameters in multivariable linear prediction and filtering theory," Republic Aviation Corp., Rept. RAC 907(ARD-947-450) (September 1962).

[5]Herget, P. (Univ. of Cincinnati) "The computation of orbits," private publication (1948).

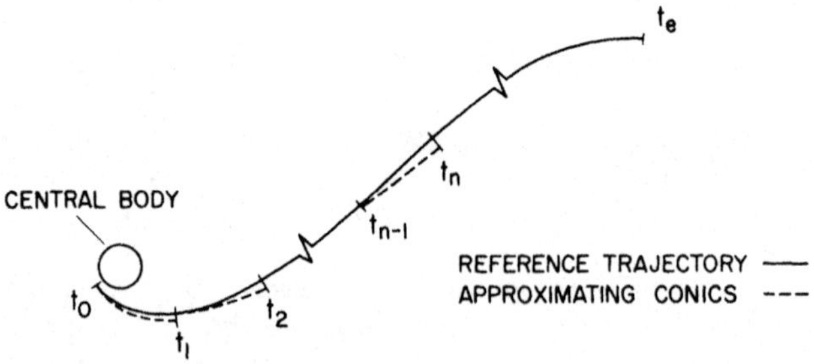

Fig. 1 Patched conic computation of matrices.

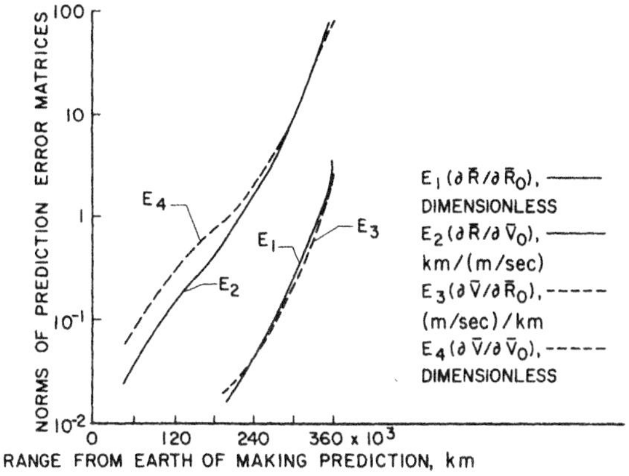

Fig. 2 Errors from integration method.

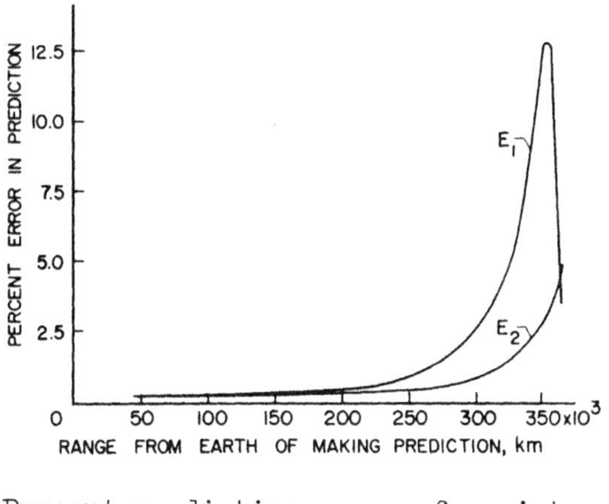

Fig. 3 Percent prediction errors from integration.

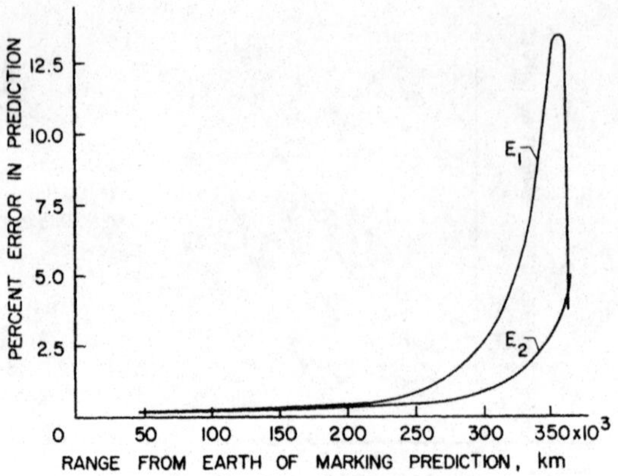

Fig. 4 Percent prediction errors from closed form.

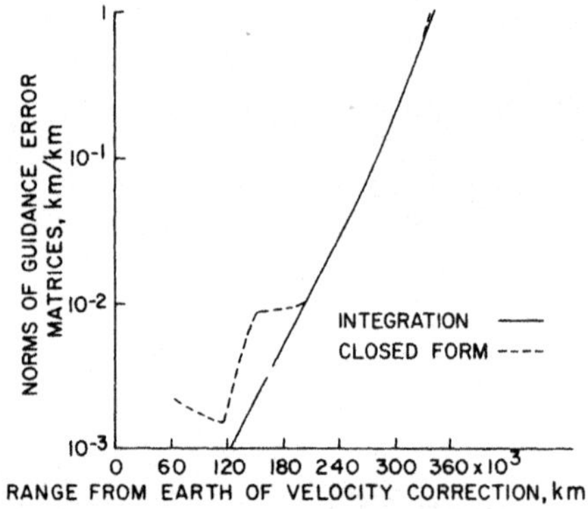

Fig. 5 Guidance errors from two-body matrices.

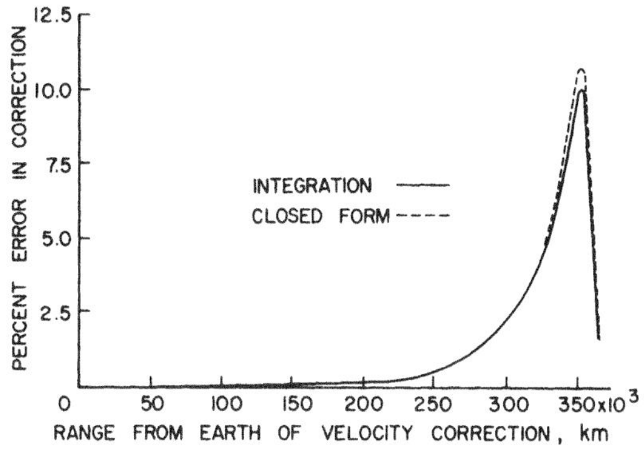

Fig. 6 Percent guidance errors from two-body matrices.

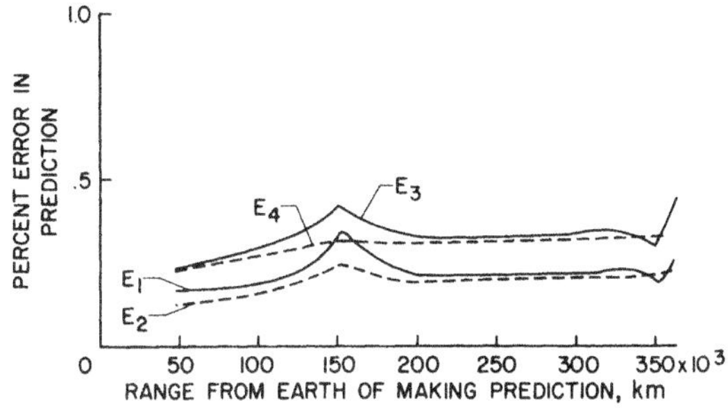

Fig. 7 Percent prediction errors from corrected matrices.

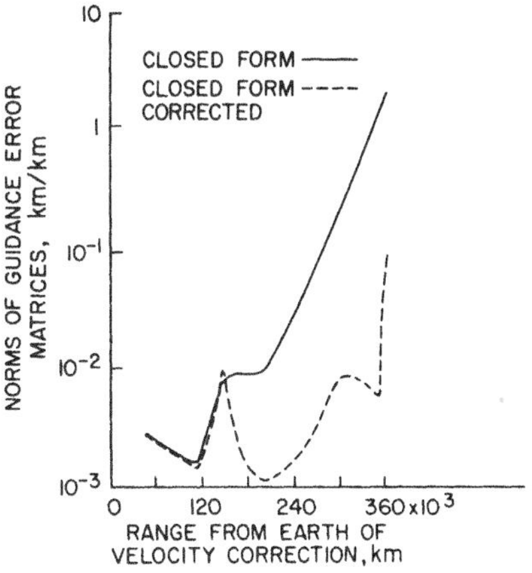

Fig. 8 Comparison of guidance errors.

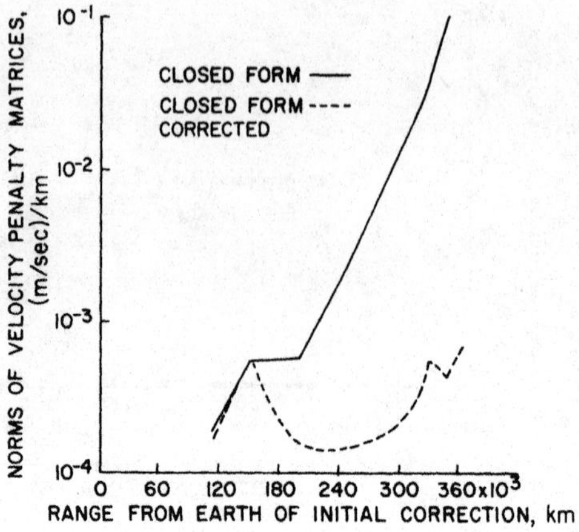

Fig. 9 Velocity penalty.

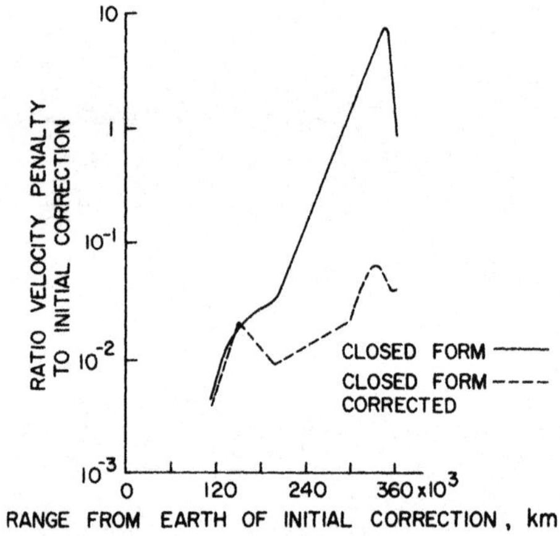

Fig. 10 Relative size of velocity penalty.

III.

ORBIT SELECTION

Orbits in the Earth-Moon Space

MINIMUM LUNAR ORBIT INCLINATION TO LUNAR EQUATORIAL PLANE FOR EARTH-LAUNCHED VEHICLE

Fred D. Breuer* and Walter C. Riddell*

General Dynamics/Astronautics, San Diego, Calif.

Abstract

The minimum inclination (without a plane change) of the lunar orbit plane of an earth-launched vehicle to the lunar-equatorial plane is analyzed. Determination of this minimum inclination identifies those lunar orbit inclinations that cannot be attained without a plane change in the immediate vicinity of the moon. Launch azimuth restrictions are examined to establish launch windows; launch azimuth limits of 72 deg to 114 deg were used. An investigation is included to show the launch window extensions that can be obtained by plane changes made in the earth parking orbit, while accelerating from parking orbit to injection velocity, or during the earth-moon transfer trajectory. The study results indicate, for the launch azimuth limits used, that the minimum lunar inclination varies between approximately plus or minus 15 deg, that there are 2 launch windows per 24-hr period of 4- to 6-hr duration each, and that launch window extension via plane change is very expensive in terms of the required plane change velocity increment.

Nomenclature

☾ = ecliptic longitude of target vector at lunar arrival, measured from equinox of date, see Fig. 5

A = right ascension of target vector at arrival

Presented as Preprint 63-403 at the AIAA Astrodynamics Conference, New Haven, Conn, August 19-21, 1963.

*Design Specialist.

A_C = right ascension of launch site at launch

A_G = right ascension of Greenwich at launch

A_{PC} = right ascension of point of trajectory plane change

A_S = right ascension of mean sun at beginning of launch day

A_S^* = ecliptic longitude of mean sun at beginning of launch day, measured from equinox of date

A_Z = launch azimuth measured east from north

A_Z^* = launch azimuth for case when trajectory plane change is to be made

b = angle subtended by target vector and ecliptic

d_L = number of days from the epoch to launch time

e = earth-moon transfer orbit eccentricity

h_L = number of hours from the beginning of the launch day to launch time, Greenwich Mean Time

i = inclination of the lunar plane to the ecliptic

i_{EL} = inclination of the lunar-equatorial plane to the lunar plane

$i_{PC_{inj}}$ = angle of trajectory plane change made while accelerating from parking orbit to injection velocity

i_{PC_P} = angle of trajectory plane change made in parking orbit

i_{PC_T} = angle of trajectory plane change made while in transfer trajectory

i_V = inclination of the vehicle trajectory plane to the earth-equatorial plane

i_{VM} = angle subtended by the V_∞ vector and the lunar-equatorial plane

i_1 = inclination of the vehicle trajectory plane to the lunar plane

I_{A_Z} = geocentric unit vector parallel to the launch azimuth vector

I_L = geocentric unit vector directed to the launch site at launch

$I_{L_{PC}}$ = geocentric unit position vector to point of trajectory plane change

I_M = geocentric unit vector normal to the lunar plane

I_T = geocentric unit target vector

I_V = geocentric unit vector normal to the vehicle trajectory plane

I^*_V = geocentric unit normal to arrival trajectory plane, i.e., after plane change

$I_{V_{PC}}$ = geocentric unit normal to launch trajectory plane, i.e., prior to plane change

I_Z = unit vector normal to lunar-equatorial plane

I_∞ = unit vector of V_∞

I_Ω = geocentric unit vector along the ascending node of the lunar plane on the ecliptic

K_1, K_2, K_3 = constant terms of ephemeris equations, see Eqs. (1-3)

ℓ = latitude of launch site

ℓ_{PC} = declination of point of plane change

R_{inj} = geocentric radius to transfer trajectory injection point

Δt = launch window extension time increment, see Fig. 11

t_T = cislunar transfer time, hr

V = velocity of the vehicle at the moon with respect to the earth

V_C = earth parking orbit circular velocity

V_{inj} = transfer trajectory injection velocity

V_M = velocity of the moon with respect to the earth

V_l = local vehicle velocity on the transfer trajectory

V_∞ = approach asymptotic velocity vector with respect to the moon

ΔV = velocity increment needed to execute trajectory plane change

ΔV_T = velocity increment needed to simultaneously accelerate from parking orbit velocity to injection velocity and execute a trajectory plane change

α = an arbitrary parameter used on Fig. 1

β = angle subtended by the target vector and the launch trajectory plane

δ = declination of target vector

ϵ = inclination of earth-equatorial plane to the ecliptic

η = true anomaly of the cislunar transfer trajectory at lunar arrival

η_1 = true anomaly of plane change point (for the case of plane change in the transfer trajectory)

γ = flight path angle of the V vector with respect to the local earth horizontal

γ_1 = local transfer trajectory flight path angle referred to the local earth horizontal

μ = earth gravitational field constant

Ω = angle in the ecliptic plane eastward from the vernal equinox to the ascending node of the lunar plane on the ecliptic, see Fig. 5

Ω_{EL} = angle in the lunar plane counter-clockwise from a target vector extension to the ascending node of the lunar-equatorial plane on the lunar plane, see Fig. 16

Ω_L = longitude eastward from Greenwich to the launch site

ϕ = geocentric angle subtended by launch position and target vectors

ϕ_{PC} = geocentric angle subtended by launch and plane change position vectors

ϕ_1 = geocentric angle subtended by the plane change position vector and the target vector

ψ = geocentric angle subtended by launch position vector and line of apsides of transfer trajectory

ρ_1 = earth central angle traversed by vehicle during boost from launch to earth parking orbit

ρ_2 = earth central angle traversed by vehicle during acceleration from parking orbit velocity to transfer trajectory injection velocity

θ = earth parking orbit coast angle

Subscripts

ec. = geocentric ecliptic coordinate system

eq. = geocentric earth-equatorial coordinate system

ℓ.eq. = selenocentric lunar-equatorial coordinate system, x axis along the ascending node of the lunar-equatorial plane on the lunar plane, and z axis coincident with the lunar North Pole

ℓ.p. = selenocentric lunar plane coordinate system, x axis outward from earth along the target vector and z axis northward

z = component of a vector

Introduction

In the planning of lunar missions, it is essential to know the orientation of the plane of the ballistic arrival trajectory with respect to the moon's surface. Certain missions require that a vehicle orbit be established in the equatorial plane of the moon. For this reason one of the parameters describing the orbit orientation, viz., the inclination of the arrival trajectory plane to the equatorial plane, is of prime importance, since it indicates the relative cost of maneuvering from the arrival trajectory to the equatorial orbit. It is known from other lunar trajectory studies that a minimum inclination (without a plane change) to any given plane can be found.[1,2] The minimum inclination that can be attained ballistically depends upon earth-moon relative geometry. The relative configuration of the several pertinent earth and moon planes and the ecliptic is constantly changing so that the minimum inclination available varies with time. Lunar mission studies given in the literature have not dwelt upon the derivation of this inclination, although Tolson has touched upon it.[2]

The purpose of this study is to determine this minimum inclination of the arrival, moon-referenced, trajectory plane with respect to the lunar-equatorial plane as a function of launch time and to determine those launch dates or periods at which the minima occur. Launch windows are used to identify the data that are practicable. These data are presented to show their variation throughout lunar months and throughout the 18.6-yr regression period of the lunar plane-ecliptic nodal line. The launch windows were established by range safety restrictions on launch azimuth at the Cape Kennedy launch site. Limits on launch azimuth of 72 deg to 114 deg were used in the study.

Included in the study is an analysis of launch window extension by means of trajectory plane changes made in the earth parking orbit, at injection, or during transfer to the moon. The velocity increment needed to execute the plane change is used to present the data.

Discussion

Earth-Moon Geometry

As viewed from the earth along the earth-moon line, the orientation of the lunar-equatorial plane is continuously changing because of the orbital motion of the moon, as shown in Fig. 1. The "target plane", i.e., the lunar-equatorial plane, has an apparent rotary, wobbling motion with a period of one nodal lunar month so that the target plane is both wobbling and moving. An earth-moon-ecliptic geometry fact to be noted is that the nodal line of the lunar-equatorial plane on the lunar plane is parallel to the nodal line of the lunar plane on the ecliptic. The ascending node of one points in the direction of the descending node of the other.

The rather complicated earth-moon dynamic geometry dictates the coding of what amounts to an ephemeris of planar orientations and moon positions. The long period before a repetition of planar configurations indicates the practicality of a calendar-type display of results. The large number of cases which must be computed to devise such a calendar demands an economical scheme of trajectory synthesis. Fortunately, the two-body conic section method described below provides such a process.

The simple form yielded by the two-body conic section method is well suited to this study.[3] Its accuracy has been demonstrated to be more than adequate.[1,4-6] In this trajectory synthesis, an earth-referenced conic representing the transfer trajectory is determined to intersect the center of the moon, with the moon being considered as a massless point. The perigee altitude, the flight time from perigee to moon, and position of the moon at the time of arrival are specified parameters of the conic. The conic lies in a plane containing the center of the earth, the earth launch site, and the moon's center at the time of arrival. The moon's center is moving slowly compared to the motion of the launch site. Thus, the trajectory plane may be visualized as rotating about the earth-moon line so as to intersect the launch site in its daily motion (Fig. 2). The fact that the target point would not be the moon's center (except for direct hits) but would be a point offset from the center is disregarded, since the difference in cislunar trajectory inclination results is insignificant.

From the velocity vector of the conic at the moon's center is subtracted the velocity vector of the moon with respect to the earth. The resulting relative velocity vector is assumed to be the asymptotic vector of the arrival, moon-referenced hyperbola (Fig. 3). (A similar scheme has been suggested by Egorov in his classical work.[7]) This asymptotic vector is essentially the velocity of the unperturbed trajectory relative to the center of the moon. This vector, directed through the moon's center, has been defined as the V_∞ vector. The plane of the hyperbola must contain this vector. This plane, which is the vehicle arrival, moon-referenced, trajectory plane, may be rotated about the V_∞ vector by small injection and/or midcourse corrections. The least inclination that the trajectory plane may take to any reference plane at the moon is the inclination of the V_∞ vector to that reference plane. It may take any steeper inclination however. The minimum inclination sought is simply the inclination of the V_∞ vector to the lunar-equatorial plane (Fig. 4). It is illustrative to note that the intersection of this V_∞ vector with the lunar surface is the point of impact for a direct, vertical, lunar approach.

Earth-Moon-Sun Model

The moon was assumed to be in a mean circular orbit for the analysis. This approximation has little or no effect on the trends or magnitudes of the results; the only effect is a slight shift of the results along the time scale. All librations of the moon, except the apparent wobbling of the lunar equator (mentioned earlier), were ignored.

Ecliptic longitude values of the target vector and the lunar plane-ecliptic nodal line at the time of lunar arrival were determined in terms of launch time from the respective expressions:

$$☾ = K_1 + 13.176397\left(d_L + \frac{h_L + t_T}{24}\right) \qquad (1)$$

$$\Omega = K_2 - 0.052953922\left(d_L + \frac{h_L + t_T}{24}\right) \qquad (2)$$

The ecliptic longitude of the earth-sun line at launch time was obtained from

$$A_S^* = K_3 + 0.98564734\, d_L \qquad (3)$$

Values for the K_1, K_2, and K_3 coefficients depend upon the epoch used in preparing the data. In the analysis, it is assumed that the epoch selected begins at midnight, Greenwich Mean Time. Equations (1-3) were drawn from Ref. 8; higher order terms were dropped as the additional accuracy would not appreciably affect the results.

Analysis

Only a brief analytic treatment is described here since the Appendix contains the detailed derivation of the equations as do Refs. 9-11. Figure 5 will be helpful in following the development.

Minimum lunar inclination

The ephemeris relationships of Eqs. (1) and (3) locate the target vector and the launch position vector. These two vectors establish the required vehicle trajectory plane. Orientation of the V vector within this plane was determined from two-body conic results that yielded the radial and azimuthal components of V at the moon with respect to the earth. Vector subtraction of the moon's velocity with respect to the earth leads to an expression for V_∞ which then permits the minimum lunar inclination to be defined. Having the trajectory plane located, the required launch azimuth can be determined. Limiting values on the launch azimuth establish launch windows that can be used to identify the data that are practicably useful.

The vehicle trajectory plane can be described by its unit normal I_V, which can be obtained from the cross-product of the launch position and target vectors; succinctly,

$$I_V = I_L \times I_T / \sin\phi \tag{4}$$

where the transfer angle ϕ follows from the dot product

$$\phi = \cos^{-1}(I_L \cdot I_T) \tag{5}$$

Since the launch azimuth vector lies in the trajectory plane, a second expression for the unit normal I_V can be formed as

$$I_V = I_L \times I_{A_Z} \tag{6}$$

A comparison of the components of the expanded forms of Eqs. (4) and (6) leads to a formulation of launch azimuth:

$$A_Z = \sin^{-1}\left[\frac{\cos\delta \sin(A - A_C)}{\sin\phi}\right] \tag{7}$$

The inclination i_1 of the trajectory plane to the lunar plane can be obtained from the unit normals to these two planes by use of the equation

$$i_1 = \cos^{-1}(I_V \cdot I_M) \tag{8}$$

Having the transfer trajectory plane-lunar plane inclination and the radial and azimuthal components of V with respect to the earth permits V_∞ to be referred to the moon in lunar plane coordinates by (using matrix notation)

$$V_{\infty_{\ell.p.}} = \begin{pmatrix} V \sin \gamma \\ V \cos \gamma \cos i_1 - V_M \\ V \cos \gamma \sin i_1 \end{pmatrix} \tag{9}$$

The angle subtended by the V_∞ vector and the lunar-equatorial plane defines the minimum inclination - the desired study result. Using coordinate transformations to express V_∞ in a selenocentric, lunar-equatorial system and defining a unit V_∞ vector as I_∞, leads to a simple expression for this minimum inclination, viz.,

$$i_{VM} = \sin^{-1} \left[(I_{\infty_{\ell.eq.}})_z \right] \tag{10}$$

Trajectory plane change

The possibility of extending the launch window was examined by considering trajectory plane changes made in the earth parking orbit, during acceleration from parking orbit to transfer trajectory injection, and while in the transfer trajectory. A convenient parameter in describing the plane change is the velocity increment needed to execute the maneuver. The point of plane change was specified by the geocentric angle ϕ_1 subtended by the target vector and the vehicle position vector at the time of plane change. Figure 6 is a sketch showing the launch trajectory plane (from launch to plane change) and the arrival trajectory plane (from plane change to lunar arrival).

For plane change in the earth parking orbit, the plane change angle i_{PC} can be related to the velocity increment ΔV

by

$$i_{PC_P} = 2 \sin^{-1} (\Delta V / 2 V_C) \tag{11}$$

The total velocity increment required to accelerate from circular parking orbital velocity to transfer trajectory injection velocity and simultaneously make a plane change can be obtained from

$$\Delta V_T = \sqrt{V_C^2 + V_{inj}^2 - 2 V_C V_{inj} \cos i_{PC_{inj}}} \tag{12}$$

which, with the definition,

$$\Delta V = \Delta V_T - (V_{inj} - V_C) \tag{13}$$

leads to

$$i_{PC_{inj}} = \frac{|\Delta V|}{\Delta V} \cos^{-1} \left[1 - \frac{|\Delta V|}{V_C} + \frac{|\Delta V|}{V_{inj}} - \frac{(\Delta V)^2}{2 V_C V_{inj}} \right] \tag{14}$$

where the sign of ΔV (as a computer program parameter) is used to determine the sign of the plane change angle. A change of plane during transfer trajectory can be written in terms of ΔV

$$i_{PC_T} = 2 \sin^{-1} \left[(\Delta V / 2 V_1 \cos \gamma_1) \right] \tag{15}$$

Use of conic orbit relationships permits Eq. (15) to be rewritten as

$$i_{PC_T} = 2 \sin^{-1} \left[\frac{\Delta V R_{inj} V_{inj}}{2 \mu (1 + e \cos \eta_1)} \right] \tag{16}$$

The launch azimuth required when a plane change is to be made follows from a definition of the launch plane miss angle β. From the dot product of the normal to the launch plane and the target vector, the sine of this angle is

$$\sin \beta = I^*_V \cdot I_T \tag{17}$$

which leads to

$$\sin \beta = \sin A^*_Z \left[\sin \zeta \cos \ell - \cos \delta \sin \ell \cos (A - A_C)\right] - \cos A^*_Z \cos \delta \sin (A - A_C) \tag{18}$$

An iterative solution of Eq. (18) yields the launch azimuth.

To get the unit normal to the arrival trajectory plane, use is made of the cross product of the unit position vector at the time of plane change and the target vector, viz.,

$$I_{V_{PC}} = \frac{I_{L_{PC}} \times I_T}{\sin \phi_1} \tag{19}$$

Having the arrival trajectory plane normal $I_{V_{PC}}$, the remainder of the analysis with a plane change is identical to that of Eqs. (1-10) with one exception: I_V in the cited section should be replaced with $I_{V_{PC}}$ of the foregoing Eq. (19).

Results

The equations just presented and those in the Appendix were coded for IBM 7090 digital computer solution. References 9-11 contain the complete study data that have been obtained. Typical and summary results for an earth-moon transfer time of 60 hr are discussed herein.

Results of previous two-body conic section studies were summarized for this analysis in terms of the transfer time variation of the radial and azimuthal velocity components at the moon and of the true anomaly of the target vector. Launch azimuth limits of 72 deg to 114 deg were used to establish exemplary launch windows for launch from Cape Kennedy, Florida.

Minimum Lunar Inclination

Presented below are the minimum inclinations for typical launch windows, a representative envelope of these minima showing their variation in a lunar month, and a plot showing the variation of certain maxima and minima of the minimum inclination envelope during the 18.6-yr-lunar regression period.

Figure 7 shows a plot of launch azimuth for a typical 24-hr period during June 13, 1967.

This figure indicates two launch windows, each of approximately 5 to 6 hr in duration, for the 24-hr period shown. It also presents the minimum lunar orbit inclination available without a plane change, as well as the required earth parking orbit coast angle associated with launch during this period. In computing the coast angle, a 15 deg value was assumed (based on other analyses) for each of the boost angles, launch-to-parking-orbit, and parking-orbit-to-injection (the respective angles ρ_1 and ρ_2 of Eq. (46)). The cyclic variation of the minimum inclination during a 24-hr period is caused by the plane of the earth-referenced trajectory rotating about the earth-moon target line as the launch site moves with the earth's rotation. Since the inclination of the moon-referenced trajectory to the lunar-equatorial plane depends upon the orientation of the earth-referenced trajectory plane, the variation of minimum inclination with hour-of-the-day is reasonable.

The curves of Fig. 8, for a one-week period, show how envelopes can be formed for the coast-angle requirements and minimum lunar inclination. These envelopes provide a convenient means for examining the cycles for various periods of interest in earth-moon geometry. Coast angle and minimum inclination angles are shown in Fig. 9 for a 60-day period; this plot illustrates the periodicity of these angles with respect to a lunar month. The one-week period of Fig.8 is indicated on Fig. 9 by the shading. Figure 9 shows, for the 60-day period considered, that the minimum lunar orbit inclination to the lunar-equatorial plane varies between approximately $\pm$ 12 deg within the launch window specified.

It should be remembered that only discrete line segments within the envelopes are valid data as was shown in Fig. 8; furthermore, the minimum inclination envelope contains the segments for both daily launch windows whereas the coast angle envelopes each contain only one of the two daily segments. The cyclic variation of the minimum inclination during a lunar month is due to two factors. Throughout a lunar month the earth-moon target line varies with respect to the earth-equatorial plane, the monthly declination cycle. This target line is also varying with respect to the moon-equatorial plane. To gain some insight into the variation of the minimum inclination during a lunar month, consider Fig. 1 and the (projected) V_∞ vector shown at two diametrically opposite positions in the moon's orbit about the earth. On Fig. 1 the two projections of V_∞ (onto the plane normal to the lunar plane and containing the earth-moon line) are assumed fixed and inclined to the lunar plane at the angle α . With this premise, the cited figure illustrates, qualitatively, how the minimum inclination takes a cyclic variation during a lunar month.

The particular earth-moon-ecliptic configuration of Fig. 1 roughly depicts conditions that will exist in 1978. At that time, the node of the lunar plane on the ecliptic coincides with its node on the earth-equatorial plane and both coincide with the Vernal Equinox. The node of the lunar-equatorial plane on the lunar plane is parallel to the vernal equinox, at that time, and the lunar plane is at its least inclination to the earth-equatorial plane.

An entire 18.6-yr period was investigated to cover all possible relative positions of the earth and moon. The 18.6-yr time period encompasses one regression cycle of the lunar plane-ecliptic nodal line. Sixty-day periods separated by 365 days were considered in order to get a complete sampling of data. The results obtained were quite similar to those shown in Figs. 7-9. The excursions of certain maxima and minima for the minimum lunar inclination data are shown in Fig. 10 for a typical 18.6-yr period; the inset on this figure identifies the maximum and minimum points under consideration. Figure 10 indicates that the minimum inclination envelopes tend to flatten during the 2- or 3-yr period leading to the minimum lunar declination that occurs in 1978. This flattening is perceivable since the variation in lunar declination within a lunar month is smaller during this period.[12, 13] As a help in understanding the regressional period variation of the minimum inclination, consider the moon to be fixed in its orbit. Then the nodal line of the lunar-equatorial plane on the lunar plane is slowly rotating clockwise with the regressional period. The declination of the moon is varying with the same period (unless it is zero) since the inclination of the lunar plane to the earth-equatorial plane is changing. These latter two variations are superimposed on all others and account for the need, as filled by this study, of an automatic digital computing machine program and the covering of at least an 18.6-yr period.

Launch Window Extension

Results showing the effect of trajectory plane change on launch window are presented in Figs. 11 and 12. The data of Fig. 11 show the change in launch azimuth for a plane change made at ϕ_I = 20 deg using velocity increment values of $\pm$ 500 fps. The launch window extension Δt that results from this plane change is identified on the figure. It was found that Δt was nearly linear with ΔV for ΔV and ϕ_I values of interest, viz.,

$$|\Delta V| \leq 500 \text{ fps}$$
$$\phi_I = 2 \text{ to } 270 \text{ deg}$$

Figure 12 is a summary plot of the Δt - ΔV - ϕ_I relationship for a transfer time of 60 hr. This figure shows that the launch window can be most effectively extended by plane changes made at ϕ_I values of approximately 15 deg. The earth radial distance indication on the curve demonstrates that this corresponds to about three-tenths of the distance to the moon which is in the same general region where midcourse corrections are likely to be made. The shape of the curve of Fig. 12 can be explained by the following: From a geometric standpoint, plane changes at ϕ_I = 90 deg or 270 deg are the most effective, whereas plane changes for ϕ_I values near 0 deg or 180 deg are quite ineffective. However, at ϕ_I = 270 deg, the vehicle is in an earth parking orbit with the associated circular orbital velocity; this high circular velocity offsets the advantages of the geometric effectiveness. At ϕ_1 = 90 deg, the vehicle, on the transfer trajectory, still has a high velocity. Although the plane change can be combined with the acceleration to injection at ϕ_1 values near 180 deg, the geometric disadvantage in this vicinity more than offsets the benefit of the combination. The study data of Fig. 12 indicate that the lower velocity on the transfer trajectory and the geometric effectiveness combine to yield optimal results for a ϕ_1 value near 15 deg.

Conclusions

Study results have just been presented for an entire 18.6-yr lunar regressional period, illustrating the minimum lunar inclination (without a plane change) and the associated earth parking orbit coast angle. Study premises included an earth-moon transfer time of 60 hr and launch azimuth restrictions of 72 deg to 114 deg for the Cape Kennedy launch site. Additional results were presented showing the launch window extensions available via trajectory plane change. These results permit the following conclusions to be drawn:

1) There are two launch windows per 24-hr period, each of 4 to 6 hr in duration.

2) The minimum lunar inclination varies between approximately $\pm$ 15 deg.

3) There are at least two times during each lunar month when a near lunar-equatorial orbit can be obtained without a plane change.

4) Generally, there is at least one direct ascent possibility during each lunar month (with a few possible exceptions). There may be up to five or six direct ascents possible within a lunar month.

5) Launch window extension via trajectory plane change is quite expensive in terms of the required velocity increment.

Further study efforts for other earth-moon transfer times up to 95 hr have revealed data trends and magnitudes quite similar to those reported herein.

Appendix

This appendix presents a detailed derivation of the equations and terms necessary to evaluate the summary expressions given in the main body of the paper. Matrix notation will be used where appropriate.

A geocentric unit target vector can be written in the ecliptic coordinate system directly from Fig. 5 as

$$I_{Tec} = \begin{pmatrix} \cos ☾ \cos b \\ \sin ☾ \sin b \\ \sin b \end{pmatrix} \tag{20}$$

The angle b subtended by the target vector and the ecliptic is given by a right spherical trigonometric relationship:

$$b = \sin^{-1} \left[\sin i \sin (☾ - \Omega) \right] \tag{21}$$

The angle b will be a first or fourth quadrant angle accordingly as the inverse sine argument of Eq. (21) is positive or negative; the lunar plane-ecliptic inclination i is assumed always positive. A coordinate rotation about the x axis through the angle ϵ can be used to express the target vector in the earth-equatorial system as

$$I_{Teq.} = \begin{pmatrix} \cos ☾ \cos b \\ \cos \epsilon \sin ☾ \cos b - \sin \epsilon \sin b \\ \sin \epsilon \sin ☾ \cos b + \cos \epsilon \sin b \end{pmatrix} \tag{22}$$

A second expression for the target vector can be written directly from Fig. 5:

$$I_{Teq.} = \begin{pmatrix} \cos A \cos \delta \\ \sin A \cos \delta \\ \sin \delta \end{pmatrix} \tag{23}$$

Equating the x components of Eqs. (22) and (23) gives

$$\delta = \cos^{-1}\left(\frac{\cos \leftmoon \cos b}{\cos A}\right) \tag{24}$$

and equating the y/x quotients yields

$$A = \tan^{-1}\left(\frac{\cos \epsilon \sin \leftmoon \cos b - \sin \epsilon \sin b}{\cos \leftmoon \cos b}\right) \tag{25}$$

From Fig. 5, it can be noted that A is a first or fourth quadrant angle when $\leftmoon$ is in the first or fourth and is a second or third quadrant angle when $\leftmoon$ is in the second or third; A may not be in the same quadrant as $\leftmoon$, however. The quadrant of δ is first or fourth as the sign of

$$\sin \delta = \sin \epsilon \sin \leftmoon \cos b + \cos \epsilon \sin b \tag{26}$$

is plus or minus; Eq. (26) results from equating the z components of Eqs. (22) and (23).

The launch position vector can be analytically defined by finding the right ascension of the mean sun. Using Eq. (3) and Fig. 13, this is

$$A_S = \tan^{-1}(\tan A_S^* \cos \epsilon) \tag{27}$$

where A_S and A_S^* are always in the same quadrant. Since the mean sun of Eq. (3) is midnight with respect to Greenwich, the right ascension of Greenwich at launch is

$$A_G = A_S + 180 + 15\, h_L \tag{28}$$

so that the launch site right ascension follows as

$$A_C = A_G + \Omega_L \tag{29}$$

An expression for the launch position vector can now be written directly from Fig. 14 as

$$I_{L\,eq.} = \begin{pmatrix} \cos \ell \cos A_C \\ \cos \ell \sin A_C \\ \sin \ell \end{pmatrix} \tag{30}$$

The vehicle trajectory plane normal relationship of Eq. (4) can be expanded using Eqs. (30) and (22) to yield

$$I_V = \frac{1}{\sin \phi} \begin{pmatrix} \sin A_C \cos \ell \sin \delta - \sin A \sin \ell \cos \delta \\ \cos A \sin \ell \cos \delta - \cos A_C \cos \ell \sin \delta \\ \cos \ell \cos \delta \sin (A - A_C) \end{pmatrix} \tag{31}$$

where ϕ can be obtained from Eqs. (5, 30, and 22) as the form

$$\phi = \cos^{-1} \left[\cos \ell \cos \delta \cos (A_C - A) + \sin \ell \sin \delta \right] \tag{32}$$

Referring to Fig. 15 and recalling the eastward launch assumption, leads to the quadrant determination of ϕ . If the z component of Eq. 31 is positive, ϕ is a first or second quadrant angle whereas if this z component is negative, ϕ is in the third or fourth.

Equation (6) can be defined in detail by the following: A geocentric unit vector parallel to the launch azimuth vector can be written directly from Fig. 14 in the double-primed system as

$$I''_{A_Z} = \begin{pmatrix} 0 \\ \sin A_Z \\ \cos A_Z \end{pmatrix} \tag{33}$$

A rotation of x'' about y'' from the launch site to the earth-equatorial plane followed by a rotation of x' about y' from the launch-site right ascension to the vernal equinox leads to

$$I_{A_Z} = \begin{pmatrix} -\cos A_C \cos A_Z \sin \ell - \sin A_Z \sin A_C \\ -\sin A_C \cos A_Z \sin \ell + \sin A_Z \cos A_C \\ \cos A_Z \cos \ell \end{pmatrix} \tag{34}$$

Substitution of Eqs. (30) and (34) into (6) gives the result

$$I_V = \begin{pmatrix} \sin A_C \cos A_Z - \cos A_C \sin \ell \sin A_Z \\ -\cos A_C \cos A_Z - \sin A_C \sin \ell \sin A_Z \\ \cos \ell \sin A_Z \end{pmatrix} \tag{35}$$

Determination of the launch azimuth given by Eq.(7) quadrant is possible by establishing the sign of $\cos A_Z$. Equating the x components of Eqs. (31) and (35) yields

$$\cos A_Z = \frac{\sin A_C \cos \ell \sin \delta - \sin A \sin \ell \cos \delta}{\sin \phi \sin A_C} + \frac{\cos A_C \sin \ell \sin A_Z}{\sin A_C} \tag{36}$$

where $\sin A_Z$ can be obtained from Eq. (7).

Expansion of Eq. (8) for the inclination of the trajectory plane to the lunar plane leads to a lengthy expression for which no convenient analytic simplification was found. Consequently, in using this equation, the vector components were numerically evaluated prior to forming the dot product. The lunar plane unit normal I_M can be found from the cross product

$$I_M = \frac{I_\Omega \times I_T}{\sin (☾ - \Omega)} \tag{37}$$

From Fig. 5, in the ecliptic system,

$$I_{\Omega_{ec}} = \begin{pmatrix} \cos \Omega \\ \sin \Omega \\ 0 \end{pmatrix} \tag{38}$$

Conversion of Eq. (38) to the earth-equatorial system and substitution into Eq. (37) together with Eq. (22) results in

$$I_{Meq} = \frac{1}{\sin (☾ - \Omega)} \begin{pmatrix} \sin \Omega \sin b \\ -\cos \epsilon \cos \Omega \sin b + \sin \epsilon \cos b \sin(\Omega - ☾) \\ -\sin \epsilon \cos \Omega \sin b - \cos \epsilon \cos b \sin(\Omega - ☾) \end{pmatrix} \tag{39}$$

The quadrant of i_1 of Eq. (8) is needed to indicate whether the vehicle, in arriving at the moon, approaches the lunar plane from above or below. Recalling the eastward launch assumption, define a first or fourth quadrant i_1 to indicate, respectively, that the vehicle approaches the lunar plane from below or above. Then, since I_T is positive in the earth-to-moon direction, and since the cross-product $I_M \times I_V$ lies along the I_T vector (but not necessarily in the same direction), it can be concluded that i_1 has the sign of $\left[(I_M \times I_V) \cdot I_T\right]$.

The unit vector I_∞ is defined by

$$I_\infty = V_\infty / |V_\infty| \tag{40}$$

Expressing I_∞ in a selenocentric, lunar-equatorial system requires two coordinate rotations. Use is made of Fig. 16 and of the fact that the ascending node of the lunar-equatorial plane on the lunar plane is parallel to the ascending node of the lunar plane on the ecliptic. The first rotation is of x about z from its target vector orientation to the lunar equatorial plane ascending node giving the x'-y'-z' system of Fig. 16. Then, a rotation of y' about x' from the lunar plane to the lunar-equatorial plane makes z' coincident with the lunar North Pole. Noting that

$$\Omega_{EL} = 180 - ☾ + \Omega \tag{41}$$

the desired form of the unit vector I_∞ is

$$I_{\infty_{\ell_{eq.}}} = \begin{pmatrix} \cos\Omega_{EL} & \sin\Omega_{EL} & 0 \\ -\sin\Omega_{EL}\cos i_{EL} & \cos\Omega_{EL}\cos i_{EL} & \sin i_{EL} \\ \sin\Omega_{EL}\sin i_{EL} & -\cos\Omega_{EL}\sin i_{EL} & \cos i_{EL} \end{pmatrix} \frac{V_{\infty_{\ell.p.}}}{|V_{\infty_{\ell.p.}}|} \tag{42}$$

Further analytic expansion of Eq. (42) is not convenient; accordingly, numerical evaluation was used to obtain the z component required by Eq. (10). The expression of Eq. (10) follows directly from

$$\sin i_{VM} = \cos(90 - i_{VM}) = (I_{\infty \ell.eq.}) \cdot (I_{Z \ell.eq.}) = (I_{\infty \ell.eq.})_3 \qquad (43)$$

since

$$I_{Z \ell.eq.} = \begin{pmatrix} 0 \\ 0 \\ 1 \end{pmatrix} \qquad (44)$$

The minimum lunar orbit inclination i_{VM} is a first or fourth quadrant angle, accordingly, as Eq. (43) is positive or negative.

In determining the earth parking orbit coast angle, the perigee of the earth-moon trajectory was assumed as a nominal injection point. From Fig. 15, the angle Ψ from launch to the injection point is

$$\Psi = \phi - \eta \qquad (45)$$

where the transfer trajectory true anomaly η is taken from the two-body conic results cited earlier. Then, from Fig. 15, the earth parking orbit coast angle θ can be written directly as

$$\theta = \Psi - (\rho_1 + \rho_2) \qquad (46)$$

The angles ρ_1 and ρ_2 are, respectively, the boost angles from launch to parking orbit and from parking orbit to injection. Since the coast angle cannot be negative, any negative value from Eq. (46) must have 360 deg added to it.

Equations (11 - 16) adequately explain the relationships between the plane change velocity increment and the plane change angle.

The angle β of Eq. (17) can be obtained by using a spherical trigonometric relationship (and Fig. 5) to give

$$\beta = \sin^{-1} (\sin i_{PC} \sin \phi_1) \qquad (47)$$

The launch plane unit normal I_{V*} of Eq. (17) is given by the equation for I_V, viz., Eq. (35) whereas the unit target vector is given by Eq. (23). Substitution of Eqs. (23) and (35) into Eq. (17) yields Eq. (18).

With ϕ, ϕ_1, and β known, the angle ϕ_{PC} subtended by the launch position vector and the plane change position vector can be obtained from spherical trigonometric considerations as

$$\phi_{PC} = \cos^{-1}\left(\frac{\cos\phi}{\cos\beta}\right) - \cos^{-1}\left(\frac{\cos\phi_1}{\cos\beta}\right) \tag{48}$$

In this equation, the angle given by the first inverse cosine term must be of the same quadrant as ϕ while the angle represented by the second inverse cosine term is of the same quadrant as ϕ_1.

The unit position vector at the time of plane change ($I_{L_{PC}}$), see Eq. (19), can be defined in terms of the declination and right ascension of the plane change point. Using spherical trigonometric relationships, the declination is

$$\ell_{PC} = \sin^{-1}(\sin\ell\cos\phi_{PC} + \cos\ell\cos A_Z^* \sin\phi_{PC}) \tag{49}$$

whereas the right ascension can be obtained from

$$\sin(A_{PC} - A_C) = \frac{\sin\phi_{PC}\sin A_Z^*}{\cos\ell_{PC}} \tag{50}$$

and

$$\cos(A_{PC} - A_C) = \frac{\cos\phi_{PC} - \sin\ell_{PC}\sin\ell}{\cos\ell_{PC}\cos\ell} \tag{51}$$

Both Eqs. (50) and (51) are needed for quadrant determination of A_{PC}. Having ℓ_{PC} and A_{PC}, the plane change unit position vector can be written directly from Fig. 6 as

$$I_{L_{PC}} = \begin{pmatrix} \cos A_{PC} \cos \ell_{PC} \\ \sin A_{PC} \cos \ell_{PC} \\ \sin \ell_{PC} \end{pmatrix} \qquad (52)$$

Substitution of Eqs. (23) and (52) into Eq. (19) leads to a formulation of the normal to the arrival trajectory plane:

$$I_{V_{PC}} = \frac{1}{\sin \phi_I} \begin{pmatrix} \sin A_{PC} \cos \ell_{PC} \sin \delta - \sin A \sin \ell_{PC} \cos \delta \\ \sin \ell_{PC} \cos A \cos \delta - \cos A_{PC} \cos \ell_{PC} \sin \delta \\ \cos \ell_{PC} \cos \delta \sin (A - A_{PC}) \end{pmatrix} \qquad (53)$$

This completes the detailed derivation needed in the analysis.

References

[1] Frimtzis, R., "Lunar vehicle guidance study," Aeronautical Systems Div. -TDR-62-207 (March 1962).

[2] Tolson, R. H., "Maximum latitude of lunar landing sites," ARS J. 32, 1091 - 1092 (1962).

[3] Mickelwait, A. B., "Lunar trajectories," ARS J. 29, 905- 914 (1959).

[4] Michael, W. H., Jr. and Tolson, R. H., "Three-dimensional lunar mission studies," NASA Memo. 6-29-59L (June 1959).

[5] Michael, W. H., Jr. and Tolson, R. H., "Effect of eccentricity of the lunar orbit, oblateness of the earth, and solar gravitational field on lunar trajectories," NASA TN D-227 (June 1960).

[6] Renyl, J., "Conic approximations for Atlas/Agena/Ranger lunar trajectories, 1962-1964," Jet Propulsion Lab., JPL EPD-95 (June 1962).

[7] Egorov, V.A., "Certain problems of moon flight dynamics," The Russian Literature of Satellites (International Physical Index, New York, 1958), Part 1, pp. 115-175.

[8] The American Ephemeris and Nautical Almanac for the Year 1960 (U.S. Government Printing Office, Washington, D.C., 1958), pp. 501 - 502.

[9] Breuer, F. D. and Riddell, W.C., "Minimum inclination of lunar orbit plane to lunar equatorial plane for earth-launched vehicles," General Dynamics/Astronautics, AE62-0965 (December 1962).

[10] Breuer, F.D., "Launch windows for earth-moon trajectories," General Dynamics/Astronautics, AE63-0119 (March 1963).

[11] Breuer, F.D. and Riddell, W.C., "Effect of plane change on earth-moon launch window," General Dynamics/Astronautics, GD/A 63-0103 (May 1963).

[12] Ehricke, K.A., Space Flight, Vol. II, Dynamics (D. Van Nostrand Co., Inc., Princeton, N.J., 1962), pp. 449-451, 930-935.

[13] Woolston, D.S., "Declination, radial distance, and phases of the moon for the years 1961 to 1971 for use in trajectory considerations," NASA TN D-911 (August 1961).

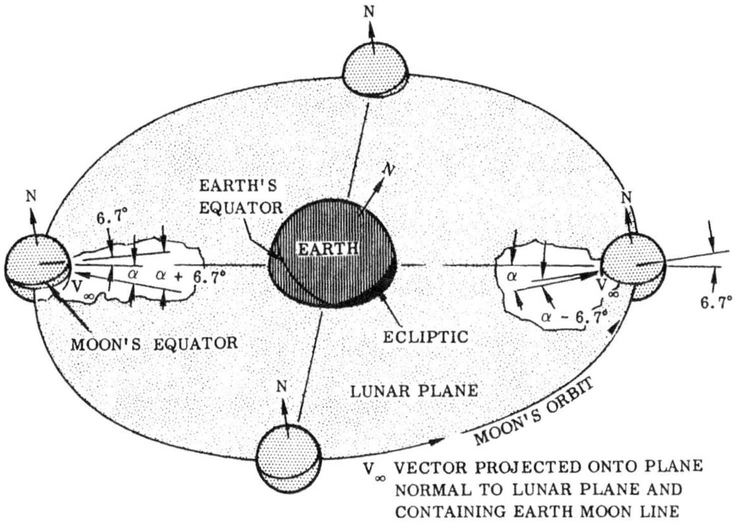

Fig. 1 Typical earth-moon geometry.

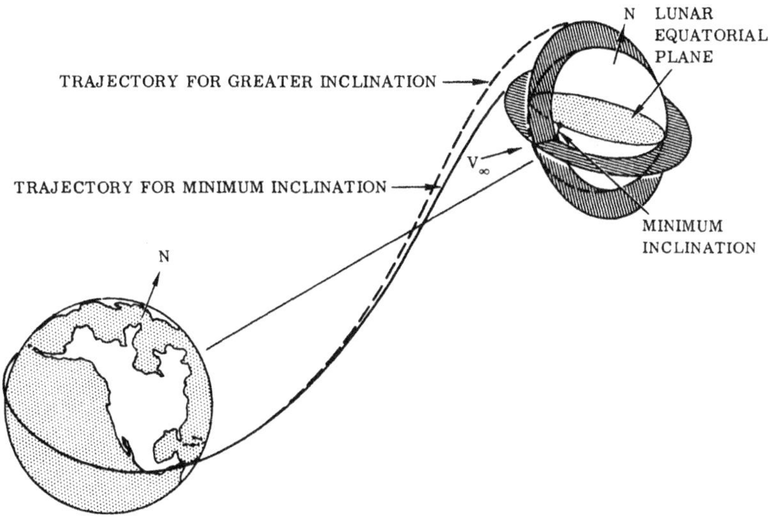

Fig. 2 Vehicle lunar-orbit inclination.

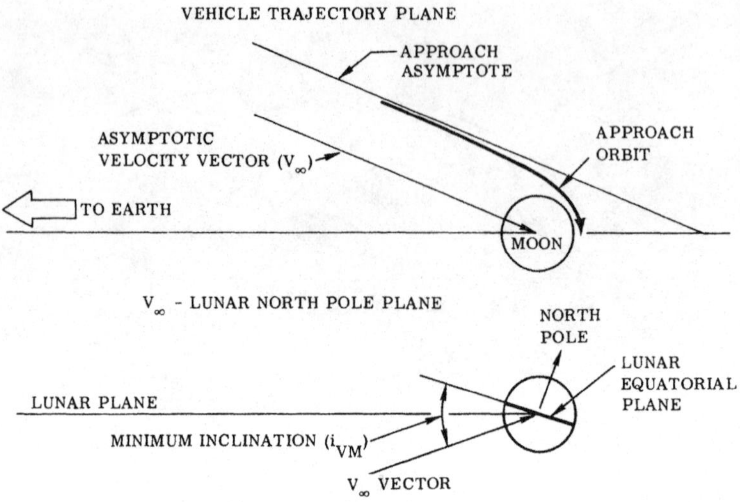

Fig. 3 Lunar approach geometry.

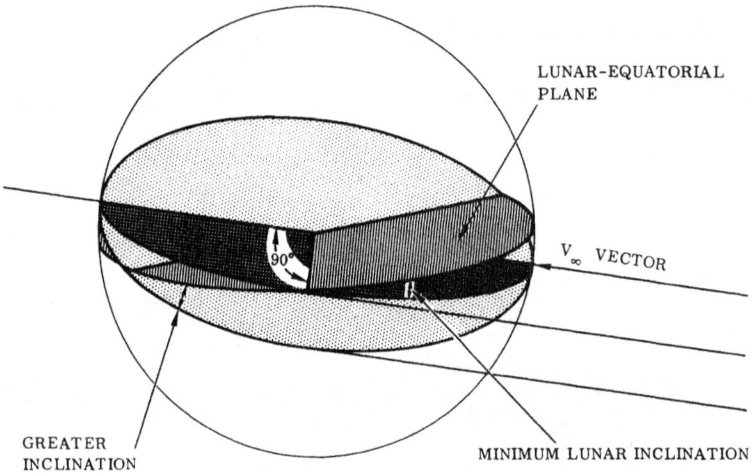

Fig. 4 Inclination of vehicle orbit to lunar-equatorial plane.

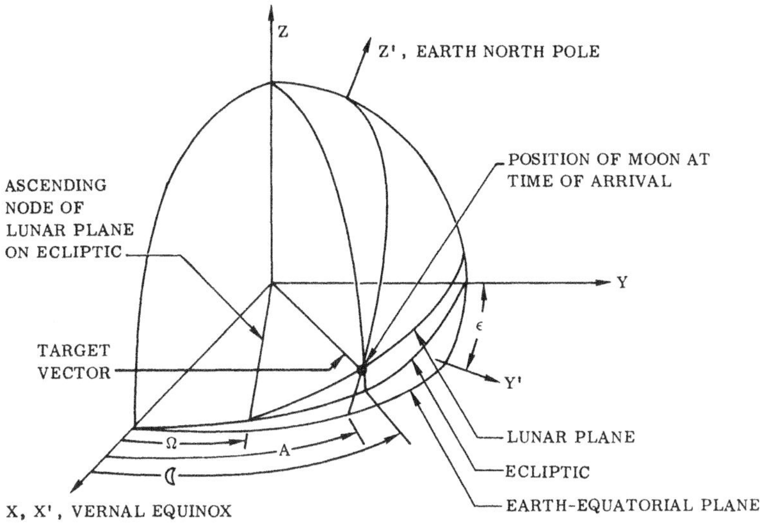

Fig. 5 Geocentric earth, moon, and ecliptic geometry.

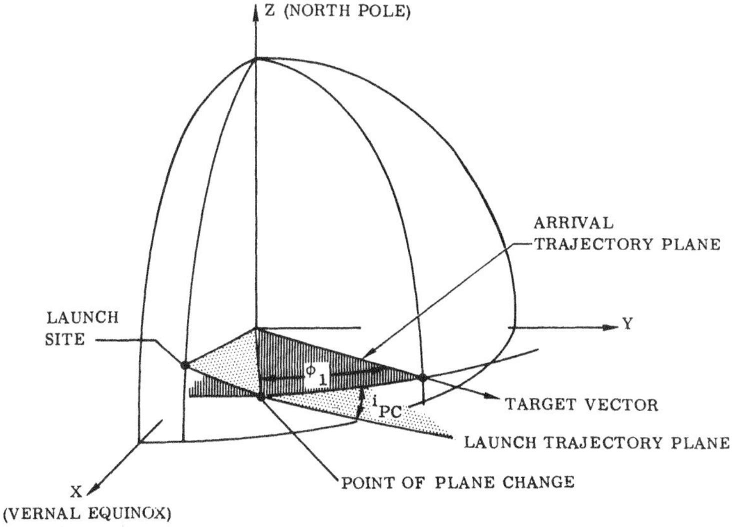

Fig. 6 Trajectory plane change geometry.

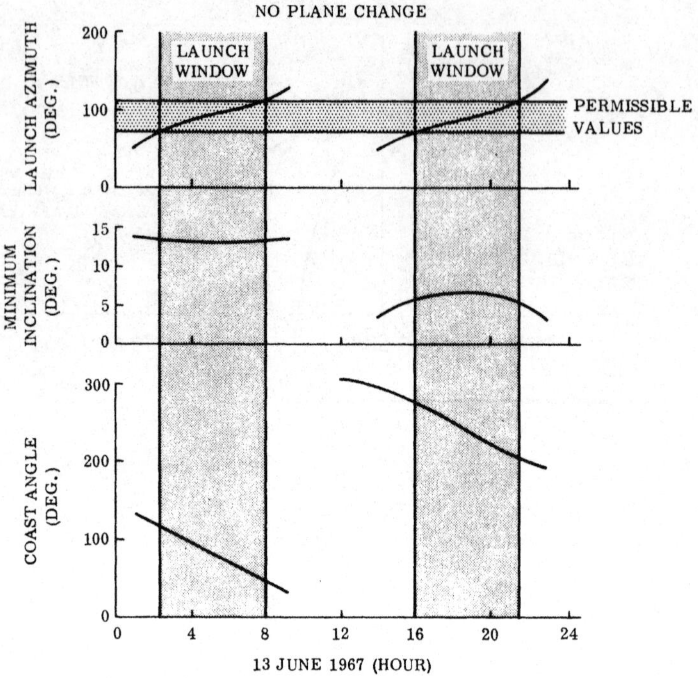

Fig. 7 Earth-moon launch window, minimum lunar inclination, and earth parking orbit coast angle.

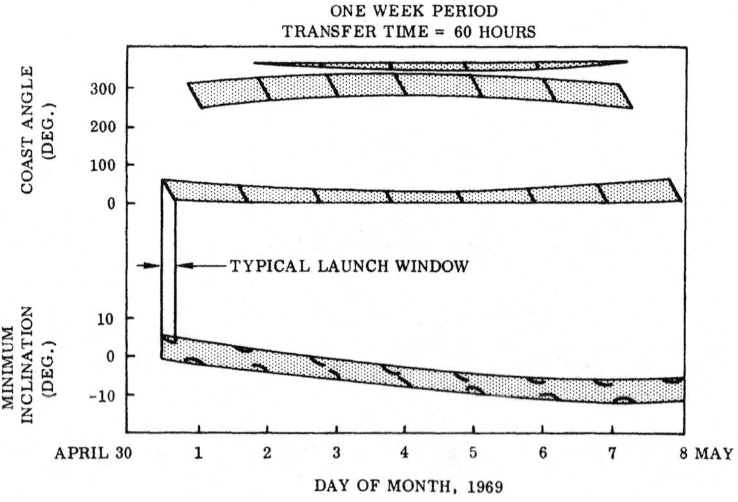

Fig. 8 Earth parking orbit coast angle and minimum lunar inclination.

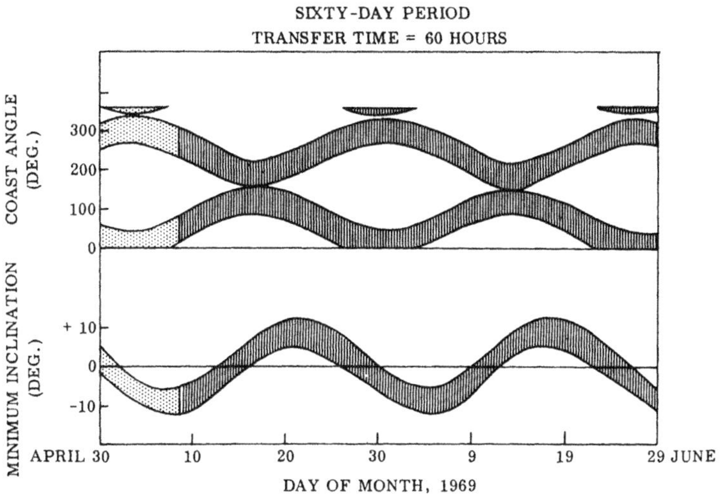

Fig. 9 Earth parking orbit coast angle and minimum lunar inclination.

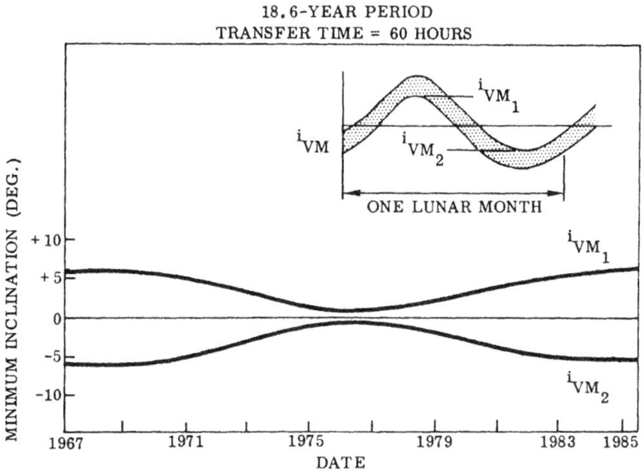

Fig. 10 Excursions of certain maxima and minima of minimum lunar inclination.

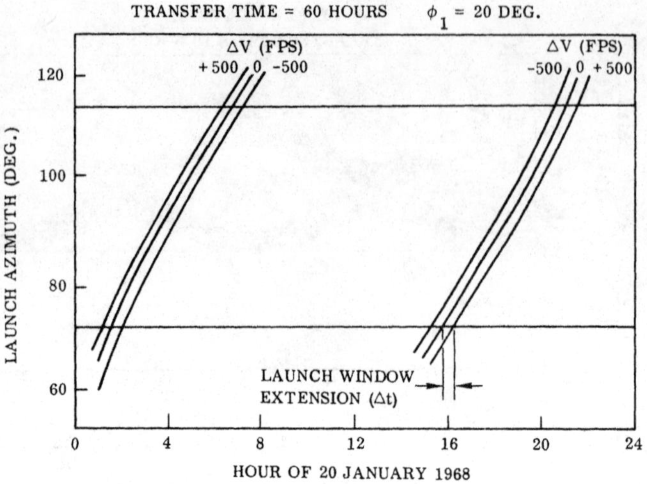

Fig. 11 Effect of plane change velocity increment on launch azimuth.

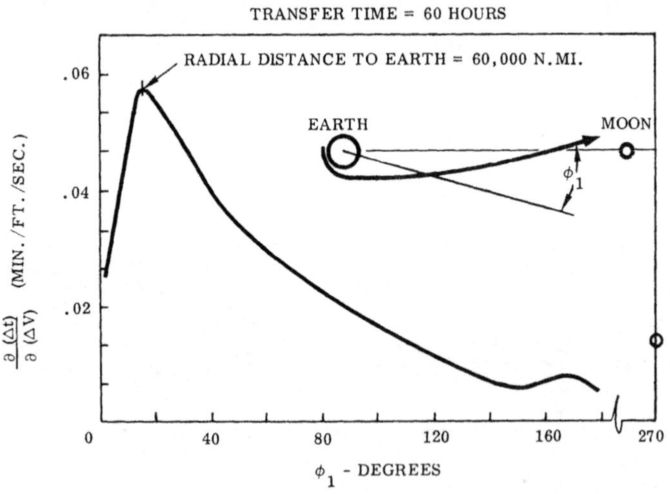

Fig. 12 Launch window extension.

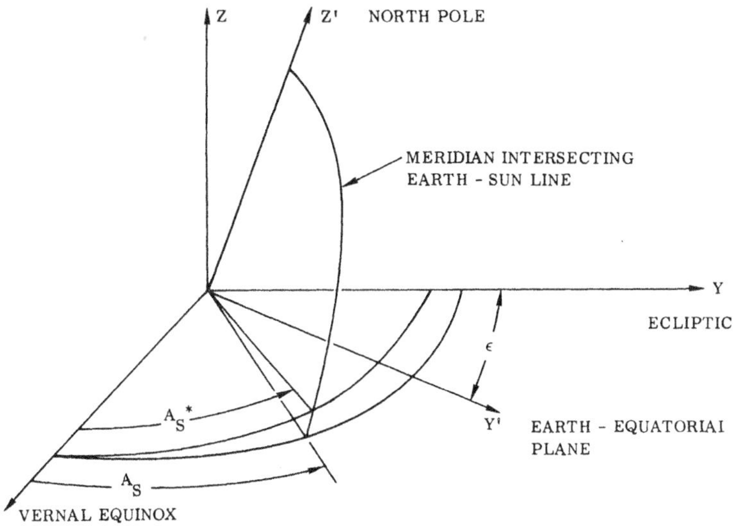

Fig. 13 Right ascension of mean sun.

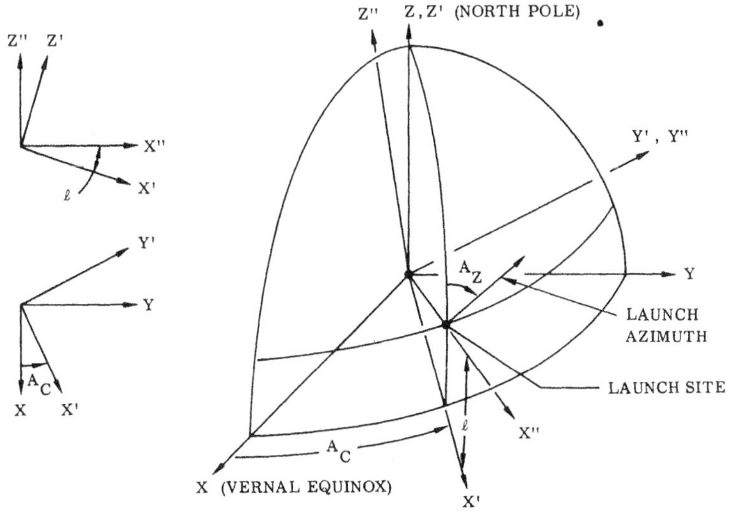

Fig. 14 Geometry of launch position vector and launch azimuth.

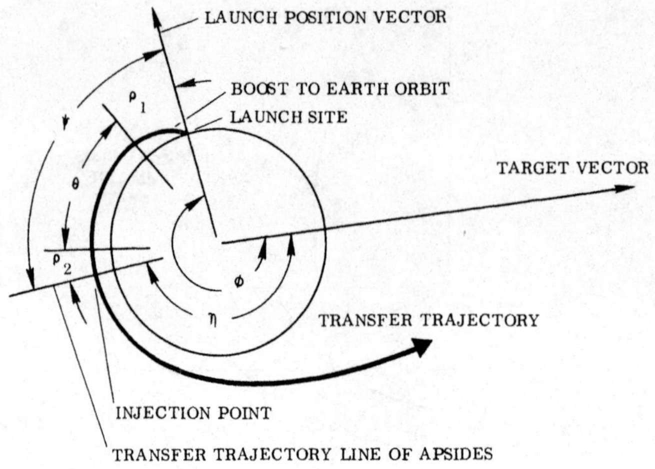

Fig. 15 Boost, coast, and transfer trajectory geometry.

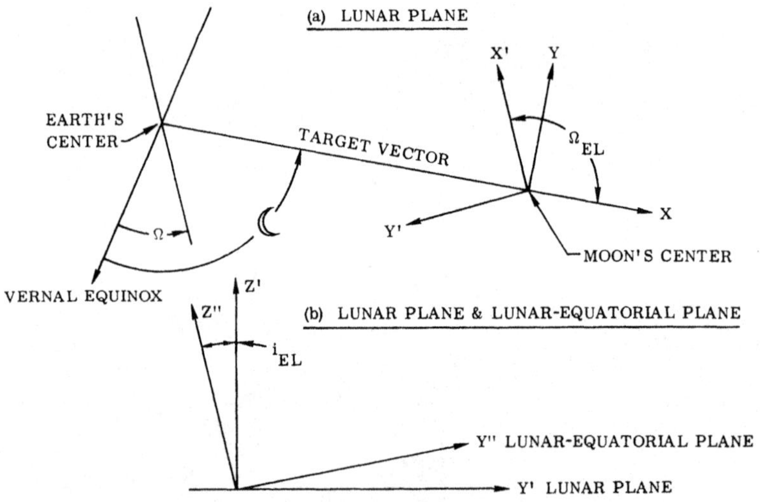

Fig. 16 Lunar coordinate systems.

GENERAL CHARACTERISTICS OF THE LAUNCH WINDOW FOR ORBITAL LAUNCH TO THE MOON

Herbert Reich*

Northrop Space Laboratories, Hawthorne, Calif.

Abstract

The analysis of the gross aspects of the flight mechanics for lunar trajectories from orbital launch is presented. An attempt is made to define the launch requirements from an earth orbit, the geometrical relationships existing between an Orbital Launch Facility (OLF) in earth orbit and the moon in its orbit, the departure trajectory sensitivity to energy and time requirements, the launch-on-time problem and the orbital "launch window."

Nomenclature

i = OLF orbit plane inclination with equatorial plane, deg

η = Lunar plane inclination with equatorial plane, deg

ϕ = Inclination angle between orbit plane and lunar plane, deg

λ_N = Longitude of orbit plane - equatorial plane line of nodes measured counterclockwise from lunar descending equatorial node, deg

$\bar{L}$ = Position vector of vehicle at launch (from geocenter)

$\bar{S}$ = Position vector of moon at arrival (from geocenter)

Presented as Preprint 63-405 at the AIAA Astrodynamics Conference, New Haven, Conn., August 19-21, 1963. The author wishes to acknowledge the invaluable assistance and contributions of Mr. Richard Wolpert (now with Space Technology Laboratories) to the study program from which this paper has been extracted.

*Chief, Advanced Systems Design.

β = In-plane central angle of transfer trajectory between $\overline{L}$ and $\overline{S}$, deg

μ_L = Plane change angle between orbit plane and transfer trajectory plane, deg

α_L = Right ascension of the vehicle position vector projection on lunar plane measured counterclockwise from the lunar ascending node, deg

α_S = Right ascension of the moon position vector measured counterclockwise from the lunar ascending node, deg

δ_L = Declination of the vehicle position vector, deg

Ω = OLF orbit plane precession rate, deg/day

h = OLF orbit altitude, naut miles

ΔV = Velocity increment, fps

ΔV_{ADD} = Additional velocity increment or penalty, fps

V_{BO} = Burnout velocity, fps

T_b = Trip time, hr

γ = Flight path angle, deg

$\hat{\theta}$ = True anomaly, deg

a = Semimajor axis of transfer ellipse

e = Eccentricity of transfer ellipse

Σ_L = Launch azimuth, deg

$\frac{N}{m}$ = Rendezvous compatible orbit synchronous ratio

q = Reference parameter equal to the difference between i and η

Δ = OLF plane - lunar plane node line position angle measured in the lunar plane, from the equatorial plane - lunar plane node line

1. Introduction

This paper presents the preliminary results of an analytical investigation into the fundamental aspects of the optimum launch conditions from an earth orbit. Inasmuch as the precise definition of launch conditions is strongly dependent on a knowledge of the physical characteristics of the launch vehicle and the specific requirements of a particular lunar mission, only a gross analysis is presented which can provide a mission analyst with the tools and data to define a performance envelope for the vehicle designer.

The chief objectives of the orbital launch window analysis are: 1) to define the earth orbit launch criteria for lunar trajectories, 2) to establish the departure trajectory sensitivities to orbital parameters, 3) to establish the mission and vehicle sensitivities to time and energy requirements, 4) to establish the analytical technique for determining optimum launch conditions, and 5) to provide the basis for the preparation of an orbital launch timetable compatible with lunar mission and earth launch requirements.

The study effort results related to each of the objectives are presented in the following sections. Pertinent assumptions are identified and conclusions and recommendations indicated where appropriate. Technical analyses in support of the data presented are included in the paper insofar as is feasible.

The scope of the work includes the definition of an operational orbit based on previous studies performed at Northrop. This orbit is defined by the earth launch site chosen (Atlantic Missile Range), the use of the concepts of rendezvous compatibility and minimum orbit maintenance, and the altitude limit due to the lower Van Allen Radiation Belt. The orbit thus defined is at an altitude of 263 nautical miles and an inclination of 30° to the equator.

2. Orbital Launch Criteria

Launch from orbit is a function of many variables, both geometric and dynamic. A clear understanding of the three dimensional geometric relationship of the satellite-earth-moon system is necessary to visualize the problem. The illustration on Fig. 1 is a display that correlates the angular and rotational variables of the three planes of interest.

The angle η between the equatorial and lunar planes varies slowly between 18.5° and 28.5° over an 18.6 year cycle.

Because of the relatively slow variation in η when compared to the real time requirements for any single lunar mission, this angle can be considered a constant in the analysis once a launch or mission date is specified. The OLF plane maintains a constant inclination to the equatorial plane as it precesses (because of earth oblateness) around the earth in a retrograde direction (from east to west). This precession rate is a function of OLF altitude and orbit plane inclination. For the chosen operational altitude and inclination (263 naut miles and 30° inclination), this rate is approximately 6.7 deg/day. The significance of orbit precession is analyzed in a later paragraph.

The angle ϕ between the OLF and lunar planes is a function of i, η, and the position of the satellite plane at any instant. This position is defined by the equatorial angle λ_N. The trigonometric relationship of these four angles defines the spatial relationship of the three planes at any time and is stated thus:

$$\phi = \cos^{-1} (\cos i \cos \eta - \sin i \sin \eta \cos \lambda_N) \qquad (1)$$

Since λ_N is changing at the precessional rate of the OLF plane (6.7 deg/day), ϕ will have a 54 day cycle (one complete rotation of the OLF plane). This variation of ϕ is illustrated in Fig. 2. The bandwidth for $i = 30°$ illustrates the effect of a two year variation in η. It will be shown later that the launch-on-time restrictions and lunar arrival window are a strong function of the angle ϕ.

3. Assumptions

The following assumptions were used in this analysis: 1) the OLF orbit plane is based on launch from AMR, 2) the OLF is in a circular earth orbit, 3) the moon's orbit is also circular, 4) plane changes occur only at launch, and 5) burnout for the orbital launch occurs at perigee of the transfer conic.

4. Optimum Launch Conditions

To define the optimum conditions, only the positional variables of the OLF and moon are significant. Projection of the OLF orbit, orbital launch vehicle transfer trajectory, and lunar orbit plane into a geocentric sphere results in the orbital launch geometric parameter relationship shown on Fig. 3.

It is of interest to correlate the four basic parameters shown in Fig. 3, to determine the extent that launches other than in-plane trajectories are feasible, to define launch window size, to discuss the launch-on-time problem, and to determine the penalties of late launch.

In Fig. 3, the out-of-plane launch angle μ_L is determined by the plane of the flight path. This plane, in turn, is determined by the geocenter and the position vectors $\bar{L}$ and $\bar{S}$. If $\bar{L}$ and $\bar{S}$ are treated as unit vectors, then

$$\left.\begin{aligned} \bar{S} &= iS_x + jS_y + kS_z \\ \bar{L} &= iL_x + jL_y + kL_z \end{aligned}\right\} \qquad (2)$$

where i, j, and k are unit vectors in the x, y, z coordinate system shown. The positive x axis is the direction of the ascending node of the OLF plane with respect to the lunar plane and the x-y plane lies in the lunar plane. In such a system

$$\left.\begin{aligned} S_x &= \cos \alpha_S \\ S_y &= \sin \alpha_S \\ S_z &= 0 \end{aligned}\right\} \qquad (3)$$

$$\left.\begin{aligned} L_x &= \cos \alpha_L \cos \delta_L \\ L_y &= \sin \alpha_L \cos \delta_L \\ L_z &= \sin \delta_L \end{aligned}\right\} \qquad (4)$$

By taking the dot product of $\bar{L}$ and $\bar{S}$, the following relationship is obtained:

$$\beta = \text{arc cos}\left[\cos \delta_L \cos(\alpha_S - \alpha_L)\right] \qquad (5)$$

From spherical trigonometry

$$\tan \delta_L = \sin \alpha_L / \cos \phi \qquad (6)$$

$$\sin \mu_L = \sin \alpha_S \sin i / \sin \beta \qquad (7)$$

For a specific ϕ, and with α_S & α_L as variables, calculations based upon the above relationships will yield the entire spectrum of possible trajectories between the OLF circular orbit and the lunar orbit. For an OLF plane inclination, i = 30°, and a lunar plane inclination, η = 23° (this η is representative of the 1964 time period), ϕ will range between 53° and 7° over its 54-day cycle. For discussion and illustration, these boundary values of ϕ (53° and 7°) are chosen as typical and are used for more detailed analysis.

The entire spectrum for $\phi = 53°$ and 7° is displayed in Figs. 4 and 5 where the central angle and plane change requirement can be determined for any position of the moon at arrival and satellite at launch. On these figures are superimposed lines of constant central angle β (light lines) and constant plane change μ_L (heavy lines). The $+\mu_L$ and $-\mu_L$ lines represent plane changes to the left and right of the OLF orbit plane. Each point on the α_S vs α_L grid defines a unique pair of transfer paths. One path (the indicated μ_L and unbracketed β's) requires a $\mu_L \leq 90°$. The alternate route is the explement of the latter path and requires a launch "backwards" ($\mu_L > 90°$). The $\mu_L > 90°$ trajectories are certainly impractical and are pointed out merely for interest.

The optimum launch time for a specified central angle occurs when the vehicle can launch with a minimum amount of expended energy. Since a plane change at the time of launch requires additional energy, the optimum launch situation exists if the transfer trajectory remains in the plane of the OLF ($\mu_L = 0$) thus taking maximum advantage of the orbiting velocity of the OLF. With the transfer orbit plane and the OLF plane coplanar, the vehicle will arrive at the moon at the line of nodes of the OLF and lunar planes. Therefore, the ideal transfer trajectory for a given central angle occurs when the moon is at either node (i.e., α_S is 0° or 180°) at vehicle arrival. Since the moon is rotating in its orbit at a mean rate of 13.2 deg/day, and the OLF plane is precessing an average of 6.7 deg/day in a direction opposite to the lunar motion, the moon is moving "effectively"* 19.9 deg/day relative to the OLF - lunar plane node line. At this rate, an optimum launch situation presents itself every 180/19.9 = 9.05 days which is the time required for the moon to travel from the ascending node to the descending node using a linear average rotation rate for the OLF plane-lunar plane node line. For gross determinations of the cyclic

---

*"Effectively" is used because a specific determination of the relative motion between the moon and the line of nodes would have to account for 1) what portion of its elliptical orbit the moon is in, 2) the inclination ϕ, 3) the sinusoidal variation in satellite precession rate, and 4) the nonlinear rotation rate of the node line between the OLF and lunar planes, which has a significant effect on the launch window cycle.

occurrence of the launch window, the linear rate is useful. A discussion of the effect of introducing the nonlinear motion of the node line is contained in Sec. 7 and must be considered in any accurate determination of the interval between escape windows.

Figure 6 shows schematically the mechanics of launch window occurrence for each time the moon crosses the node line. This figure shows the first window occurring when the moon is at First Quarter (for illustration purposes only). The size of the launch window is determined by the value of ϕ at the time of launch. Specification of ϕ for any i, h, or η determines the rotational rate of the node line. The second window occurs when the counterclockwise motion of the moon and the clockwise movement of the node line causes the moon to again intersect the line of nodes. The size of the second window differs from the first because of the changing nature of ϕ during the interval between windows. The dependence of window size upon ϕ will be described in a later paragraph.

5. Trajectory Analysis

5.1 Description of Computer Program

The digital computer program for automatic trajectory computation in the study is a restricted three-body program based upon a modified Buckheim model of the earth-moon system (Ref. 1). The program ignores the following perturbing forces: 1) effect of the sun and other planets, 2) solar radiation pressure, 3) eccentricity of the moon's orbit, and 4) oblateness of the earth and the moon.

The effects of these perturbations upon a lunar trajectory are listed: 1) the presence of the sun can alter the required burnout velocity as much as 10 fps, 2) the effect of radiation pressure is less than that of the sun, 3) the velocity difference between launching to the moon when it is at apogee from the value at perigee is a 50 fps increment. A precise figure for the perturbing effect of oblateness was not determined; however, it is of a magnitude comparable to the sun's influence.

For the purpose of this analysis, interest lies mainly in general analysis, and comparisons of various trajectories. The restricted three body program is adequate (and in some cases desirable) for this type of computation. Such a simplified program is faster and less complex to operate than the more sophisticated routines. Calculations were performed on an IBM 7090 computer.

5.2 Two-Dimensional Analysis

Using the above mentioned IBM program, a two-dimensional analysis of the trajectory conic was performed in the plane of the lunar orbit to determine the relationship between the various in-plane dynamic parameters displayed on Fig. 7. The results are shown on Fig. 8 where the trip time, central angle, launch burnout velocity, and vehicle arrival velocity at pericynthion≠ are correlated. The position of the moon at launch was timed such that the pericynthion altitude of the transfer trajectory was 50 naut miles and pericynthion velocity could be correlated with launch burnout velocity. Launch altitude for this study was 263 naut miles (303 statute miles) and burnout flight path angle was specified as zero (burnout at perigee of the transfer conic). The information on Fig. 8 was then applied to determine in-plane data for all trajectories in this study, regardless of their inclination to the equator. This assumption is within the accuracy of the computer program employed. Fig. 8 indicates that central angle and trip time increase as burnout velocity decreases. As β approaches V_{BO} minimum ($\beta = 180°$), the velocity reduction becomes insignificant compared to the increased trip time.

If circular orbit about the moon is desired, an incremental retrothrust must be imposed on the pericynthion velocity in order to achieve lunar circular orbit velocity (5380 fps) at the 50 naut mile altitude. The velocity increment described by the equation below includes the retrothrust requirements at the moon plus the in-plane velocity that must be added to the circular orbit velocity at earth to launch the transfer trajectory.

$$\Delta V_\beta = (V_{BO} - V_{circ_e}) + (V_{p_m} - V_{circ_m}) \qquad (8)$$

If the moon at arrival is not at the node line, then a plane change angle μ_L must be negotiated at the time of launch to accomplish lunar rendezvous. Thus, an in-plane angle β requires the V_{BO} described in Fig. 8; if the plane of the trajectory is inclined to the OLF plane by μ_L, then an additional velocity increment must be added to obtain the total energy

≠Pericynthion is the point of closest approach of a trajectory to the lunar surface.

requirement i.e., $\Delta V_{total} = \Delta V_B + \Delta V_{\mu_L}$. ΔV_{μ_L} is calculated from:

$$\Delta V_{\mu_L} = \left[V_{circ_e}^2 + V_{BO}^2 - 2V_{circ_e} V_{BO} \cos\mu_L\right]^{\frac{1}{2}} - \left[V_{BO} - V_{circ_e}\right] \quad (9)$$

6. Launch Window

It has been indicated that launch should occur when the resulting transfer trajectory can cause rendezvous with the moon at the line of nodes. If launch does not occur on time, an adjustment must be made in trajectory central angle β or the launch plane change angle μ_L or both. These remedies require additional energy, the amount of which increases with late time. An area exists (shown shaded in Fig. 3) which extends on either side of the line of nodes that can be reached with feasible alterations in the original programmed trajectory. The time required for the moon to pass through this area of feasibility defines what can be termed a geometric window. More important than the geometric window are the adjustments made at the launch point for the purpose of providing a "launch window." A precise determination of the window size or length of time which is available for launch when the moon is in this area must take into account the following: 1) OLF-lunar plane geometry, 2) the launch technique, and 3) the amount of additional energy (called ΔV_{ADD}) that is carried to provide a launch window. Detailed discussion of each of these items follows.

6.1 OLF-Lunar Plane Geometry

Arbitrary limits of $-10° \leq \mu_L \leq +10°$ and $150° \leq \beta \leq 180°$ are applied for definitive and display purposes. The energy or velocity requirement in this range can be determined by adding the burnout velocity corresponding to a given β directly to that ΔV specified by μ_L. The arbitrary limits are seen to be practical limitations as well since a 10° plane change requires 1200 fps additional velocity, and a reduction in β from 180° to 150° increases the in-plane burnout velocity 1284 fps (from Fig. 8). The sum of these two results is approximately a 25% velocity increase over a more ideal case (e.g. $\beta = 170°$, $\mu_L = 0°$).

The range enclosed by the imposed limitations is shown shaded in Figs. 4 and 5 and enlarged in Figs. 9 and 10. When the moon is outside this range, launch becomes inpractical because of excessive energy requirements. It is noticed by comparing Figs. 9 and 10 that the launch window increases in size as ϕ decreases and β increases. At $\phi = 53°$ the maximum window size, for the average relative motion rate considered, occurs at $\beta = 150°$. This window increases considerably for $\phi = 7°$ under the same energy and plane change limitations, as shown in Fig. 10. The maximum window possible occurs when $\phi = 0°$, i.e., the OLF and lunar planes are coplanar. This condition cannot occur for $i = 30°$ and $\eta = 23°$ as used in the illustrative figures. It could possibly occur in 1969 with both i and η equal to 28.5°, and proper positioning of the OLF orbit plane. The effect on window size of the nonlinear relative motion rate is discussed in Sec. 7.

6.2 Launch Technique

Two general approaches that exist for the accomplishment of lunar missions from an OLF are 1) trip time may be held constant(β preselected and fixed), and 2) trip time may be flexible (β variable and determined by lunar position at the time of launch). Trip time (T_b) is defined as that time elapsed between burnout at earth and arrival at the moon. Each approach requires separate analysis for the launch-on-time consideration. Holding the trip time constant has all the advantages consistent with guidance and tracking coordination, predicting crew necessities, return trip and reentry conditions, etc. However it will be pointed out that fixed trip time carries severe late launch velocity penalties for the high ϕ inclinations. This situation may be relieved if a certain amount of trip time tolerance is introduced. A discussion of the forementioned approaches follows.

6.2.1 Fixed trip time; Specification of constant trip time and the assumption that the moon is moving with constant angular rate about the earth implies a fixed relationship between the position of the moon at the time of launch and its later position at the time of arrival. This is illustrated in Fig. 11. For this reason the arrival geometry can be used to describe launch conditions. Since constant trip time implies constant β, compensation for late (or early) launch must be made by a plane change adjustment μ_L. Figs. 9 and 10 show the plane change and central angle requirements as a function of the positions of the satellite and moon. The feasible situations within the arbitrarily chosen constraints occur each time the lunar position is in the vicinity of the node line ($\alpha_S = 180°$ or $0°$).

The time it takes the moon to pass through the launch window is a function of the relative motion of the moon and the OLF-lunar plane node line. If an average rate is used for node line precession as previously indicated, it is possible to calculate the time of passage of the moon beyond the line of nodes (late launch) for various values of ϕ. For data presentation, a nominal point for reference should be selected. In the selection of this point, an attempt should be made to pick a realistic, optimum transfer trajectory. A nominal value for β (170°) was selected which is large enough to be in the minimal portion of the β vs V_{BO} curve (Fig. 8), yet not in the excessive trip time portion of the T_b vs V_{BO} curve (Fig. 8). A zero nominal plane change angle was also selected (since this is an optimum figure). An arrival altitude at the moon of 50 naut miles is also used for reference. Such a nominal trajectory would have the following energy characteristics: 1) $\beta = 170°$, $V_{BO} = 35{,}110$ fps, 2) $\mu_L = 0°$, $\Delta V_{ADD} = 0$ fps, 3) arrival velocity at the 50 naut mile lunar altitude = 8530 fps, and 4) a lunar retrothrust increment of 8530 - 5380 = 3150 fps must be supplied to achieve lunar orbit.

The total nominal velocity increment that the reference trajectory requires for a lunar orbit mission is obtained by adding the various components,

$$\Delta V_{nom} = (35{,}110 + 0 + 3150) \approx 38{,}260 \text{ fps}$$

Actually, since the OLF at the specified altitude already has a circular velocity of 24,999 fps, the total additional energy to escape earth orbit and capture into lunar orbit is (38,260 - 24,999) = 13,261 fps for the nominal case mission. The additional energy packages ΔV_{ADD} that are used to parameterize the window analysis are increments in addition to the nominal amount of 38,260 fps and represent fuel carried for deviations from the nominal. The fixed trip time launch windows for various ΔV_{ADD} are shown in Fig. 12. Window size, which includes early launch time as well, is determined by doubling the time for a given velocity increment. It can be shown that the window size decreases with decreasing β and this factor is important in the overall mission analysis to emphasize the effect of any changes of the escape condition on the lunar end of the mission, with the resulting effect on launch window size.

Since maximum window size will occur near minimum ϕ, orbital launch missions will utilize this fact by synchronizing the launch window near minimum ϕ. Minimum attainable ϕ equals the inclination of the lunar plane to the equator minus the latitude of the launch site (AMR latitude = 28.5°). In this study, for the OLF orbit selected, minimum ϕ is 1.5°.

6.2.2 Variable trip time: While the fixed trip time approach involves holding the transfer trajectory central angle constant and making a plane change to allow for late or early launch, another possibility for adjustment exists which would not involve any plane change, requiring instead a reduction in the central angle. This is described in Fig. 13. Although the picture shows the moon at arrival fixed in space, it should be remembered that the transfer plane is moving in space (coplanar with satellite plane) so that the moon at arrival is moving at an angular rate equal to the precessional motion of the satellite plane. A reduction in central angle results in a faster transfer. This is a more optimum procedure because an adjustment in central angle for a given increment of tardiness involves less additional energy than a plane change. However, the variable trip time technique that is considered here actually involves a combination of plane change and central angle change. That is, for a given position of the moon at the time of launch, there exists a combination plane change and central angle which requires a minimum energy mission velocity. Also to be considered in the variable trip time approach is the fact that, as central angle decreases, inplane burnout velocity and arrival velocity at the moon increase. A retro-thrust increment is anticipated at the moon for landing or orbiting. This increment will increase as the central angle decreases and is accounted for when comparing the fixed and variable trip time approaches. Using data from plots for various values of ϕ between 7° and 53° the variable trip time window size for different values of ΔV_{ADD} is shown in Fig. 14.

6.3 Comparison of Techniques

A comparison of the results of the variable and fixed trip time approaches (Figs. 12 and 14) illustrates the advantage that the variable trip time technique holds over the fixed trip time approach. It should be emphasized that any lunar shot from orbit would most likely be synchronized to coincide with near minimum ϕ values. Only when the initial attempt fails and the satellite plane is committed to a given ϕ cycle, do the higher ϕ inclinations become pertinent. Values of ϕ for a typical series of launch opportunities were shown in Fig. 2. The change in ϕ during waiting periods is apparent.

7. Adjustment for Nonlinear Nodal Precession

The discussion in the previous sections and related figures has assumed a linear or constant relative motion between the moon and the OLF Plane-lunar plane node line. Actually, depending on the values of i and η, the node line position, defined by Δ, is a nonlinear function of λ_N. This nonlinearity is evident in the following analysis and can be calculated for any value of i and η.

From Fig. 1, the following trigonometric relationships are obtained.

$$\cos i = \cos \eta \cos \phi + \sin \eta \sin \phi \cos \Delta \tag{10}$$

or

$$\cos \Delta = \frac{\cos i - \cos \eta \cos \phi}{\sin \eta \sin \phi} \tag{11}$$

From spherical trigonometry

$$\frac{\sin i}{\sin \Delta} = \frac{\sin \phi}{\sin \lambda_N} \quad \text{or} \quad \sin \phi = \frac{\sin i \sin \lambda_N}{\sin \Delta} \tag{12}$$

substituting (12) in (11)

$$\cos \Delta = \frac{\cos i - \cos \eta \cos \phi}{\sin \eta \left[\dfrac{\sin i \sin \lambda_N}{\sin \Delta}\right]} \tag{13}$$

which reduces to

$$\tan \Delta = \frac{\sin \eta \sin i \sin \lambda_N}{\cos i - \cos \eta \cos \phi} \tag{14}$$

Substituting the expression for $\cos \phi$ [Eq. (1)] from Sec. 2 in Eq. (14), the expression for $\tan \Delta$ reduces to

$$\tan \Delta = \frac{\sin i \sin \lambda_N}{\cos i \sin \eta - \sin i \cos \eta \cos \lambda_N} \tag{15}$$

From (15), with i and η constant, Δ is a nonlinear function of λ_N and $d(\Delta)/d(\lambda_N)$ is not a constant. This is illustrated in Fig. 15 for an i = 30 and various η's. Since the interval between window occurrence is a function of the rate of change of α_S, this interval will vary depending on i, η, and the position of the previous window. The variation in this interval between windows is illustrated in Table 1 for various η's.

Table 1 Interval between windows (i = 30 deg)

η	Window	α_s = const	α_s = nonlinear
	0	0 days	0 days
28.5	1	9.05 days	5.63 days
	2	9.05 days	10.7 days
	3	9.05 days	10.6 days
	0	0 days	0 days
27.0	1	9.05 days	6.0 days
	2	9.05 days	10.5 days
	3	9.05 days	10.45 days
	0	0 days	0 days
25.3	1	9.05 days	6.2 days
	2	9.05 days	10.35 days
	3	9.05 days	10.40 days

The above table assumes that the first window (window 0) occurs at the condition of ϕ minimum. It should be noted that the interval between can be adjusted depending on when the ϕ cycle is entered.

The effect of this nonlinear geometrical condition on the window size must also be considered. Since window size is essentially dependent on both the geometric conditions as well as the vehicle constraints of allowable energy packages, any variation in the rate of change of the geometric conditions will adjust the window size.

Because of the condition of rapidly changing variables in the ϕ_{min} region, the graphical technique employed for adjusting window sizes is not sufficient and any conclusive statement concerning this region cannot be made at this time. A more detailed study should be made which should include programming the window calculations on a digital computer. The effect at small ϕ's is sufficiently large to indicate that the conditions producing these values might best be avoided and launch should occur a little before or after the ϕ_{min} condition and at values of ϕ greater than $\sim$ 5 deg. This situation results from the fact that, although the geometrical size of the window is large, the rate at which the node line passes through the window is so high as to nullify the physical condition.

A limiting condition exists when $\phi = 0$. This condition is met for i = 28.5° and η = 28.5°. When this occurs, as noted before, a very large window exists because launch can be made in the lunar plane.

The effect of the nonlinear motion of the node line is not altogether detrimental. By entering the ϕ cycle (establishing the OLF orbit plane) such that a launch opportunity occurs at the "knee" in the curves of Fig. 15, a short interval between windows is achieved. An interval as small as 1.3 days on either side of the ϕ_{min} point can be obtained, thus providing a second launch opportunity 2.6 days from the first. This is a one-time occurrence and can be utilized should the stay in orbit be limited by operational factors.

8. Initial Positioning of Orbit Plane and Orbital Launch Procedure

Once a satellite plane is in orbit, the rate at which the orbit precesses, and thus the rate of change of the satellite plane inclination ϕ to the lunar plane, is fixed. Although the cyclic nature of ϕ cannot be controlled once the orbit plane is established, the initial value of ϕ can be preselected by judiciously choosing the time of launch from earth surface. This amounts to choosing the point at which the ϕ cycle is entered. Since, for maximum launch window, it is desirable to have a small ϕ at the time of orbital launch to the moon, the initial ϕ is determined by allowing for the time T required to rendezvous with an orbital vehicle. The ϕ cycle is then entered T hours before it reaches the minimum range. If the orbital rendezvous proceeds successfully, the vehicle will be checked out and ready for launch at that time when ϕ is a near minimum and the moon is positioned to eventually coincide with the calculated launch window. Two considerations are important here. Not only is it desired that the OLF plane be situated for minimum ϕ thus defining the planar relationship, but also that the moon at that time will be in the proper location within its plane as defined by its monthly ephemeris. The OLF plane is positioned correctly by launching at the proper time of day. The position of the moon is relatively established by launching the satellite at the correct time of the month.

Vehicle rendezvous would preferably be completed early and launch would await orbit plane precession into the optimum launch position. If for some reason there is a launch delay, then the vehicle waits for the next pass (or orbit), and the next, etc. As the delay time increases, the velocity penalty mounts. The rate of increase of this penalty is a function of ϕ and the central angle β (or trip time) which has been selected for the cislunar mission. If it is not necessary to hold β constant, the variable trip time approach may be used and the launch-on-time penalties are relieved somewhat. Once the late time becomes so large that the velocity penalty exceeds that reserve which has been set aside for this purpose, the mission would be postponed until the OLF plane and moon position are again in position for optimum launch. However, some days later, ϕ will have changed from its minimum to a higher value. For this second opportunity there will be less time available for launch (smaller launch window).

9. Effect of Changes in Nominal Parameters

This analysis has assumed a specific altitude and inclination for the OLF as well as a nominal transfer trajectory and lunar orbit altitude. It is the intent of the following section to indicate the extent which deviations in these parameters have on the launch window calculations.

Data from Ref. 3 illustrate the relationship between burnout velocity, retrothrust velocity increment at the moon, central angle, and trip time for 150 and 300 naut mile earth orbit altitudes. This altitude selection is representative for this study because higher than 300 naut miles borders on the Van Allen radiation belt while an altitude lower than 150 naut miles enters the fringe of the atmosphere with high drag implications.

Using the data from Ref. 3, energy comparisons were made for two altitudes of the OLF by employing the following relationships:

$$\Delta V_{total} = \left[V_{BO\ nom} - V_{BO}\right]_{\beta} + \left[V_{R_{nom}} - V_R\right] + \Delta V_{\mu_L} \tag{16}$$

where ΔV_{total} is the energy package to provide the escape window, $\left[V_{BO_{nom}} - V_{BO}\right]_{\beta}$ is the velocity difference due to a change in central angle between the selected transfer trajectory and the nominal, $\left(V_{R_{nom}} - V_R\right)$ is the lunar retrothrust velocity difference due to a change in central angle between the selected transfer trajectory and the nominal, and ΔV_{μ_L} is the velocity supplied to change planes at the time of launch. The change in mission energy with orbit altitude for a given nominal travel time of 66 hr. is obtained from Table 2.

Table 2. Nominal Lunar Mission Energy Requirements

h naut. mi.	β deg	$V_{BO_{nom}}$ fps	$V_{R_{nom}}$ fps	$V_{BO} + V_R$ fps
150	170	35,800	3,200	39,000
300	170	35,070	3,200	38,270

The lower altitude mission requirement is seen to be higher by 730 fps. Thus, the nominal point velocity is affected by altitude. The influence of altitude upon window size is assessed by comparing its effect upon the ΔV_{total} from Eq. 16. A typical 5 deg. deviation (β = 165) from the 170 deg. nominal central angle is assumed and results shown in Table 3.

Table 3. Mission Energy Variation with Altitude

h	V_{BO}		ΔV_{BO}	V_R		ΔV_R
naut. mi.	β = 170°	β = 165°		β = 170°	β = 165°	
150	35,800	35,970	170	3,200	3,800	600
300	35,070	35,230	160	3,200	3,750	550
			10 fps			50 fps

The resulting ΔV differential due to altitude is 60 fps. This differential is small because the data from Ref. 3 indicate changes in net total energy requirements to be almost independent of V_{BO} and β. The difference in ΔV_{μ_L} for the altitudes being compared, for a 10° plane change, is also small and of a magnitude less than 10 fps. Therefore, the ΔV calculations can be generally considered independent of altitude. Thus the launch window can also be considered nearly independent of altitude in the altitude range of interest.

However, for the general case, this is not true since large changes in altitude will change the OLF orbit precession rate. If the precession rate is increased or decreased because of changes in the orbit parameters, then the frequency with which the launch window occurs as well as the time in a given window will be increased or decreased accordingly.

The most important effect of a change in OLF plane inclination is to influence the OLF plane-lunar plane node line precession rate as discussed in Sec. 7.

10. "Pushbutton Error"

The discussion preceding has been concerned with delays exceeding one orbit period. Of interest also is the allowable delay from the correct instant of orbital launch in any single orbit. This delay could be due to instrumentation threshold limits as well as to random causes that would result in a launch hold beyond the instrumentation tolerances.

Since the OLF is orbiting at an angular rate (for the 263 naut mile altitude) of 3.83 deg/min, any delay from the correct launch point would result in a decrease in the central angle β between the position vectors $\overline{L}$ and $\overline{S}$ in the transfer trajectory plane (Fig. 7). Assuming that information and equipment is available to sense this change and allowance made in the burnout parameters, the effect on the transfer trajectory would be to shorten the trip time. Using a $\beta = 170°$ ($T_b = 66.8$ hours) as the nominal point for a correct launch, a 3-min delay would amount to decreasing β to $\sim 158°$ with an equivalent trip time of 40 hours. In a 3 min period, the moon's position can be considered as relatively fixed. Therefore, if launch occurred in-plane, the vehicle would arrive at the moon's orbit 27 hr ahead of the moon. Thus, to intersect the moon, it would be necessary to make a plane change either at launch or midcourse. If the plane change were made at launch, the change required would be a function of ϕ existing at launch. Assuming the best conditions or a reasonably small ϕ, the plane change would be about 10° for $\phi = 10°$ and less for smaller ϕ's. But with a $\phi = 50°$, the plane change would be as high as 40°.

Figure 16 illustrates the velocity penalty at various ϕ's for "pushbutton" error up to 3 min in any one orbit. This velocity penalty is the combined effect of the plane change requirement at launch and the resultant increase in burnout velocity from the nominal value at $\beta = 170°$.

It is apparent upon examining Fig. 16, that the penalties are severe for even a delay of one minute. The alternative technique of making a plane change at some midcourse point decreases this penalty (how much depends on the vehicle design, guidance equipment, and precision trajectory data). Delaying launch to the next orbital pass is considered the most practical as well as economical approach.

11. Conclusions

Because of the relative motion between the moon and the precessing OLF orbit plane, feasible launch opportunities (launch windows) occur several times in one month. The interval between launch opportunities is determined by the relative positions of the OLF and lunar planes. These positions can be controlled to a certain degree by timing the first launch window occurrence. An arrangement between planes for this first opportunity might be selected which provides the largest possible launch window, or the arrangement might be such that the first two launch windows follow one another in a short interval (second window occurring about 3 days after first) with the third window occurring 9 or 10 days later. As might be suspected, the largest window and short interval sequence are not synonymous but by judicious orbit plane determination, a little of both can be obtained. Precise definition as to the degree which the "little of both" can be accomplished involves a detailed study in a range where the precession rate of the OLF-lunar planes node line is extremely nonlinear. The graphical techniques employed in the analysis to date are approximate but not sufficient for proper investigation of this range. A digital computer program should be used to perform the calculations accurately.

The launch window size or time available for launch is a function of three general factors: 1) geometrical relationship between planes, 2) extra propellant allowed for provision of the window, and 3) launch technique employed. The first concerns the correlation between the position parameters of the moon and OLF in their respective planes and the transfer trajectory parameters. Optimum launch situations arise when the position of the moon at arrival coincides with intersection of the node line of the OLF-lunar plane at launch and the lunar orbit. The size of the window when this coincidence occurs is a function of the inclination ϕ between the two planes at the time of launch.

Since adjustments made for early or late launch involve changes in the launch and trajectory parameters that require expenditure of additional propellant, the more propellant available the longer delay times may be tolerated.

Analysis of two techniques for altering the launch conditions to insure lunar rendezvous shows the launch window can be widened considerably if a trip time tolerance is introduced. Allowing a decrease in central angle of 3° (decrease nominal trip time by 10 hr) coupled with a 500 fps additional energy provision leads to a window several times larger than that which results from a fixed trip time restriction.

The bulk of the launch window analysis was performed for a particular OLF altitude and inclination. However, it is shown that the results are relatively independent of altitude and may be generally applied for any case within the operational orbit limits of an Orbital Launch Facility.

References

[1] Buckheim, R. W., "Artificial satellites of the moon," Rand Corp., Research Memo 1951 (June 1956).

[2] Woolston, Donald S., "Declination, radial distance and phases of the moon for the years 1961 to 1971 for use in trajectory considerations," NASA TN D-911, Langley Research Center (August 1961).

[3] Weber, R. J., Pauson, W. M., & Burley, R. R., "Lunar trajectories," NASA TN-D-866, Lewis Research Center (August 1961).

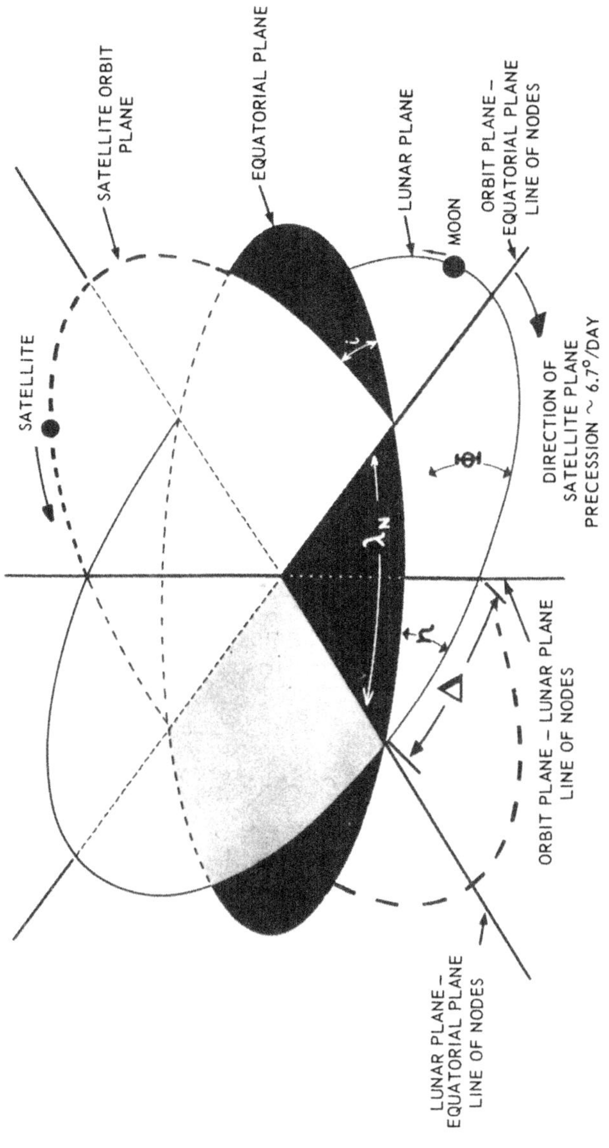

η = INCLINATION BETWEEN LUNAR AND EQUATORIAL PLANES
i = INCLINATION BETWEEN SATELLITE AND EQUATORIAL PLANES
ϕ = INCLINATION BETWEEN SATELLITE AND LUNAR PLANES
λ_N = LONGITUDE OF OLF–EQUATOR LINE OF NODES MEASURED FROM LUNAR DESCENDING EQUATORIAL NODE

Fig. 1 Geometric relationship for satellite-earth-moon system.

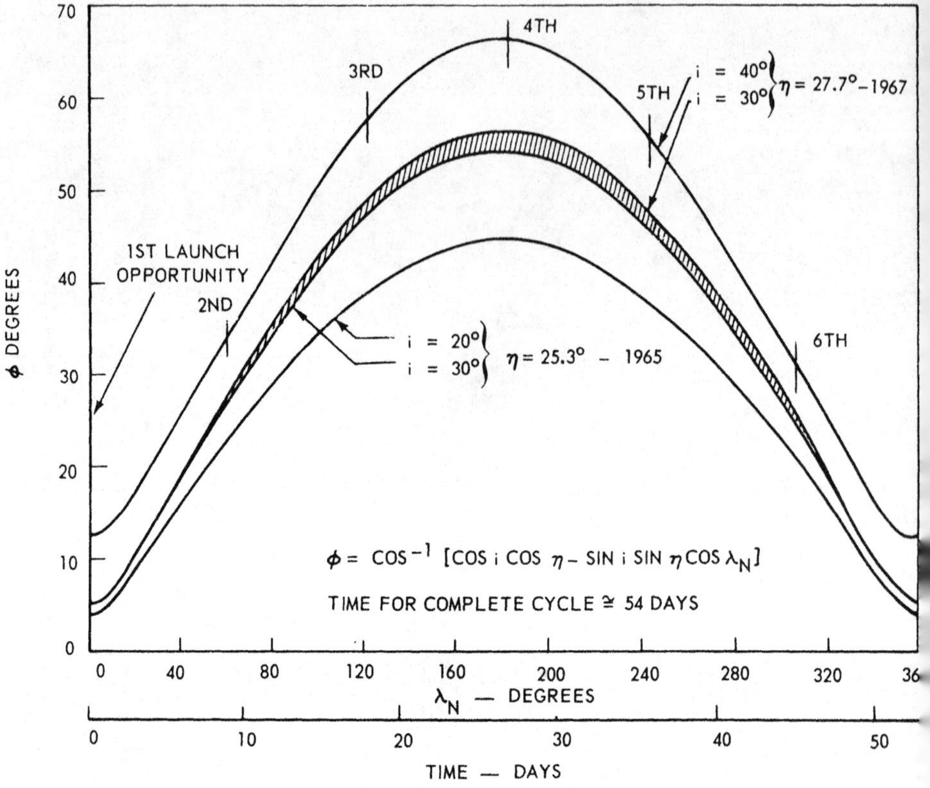

Fig. 2 OLF plane-lunar plane inclination cycle, (ϕ vs λ_N).

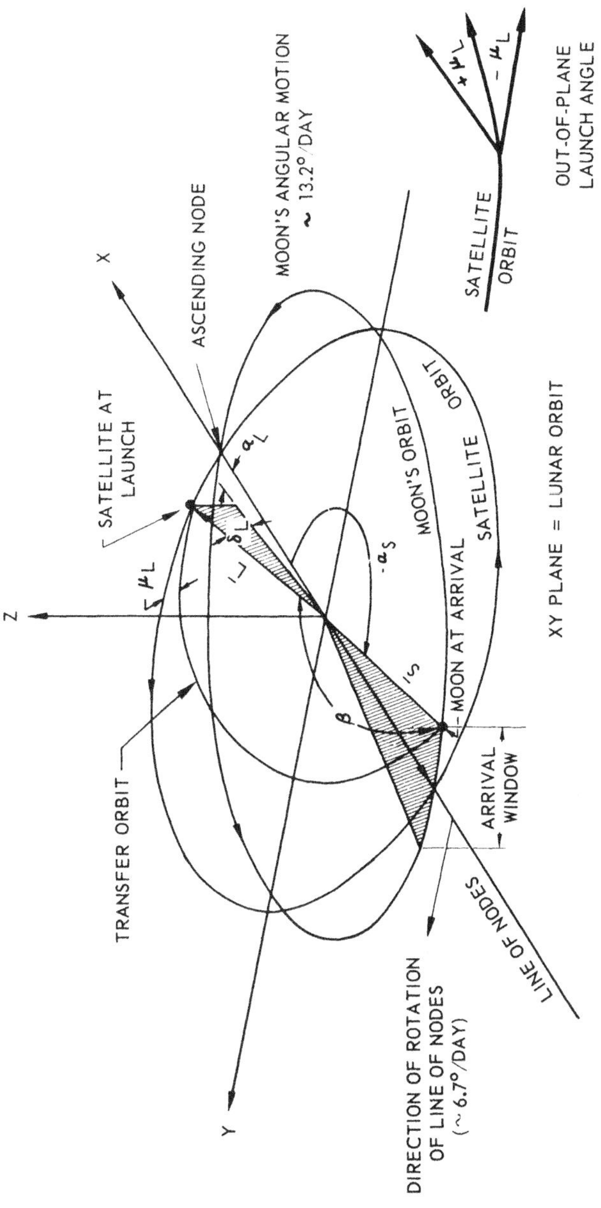

Fig. 3 Orbital launch geometric parameters.

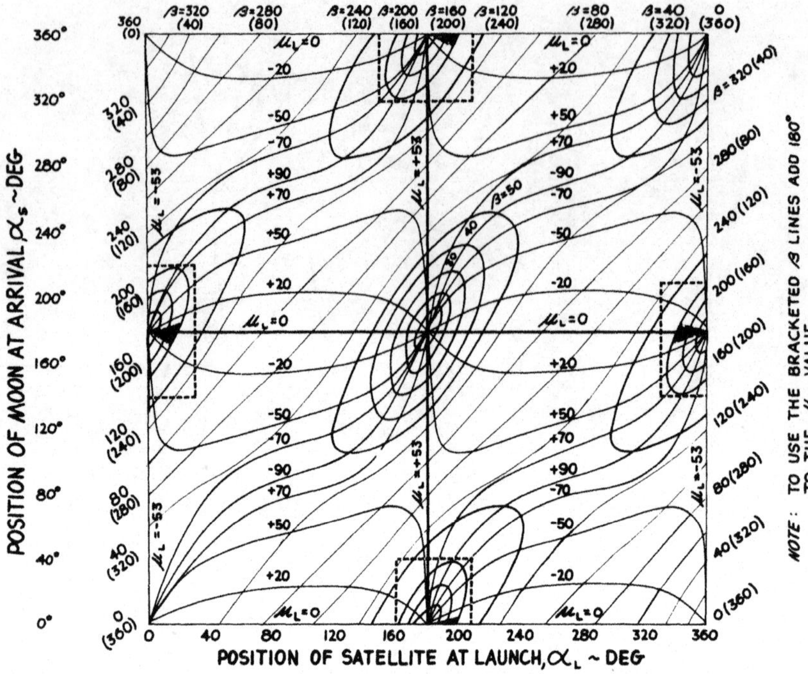

Fig. 4 Complete display of geometric parameters, $\phi = 53°$.

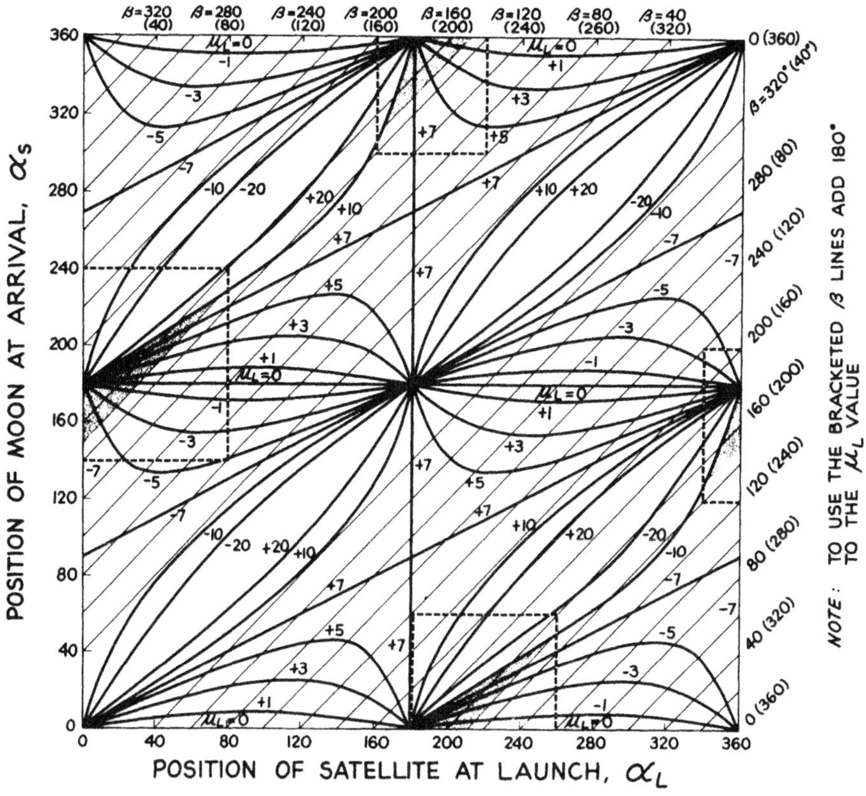

Fig. 5 Complete display of geometric parameters, $\phi = 7°$.

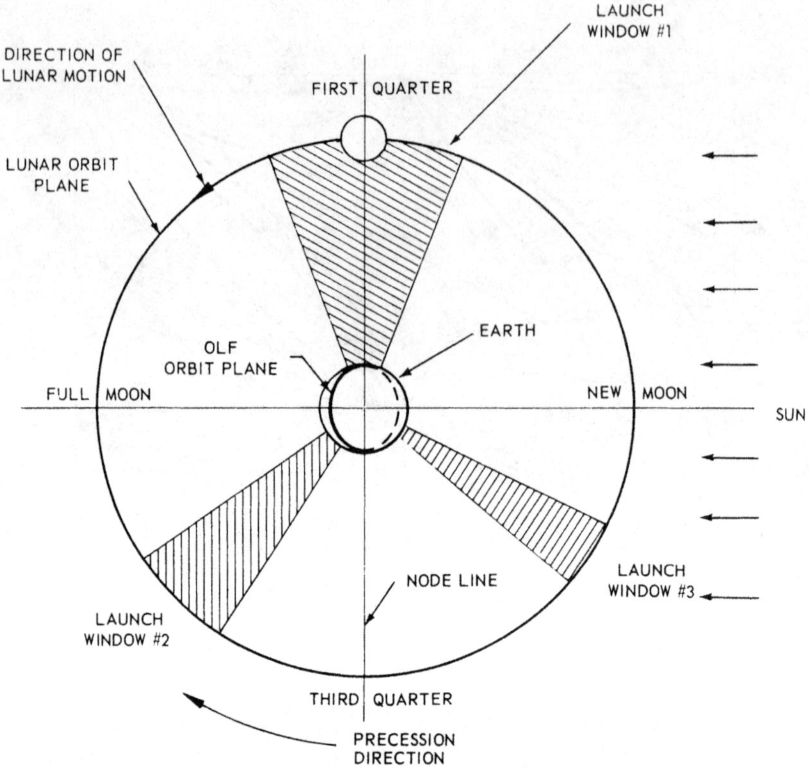

Fig. 6 Orbital launch windows.

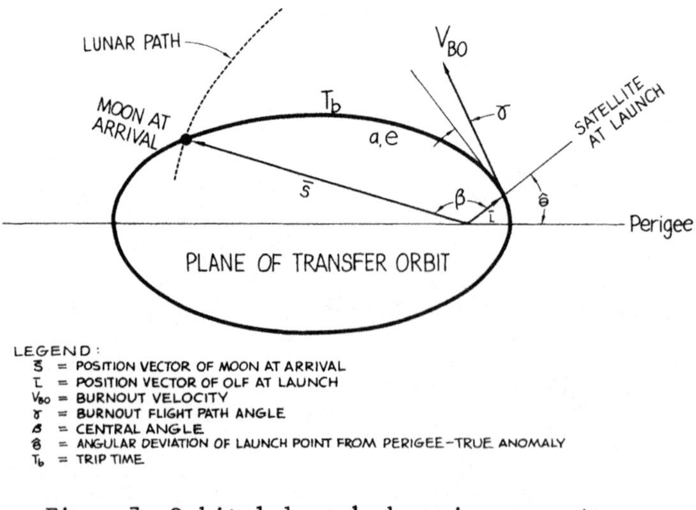

Fig. 7 Orbital launch dynamic parameters.

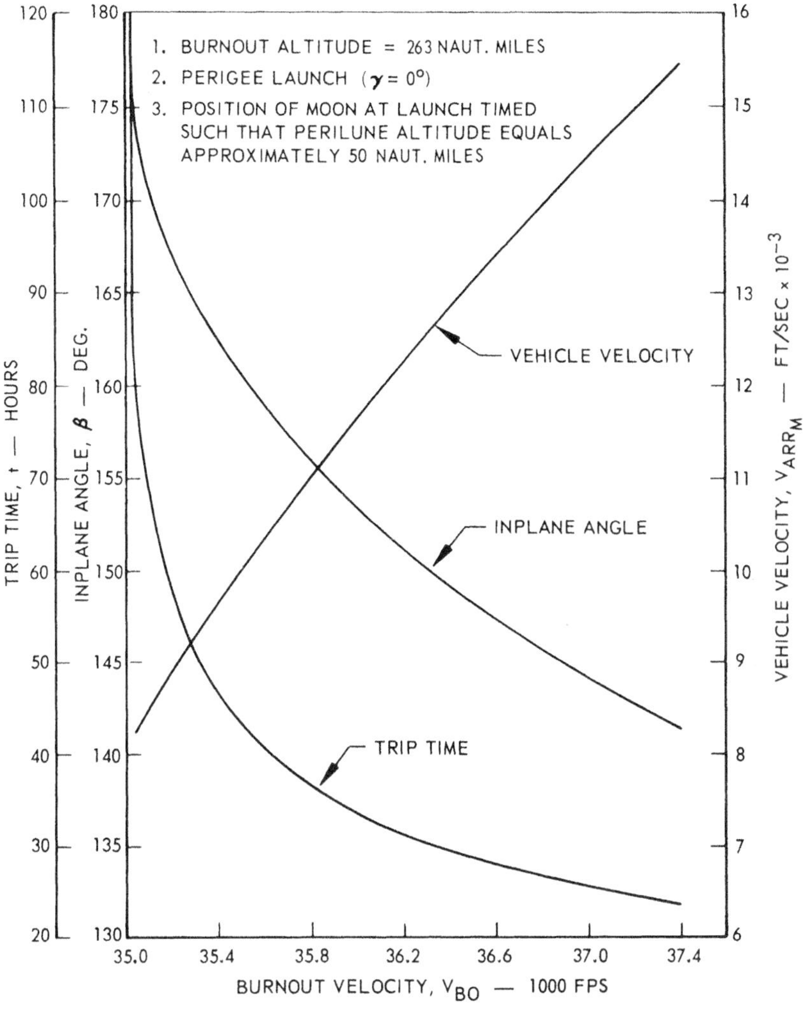

Fig. 8 Dynamic parameters display.

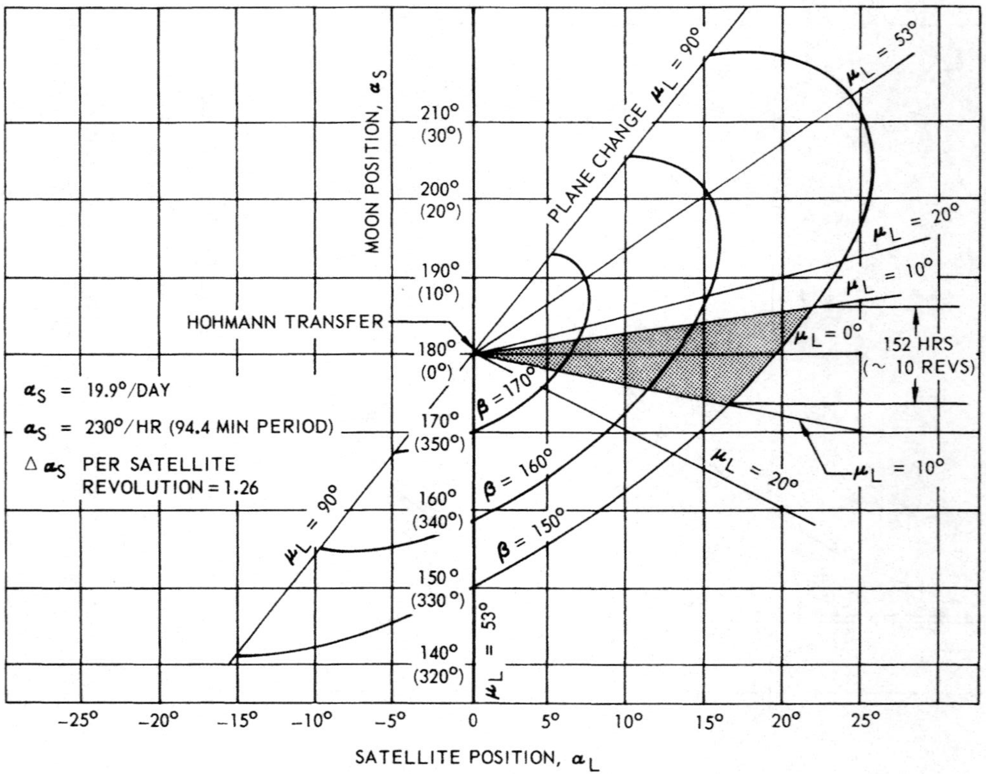

Fig. 9 Enlarged partial display of geometric parameters, $\phi = 53°$,

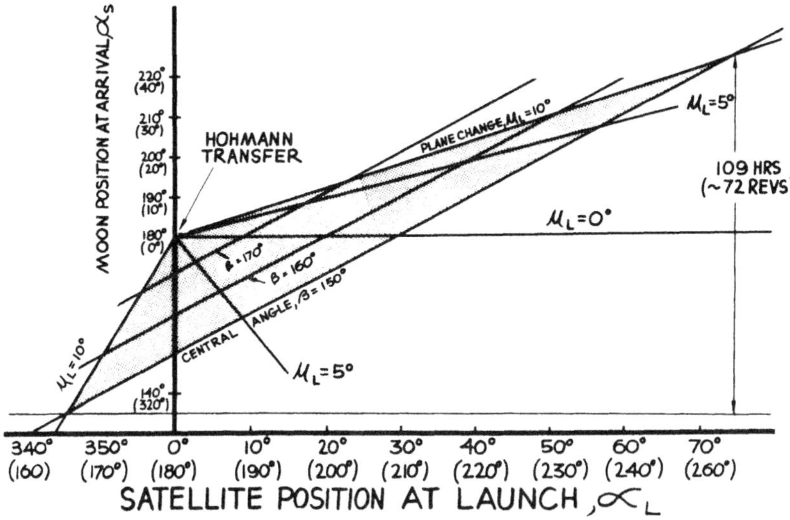

Fig. 10 Enlarged partial display of geometric parameters, $\phi = 7°$.

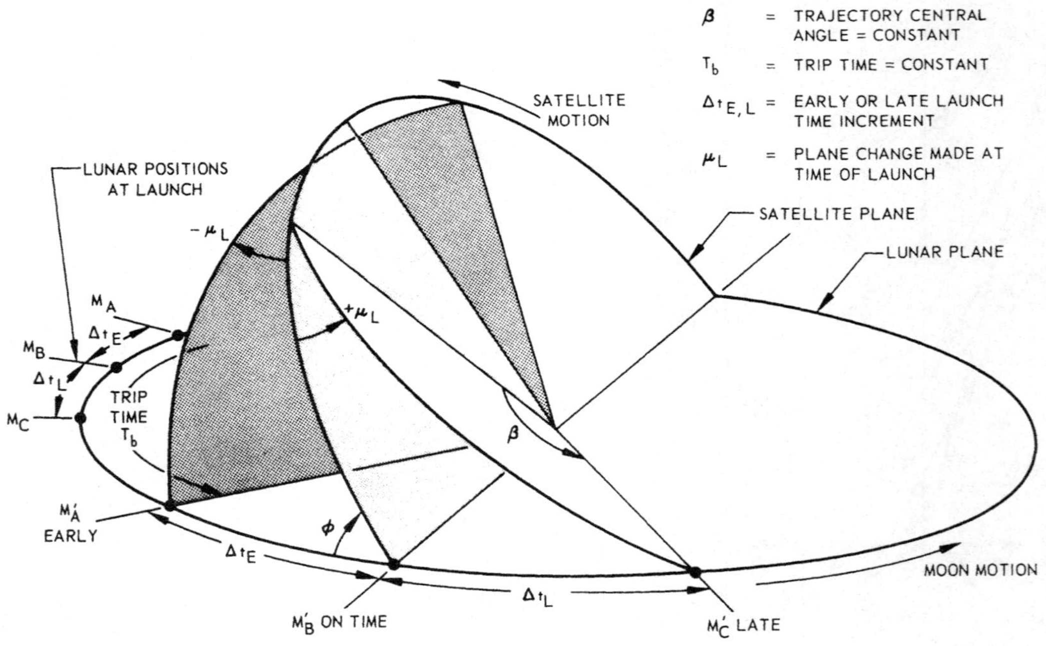

Fig. 11 Fixed trip time trajectory illustration.

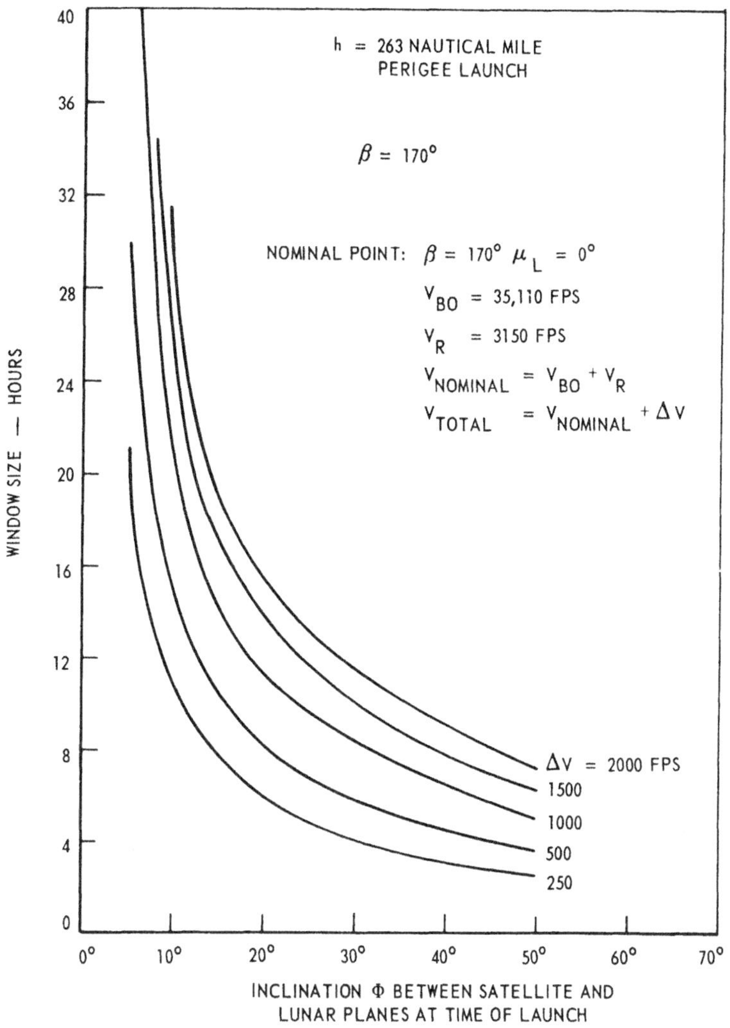

Fig. 12 Fixed trip time window size.

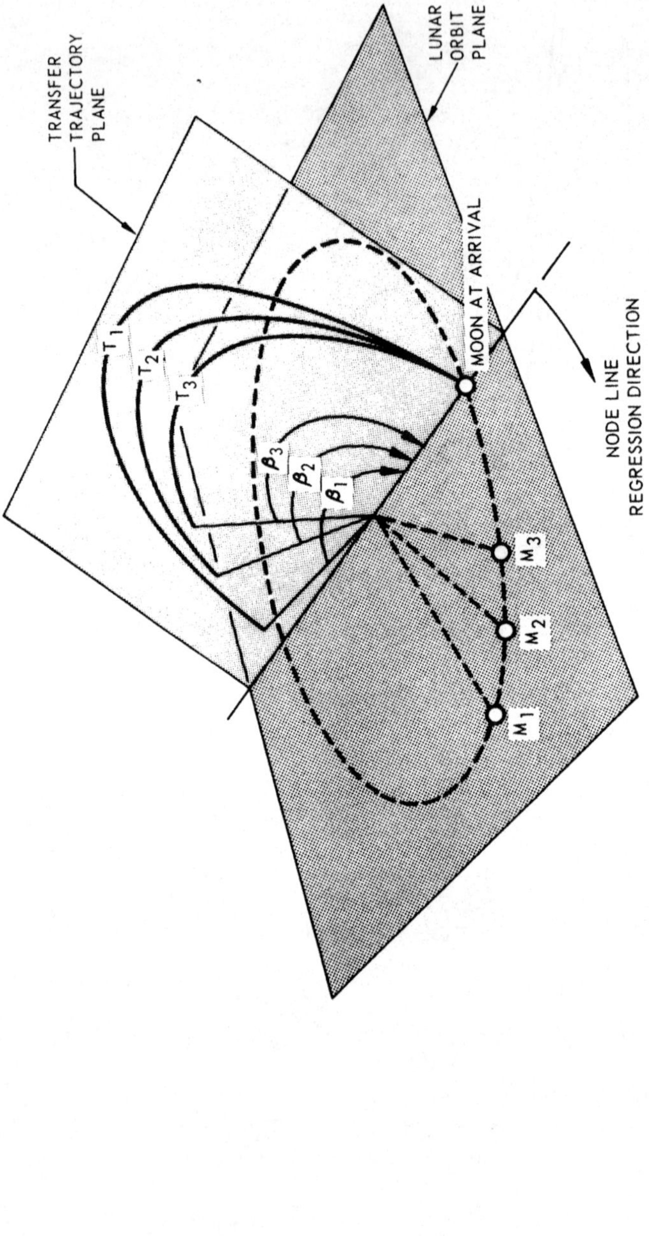

Fig. 13 Variable trip time trajectory illustration.

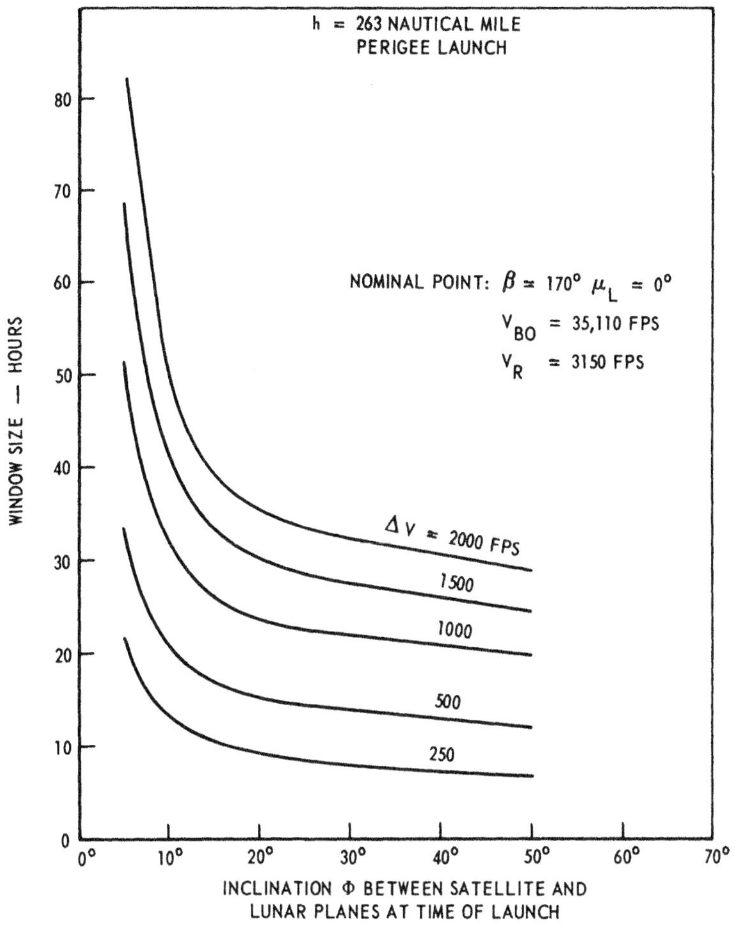

Fig. 14 Variable trip time window size.

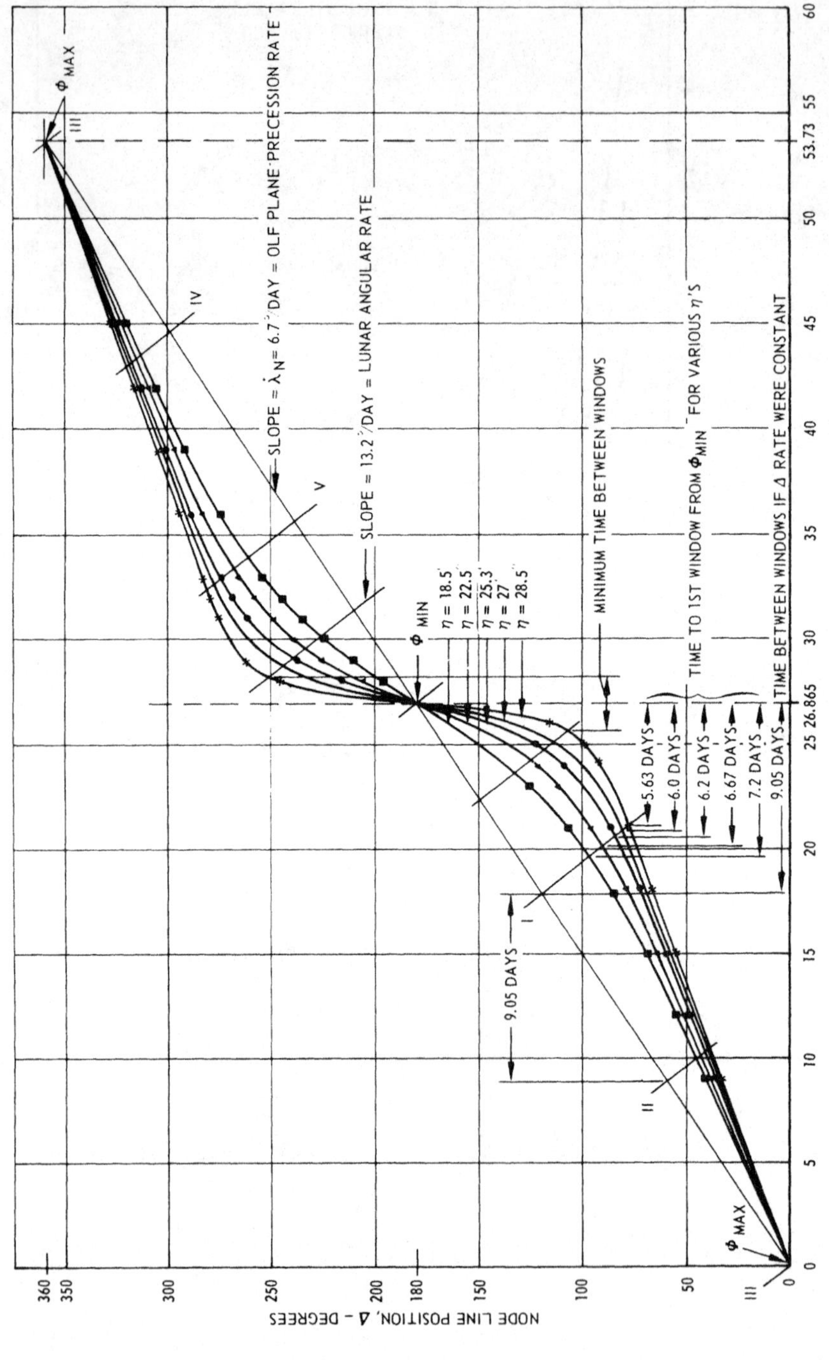

Fig. 15 Node line position variation with time.

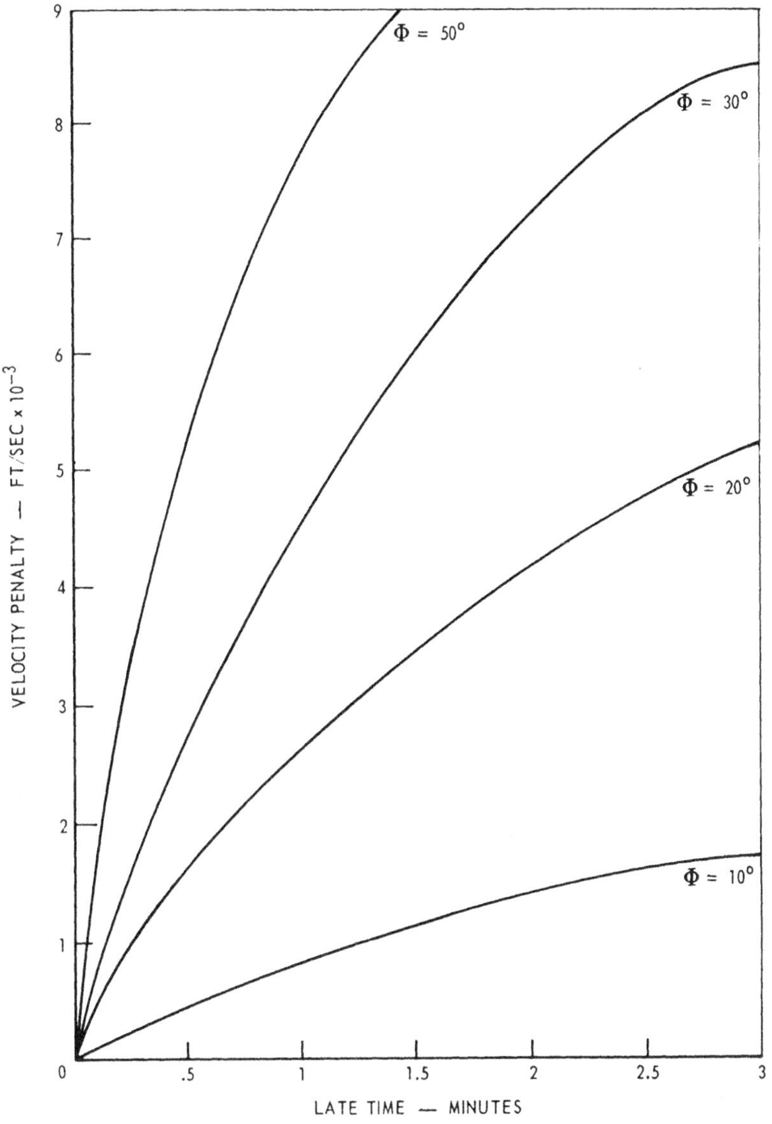

Fig. 16 Pushbutton error velocity penalty.

LAUNCH WINDOWS FOR HIGHLY ECCENTRIC ORBITS

Barbara E. Shute*

NASA Goddard Space Flight Center, Greenbelt, Md.

Abstract

Launch windows of a few hours daily are obtained for highly eccentric orbits in order that lunar and solar perturbations will not lower perigee into the atmosphere for a specified period after launch, usually a year. This paper discusses the characteristics of the perturbations, the methods employed in numerically integrating the perturbations, the construction of a launch window map, and the characteristics of the launch window map.

Discussion

Orbits of high eccentricity are useful for satellites conducting experiments at distances of several earth radii. They are feasible because the energy required to send a satellite to large distances is minimized with a near earth injection. This results in perigee close to the earth and an apogee distance that is sensitive to the magnitude of the velocity vector at injection. The satellite will spend a large percentage of the orbital period at the desired distances.

Solar and lunar perturbations may force the perigee into regions of dense atmosphere before expiration of the instrument's design lifetime. In order to ensure an orbital lifetime of at least the specified time, usually a year, a grid of possible orbits is studied by numerical integration. The grid is constructed by considering that the position and velocity coordinates of injection are fixed

---

Presented as Preprint 63-412 at the AIAA Astrodynamics Conference, New Haven, Conn., August 19-21, 1963.

*Space Scientist, Theoretical Division.

with respect to the earth's surface; the variable that is incremented is the injection time, both in hours of the day and in days of the year. The variations in perigee height during a year of satellite flight time are obtained for each value of injection time by numerically integrating the perturbations using IBM 7094 computer programs. The computer output is examined to determine if perigee went close to or below the surface of the earth.

Explorer XII is the first satellite to be launched at a time that was preselected for a minimum orbital lifetime of a year. The changes in perigee height are shown in Fig. 1. The curve was computed by ITEM,[1] a numerical integration program described later, using an injection vector derived from data reduction. The points are observations, smoothed over intervals of a week. The graph ends when the radio ceases to function, 4 months after launch. However, it can be extrapolated that the maximum perigee height is 1000-1100 km, after which the perigee height will start to decrease. The orbit will be terminated by atmosphere drag in the summer of 1963, when the perigee has been lowered into dense regions of the atmosphere.

In general, the magnitude of the change in perigee height is greater for larger eccentricities and semimajor axes. This magnitude may be as large as 600 km/orbit for a satellite with semimajor axis 145,000 km (1/3 lunar distance) and 0.95 eccentricity. A decrease in perigee height of this magnitude will cause a collision with the earth on the first return to perigee if the injection height is 200 or 300 km.

Two types of perturbations by the sun and moon on the satellite's orbital elements may be distinguished in the numerical integration results. "Long period" is used here to denote the major trend of orbital changes, which depend on the orientation of the satellite orbit with respect to the orbital plane of the disturbing body. The period cannot be correlated at present with any physical parameter but is found by numerical integration results to be on the order of 10 yrs in the earth-moon system. This type of variation is demonstrated on the eccentricity in Fig. 2. The sun has roughly the same effect as the moon in producing a long period variation. When each force is considered independently, the frequency of the variation is shorter than when both forces are used in the calculation. "Short period" is used to denote variations in the orbital elements that depend on the position of the sun or the moon.

The short period effect has a period of 1/2 the orbital period of the disturbing body (1/2 year and 1/2 month) and superposes sinusoidal oscillations on the long period variation. The amplitudes of these fluctuations are usually small compared to the total amplitudes of the long period variations. However, short period fluctuations might cause the satellite to hit the earth for low perigee heights soon after launch if the long period variation does not produce a large enough upward change to make them ineffective. The short period moon term is apparent in Fig. 1. The short period sun effect is shown in Fig. 3 as an oscillation about the long period rise in the inclination.

The oblateness does not directly perturb the eccentricity, but there is an indirect perturbation because the oblateness changes the orientation of the orbit with respect to the planes of the moon and sun.

Atmospheric drag is usually omitted in determining the approximate lifetime of a high eccentricity orbit since changes in perigee height, up or down, are large compared to the effect of drag at one perigee level, that is, perigee will move rapidly into regions of greater or lesser atmospheric density. If perigee is low enough so that the drag perturbation predominates, the satellite orbit will decay in a very few revolutions. Fig. 4 shows the reduction in semimajor axis per perigee pass as a function of perigee altitude. It will be seen that above 250 km, the reduction of a 50,000 km semimajor axis (eccentricity = 0.85) is less than 1 km/orbit ($A/m = 0.23\ cm^2/g$).

When the injection conditions with respect to the earth are fixed, a selection of orientations of the satellite orbit with respect to the celestial sphere is available every day because of the rotation of the earth. Under these conditions, all orbital elements are specified except for the right ascension of the ascending node. The lifetime of the orbit may be determined as a function of this angle independent of the time of year (except for those instances where the short period effect due to the sun or the moon on perigee height are comparable, e.g., soon after launch), and the node may be converted into time of day for any given day. A "launch window map" is constructed by plotting the acceptable hours of the day vs day of year (Fig. 5).

The upper boundary is formed by those orbits that last exactly a year. This is caused by a change in phase in the long period variation; therefore, the same orientation with respect to the ecliptic and moon's orbital plane will produce the same orbital lifetime. To obtain this orientation requires a fixed value of the right ascension of the ascending node. A fixed value of the right ascension of the ascending node, using a set of earth-fixed injection conditions, is obtained by varying the launch time by 4 min/day to compensate for the rotation of the earth around the sun. Orbits between the two lines exist several years before perigee becomes too low. The lower boundary is formed between orbits that have an initial downward trend of perigee and initial upward trend of perigee. The reversals in curvature are due to the short period effect of the sun; therefore, the pattern repeats itself every 1/2 yr. The Halphen method,[2] which provided the numerical results used in constructing the launch window map, averages over the orbit of the moon and, therefore, does not provide information on the short period effect of the moon. Early orbits must be recalculated by the ITEM (or similar) program if the short period effect of the moon is large enough to destroy the orbit.

Requirements, other than lifetime, such as those imposed by the experiments, may reduce possible launch times. Some of the requirements for the present generation of satellites include upper limits on the time in shadow, a lower and an upper limit on the angle between the satellite's spin axis and the sun, and specified values of the initial angle. For example, the requirement that a spacecraft may not be injected into orbit in darkness adds nearly horizontal lines of constraint to a launch window map.[3]

Two computer programs are currently employed in the construction of launch window maps. The corrected Halphen method with modifications by Peter Musen was programmed by Arthur Smith, Jr. This method requires less than one minute of IBM 7090 machine time per year of satellite flight time. This computational speed is possible because the method is semianalytic. The disturbing force of the moon is numerically averaged over both the satellite's orbit and the moon's orbit, that of the sun is expressed with the second Legendre polynomial. The effect of the earth's second harmonic on the right ascension of the ascending node and the argument of perigee is expressed analytically. The perturbations on the satellite elements are numerically integrated using integration intervals of 10 days. Typical cases computed by this method are recomputed by the ITEM program, and the results are carefully compared. The ITEM program is based on a modified Encke method. It includes perturbations due to the sun, moon, other planets and atmospheric drag and radiation pressure. One half hour of IBM 7090 machine time is used in computing the satellite's orbital history for 1 yr.

This cross checking is necessary to ensure against program errors and against omission of significant theoretical effects. For example, as mentioned earlier, the short period moon effect may cause the satellite to hit the earth during early phases of orbital flight. The effect of the amplitude of the variation in the eccentricity due to this perturbation may be obtained from the ITEM program and plotted as perigee height vs time and observing the oscillatory differences between the two curves (Fig. 6).

Further studies in this area may lead to deeper insights in celestial mechanics as well as provide practical results for satellite technology.

References

[1]Shaffer, F., Squires, R. K., and Wolf, H., "Interplanetory trajectory Encke method, program manual," Goddard Space Flight Center, Rept. X-640-63-71 (May 1, 1963).

[2]Musen, P., "A discussion of Halphen's method of secular perturbations and its application to the determination of long-range effects in the motion of celestial bodies," Rev. Geophys. 1, 85-122.

[3]Montgomery, H. E., "EGO launch window study," Goddard Space Flight Center Rept. X-643-62-225 (Jan. 1, 1963).

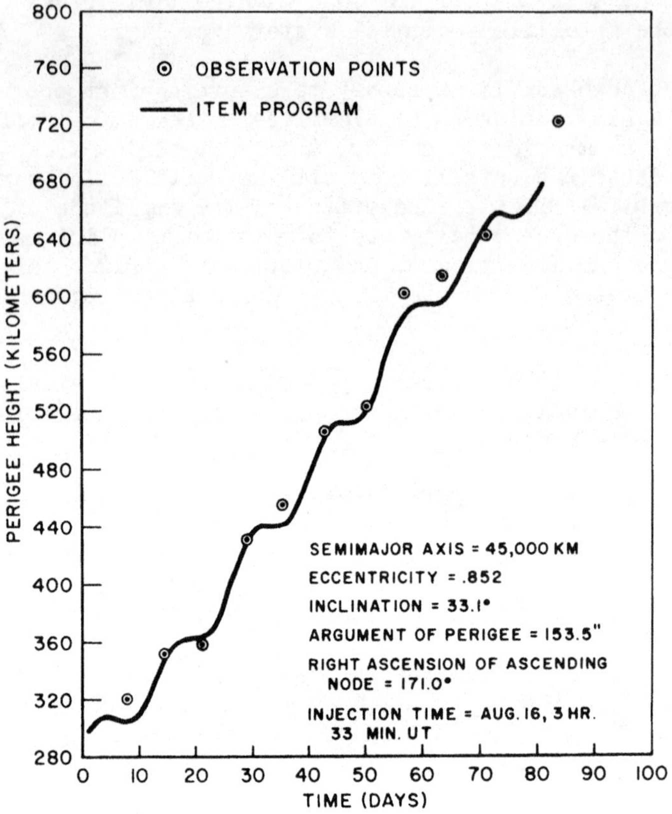

Fig. 1 Perigee rise of actual Explorer XII orbit.

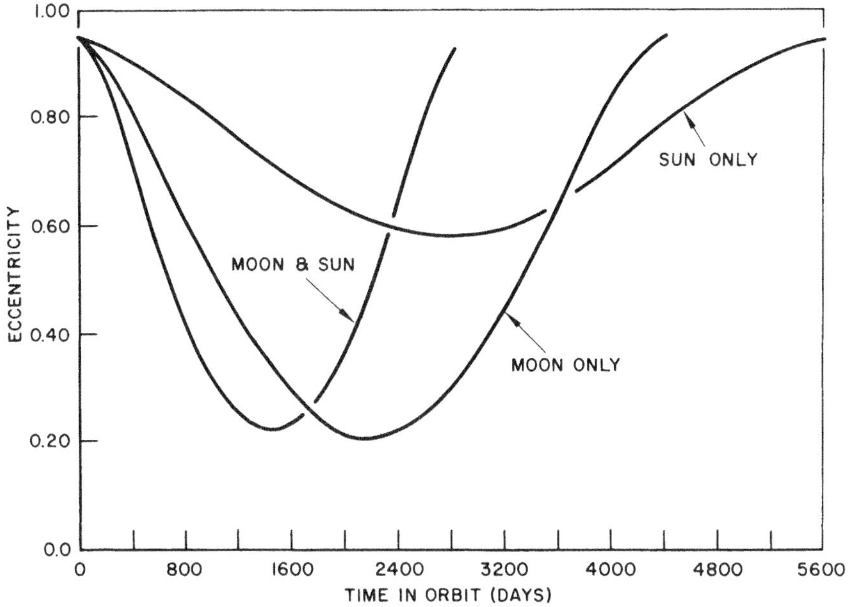

INITIAL ORBITAL ELEMENTS:
SEMIMAJOR AXIS = 145,000 KM
ECCENTRICITY = .956
INCLINATION = 33.0°
ARGUMENT OF PERIGEE = 270.0°
RIGHT ASCENSION OF ASCENDING NODE = 270.0°
(ANGLES MEASURED WITH RESPECT TO EARTH'S EQUATOR)

Fig. 2 Long period variation of the eccentricity Halphen method used for computation.

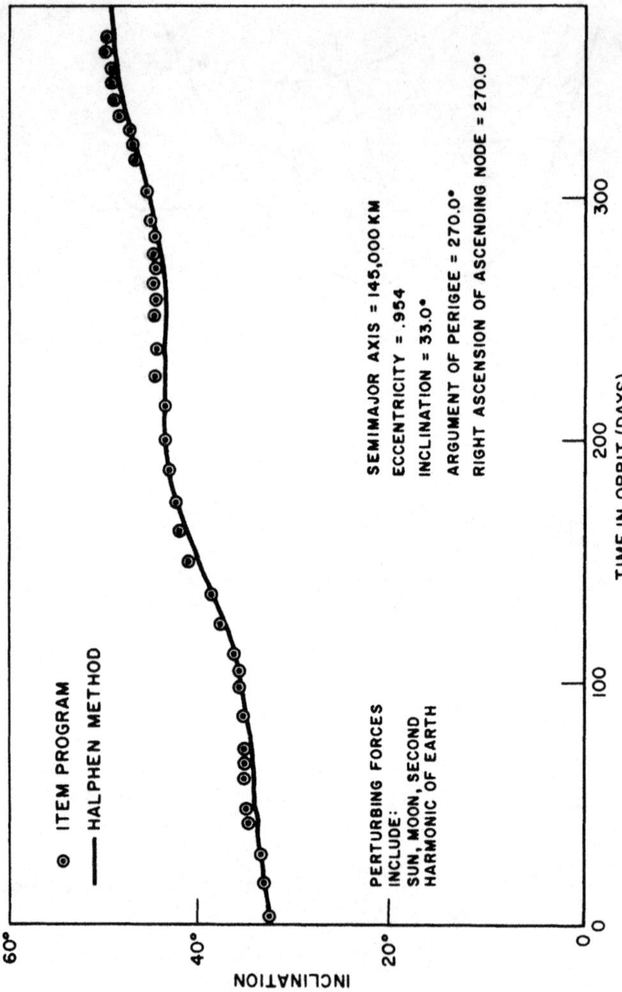

Fig. 3 "IMP" inclination vs time.

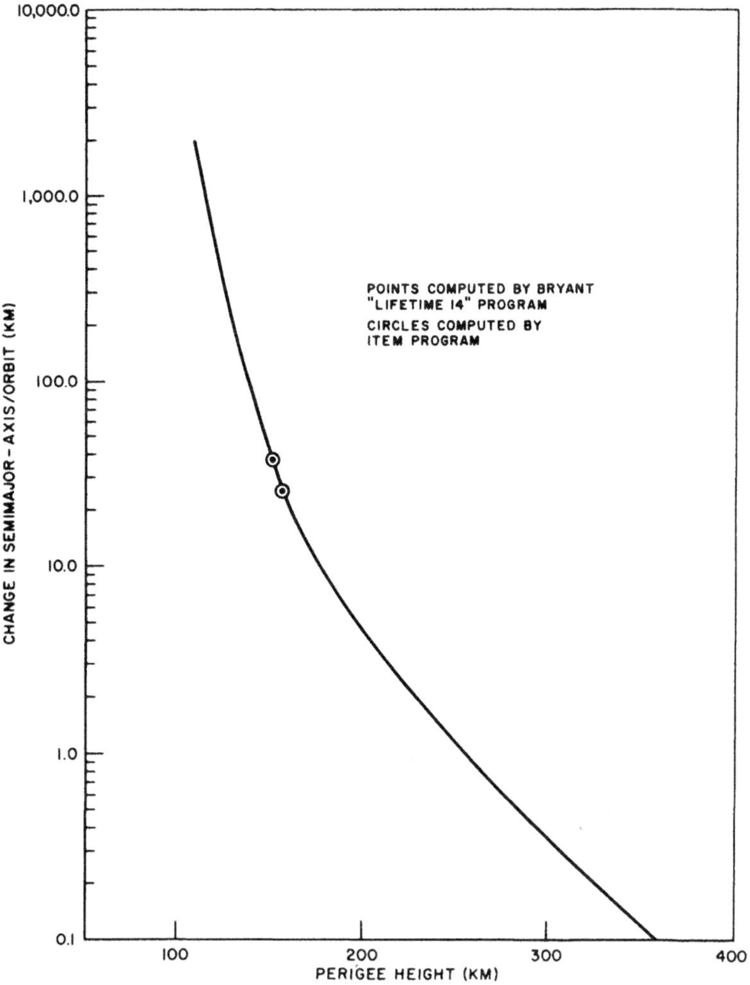

Fig. 4 Reduction in semimajor axis due to atmospheric drag per perigee pass vs perigee height.

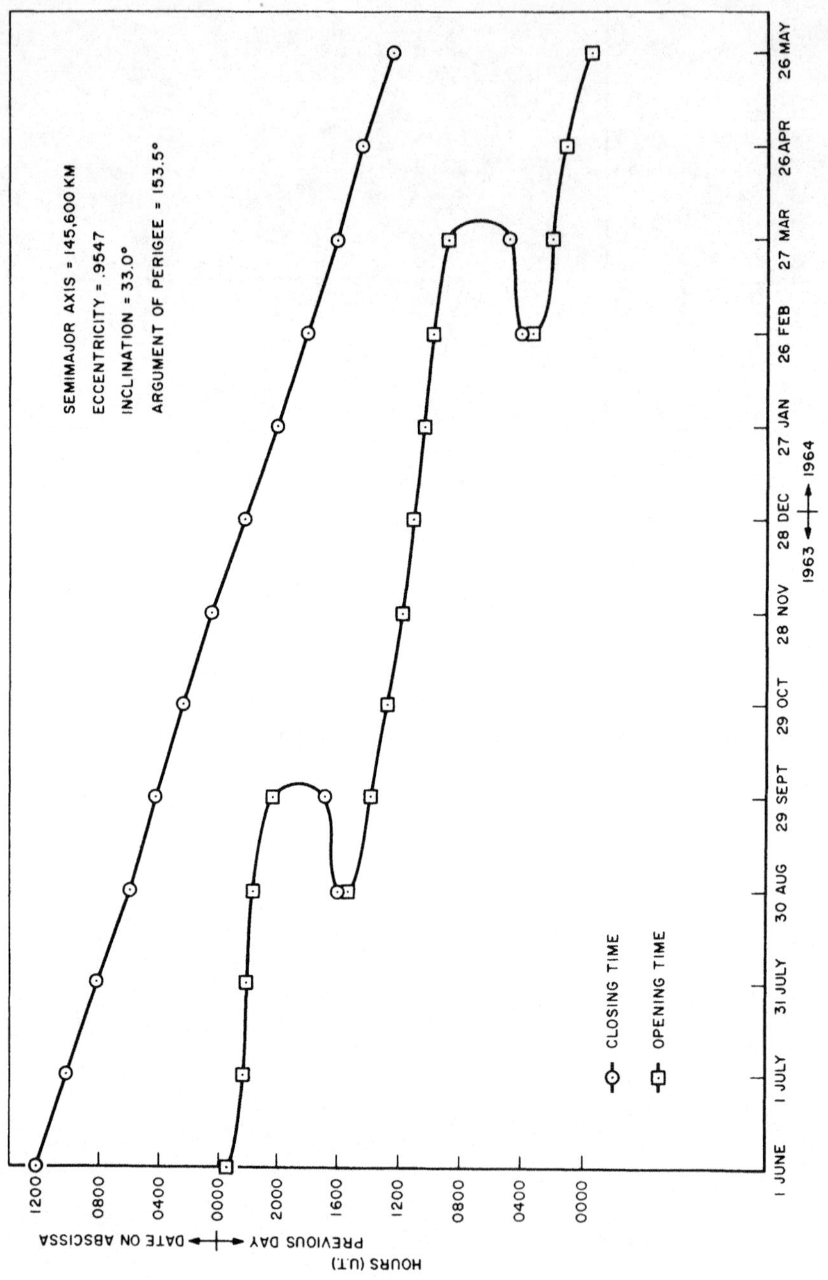

Fig. 5 Launch window map.

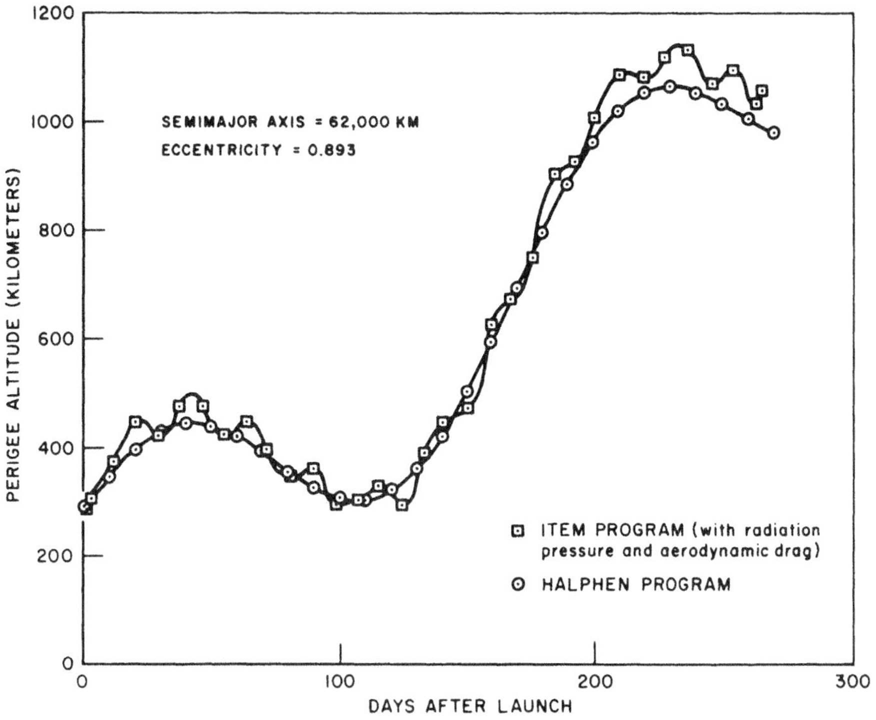

Fig. 6 Comparison of computer runs perigee altitude vs days after launch.

III.

ORBIT SELECTION

Trajectories for Return to Earth

MOON-TO-EARTH TRAJECTORIES

Saterios S. Dallas*

Jet Propulsion Laboratory,
California Institute of Technology, Pasadena, Calif.

Abstract

The design and characteristics of trajectories that initiate on the surface of the moon and terminate at a specific landing site on the surface of the Earth are presented, along with computing techniques for such trajectories. Results from an analytic trajectory-computing program are used for the qualitative discussions, and results from a precision integrating program are used for the quantitative discussions.

Nomenclature

Symbols

$\underline{(\ )}$ = vector (indicated by boldface type in figures)

a = semimajor or semitransverse axis

A = inertial azimuth measured east of true north

$\underline{B}$ = impact parameter (defined in Sec. 3. 3)

$\underline{B} \cdot \underline{R}$ = projection of the impact parameter $\underline{B}$ upon the vector $\underline{R}$

Presented as Preprint 63-402 at the AIAA Astrodynamics Conference, New Haven, Conn., August 19-21, 1963. This paper presents the results of one phase of research carried out at the Jet Propulsion Laboratory, California Institute of Technology, under Contract No. NAS 7-100, sponsored by NASA. The author thanks V. C. Clarke, Jr. for his assistance in the theory of lunar-return trajectories and D. Richardson for his assistance in the search techniques of a high-speed computer and for the programming of the necessary equations.

*Research engineer.

Symbols (Cont'd)

$\underline{B} \cdot \underline{T}$ = projection of the impact parameter $\underline{B}$ upon the vector $\underline{T}$

C_3 = twice the total energy per unit mass, $C_3 = V^2 - 2Gm/r$

e = eccentricity

G = universal constant of gravitation

i = inclination of the orbit plane to the equatorial plane

m = mass of a body

p = point of closest approach

$\underline{P}$ = a unit vector from the central body to the point of closest approach

$\underline{Q}$ = a unit vector perpendicular to $\underline{P}$ and lying in the trajectory plane

$\underline{r}$ = vector directed from the center of the central body to a second body. For example, $\underline{r}_{\oplus☾}$ is the vector from the Earth to the moon and $r_{\oplus☾}$ is its magnitude

$\underline{r}_{\ell}$ = a vector directed from the center of the central body to the launch site

$\underline{R}$ = unit vector that forms the right-handed system $\underline{R}$, $\underline{S}$, $\underline{T}$. $\underline{R} = \underline{S} \times \underline{T}$

$\underline{S}$ = unit vector in the direction of the incoming asymptote for an impacting hyperbola or the outgoing asymptote for an escape hyperbola (outward radial direction). In the case of an ellipse, a pseudo-asymptote is defined, see Sec. 3.3

$\underline{T}$ = unit vector perpendicular to $\underline{S}$ which lies in the equatorial plane

Symbols (Cont'd)

t_F = time of flight from launch to impact, hr

t_i = time of injection

v = an angle measured in the plane of the trajectory between two points of the trajectory. For example, v_{sr} is the angle between the exit point s and the re-entry point r

v_s = the angle between the closest approach direction $\underline{P}$ and the outward radial direction $\underline{S}$

V = inertial speed with respect to the central body

$\underline{V}_{\oplus\Delta}$ = inertial velocity of the probe relative to the Earth

$\underline{V}_{☾\Delta}$ = inertial velocity of the probe relative to the moon

$\underline{V}_{\oplus☾}$ = inertial velocity of the moon relative to the Earth

$\underline{W}$ = unit vector that forms the right-handed system $\underline{P}$, $\underline{Q}$, $\underline{W}$. $\underline{W} = \underline{P} \times \underline{Q}$

γ = inertial flight path angle with respect to the local horizontal, measured in a positive direction away from the central body

$\delta_☾$ = declination of the moon at launch

η_r = re-entry maneuver angle. The in-plane angle between the re-entry point and the impact point

λ = longitude measured eastward in a geographic system

ϕ = geodetic latitude

$\omega_\oplus$ = rotation rate of the Earth about its axis

$\omega_{\oplus☾}$ = rotation rate of the moon about the Earth

Units of Measurement

The units used throughout this paper, except for the Earth tracks and associated data, are as follows:

Distance - km

Speed - km/sec

Angles - deg

Time - hr

Angular Momentum - km^2/sec

Energy - km^2/sec^2

Subscripts

bo = burnout point

ca = closest approach

i = impact point

ℓ = launch point

p = point of closest approach

r = re-entry point

s = point of exit from sphere of influence

Δ = object or space vehicle

$\oplus$ = quantities referring to the Earth

☾ = quantities referring to the moon

Abbreviations

DSIF = Deep Space Instrumentation Facility

LSOI = Lunar sphere of influence

1. Introduction

Moon-to-Earth trajectories, also known as lunar return trajectories, have gained in importance as a result of the recent desire to send man to the moon and back again. As a result of intensive trajectory analysis in projects such as Ranger and Surveyor, the problem of launching from the Earth to the moon has been solved. However, the reverse problem, that of launching from the moon and returning to the Earth, still remains to be thoroughly understood.

Many of the theoretical results presented are extensions of the work of V. A. Egorov[1] and P. A. Penzo[2]. To the best of the author's knowledge, Egorov was the originator of the so-called "matched conic analysis" as applied to Earth-moon space. Penzo has taken the work of Egorov and has developed it further in both theory and in trajectory computing techniques. Both the works of Egorov and Penzo are concerned with the analytical problem of moon-to-Earth trajectories. This paper presents some additional ideas in the theoretical realm and also presents the application of the theory and analytical methods in the actual design of accurate trajectories numerically integrated on a high-speed digital computer.

1.1. Definition and the Method of Solution of the Problem

The analysis presented is concerned with the geometrical and dynamical problem of launching a spacecraft from the surface of the moon so that it will return to a specific landing site on the surface of the Earth. Results from an analytic trajectory computing program using conic section theory are used for the qualitative discussions, and results from a precision integrating program are used for the quantitative discussions.

Section 2 describes the general characteristics of the lunar return trajectory. From the knowledge of these characteristics, a set of search parameters is chosen which defines an acceptable lunar return trajectory upon convergence. Section 3 presents the IBM 7090 trajectory computing programs and techniques used in designing a set of lunar return trajectories. Section 4 presents a typical set of lunar return trajectories and shows how the actual characteristics compare with the theoretical characteristics.

1.2 Summary of Results

The matched conic analysis has been used in several investigations concerned with the problem of launching from the Earth and impacting or circumnavigating the moon. In these investigations, the perturbations (accelerations) of the N-body problem are shown to be small compared with the two-body or Keplerian accelerations. A theoretical explanation for this phenomenon has not been given in previous work. Section 2 presents the explanation. The equations of motion in the matched conic analysis are shown to be of the relative motion form with the central body chosen so as to make the perturbing terms negligible and, therefore, very nearly equations of Keplerian motion. In addition, a careful study of the matched conic sections reveals that the entire moon-to-Earth trajectory problem may be thought of as a simple rotation of elements of two cones in three-dimensional space.

Analysis of the types of lunar return trajectories shows that the best type of trajectory is strongly dependent upon the mission objectives, but that the counterclockwise class of trajectories is superior to the clockwise class of trajectories. Clarke[3] has shown that the parking-orbit technique is superior to the direct-ascent technique for Earth-to-moon trajectories because of several specific geometrical properties of Earth-to-moon trajectories. Investigations of similar geometrical properties for moon-to-Earth trajectories show that superiority depends upon the mission objectives. For example, the mission objectives define a desirable region of launch sites upon the moon. Launch sites located outside this region are best utilized by using the parking-orbit technique.

Section 3 presents the search technique used in designing the best set of trajectories for a lunar return mission. The ideal set of three independent variables from the selenographic injection conditions are azimuth, velocity, and flight-path angle.

Section 4 presents a typical set of lunar return trajectories for the mission of returning a spacecraft from the surface of the moon to one of two landing sites upon the surface of the Earth. Realistic design constraints were chosen so as to determine the detailed behavior of such a set of trajectories. The information presented is intended to point out the problem areas associated with such a mission, especially in the design of the spacecraft and in the tracking system. For example, the Earth tracks for moon-to-Earth trajectories are quite different from those of Earth-to-moon trajectories.

Since the Deep Space Instrumentation Facilities are designed for missions initiating from the Earth, modifications may be necessary for the case of the lunar return missions.

2. Lunar Return Trajectory Characteristics

In the design of Ranger Earth-to-moon trajectories, a technique for understanding general geometrical relationships is used very successfully. This technique assumes that the gravitational field of the Earth-moon system consists of two independent inverse-square force fields, one associated with the Earth and one associated with the moon. Under this system, a trajectory consists of two Keplerian orbits, usually elliptic in the sphere of influence of the Earth and hyperbolic in the sphere of influence of the moon. Since the geometrical problem of a moon-to-Earth trajectory is similar to that of an Earth-to-moon trajectory, the same technique is used in this analysis.

2.1. Keplerian Presentation of the Lunar Return Geometry

The first step in an analysis of a dynamical problem is to choose a basis or frame of reference. Once a basis is chosen, the equations governing the motion of the bodies of concern relative to that basis can be derived. In most problems, a proper choice of the basis will simplify not only the equations of motion but the motion itself and, therefore, permit the investigator to obtain a better understanding of the problem. This is the case with the problem of designing moon-to-Earth trajectories. The basis is chosen initially with origin at the center of mass of the moon and then shifted to the center of mass of the Earth at a point when the return vehicle reaches a certain distance from the moon. By choosing the basis in this manner, the equations of motion are reduced to their simplest form. The classical name for the equations of motion in this form is "the equations of relative motion"(Ref. 4, p. 161). The derivation of the equations of relative motion is as follows:

$\underline{I}, \underline{J}, \underline{K}$ = inertially fixed basis

$\underline{i}, \underline{j}, \underline{k}$ = moving basis fixed to body 1

$\underline{r}_{12}$ = radius vector of body 2 relative to body 1 ($\underline{i}, \underline{j}, \underline{k}$ system)

$\underline{r}_1$ = radius vector of body 1 relative to the $\underline{I}, \underline{J}, \underline{K}$ system

r_2 = radius vector of body 2 relative to the $\underline{I}$, $\underline{J}$, $\underline{K}$ system

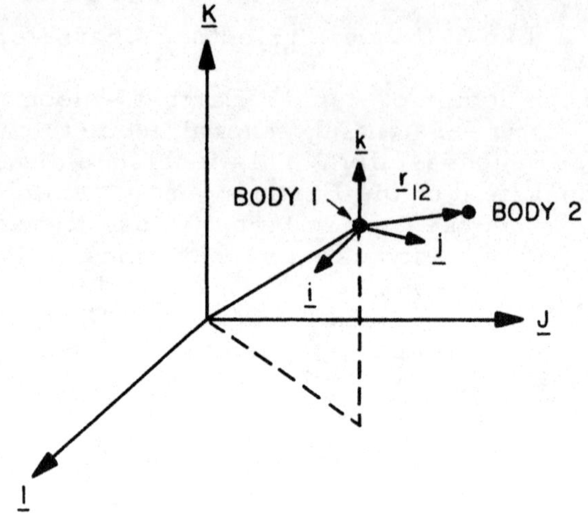

Therefore,

$$\underline{r}_{12} = \underline{r}_2 - \underline{r}_1 \Rightarrow \frac{d^2\underline{r}_{12}}{dt^2} = \frac{d^2\underline{r}_2}{dt^2} - \frac{d^2\underline{r}_1}{dt^2}$$

and

$$\underline{r}_{21} = \underline{r}_1 - \underline{r}_2 = -\underline{r}_{12} \Rightarrow r_{12} = r_{21} \tag{1}$$

From Newton's law of motion,

$$\underline{F}_i = km_i\frac{d^2\underline{r}_i}{dt^2} = m_i\frac{d^2\underline{r}_i}{dt^2} \tag{2}$$

where

i = 1, 2, . . . , n (n bodies)

$\underline{F}_i$ = force acting on body i

m_i = mass of body i

r_i = radius distance of body i

k = 1 (the units of force, mass, length, and time are chosen so that $k = 1$).

Using Newton's law of gravitation,

$$\underline{F}_{ij} = G\frac{m_i m_j}{r_{ij}^3}\underline{r}_{ij} = k^2\frac{m_i m_j}{r_{ij}^3}\underline{r}_{ij} \tag{3}$$

where

$i = 1, 2, \ldots, n$

$j = 1, 2, \ldots, n$

$\underline{F}_{ij}$ = force of attraction between bodies i and j

m_i = mass of body i

m_j = mass of body j

r_{ij} = the distance between bodies i and j

$k^2 = G$ = universal constant of gravitation.

From Newton's law of action and reaction, the force acting on body i is

$$\underline{F}_i = \sum_{j=1}^{n}{}^{*} \underline{F}_{ij} + m_i\underline{p}_i \tag{4}$$

where * indicates $j \neq i$ and where $\underline{p}_i$ represents accelerations due to thrust, drag, etc., and gravitational forces exterior to the system of n bodies or arising from their departures from homogeneity in spherical concentric layers or from Einstein effects.

From Eqs. (2-4),

$$m_i\frac{d^2\underline{r}_i}{dt^2} = \sum_{j=1}^{n}{}^{*} \underline{F}_{ij} + m_i\underline{p}_i = k^2\sum_{j=1}^{n}{}^{*} m_i m_j\frac{\underline{r}_{ij}}{r_{ij}^3} + m_i\underline{p}_i$$

where * indicates $j \neq i$. Also,

$$\frac{d^2\underline{r}_i}{dt^2} = k^2\sum_{j=1}^{n}{}^{*} m_j\frac{\underline{r}_{ij}}{r_{ij}^3} + \underline{p}_i \tag{5}$$

Letting i = 1, 2 in Eq. (5),

$$\left.\begin{aligned}\frac{d^2\underline{r}_1}{dt^2} &= k^2 m_2 \frac{\underline{r}_{12}}{r_{12}^3} + k^2 \sum_{j=3}^{n} m_j \frac{\underline{r}_{1j}}{r_{1j}^3} + \underline{p}_1 \\ \frac{d^2\underline{r}_2}{dt^2} &= k^2 m_1 \frac{\underline{r}_{21}}{r_{21}^3} + k^2 \sum_{j=3}^{n} m_j \frac{\underline{r}_{2j}}{r_{2j}^3} + \underline{p}_2\end{aligned}\right\} \tag{6}$$

From Eqs. (1) and (6),

$$\frac{d^2\underline{r}_{12}}{dt^2} = k^2\left(m_1 \frac{\underline{r}_{21}}{r_{21}^3} - m_2 \frac{\underline{r}_{12}}{r_{12}^3}\right) + k^2 \sum_{j=3}^{n} m_j \left(\frac{\underline{r}_{2j}}{r_{2j}^3} - \frac{\underline{r}_{1j}}{r_{1j}^3}\right) + \underline{p}_2 - \underline{p}_1$$

$$\frac{d^2\underline{r}_{12}}{dt^2} = -k^2(m_1 + m_2)\frac{\underline{r}_{12}}{r_{12}^3} + k^2 \sum_{j=3}^{n} m_j \left(\frac{\underline{r}_{2j}}{r_{2j}^3} - \frac{\underline{r}_{1j}}{r_{1j}^3}\right) + \underline{p}_2 - \underline{p}_1 \tag{7}$$

Equation (7) is the relative motion form of the equation of motion. The terms in the summation involving $\underline{r}_{1j}/r_{1j}^3$ and the term $\underline{p}_1$ are usually referred to as the "indirect terms" or as the indirect perturbations of motion of body 2 relative to body 1. Therefore, Eq. (7) can be written as

$$\left.\begin{aligned}\frac{d^2\underline{r}_{12}}{dt^2} = &\left[-k^2(m_1 + m_2)\frac{\underline{r}_{12}}{r_{12}^3}\right] \\ &+ \left[k^2 \sum_{j=3}^{n} m_j \frac{\underline{r}_{2j}}{r_{2j}^3} + \underline{p}_2\right] \\ &- \left[k^2 \sum_{j=3}^{n} m_j \frac{\underline{r}_{1j}}{r_{1j}^3} + \underline{p}_1\right]\end{aligned}\right\} \tag{8}$$

or

$$\left.\begin{aligned} \frac{d^2\underline{r}_{12}}{dt^2} &= \text{Keplerian acceleration} \\ &+ \text{direct perturbations} \\ &- \text{indirect perturbations} \end{aligned}\right\} . \tag{8 (Cont'd)}$$

Equation (8) implies that if the direct and indirect perturbations are very nearly equal, Keplerian motion exists. This is the case in the design of moon-to-Earth trajectories. Consider Eq. (7) with respect to the two bases defined previously ($\underline{i}, \underline{j}, \underline{k}$ system initially moon-centered and then Earth-centered):

$$\begin{aligned} \frac{d^2\underline{r}_{☾\Delta}}{dt^2} &= -k^2(m_☾ + m_\Delta)\frac{\underline{r}_{☾\Delta}}{r_{☾\Delta}^3} \\ &+ k^2\sum_{j=3}^{n} m_j\left(\frac{\underline{r}_{\Delta j}}{r_{\Delta j}^3} - \frac{\underline{r}_{☾ j}}{r_{☾ j}^3}\right) \\ &+ \underline{p}_\Delta - \underline{p}_☾ \end{aligned} \tag{9}$$

and

$$\begin{aligned} \frac{d^2\underline{r}_{\oplus\Delta}}{dt^2} &= -k^2(m_\oplus + m_\Delta)\frac{\underline{r}_{\oplus\Delta}}{r_{\oplus\Delta}^3} \\ &+ k^2\sum_{j=3}^{n} m_j\left(\frac{\underline{r}_{\Delta j}}{r_{\Delta j}^3} - \frac{\underline{r}_{\oplus j}}{r_{\oplus j}^3}\right) \\ &+ \underline{p}_\Delta - \underline{p}_\oplus \end{aligned} \tag{10}$$

Equations (9) and (10) simplify, since in cislunar space the direct and indirect terms are nearly the same for all perturbing bodies except for the perturbations due to the Earth in Eq. (9) and the moon in Eq. (10). Therefore,

$$\frac{d^2\underline{r}_{☾\Delta}}{dt^2} = -k^2(m_{☾} + m_{\Delta})\frac{\underline{r}_{☾\Delta}}{r^3_{☾\Delta}} + k^2 m_{\oplus}\left(\frac{\underline{r}_{\Delta\oplus}}{r^3_{\Delta\oplus}} - \frac{\underline{r}_{☾\oplus}}{r^3_{☾\oplus}}\right) \tag{11}$$

and

$$\frac{d^2\underline{r}_{\oplus\Delta}}{dt^2} = -k^2(m_{\oplus} + m_{\Delta})\frac{\underline{r}_{\oplus\Delta}}{r^3_{\oplus\Delta}} + k^2 m_{☾}\left(\frac{\underline{r}_{\Delta☾}}{r^3_{\Delta☾}} - \frac{\underline{r}_{\oplus☾}}{r^3_{\oplus☾}}\right) \tag{12}$$

Equations (11) and (12) are the equations of motion representing a space vehicle in free flight. Furthermore, if Eq. (11) is defined as applicable only to motion of the vehicle near the center of mass of the moon, that is, where $\underline{r}_{\Delta\oplus}/r^3_{\Delta\oplus} \cong \underline{r}_{☾\oplus}/r^3_{☾\oplus}$, and if Eq. (12) is defined as applicable only to motion of the vehicle near the center of mass of the Earth, that is, where $\underline{r}_{\Delta☾}/r^3_{\Delta☾} \cong \underline{r}_{\oplus☾}/r^3_{\oplus☾}$, then Eqs. (11) and (12) reduce to Keplerian motion

$$\frac{d^2\underline{r}_{☾\Delta}}{dt^2} = -k^2(m_{☾} + m_{\Delta})\frac{\underline{r}_{☾\Delta}}{r^3_{☾\Delta}} \tag{13}$$

and

$$\frac{d^2\underline{r}_{\oplus\Delta}}{dt^2} = -k^2(m_{\oplus} + m_{\Delta})\frac{\underline{r}_{\oplus\Delta}}{r^3_{\oplus\Delta}} \tag{14}$$

Equation (13) is used to represent the motion of the space vehicle within the sphere of influence of the moon*, and Eq. (14) is used to represent the motion of the space vehicle exterior to the sphere of influence of the moon. This is equivalent to saying that the basis is initially fixed to the center of mass of the moon and then shifted to the center of mass of the Earth at the time the space vehicle passes through the sphere of influence of the moon.

---

* The locus of points about body 1 at which the ratio between the force with which body 2 perturbs the motion of body 3 and the force of attraction of body 1 is equal to the ratio between the force with which body 1 perturbs the motion of body 3 and the force of attraction of body 2. The radius of this sphere for the moon is equal to $0.87 r_{☾\oplus}(m_{☾}/m_{\oplus})^{2/5}$.

The approximation of Eqs. (13) and (14) to the actual motion of the space vehicle is least accurate at distances from the moon equivalent to the radius of the sphere of influence of the moon (the point of basis shift).

Figure 1 presents a three-dimensional drawing of the Keplerian presentation of the lunar return trajectory geometry. For any given set of re-entry conditions at the Earth, two solutions exist. The first solution is a counterclockwise trajectory shown as a solid line and the second solution is a clockwise trajectory shown as a dashed line. The lunar return trajectory is approximated by two joined conic sections. The first conic section is drawn in a moon-centered inertial system and is a hyperbolic orbit up to the piercing point s of the lunar sphere of influence (LSOI). At this point, the reference frame is changed to the Earth-centered inertial system by adding the velocity of the moon relative to the Earth, $\underline{V}_{\oplus☾}$, to the velocity of the probe relative to the moon, $\underline{V}_{☾\Delta}$, and obtaining the velocity of the probe relative to the Earth $\underline{V}_{\oplus\Delta}$. The second conic section is drawn in the Earth-centered inertial system and is an elliptical orbit that terminates at the point of entry into the atmosphere of the Earth. The apparent discontinuity of the trajectory at the exit point of the LSOI is a result of drawing the first phase of the trajectory as seen in a moon-centered inertial system and the second phase of the trajectory as seen in an Earth-centered inertial system. If a single frame of reference is used, the actual shape of the trajectory with respect to the frame of reference chosen is presented and the discontinuity is removed, but now the trajectory is not presented as two joined conics and the general trajectory relationships can not be easily discussed. The objective of Fig. 1 is to present the general trajectory relationships and not the actual shape of the trajectory.

It has been empirically determined that the velocity vector $\underline{V}_{☾\Delta}$ is very nearly equal to the hyperbolic excess velocity vector at the piercing point of the LSOI. As a result, the direction of $\underline{V}_{☾\Delta}$ closely represents the outward radial direction or outgoing asymptote of the selenocentric hyperbola. Since $\underline{V}_{\oplus☾}$ points to the lunar west and is approximately in the Earth-moon plane perpendicular to the Earth-moon line, every outward radial direction that describes a lunar-return trajectory must point toward the lunar east so that the velocity with respect to the Earth, at the piercing point of the LSOI, points toward the Earth as shown in Fig. 1.

The exit points of the outward radial directions for a constant Earth-to-moon distance, time of flight, and re-entry path angle form a circular locus at the LSOI with the exit point of the vertical impact outward radial direction as the center. The position and size of the locus of exit points relative to the Earth-moon line depend upon three parameters: the Earth-to-moon distance, the flight time, and the re-entry angle. First, as the Earth-to-moon distance increases, the angular distance of the locus of exit points from the Earth-moon line decreases. Second, as the flight time increases, the angular distance of the locus from the Earth-moon line and the angular diameter of the circular locus increase. Finally, as the re-entry angle increases, the angular diameter of the locus decreases.

Figure 2 presents the locus of asymptote piercing points or exit points for lunar return trajectories having a flight time of 70 hr and a re-entry flight-path angle of a negative 6° for a mean Earth-to-moon distance. The outward radial direction for the lunar return trajectory which impacts upon the Earth vertically is at an angle of about 52.5° with the Earth-moon line and is in the Earth-moon plane. The outward radial direction for a clockwise re-entry trajectory is to the right of this at an angle of 57.0° with the Earth-moon line and requires a maximum $\underline{V}_{☾\Delta}$ (Ref. 2, pp. 58-59). The outward radial direction for a counterclockwise re-entry trajectory is to the left of the vertical impact outward radial direction at an angle of 48.0° with the Earth-moon line and requires a minimum $\underline{V}_{☾\Delta}$. These two cases represent the worst and the best trajectories, respectively, from the viewpoint of a required lunar boost impulse. The remaining outward radial directions exhibit impulses between these two. Consequently, the magnitude of $\underline{V}_{☾\Delta}$ is a function of the Earth-to-moon distance, the flight time, the re-entry angle, and the direction of the trajectory (clockwise or counterclockwise).

The locus of exit points may be thought of as the locus formed by the point s as the outward radial direction is rotated about the vertical impact outward radial direction by controlling $\underline{V}_{☾\Delta}$. Figure 1 shows that the vector from the center of the Earth to the point s, $\underline{r}_{\oplus s}$, rotates about the vertical impact trajectory as the locus of exit points is formed. The vertical impact trajectory is very nearly the direction of the moon at launch since the angular momentum of the probe relative to the Earth is cancelled by the angular momentum of the moon relative to the Earth during transit through the LSOI. The rotation of $\underline{r}_{\oplus s}$ about the vertical impact trajectory forms a cone with its apex at the center of the Earth. The trajectory plane contains $\underline{r}_{\oplus s}$ and the center of the Earth and, therefore, rotates about the vertical impact trajectory as the locus of exit points is formed. The in-plane angle between the exit point s and the point of re-entry r, v_{sr}, is a function of the

Earth-to-moon distance (launch date), the time of flight (energy $C_{3\oplus}$), and the re-entry flight path angle (Ref. 2, p. 43). Since these three parameters are defined as constant, v_{sr} is constant. By defining a constant re-entry maneuver angle η_r, the total inplane angle v_{si} is constant. Therefore, the impact radius vector $\underline{r}_i$ in the trajectory plane is at a constant angle of v_{si} from the vector $\underline{r}_{\oplus s}$. As the trajectory plane rotates about the vertical impact trajectory, the vector $\underline{r}_i$ forms a second cone with its apex at the center of the Earth and the direction of the vertical impact trajectory as its axis. The intersection of this cone with the surface of the Earth forms a circular locus of impact points. Since the outward radial direction, the vector $\underline{r}_{\oplus s}$, and the vector $\underline{r}_i$ are all slaved together, the locus of exit points and the locus of impact points are formed simultaneously as the trajectory plane rotates about the vertical impact trajectory according to the variation of the velocity vector $\underline{V}_{☾\Delta}$. Hence, for each exit point on the locus of exit points there is a corresponding impact point on the locus of impact points. Figures 1 and 2 show that as the locus of exit points is traversed from maximum $\underline{V}_{☾\Delta}$ to minimum $\underline{V}_{☾\Delta}$ in a clockwise fashion, the azimuth of impact varies in an increasing fashion. Therefore, given an Earth-to-moon distance, a time of flight, a re-entry angle, and a maneuver angle, the circular locus of impact points is defined. In addition, if a specific impact latitude is defined, then the circular locus reduces to two impact points. The first impact point is associated with a counterclockwise re-entry having an impact azimuth between 0° and 180°, and the second impact point is associated with a clockwise re-entry having an impact azimuth between 180° and 360°. The two points would differ in longitude by a $\Delta\lambda$, the angular diameter of the impact cone (Fig. 1). If a specific impact longitude is defined, there are still two possible trajectories, one counterclockwise and one clockwise. The correct impact longitude for each of the cases is obtained by varying the launch time, which is directly related to impact longitude. Varying the launch time an hour or so in order to correct impact longitude does not significantly affect the remaining trajectory characteristics.

If the re-entry angle is varied from shallow to steep angles, the angular diameter of the impact cone will increase to a maximum of 180° and then decrease on the moon side of the Earth to a point corresponding to the vertical impact point, which is at a latitude equal to the declination of the moon at the time of exit from the lunar sphere of influence.

Figure 1 presents a typical manned lunar return trajectory from the lunar injection point to the impact point upon the Earth (shown as a solid line). The ascent trajectory is not typical of a manned lunar return trajectory; however, the important near-Earth characteristics are independent of the ascent trajectory and are typical. Clearly, from the previous discussion, the trajectory from the exit point s to the re-entry point r is dependent only upon the conditions at the exit point and not on how these conditions arise The trajectory after exit from the LSOI consists of a highly elliptical orbit characteristic of a flight time of 70 hr. For a flight time of 45 hr, this phase of the trajectory would be a hyperbolic conic. The clockwise and counterclockwise re-entry trajectories are shown. The choice of which of the two solutions is better depends upon two criteria, the amount of return-boost impulse required and the amount of re-entry retrorocket impulse required. The previous discussions of Figs. 1 and 2 indicated that the amount of return-boost impulse required is less for the counterclockwise re-entry trajectory. In addition, the amount of retrorocket impulse required at re-entry is less for the counterclockwise re-entry trajectory, since the inertial velocity of the probe relative to the rotating atmosphere in inertial space is less. For these reasons, the counterclockwise trajectory is better and is the one primarily used and discussed in this paper.

Penzo[2] points out several interesting characteristics of the geocentric phase of the lunar return trajectory, which are apparent in Sec. 4 of this paper. These characteristics are summarized below:

1) For a fixed flight time, vehicles launched on those days when the moon is farthest from the Earth must have higher energies than those launched when the moon is closest to the Earth.

2) The effects of the re-entry path angle γ_r, the impact site, and the direction of the trajectory (clockwise or counterclockwise) on the inertial re-entry velocity are negligible. The re-entry velocity is primarily dependent upon the flight time and the Earth-to-moon distance.

3) The energy of the geocentric conic is a function of the Earth-to-moon distance, the flight time, and the re-entry path angle. Trajectories having steep re-entry path angles have lower energies than trajectories with the same flight time but with shallower re-entry path angles. The trajectories with steep re-entry angles re-enter on the side of the Earth facing the moon and require less distance to travel than the trajectories with shallow re-entry angles which re-enter on the far side of the Earth.

4) The in-plane angle ν_{sr} between the exit point of the LSOI and the re-entry point at the Earth is dependent upon the Earth-to-moon distance, the flight time, and the re-entry angle. The re-entry radius distance is fixed at a radius distance of 6500 km. The angle ν_{sr} can also be represented as a function of the Earth-to-moon distance $r_{\oplus ☾}$, the energy $C_{3\oplus}$, and the re-entry flight path angle γ_r. The energy $C_{3\oplus}$ is a function of the Earth-to-moon distance, the flight time, and the re-entry angle. The parameter C_3 is used rather than velocity because it describes the dynamical nature of the trajectory completely. It is analogous to the eccentricity, which describes the geometrical nature of the trajectory completely. In addition, C_3 is a measure of a launch vehicle's capability and, therefore, can be directly related to payload capabilities of various vehicles given a prescribed trajectory. From Fig. 1 and Ref. 2, $C_{3\oplus} \triangleq V^2_{\oplus\Delta} - (2\mu_\oplus/r_{\oplus s})$ and $a = \mu_\oplus/C_{3\oplus}$, where $\mu_\oplus \triangleq Gm_\oplus$, $r_r \triangleq$ 6500 km (radius distance of point of re-entry), and $r_{\oplus s} \cong r_{\oplus ☾} - 0.87(r_{\oplus ☾})(m_☾/m_\oplus)^{2/5}$. Using the conservation of angular momentum and the definition of $C_{3\oplus}$, we have the relations $r_r V_r \cos\gamma_r = r_{ca} V_{ca}$, $V^2_{ca} = C_{3\oplus} + (2\mu_\oplus/r_{ca})$, and $V^2_r = C_{3\oplus} + 2\mu_\oplus/r_r$, so that

$$r_r^2 V_r^2 \cos^2\gamma_r = r_{ca}^2 V_{ca}^2$$

$$r_r^2\left(C_{3\oplus} + \frac{2\mu_\oplus}{r_r}\right)\left(\cos^2\gamma_r\right) = r_{ca}^2\left(C_{3\oplus} + \frac{2\mu_\oplus}{r_{ca}}\right)$$

$$r_{ca}^2(C_{3\oplus}) + r_{ca}(2\mu_\oplus) - r_r^2(\cos^2\gamma_r)\left(C_{3\oplus} + \frac{2\mu_\oplus}{r_r}\right) = 0$$

$$r_{ca}^2 + \left(\frac{2\mu_\oplus}{C_{3\oplus}}\right) r_{ca} - \left(\frac{r_r^2\cos^2\gamma_r}{C_{3\oplus}}\right)\left(C_{3\oplus} + \frac{2\mu_\oplus}{r_r}\right) = 0$$

$$r_{ca} = -\frac{\mu_\oplus}{C_{3\oplus}} + \sqrt{\frac{\mu_\oplus^2}{C_{3\oplus}^2} + \left(\frac{r_r^2\cos^2\gamma_r}{C_{3\oplus}}\right)\left(C_{3\oplus} + \frac{2\mu_\oplus}{r_r}\right)} \qquad (15)$$

Equation (15) shows that $r_{ca} = f_1(C_{3\oplus}, \gamma_r)$. In addition, $e = 1 - (r_{ca}/a)$ and ν_r, the true anomaly of re-entry, is equal to

$$\cos^{-1}\left[\frac{r_{ca}(1 + e)}{er_r} - \frac{1}{e}\right]$$

where $\pi \leq v_r \leq 2\pi$ (re-entry always occurs before closest approach) and

$$v_{rp} = 2\pi - v_r \tag{16}$$

From Eq. (16) and the discussion above

$$v_{rp} = f_2(C_{3\oplus}, \gamma_r)$$

Also

$$v_{sp} = \cos^{-1}\left[\frac{b^2 + r_{\oplus s}a}{r_{\oplus s}(r_{ca} - a)}\right] \tag{17}$$

where

$$b^2 = r_{ca}(r_{ca} - 2a) \qquad 0 \leq v_{sp} \leq \pi$$

From Eq. (17)

$$v_{sp} = f_3(r_{\oplus☾}, C_{3\oplus}, \gamma_r) \tag{18}$$

and finally

$$v_{sr} = v_{sp} - v_{rp}$$

$$v_{sr} = f_4(r_{\oplus☾}, C_{3\oplus}, \gamma_r)$$

Figure 3 presents v_{sr} vs t_F and $C_{3\oplus}$ for a re-entry angle of -6.0°.

5. The declination and right ascension of the exit point s are within 1.5° of the moon at launch as seen from the Earth. The angular diameter of the LSOI is about 17° as seen from the Earth (Fig. 1 is obviously not to scale).

2.2 Types of Lunar Return Ascent Trajectories

There are two basic types of lunar return ascent trajectories, the direct-ascent return and the parking-orbit return. The direct-ascent return consists of a continuous powered flight from liftoff to injection into the transfer orbit.

In the case of a multiple-stage lunar return vehicle, the direct-ascent return consists of consecutive powered flights separated by short coast periods known as staging periods. Figure 4 presents the altitude vs range profile of a direct-ascent return trajectory. Launch occurs at point A_0 and is followed by injection into the transfer orbit at point A_1.

The parking-orbit return consists of two powered flight phases separated by a coasting phase that is usually longer than a staging period. Figure 4 presents the altitude vs range profile of a parking-orbit return trajectory. Launch occurs at B_0 and is followed by the first powered flight, which places the return vehicle at point B_1, injection into a low circular satellite orbit about the moon. The return vehicle coasts in this orbit to a prescribed location, point B_2, where ignition of the second powered flight occurs. The second powered flight places the return vehicle at point B_3, injection into the transfer orbit to the Earth.

Clarke[3] has shown that the parking-orbit technique is superior to the direct-ascent technique for Earth-to-moon trajectories because of several specific geometrical properties of Earth-to-moon trajectories. Investigations of similar geometrical properties for moon-to-Earth trajectories show that superiority depends upon the mission objectives.

Three geometrical properties that strongly influence the choice between the two lunar return techniques are the available launch sites, outward radial directions, and launch azimuths. The launch site is dependent upon the mission objectives, since any point upon the lunar surface may be a landing site for an Earth-to-moon trajectory and, hence, a launch site for a moon-to-Earth trajectory. The outward radial direction is dependent upon the launch date, the re-entry angle, and the flight time, which are mission dependent. Once these three parameters are chosen, the outward radial direction and the injection energy are specified. The launch azimuth is not restricted in any apparent manner since both atmosphere and life are nonexistent upon the moon. Consequently, depending upon mission objectives, a considerable amount of flexibility can exist in these three geometrical properties in contrast to the Earth-to-moon trajectories.

The importance of flexible launch sites, outward radial directions, and launch azimuths is shown in Figs. 5-7. Figure 5 presents the launch geometry of a typical lunar-return trajectory for a particular launch day, re-entry angle, and flight time. The outward radial direction $\underline{S}$ and the injection energy are uniquely defined. Furthermore, the direction of closest approach $\underline{P}$ is defined relative to $\underline{S}$ by an angle v_s, where v_s is a function of the closest approach distance to the moon and the injection energy $C_{3☾}$. The injection energy $C_{3☾}$ is a function of the Earth-to-moon distance, flight time, and re-entry angle.

From Fig. 1 and Ref. 2,

$$C_{3☾} \triangleq V^2_{☾\Delta} - \frac{2\mu_☾}{r_{☾s}}$$

and (19)

$$a = -\frac{\mu_☾}{C_{3☾}}$$

where $\mu \triangleq Gm_☾$ and $C_{3☾}$ is constant for fixed launch day, re-entry angle, and flight time. In addition,

$$e = 1 - (r_{ca}/a) \tag{20}$$

and

$$v_s = \cos^{-1} -(1/e) \tag{21}$$

Using the conservation of angular momentum and the definition of $C_{3☾}$, we have the relations $r_{bo}V_{bo}\cos\gamma_{bo} = r_{ca}V_{ca}$, $V^2_{ca} = C_{3☾} + (2\mu_☾/r_{ca})$, and $V^2_{bo} = C_{3☾} + 2\mu_☾/r_{bo})$, so that

$$r^2_{bo}V^2_{bo}\cos^2\gamma_{bo} = r^2_{ca}V^2_{ca}$$

$$r^2_{bo}\left(C_{3☾} + \frac{2\mu_☾}{r_{bo}}\right)\left(\cos^2\gamma_{bo}\right) = r^2_{ca}\left(C_{3☾} + \frac{2\mu_☾}{r_{ca}}\right)$$

$$r^2_{ca}(C_{3☾}) + r_{ca}(2\mu_☾) - r^2_{bo}\left(\cos^2\gamma_{bo}\right)\left(C_{3☾} + \frac{2\mu_☾}{r_{bo}}\right) = 0$$

$$r^2_{ca} + \frac{2\mu_☾}{C_{3☾}}r_{ca} - \left(\frac{r^2_{bo}\cos^2\gamma_{bo}}{C_{3☾}}\right)\left(C_{3☾} + \frac{2\mu_☾}{r_{bo}}\right) = 0$$

$$r_{ca} = -\frac{\mu_☾}{C_{3☾}} + \sqrt{\frac{\mu^2_☾}{C^2_{3☾}} + \left(\frac{r^2_{bo}\cos^2\gamma_{bo}}{C_{3☾}}\right)\left(C_{3☾} + \frac{2\mu_☾}{r_{bo}}\right)} \tag{22}$$

Equation (22) shows that

$$r_{ca} = f(r_{bo}, \gamma_{bo}) \tag{23}$$

Assuming that optimum flight is desired for a given boost vehicle, injection occurs at the closest approach distance of the selenocentric hyperbola. The powered flight characteristics of importance here are the burning arc $v_{\ell bo}$ required to achieve the injection energy $C_{3☾}$ and the resulting burnout radius distance r_{bo}. Since injection occurs at the closest approach distance, $\gamma_{bo} = 0$ and $r_{ca} = r_{bo}$. Figure 6 presents the variation of v_s with flight time t_F and injection energy $C_{3☾}$ for a closest approach distance of 1768.6 km. Figure 5 shows that the sum of v_s and $v_{\ell bo}$ forms an angle $v_{\ell s}$ between the launch site vector $\underline{r}_{\ell}$ and the outward radial direction $\underline{S}$ in the trajectory plane, which is defined by the outward radial direction, the injection point (point of closest approach), and the launch site. For a flight time of 70 hr, a re-entry angle of -6.0°, and a mean Earth-to-moon distance, Fig. 6 shows that v_s is equal to 134° for an assumed burnout radius r_{bo} of 1768.6 km. Assuming that the boost vehicle requires a burning arc $v_{\ell bo}$ of 8.0° to achieve the required $C_{3☾}$, $v_{\ell s}$ results in a value of 142°. Using this value of $v_{\ell s}$, the locus of possible launch sites for a direct-ascent trajectory is obtained by rotating $\underline{r}_{\ell}$ about $\underline{S}$ at an angle $v_{\ell s}$ of 142°. This locus is presented in Fig. 7. Each point on the locus represents a possible launch site. Associated with each point are a unique launch azimuth and a unique orbit inclination. If the mission objectives require launch sites other than those described by this locus, then nonoptimum powered flight must occur. A certain amount of variation in the locus of launch sites may be obtained by varying flight time and, therefore, varying $v_{\ell s}$. However, mission constraints usually require a relatively fixed flight time.

Launch sites located within the circular locus require very inefficient direct-ascent powered flights, since injection occurs before the closest approach distance of the selenocentric hyperbola. These launch sites could best be utilized in the parking-orbit technique.

Launch sites located outside the circular locus require inefficient direct-ascent powered flights to a lesser extent. Equation (23) shows that r_{ca} is a function of r_{bo} and γ_{bo}. For a given lunar return boost vehicle, the burnout radius r_{bo} and the burning arc $v_{\ell bo}$ vary little with variations in pitch control. Therefore, r_{ca} is essentially a function of burnout flight path angle γ_{bo}. If burnout occurs at an angle γ_{bo}, $v_{\ell s}$ is no longer the sum of v_s and $v_{\ell bo}$ but instead the sum of v_s and $v_{\ell bo}$ minus the true anomaly angle of burnout v_{bo} where

$$v_{bo} = \cos^{-1}\left[\frac{b^2 + r_{bo}a}{r_{bo}(r_{ca} - a)}\right]$$

and

$$b^2 = r_{ca}(r_{ca} - 2a)$$

Figure 8 presents the geometry and a typical locus of launch sites.

There is an extreme set of r_{bo} and γ_{bo} which achieves the required $C_{3☾}$ for a given lunar boost vehicle using a direct-ascent trajectory. The extreme combination places a minimum value that may be obtained for $v_{\ell s}$. The maximum value of $v_{\ell s}$ will be determined by trajectories that inject at the closest approach distance. As a result, a region of launch sites is defined such that $v_{\ell s(min)} \leq v_{\ell s} \leq v_{\ell s(max)}$ for a given flight time, as shown in Fig. 8. Launch sites located outside this region are best utilized by using the parking-orbit technique.

The previous discussion shows that the direct-ascent return trajectory is feasible depending upon the over-all mission objectives. The direct-ascent technique requires much simpler guidance and propulsion hardware, simplifies tracking and communication, and is more reliable. For this reason and the fact that the near-Earth trajectory is independent of the ascent trajectory, this paper assumes a direct-ascent trajectory rather than a parking-orbit trajectory.

2.3 Rendezvous Return

A special class of the parking-orbit return trajectory of particular interest is the rendezvous return trajectory. In this type of trajectory, the return vehicle is launched in such a manner that a rendezvous with another vehicle or spacecraft is made at parking-orbit injection. The two vehicles combine during the coast and are then accelerated together into the transfer orbit to the Earth. The rendezvous return trajectory is restricted, compared with the general parking-orbit trajectory, in that a particular parking orbit must be used as well as a set of particular launch times. The parking orbit must be that of the orbiting vehicle, and the launch time must be such that the sum of launch time and time of powered flight must equal times of injection at which the position of the orbiting vehicle corresponds to the point of injection into the parking orbit.

3. Trajectory Computing Techniques

The Analytic Lunar Return Program and the Space Trajectories Program permit the design of the best set of lunar return trajectories for a specified mission. The technique for trimming or searching-in this set of trajectories is based upon a specific set of search parameters.

3.1. Analytic Lunar Return Program[5]

The Analytic Lunar Return Program is a matched conic program. The lunar return trajectory is split into two phases, near-moon, and near-Earth, as shown in Fig. 1. The near-moon or selenocentric phase initiates upon the surface of the spherical moon and terminates at the exit point of the LSOI. The near-Earth or geocentric phase initiates at the exit point of the LSOI and terminates at the impact point upon the spherical Earth. The selenocentric phase consists of a simplified powered flight followed by free flight in a hyperbolic orbit. The powered flight portion is simulated by specifying a launch site, a burning arc, and a burnout altitude. The geocentric phase consists of an elliptical or hyperbolic orbit terminating at the re-entry point and followed by a simplified re-entry trajectory described by a re-entry altitude, a maneuver time, and an impact site.

The basic input data consist of the selenographic longitude and latitude of the launch site, the launch day, the lunar powered-flight angle from launch to lunar burnout, the burnout altitude, the re-entry maneuver angle and maneuver time to impact, the longitude and latitude of the impact site, the re-entry angle, the re-entry altitude, and the total time of flight. The output data consist of spherical position and velocity parameters and conic characteristics at selenocentric injection, geocentric re-entry, and geocentric impact.

The objective of this program is twofold. First, the simplicity of the program permits rapid computation of many trajectories and, therefore, permits a parametric study of all trajectories of interest. Second, once the ideal set of trajectories has been obtained from the parametric study, good initial conditions are available as input to the more exact Space Trajectories Program. This, of course, is because of the closeness of the moon-to-Earth trajectory to the patched conic representation as discussed in Sec. 2.1.

3.2. Space Trajectories Program[6,7]

The Space Trajectories Program computes the motion of a probe confined to the solar system and influenced by the nonspherical Earth and moon, and point masses representing the sun, Venus, Mars, and Jupiter. Any of these bodies may serve as the reference body at the injection epoch. Stepwise numerical integration of the equations of motion appropriate to either a Cowell or an Encke scheme serves to step the probe along its flight to one of the previously mentioned bodies, which then serves as a target.

The starting conditions may be entered in Cartesian or spherical coordinates based on one of four reference frames: mean equator and equinox of 1950.0, mean ecliptic and equinox of 1950.0, true equator and equinox of date, and true ecliptic and equinox of date. In addition, the Earth-fixed spherical set of starting conditions, based upon a rotating Earth, is available for the Earth as the injection body; and the selenographic or moon-fixed spherical set of starting conditions, based upon a rotating moon, is available for the moon as the injection body. The output data may consist of any of the forementioned quantities, ephemeris information expressed in any one of the Cartesian or spherical coordinate systems, elements of the osculating two-body orbit referred to one of the standard Cartesian systems, geometrical angles between the probe and the celestial bodies, and tracking characteristics from a maximum of 15 tracking stations.

The Space Trajectories Program is basically a free-flight program. A simplified powered-flight trajectory may be simulated which assumes a constant-thrust constant-burning-rate motor with thrust direction fixed in space. In addition, a simplified re-entry trajectory may be simulated as an arc that connects the re-entry point with the impact point, lies in the orbit plane, and has a re-entry radius of r_r, a maneuver angle of η_r, and a maneuver time of t_r.

3.3. Search Parameters

The search parameters for a lunar return trajectory are governed by the set of terminal conditions that must be satisfied. Such a set consists of a particular re-entry angle γ_r and an impact site defined by a latitude ϕ_i and a longitude λ_i. From these three desired terminal conditions and the general trajectory constants such as the re-entry radius, maneuver angle, etc., a set of desired search parameters can be computed which varies with burnout conditions in a fairly linear fashion. This set is based upon the impact parameter $\underline{B}$ and consists of $\underline{B} \cdot \underline{T}$, $\underline{B} \cdot \underline{R}$, t_F, and λ_{ca}.

Since the near-Earth orbit for a lunar-return trajectory is either elliptic or hyperbolic, depending upon the flight time, two definitions of the impact parameter and of the orthogonal vectors $\underline{R}$, $\underline{S}$, $\underline{T}$ are required. Parabolic orbits are excluded in this analysis because of their low probability of occurrence. First, for the elliptical case, the impact parameter $\underline{B}$ is defined as the position vector in the plane of the trajectory, originating at the center of gravity of the target body and directed normally to the perigee direction of the ellipse. Vector $\underline{S}$ is a unit vector in the direction of perigee $\underline{P}$, $\underline{T}$ is a unit vector perpendicular to $\underline{S}$ that lies in the equatorial plane as shown in Fig. 9, and $\underline{R}$ is a unit vector which forms the third vector of the dextral orthogonal system $\underline{R}$, $\underline{S}$, $\underline{T}$. Second, for the hyperbolic case, the impact parameter $\underline{B}$ is defined as the position vector in the plane of the trajectory, originating at the center of gravity of the target body and directed normally to the incoming asymptote of the hyperbola. Vector $\underline{S}$ is a unit vector in the direction of the incoming asymptote, $\underline{T}$ is a unit vector perpendicular to $\underline{S}$ that lies in the equatorial plane as shown in Fig. 9, and $\underline{R}$ is a unit vector which forms the third vector of the dextral orthogonal system $\underline{R}$, $\underline{S}$, $\underline{T}$. In both cases, $\underline{B}$ is approximately the vector miss that would occur if the target body had no mass. The search parameters $\underline{B} \cdot \underline{T}$ and $\underline{B} \cdot \underline{R}$ are the projections of the impact parameter $\underline{B}$ upon the vectors $\underline{T}$ and $\underline{R}$, respectively.

The time of flight t_F and the longitude of closest approach, λ_{ca}, complete the set of search parameters. The parameters $\underline{B} \cdot \underline{T}$, $\underline{B} \cdot \underline{R}$, and t_F determine the shape of the trajectory, whereas λ_{ca} determines the launch time. The search parameters $\underline{B} \cdot \underline{T}$, $\underline{B} \cdot \underline{R}$, and t_F are very powerful in searching-in a set of trajectories because of their linear behavior with variations in lunar burnout conditions. Figures 1 and 9 show that a perturbation on any one of the lunar burnout conditions causes direct variations in both the flight time and the impact parameter in a continuous functional manner.

The longitude of closest approach, λ_{ca}, is used to control launch time for two reasons; first, it exists for all trajectories, and second, it is directly dependent upon launch time when $\underline{B} \cdot \underline{T}$, $\underline{B} \cdot \underline{R}$, and t_F are held constant. References 7 and 8 present a detailed discussion of the equations and characteristics of the search parameters $\underline{B} \cdot \underline{T}$, $\underline{B} \cdot \underline{R}$, t_F, and λ_{ca}.

3.4 Search Technique

The search procedure consists of computing an initial trajectory based upon the selenocentric initial conditions obtained from the Analytic Lunar Return Program and then computing three perturbed trajectories. The three perturbed trajectories are obtained by independently incrementing three of the initial set of starting conditions. The size of the increments depends upon the linearity of the perturbed initial conditions with the search parameters. The variations of the search parameters are associated with the corresponding perturbations through a linearized differential correction scheme, and corrections to the initial set of starting conditions are made This process is repeated until the desired search parameters are obtained within specified convergence criteria. An adjustment is then made upon launch time and the entire process is repeated.

At the present time, the Space Trajectories Program is incapable of computing a lunar-boost powered flight trajectory. As a result, the versatility of the lunar return search program is restricted, and the computation of a set of realistic burnout or injection conditions is impossible since the injection conditions of a launch-to-impact trajectory are dependent upon the launch vehicle characteristics. For this reason, the trajectories described in this paper begin at an approximate lunar injection point. However, a booster-powered flight program is presently being added to the Space Trajectories Program to eliminate this restriction.

Figure 10 presents the search logic using the present Space Trajectories Program. An initial trajectory is computed using injection conditions obtained from the Analytic Lunar Return Program. The initial trajectory is integrated to the point of closest approach to the Earth. At this point, the desired search parameters $\underline{B} \cdot \underline{T}$, $\underline{B} \cdot \underline{R}$, and t_F are computed using parameters from the initial trajectory along with the semimajor or semitransverse axis a, the closest approach distance r_{ca}, and the inclination of the orbit plane to the equatorial plane i, obtained from the Analytic Lunar Return Program solution. These desired search parameters are held constant throughout the first search loop. Three independent search variables, A, B, and C, which are obtained from the initial set of injection conditions, are then independently incremented, and three perturbed trajectories are integrated to the point of closest approach to the Earth. The actual $\underline{B} \cdot \underline{T}$, $\underline{B} \cdot \underline{R}$ and t_F from the three perturbed and initial trajectories are compared with the desired $\underline{B} \cdot \underline{T}$, $\underline{B} \cdot \underline{R}$, and t_F. A linearized differential correction scheme computes the corrections to the selenographic independent variables required to cause these actual search parameters to approach the desired search parameters. The corrections are added to the unperturbed selenographic injection conditions, and a corrected trajectory is integrated to the point of closest approach to the Earth. The resulting search parameters are tested according to the convergence criteria specified. If the test is negative, then this search loop is repeated. New values for the desired $\underline{B} \cdot \underline{T}$ and $\underline{B} \cdot \underline{R}$ are computed using parameters a, r_{ca}, and i from the corrected trajectory. If the test is affirmative, a test for convergence is made upon λ_{ca}. If this test is affirmative, the search is completed and the trajectory is printed. If this test is negative, a correction to the launch time Δt is made. Since the lunar powered-flight portion of the trajectory is not simulated, the correction to launch time is made indirectly by fixing the moon-fixed spherical (selenographic) injection conditions and correcting the injection time by Δt. The selenographic injection conditions with the corrected injection time are integrated to the point of closest approach to the Earth. This trajectory now becomes the corrected trajectory for the search on $\underline{B} \cdot \underline{T}$, $\underline{B} \cdot \underline{R}$, and t_F. The desired search parameters are recomputed, based upon this new corrected trajectory, and the search loop is repeated. This entire procedure is repeated until all the convergence criteria are met. This search technique is known as a 3 x 3 search followed by a 1 x 1 search.

An analysis was made to determine the ideal set of three independent variables to use from the selenographic injection conditions. The results showed that azimuth, velocity, and flight-path angle composed the ideal set. Using this set in the forementioned search technique, a typical set of lunar return trajectories was designed for the month of March 1969.

4. A Typical Set of Lunar Return Trajectories

The design of a set of trajectories consists of the following steps: definition of the mission, design and programming of the required trajectory computing programs, specification of the trajectory design constraints, determination of the set of best trajectories by means of a parametric analysis, and searching-in the best set of trajectories.

A typical set of lunar return trajectories is presented for the arbitrarily chosen month of March 1969. The mission is to return a spacecraft from the surface of the moon to one of two landing sites upon the surface of the Earth. The required trajectory computing programs designed and programmed for this mission are presented in Secs. 3. 1 and 3. 2. The trajectory design constraints are presented in Sec. 4. 1. The parametric study that determined the best set of trajectories is not presented in this paper, but the searched-in trajectories are presented in Secs. 4. 2-4. 4.

As the result of the limited space available in this volume, detailed trajectory data are not presented in this paper, and only three Earth tracks are included. Complete trajectory data and the remainder of the Earth tracks may be found in Ref. 9.

4. 1. Design Constraints

The trajectory design constraints are chosen so that a manned lunar return trajectory is simulated from lunar burnout to geo-impact. The constraints are as follows: launch month, March 1969; launch site (moon), Palus Putredinis (latitude, 27. 0°, longitude (east), 0. 0°); powered flight angle, 8. 0°; time of powered flight, 25. 0 sec; burnout radius distance, 1768. 0 km; flight time, 70. 0 hr; re-entry, counterclockwise; re-entry flight path angle, -6. 0°; re-entry maneuver angle, 40. 0°; time of re-entry maneuver, 0. 2 hr; landing sites, 1) San Antonio, Texas (geodetic latitude, 29.48°, longitude (east), 261. 0°) and 2) Bourke, Australia, (geodetic latitude, -30. 09°, longitude (east), 145. 6°).

The Analytic Lunar Return Program was used to investigate all the possible trajectories satisfying the constraints just listed. The results indicated that March 8 through 16 yielded acceptable trajectories for landing site no. 1 (San Antonio, Texas) and that March 20 through 25 and 27 through 30 yielded acceptable trajectories for landing site no. 2 (Bourke, Australia). The selenographic injection conditions from the analytic program were entered into the Space Trajectories Program for each of these days. Using the search technique described in Sec. 3. 4, integrated trajectories were obtained. Figure 1 shows the general characteristics of these trajectories.

4. 2 Ascent Trajectory

The direct-ascent technique is used. Figures 11 and 12 show the lunar track of the ascent trajectories initiating at the launch site and terminating at the injection points for March 8 and 16 and for March 20 and 30, consecutively. Figure 1 shows that when the moon is at positive declination at launch, the landing site for the type of trajectory required must have negative latitude. Therefore, since the declination of the moon is positive during the period of March 20 through 30, the landing site is Bourke, Australia, at a negative latitude of 30.09°. Similarly, the declination of the moon is negative during the period of March 8 through 16; therefore, the landing site is San Antonio, Texas, at a positive latitude of 29.48°. As a result, Figs. 11 and 12 present lunar tracks for San Antonio landings and Bourke landings, consecutively. These figures show that, regardless of where the impact site is located upon the Earth, launchings must be made in the lunar-east direction for a direct-ascent trajectory. In addition, the injection points for the first and last days in each of the two launch periods are located close together, enabling direct tracking and communication during the ascent phase of flight throughout the launch periods.

Table 1 of Ref. 9 presents the following injection conditions: the lighting conditions, the visibility from the Deep Space Instrumentation Facilities (DSIF), the moon-fixed spherical coordinates, the selenocentric inertial velocity, the geocentric latitude and longitude, and the major-body angles. The lighting condition at injection is, in general, the same as the lighting condition at launch since the powered flight arc is only 8.0°. The lighting condition varies from month to month as a result of the changing relationship between the position of the moon and the quarters of the month. The visibility from the DSIF at launch is governed by the longitude of the impact site and the time of flight. For the San Antonio impact site and a time of flight of 70 hr, the DSIF station at Johannesburg, South Africa, can observe the launch. For the Bourke impact site and a time of flight of 70 hr, the DSIF station at Goldstone, California, can observe the launch. The launch period from March 20 through 30 has a launch day apparently missing, March 26. This apparent discontinuity is caused by the fact that fixing the impact site at Bourke and the flight time at 70 hr creates a geometrical situation that has no trajectory solution for this launch day. The path angle ranges in value from 45° to 60°, which indicates that this launch site (Palus Putredinis) is not optimum from the payload viewpoint. The path angle should be zero, injection at periselene.

4.3. Transfer Orbit

The transfer orbit is defined as that portion of the launch-to-impact trajectory which lies between the lunar impact point and the re-entry point into the Earth's atmosphere. Section 2.1 has shown that the transfer orbit can be expressed as two conics: a selenocentric hyperbola from the lunar injection point to the point of exit from the lunar sphere of influence and a geocentric ellipse having an eccentricity of at least 0.960 from the exit point to the re-entry point.

Table 2 of Ref. 9 presents the characteristics of the selenocentric hyperbola for each acceptable launch day in March 1969. The conic characteristics are taken as the osculating conic characteristics at the time of selenocentric injection. Table 3 presents the characteristics of the point of exit from the LSOI. Table 4 presents the characteristics of the geocentric ellipse. The conic characteristics are taken as the osculating conic characteristics at the time of closest approach to the Earth.

Figures 13-15 show the Earth tracks relative to the DSIF for the final day of flight for three acceptable launch days in March 1969; the Earth tracks for the remaining acceptable launch days in March 1969 may be found in Ref. 9.

The dependence of periselene upon injection path angle when the injection radius distance is held fixed is evident from Tables 1 and 2 in Ref. 9. The variation of the energy $C_{3☾}$ with launch day for a fixed flight time and re-entry angle is a result of the varying Earth-to-moon distance (taken at the time of re-entry). The minimum Earth-to-moon distance occurs for the launch date March 10, which exhibits the minimum value of $C_{3☾}$. Table 3 in Ref. 9 shows the changing size of the LSOI according to the varying Earth-to-moon distance. The time of transit from launch to exit from the LSOI is about 12 hr. The exit points occur at angles from the Earth-moon line varying from 35° to 48°. From Table 4, eccentricities ranging from 0.972 to 0.979 km and perigees ranging from 6426 to 6429 km characterize the geocentric ellipse.

Figures 13 through 31 in Ref. 9 show the variation of the Earth track for the last day of flight as a function of launch day. The reversal or looping of the Earth track is caused by the change in angular rate of the probe relative to the Earth. When the probe is at a large distance from the Earth (40,000 miles), it appears stationary in inertial space, similar to a star. Thus, as the Earth rotates under this apparently stationary object, an Earth track is made which appears as a fairly straight line parallel to the Earth's equator. However, as the probe nears the Earth, its angular rate relative to the Earth increases to a value greater than the angular rate of the Earth and, hence, causes the Earth track to change directions. The changing shape of the Earth tracks from launch day to launch day is a result of the changing declination of the moon at launch. As the magnitude of the declination of the moon increases, the inclination of the trajectory plane decreases, and as the magnitude of the declination of the moon decreases, the inclination of the trajectory plane increases; see Tables 1 and 4 of Ref. 9 (the geocentric latitude of injection is approximately equal to the declination of the moon at launch). The reversal of direction of the Earth track is a function of the inclination of the trajectory plane to the equatorial plane, since the angular rate of the probe relative to the Earth is dependent upon inclination. The larger inclinations cause a looping, whereas the smaller inclinations cause just a reversal. The numbers on the Earth tracks are altitudes in statute miles and times before impact in hours (units used by DSIF personnel). The solid line indicates that at least one of the three DSIF stations is able to track, whereas the dashed line indicates that none of the DSIF stations is able to track.

4.4. Re-Entry Trajectory

The re-entry trajectory is defined as that portion of the launch-to-impact trajectory which lies between the re-entry point and the impact point. The design of the re-entry trajectory is a field of its own and is not considered in any detail during the design of the over-all lunar return trajectory. Therefore, the re-entry trajectory is assumed as any path that connects the re-entry point with the impact point and has a re-entry radius of r_r, a re-entry flight path angle of γ_r, a maneuver angle of η_r, and a maneuver time of t_r, and lies in the plane of the trajectory. Figure 1 shows a typical re-entry trajectory. The design of the trajectory computing programs discussed in Sec. 3.2 and the discussion of the trajectory characteristics are based upon this assumption.

Table 5 in Ref. 9 presents the characteristics of the point of re-entry for each of the acceptable days in March 1969. Three days in each of the two launch periods exist which have daylight launches and impacts, March 8 through 10 and March 28 through 30. The re-entry date of April 1 has no solution for reasons similar to those given for the missing March 26 launch date. The Earth-fixed and inertial velocities differ by about 0.32 km/sec for these counterclockwise trajectories. If clockwise trajectories are used, about 0.6 km/sec additional retrorocket capability is required. The time of impact is equal to the time of re-entry plus 0.2 hr.

4.5. Launch-on-Time Theory

The launch-on-time problem consists of launching a space vehicle at the proper time, given a certain launch azimuth, so that a particular set of terminal conditions is met. In the case of Earth-to-moon missions, there is a very strong relationship between launch time and launch azimuth because of the rotation of the Earth. The trajectory plane must contain the outward radial direction $\underline{S}$, which is fixed in space at a particular inclination to the equatorial plane. As a result, launch must occur when the launch site passes through the spaced-fixed plane. Since the launch site is rotating with the Earth at a speed of about 0.25°/min, a delay in launch time causes the trajectory plane to rotate about the Earth's spin axis away from the desired outward radial direction.

In the case of moon-to-Earth missions, the moon is rotating at a considerably slower rate than the Earth, which simplifies the launch-on-time problem. Delays in the launch time in the order of minutes change the trajectory plane very little so that the actual outward radial direction is nearly the same as the desired outward radial direction. However, significant error occurs in the impact site longitude. If a delay of Δt occurs in the launch time, then the actual impact site is not the desired impact site as a result of the rotation of the desired impact site from the nominal location in inertial space. The angular rotation is equal to the rotation of Earth relative to the moon in time Δt, $(\Delta t)(\omega_{\oplus} - \omega_{\oplus☾})$, and is equal to about 8.0° for a firing window of 30 min. In order to achieve the desired impact site, a correction Δt_F is made for this error and corrections $\Delta(\underline{B} \cdot \underline{T})$ and $\Delta(\underline{B} \cdot \underline{R})$ are made for errors in the outward radial direction by means of a midcourse maneuver.

As a result of the importance of accuracy in the impact site, assuring the safety of the astronauts, a second correction is made during re-entry into the Earth's atmosphere. This is accomplished by controlling the re-entry trajectory by means of a retromaneuver at re-entry or by aerodynamic flying during re-entry, or by both. Figure 16 presents the nominal trajectory and the corrected trajectory as a line of arrows and a dashed line, respectively. The corrected trajectory is required because of the errors in $\Delta\lambda$ and $\Delta\phi$ not corrected by the midcourse maneuver. The corrections required are out-of-plane maneuver angles of ΔA and $\Delta\eta_r$. The re-entry maneuver is purely a precision maneuver and is not intended to correct for large errors in $\Delta\phi$ and $\Delta\lambda$, as is the midcourse maneuver.

References

[1]Egorov, V. A., "Certain problems of moon flight dynamics," Russian Literature of Satellites (International Physical Index, Inc., N. Y., 1958), Part I, pp. 107-174.

[2]Penzo, P. A., "An analysis of moon-to-earth trajectories," Rept. 8976-0008-RU-000, Space Technology Lab., Inc., Redondo Beach, Calif. (October 30, 1961); ARS Paper 105621 (November 18, 1962).

[3]Clarke, V. C., Jr., "Design of lunar and interplanetary ascent trajectories," TR 32-30, Jet Propulsion Lab., Pasadena, Calif. (March 15, 1962).

[4]Baker, R. M. L. and Makemson, M., An Introduction to Astrodynamics (Academic Press, N. Y., 1960).

[5]Penzo, P. A., Kliger, I., and Tonies, C. C., "Analytic lunar return program computer program guide," Rept. 8976-0005-MU-000, Space Technology Lab., Inc., Redondo Beach, Calif. (August 23, 1961).

[6]Holdridge, D. B., "Space trajectories program for the IBM 7090 computer," TR 32-223, Jet Propulsion Lab., Pasadena, Calif. (March 2, 1962).

[7]Dallas, S. S., "Two search routines for computing lunar return trajectories," Sec. 312, request for programming no. 72, Jet Propulsion Lab., Pasadena, Calif. (June 26, 1962).

[8]Kizner, W., "A method of describing miss distance for lunar and interplanetary trajectories," External Publication no. 674, Jet Propulsion Lab., Pasadena, Calif. (August 1, 1959).

[9]Dallas, S. S., "Moon-to-earth trajectories," TR 32-412, Jet Propulsion Lab., Pasadena, Calif. (June 1, 1963).

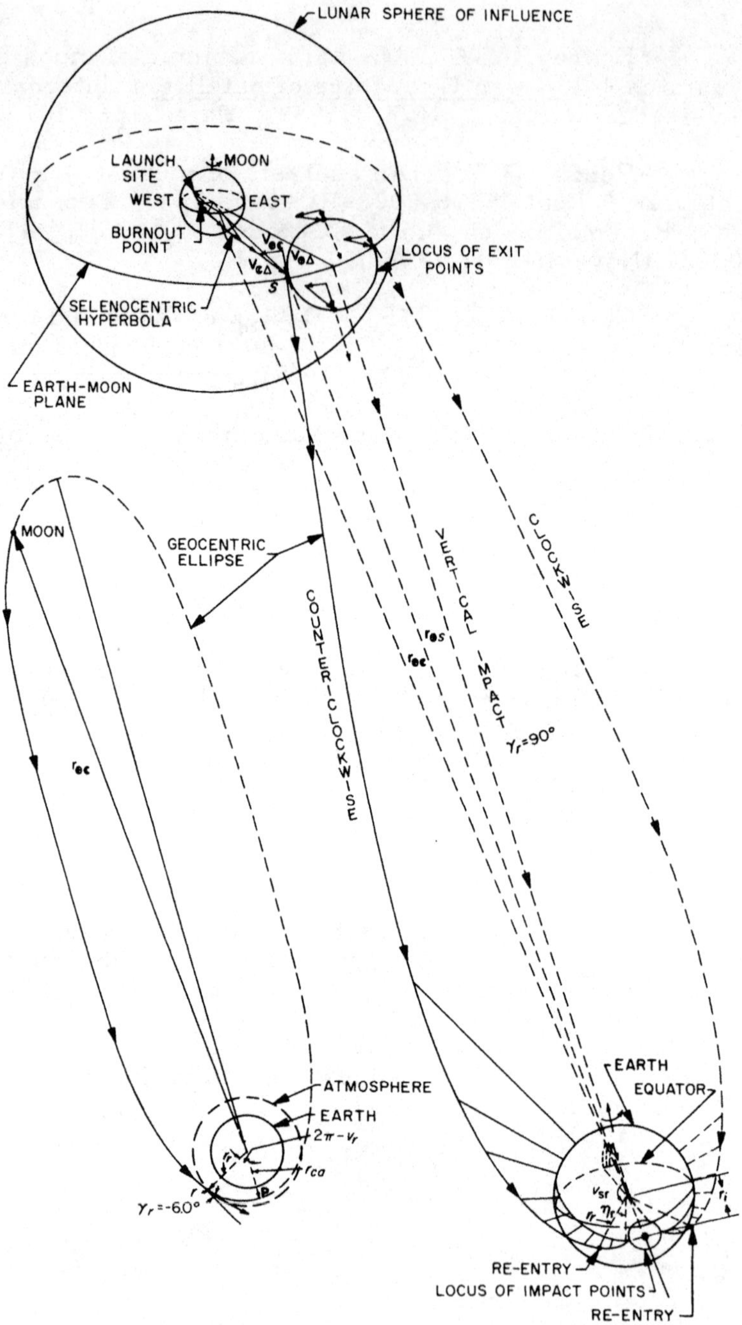

Fig. 1 Keplerian presentation of the lunar return geometry.

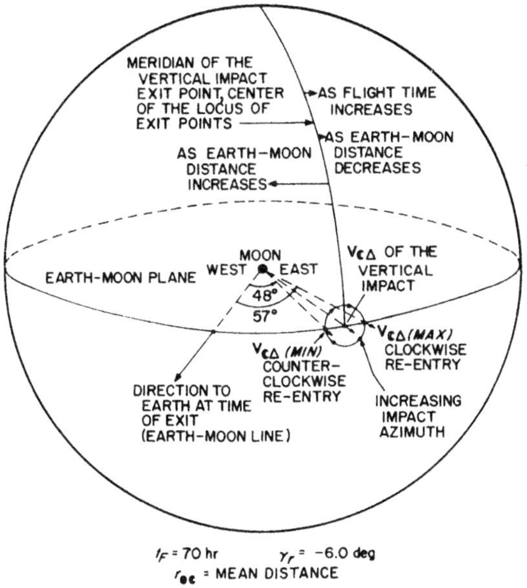

Fig. 2 Locus of asymptote piercing or exit points, lunar sphere of influence.

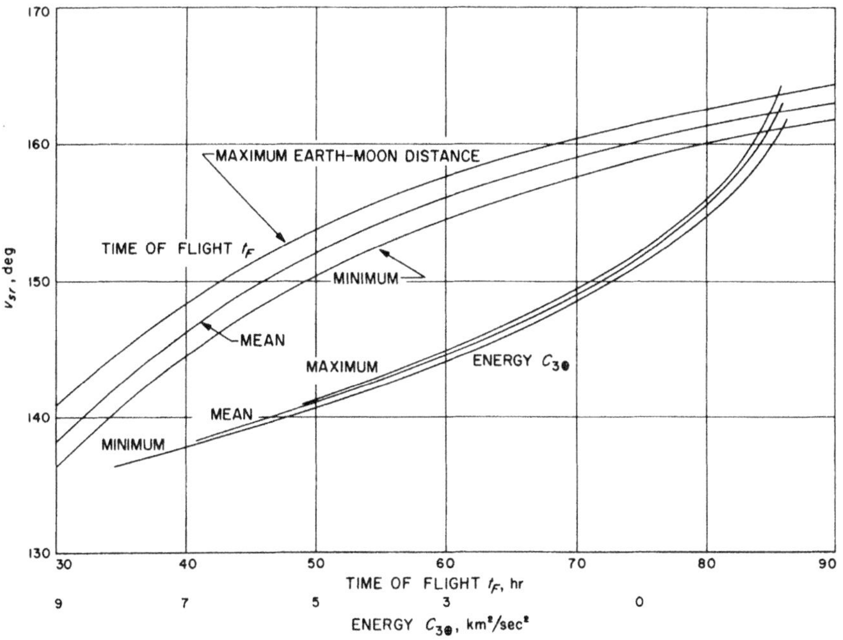

Fig. 3 In-plane angle between exit point and re-entry point vs time of flight and energy $C_{3\oplus}$.

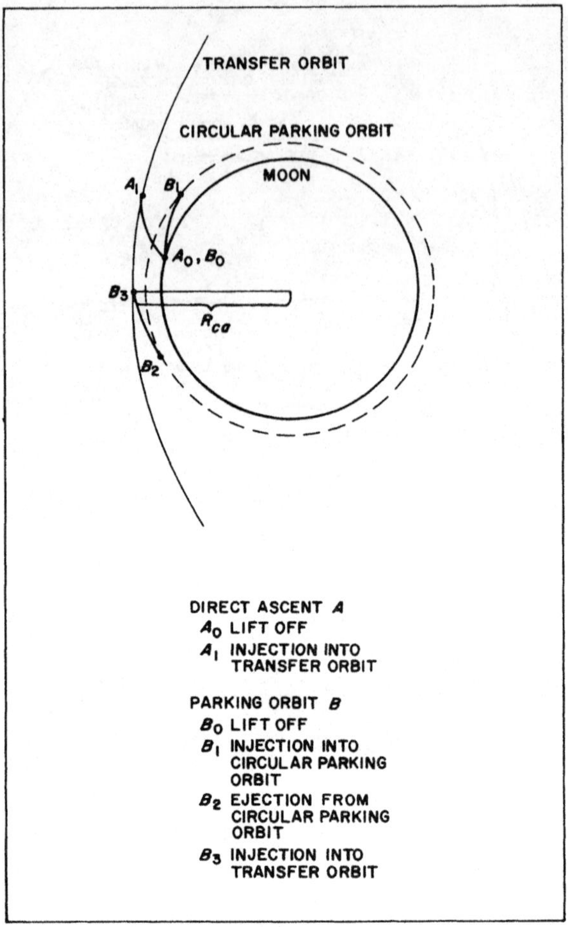

Fig. 4 Types of lunar-return trajectories.

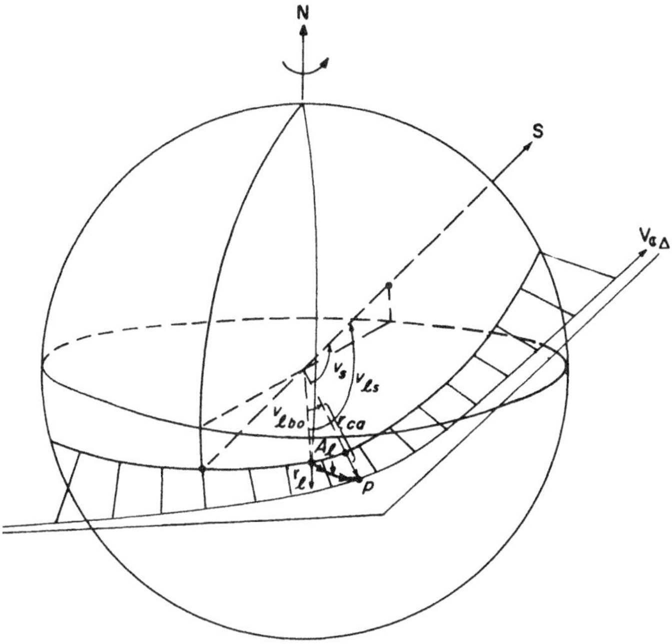

$$v_{\ell s} = v_{\ell bo} + v_s$$

Fig. 5 Launch site location referenced to the outward radial direction.

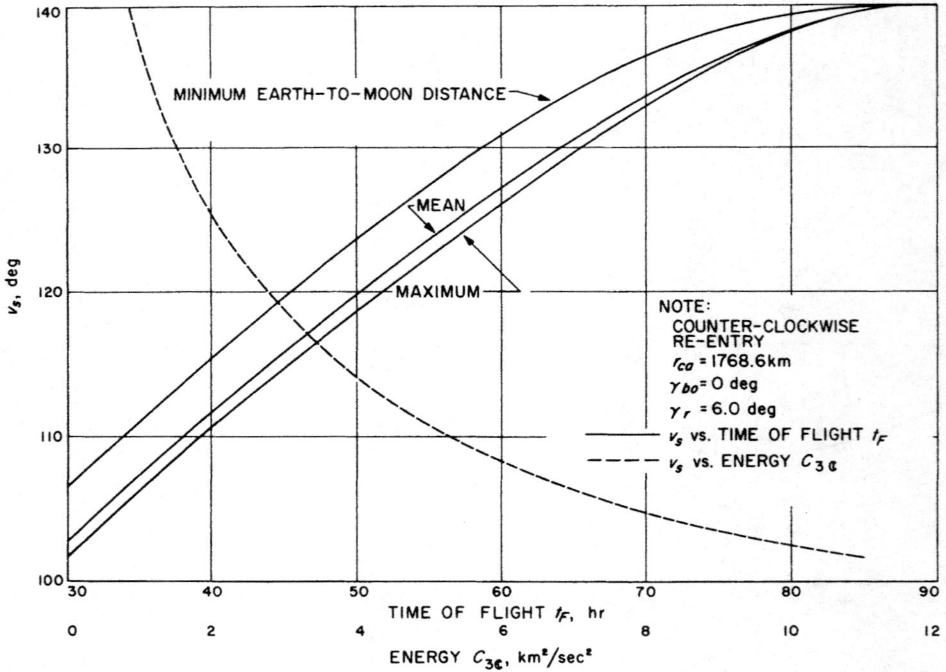

Fig. 6 Angle between outward radial direction and periselene vs time of flight and energy $C_{3☾}$.

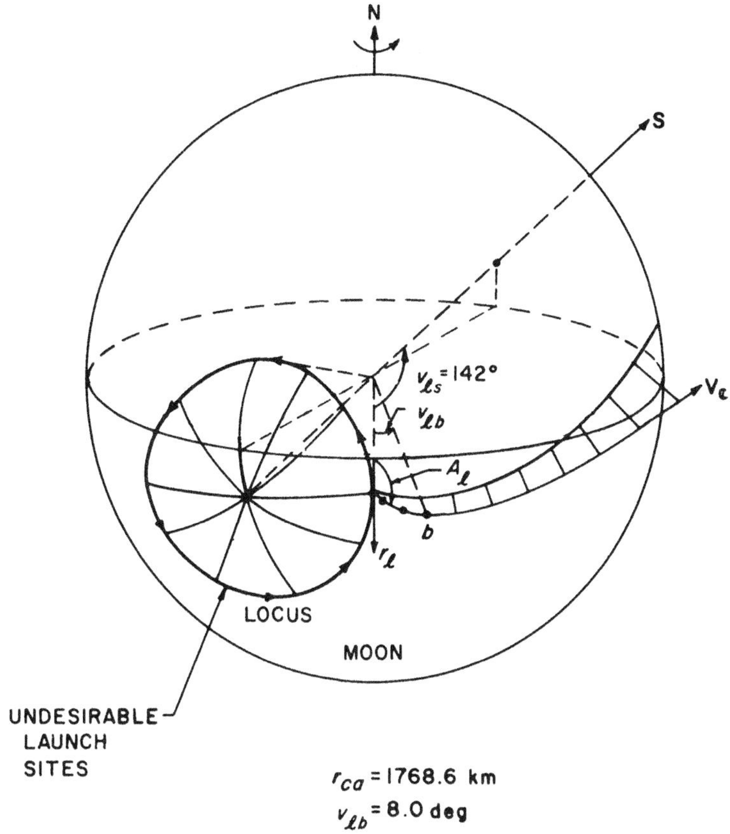

Fig. 7 Locus of launch sites for injection at closest approach distance.

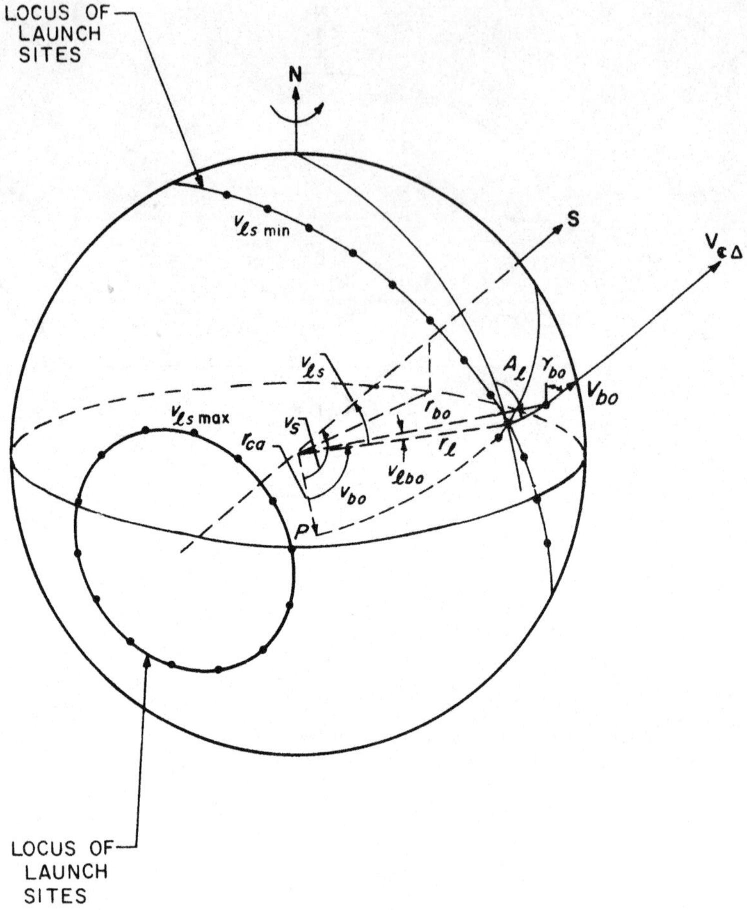

Fig. 8 Locus of launch sites for injection at positive γ_{bo}.

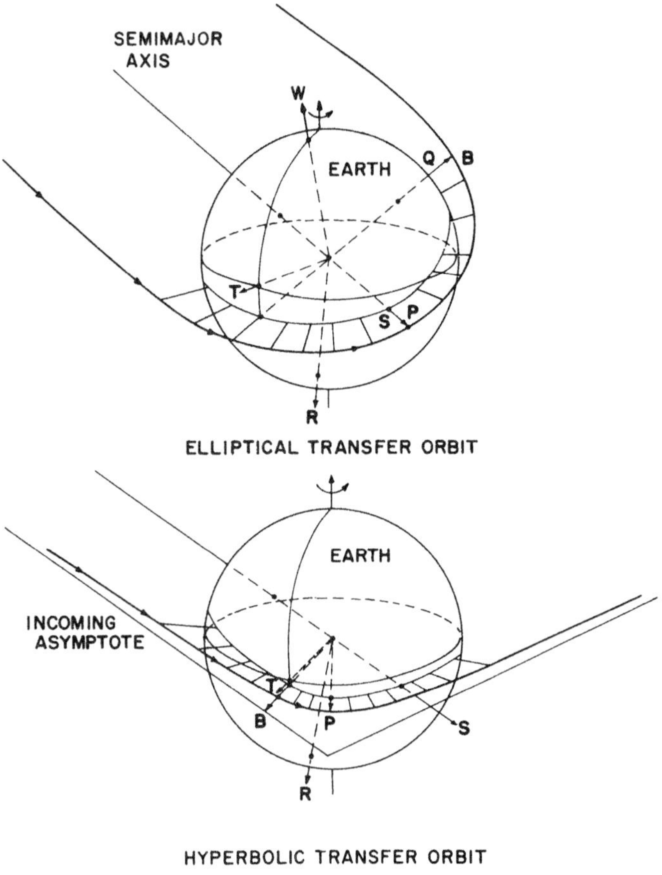

Fig. 9 Definition of the impact parameter $\underline{B}$.

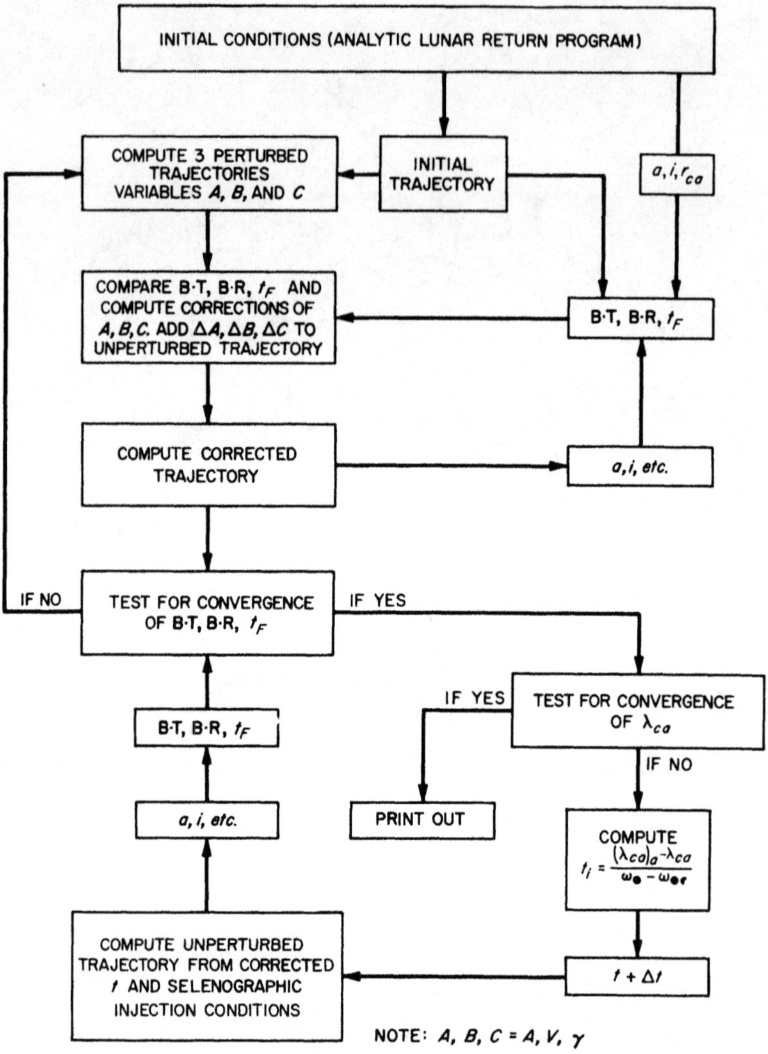

Fig. 10 Present space trajectories program search routine.

Fig. 11 Lunar tracks of the ascent trajectories, impact site, San Antonio, Texas.

Fig. 12 Lunar tracks of the ascent trajectories, impact site, Bourke, Australia.

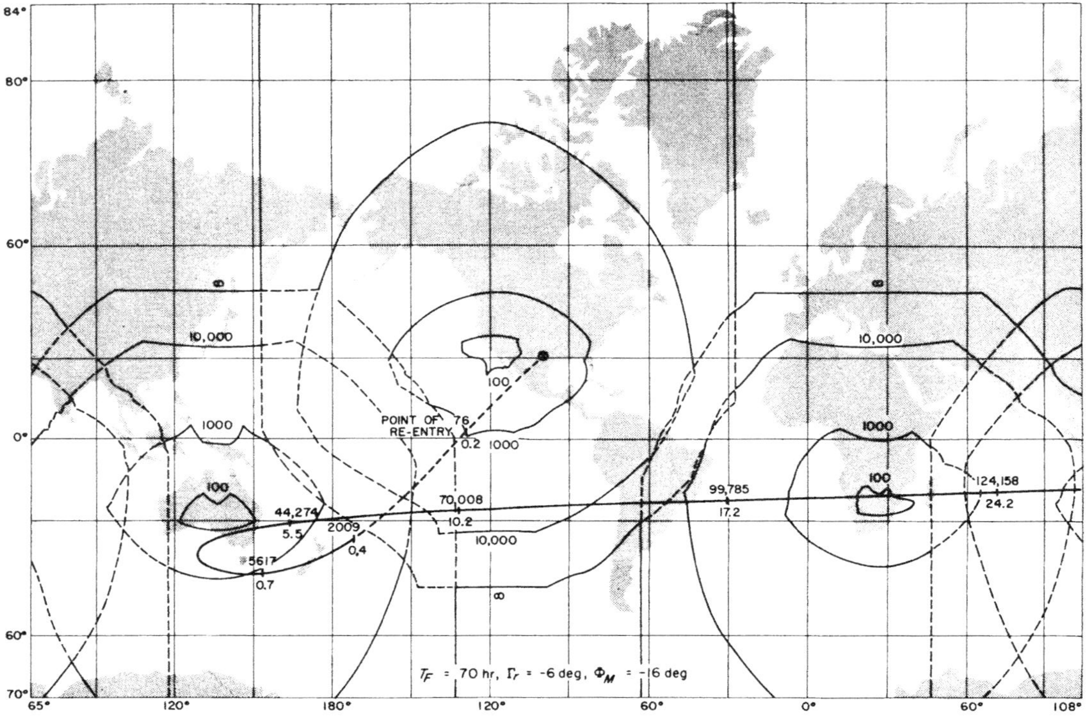

Fig. 13 Lunar return Earth track, launch date, March 8, 1969; landing site, San Antonio, Texas ($t_F = 70$ hr, $\gamma_r = -6°$, $\delta_{☾} = -16°$). For Figs. 13-15, the dashed line indicates loss of DSIF tracking; numbers above the track are altitudes in statute miles, and numbers below the track are times before impact in hours.

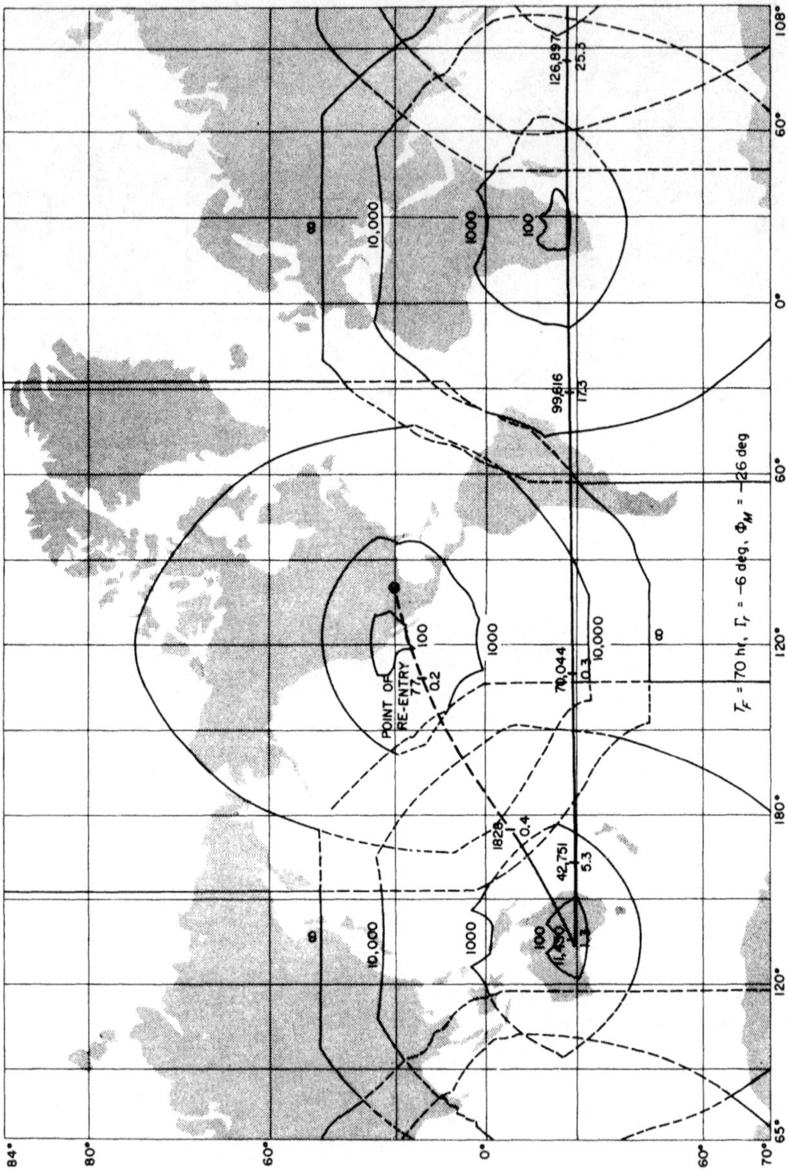

Fig. 14 Lunar return Earth track, launch date, March 10, 1969; landing site, San Antonio, Texas (t_F = 70 hr, γ_r = -6°, $\delta_{☾}$ = -26°).

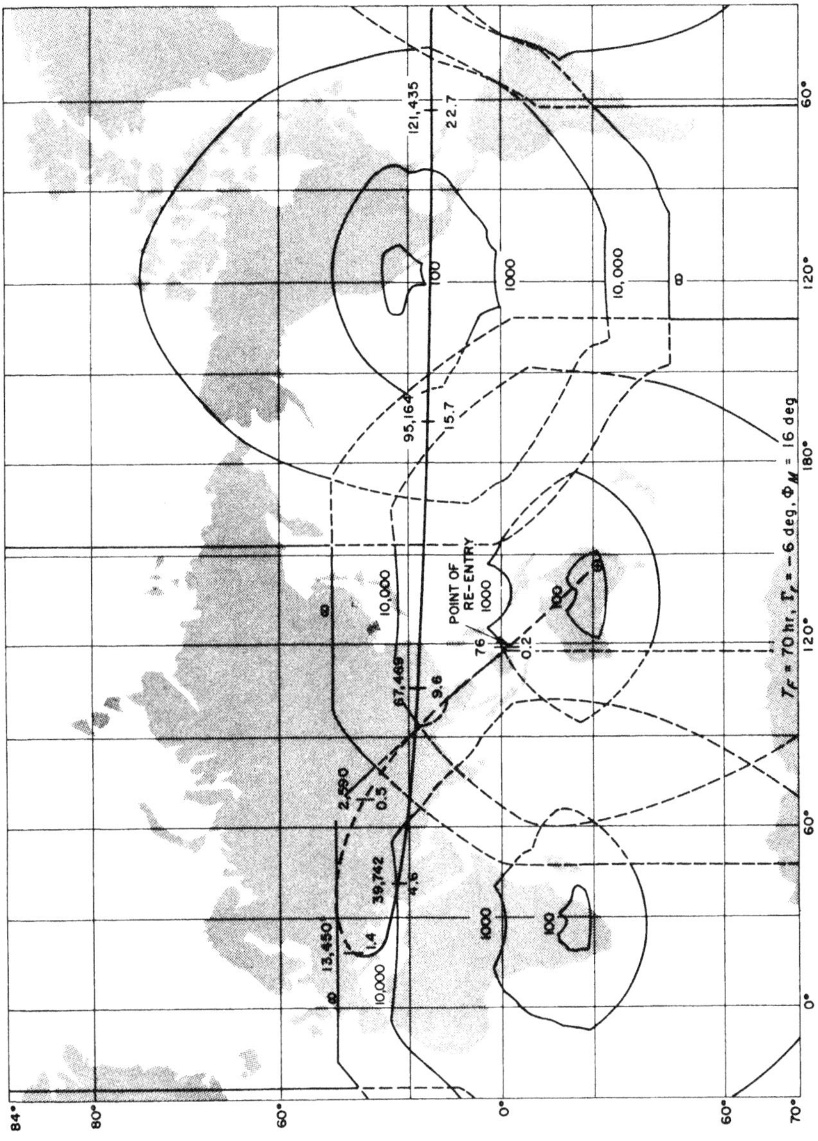

Fig. 15 Lunar return Earth track, launch date, March 30, 1969; landing site, Bourke, Australia (t_F = 70 hr, γ_r = -6°, $\delta_☾$ = 16°).

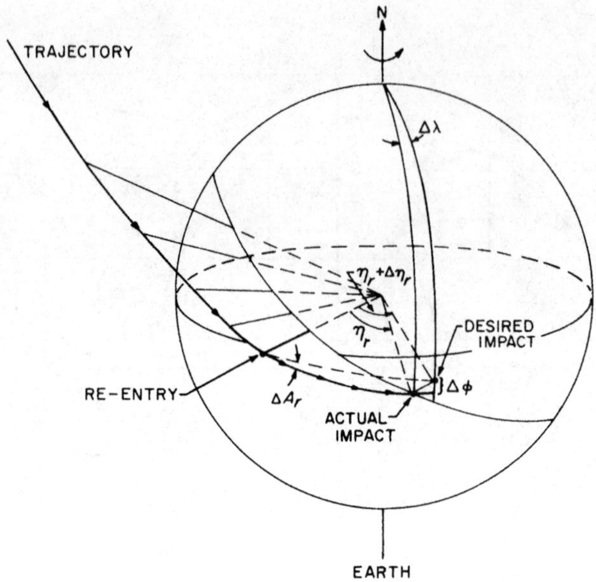

Fig. 16 Launch-on-time problem.

AN ANALYSIS OF FREE RETURN CIRCUMLUNAR TRAJECTORIES

Paul A. Penzo*

TRW Space Technology Laboratories, Redondo Beach, Calif.

Abstract

An analysis of a special class of free-flight circumlunar trajectories is presented utilizing a simplified model of motion in the earth-moon system. This model is based on the sphere-of-action concept where motion outside of this sphere (centered at the moon) is represented by an earth influenced conic, and motion within the sphere is a moon influenced conic. The conics are chosen such that the position and velocity match at the spherical boundary as well as satisfying other constraints. This study has concentrated on the class of free-flight circumlunar trajectories, which will most likely be of interest in the U.S. Space Program, i.e., launch from a circular parking orbit, very close approach to the moon, and re-entry into the earth's atmosphere at a shallow angle. A qualitative analysis of the behavior of the velocity vectors at the sphere of action is presented to provide an intuitive feeling for the relationships between time of flight, pericynthion distance, and outward and return inclinations to the moon's plane. To generate results of practical interest, a special computer program was written and some results are presented here both in a parametric form and for a specific

---

Presented as Preprint 63-404 at the AIAA Astrodynamics Conference, New Haven, Conn., August 19-21, 1963. This work was prepared for Jet Propulsion Laboratory, California Institute of Technology, Pasadena, Calif. under Contract No. 950045, under NASA Contract NAS 7-100. The author is indebted to G. W. King and G. J. Hanson for their extensive effort in programming and checking out the Analytic Circumlunar Program, which was used to generate the data in this paper. He is also grateful to L. J. Martin and R. W. Johnson for their help in processing this data and plotting the majority of the accompanying graphs. Finally, he wishes to thank E. H. Tompkins for initiating the problem and for helpful discussions during various stages of the analysis.

*Head, Trajectory Analysis Section.

time period. Also, a comparison with machine integrated circumlunar trajectories is presented, and methods by which circumlunar trajectories may be generated are indicated.

Introduction

One of the first manned lunar missions will, in all probability, be along a free-flight circumlunar trajectory. Such a trajectory is one for which no major midcourse thrust is applied from the time that it is injected from a parking orbit about the earth to the time that it re-enters the earth's atmosphere. As shown in this study, the number of such trajectories is very limited if one is interested in launching from a specified launch site with a limited range of azimuths, in coming within a fixed distance of the moon, and in re-entering the earth's atmosphere near a specified landing site. If, further, one wishes to choose a spot on the moon over which the trajectory passes, or the inclination of the return portion of the trajectory with the equator, the possibilities are very limited indeed and in many cases nonexistent.

This paper proposes to handle the problem of determining free-flight circumlunar trajectories in the following manner:

1) Choose a simple gravitational model of the earth-moon system such that trajectories in this system may, in a large part, be studied qualitatively.

2) Construct a computer program based on this model which may be used to generate data to perform a parametric analysis on those variables which are considered, from 1), to be important.

3) Determine the validity of the data by comparing the results with those generated by an integration (exact) program and determine for which trajectories the simplified gravitational model breaks down.

One of the most important products of an approach of this sort is the resulting computer program;† for such a program not only may be used to generate parametric data, but the injection conditions and the approach to the moon, both outputs of the program, may be used as first approximations in the exact program. Linear search methods may then be used to solve for the corresponding exact trajectory. This method

†A complete description of the Analytical Circumlunar Program is presented in the Preprint.

will most likely be used to generate the nominal trajectory for an actual mission.

Further, if the computer program is accurate enough, it may be used to generate sensitivity coefficients of terminal conditions with respect to injection or midcourse conditions. Such data would be very useful for general guidance studies.

The Gravitational Model

The model on which the following analysis and computer program are based was described and used by V. A. Egorov in 1956[1‡] and has since been successfully applied at TRW Space Technology Laboratories to generate moon-to-earth trajectories.[2] This model consists of enclosing the moon in a region within which only the moon's gravity is considered and outside of which only the earth's gravitational field is in effect. More specifically, the dividing surface is defined as the locus of points at which the ratio between the force with which the earth perturbs the motion of a third body and the force of attraction of the moon is equal to the ratio between the perturbing force of the moon and the force of attraction of the earth. For the earth-moon system, this surface is approximately a sphere centered at the moon having a radius of about 10 earth radii.

With such a model, all free-flight trajectories in earth-moon space will consist of a conic in the moon phase (within the sphere) and a conic in the earth phase. For circumlunar trajectories there will be two earth phase conics (see Fig. 1). These conics are such that the position and velocity match at the phase boundary. The apparent discontinuity in the velocity in Fig. 1 is due to the fact that the moon is revolving about the earth requiring a velocity translation at this point. Specifically, whenever the trajectory arrives at the moon's sphere of action, henceforth referred to as S, from either the earth phase or moon phase, it is necessary to perform the following translations:

$$\bar{x} = \bar{y} + \bar{x}_m$$

$$\bar{u} = \bar{v} + \bar{u}_m$$

where $(\bar{x}, \bar{u})$ and $(\bar{y}, \bar{v})$ are the position and velocity vectors of the spacecraft with respect to the earth and moon,

‡Superscript numbers refer to the list of references.

respectively, and $(\bar{x}_m, \bar{u}_m)$ are the position and velocity of the moon with respect to the earth at the time that the spacecraft passes through S.

Egorov discusses general circumlunar trajectories such as those which make several revolutions about the earth-moon system; however, his analysis is confined to two-dimensional motion. In contrast, this paper will treat three-dimensional circumlunar trajectories with the discussion restricted to so-called single revolution trajectories. That is, they will begin near the earth, enter the sphere of action in some manner, and then exit and return to the earth. The moon phase conic will be hyperbolic, except in very rare cases, whereas each leg of the earth phase conics may be independently elliptic, hyperbolic, or parabolic.

Classifications

In order to more easily discuss circumlunar trajectories, it is convenient to classify the trajectories in some manner and, in order to do this, some of the properties that will be derived must be anticipated. One such property is the orientation of the velocity vector at entry and exit of the moon's sphere of action. Considering the earth-centered velocity first, and limiting the trajectories to those energies where direct flights to the moon would lie between 50 and 90 hrs, then the flight path angle, i.e., the angle that the velocity vector makes with the normal, will range from 5° to 30°. This is assuming that launch is from a parking orbit and that on return the spacecraft has a very shallow re-entry, for example, 96° with the normal. Further, the moon-centered velocity direction with the earth-moon line will range from 30° to 80°.

On the basis of this behavior, it is possible to form the first major classification of circumlunar trajectories for this range of energies. Neglecting the rotation of the moon's velocity vector during the time that the spacecraft is in the moon's sphere of action, the velocity vector additions will generally look as those shown in Fig. 2. There will exist four possible velocity diagrams, and these will yield the six possible circumlunar trajectories shown. To the first two classes, the name "tight trajectories" is applied because the moon phase velocity must turn through the greatest possible angle which is accomplished only if the trajectory approaches very near the moon. In the class T-1, the spacecraft approaches the moon on the ascending leg of the outward ellipse, and leaves the moon on the descending leg of the return ellipse. In this case, the spacecraft never passes through the apogee of either transfer ellipse, resulting in a much lower flight time than in the other five classes.

The second class T-2 is also a tight trajectory; however, here the spacecraft enters the moon's sphere of action from behind or on the descending leg of the trajectory and leaves the sphere on the ascending leg and does not go "around" the back side of the moon. The major difference between this class of trajectories and the others is the fact that the spacecraft passes through apogee on both the outward and inward legs of the earth phase ellipses resulting in the longest flight time of all six classes.

The loose classes obtain their name from the fact that the moon phase velocity vector has a minimum change in direction which results only if the distance of the closest approach to the moon is considerably farther than for the tight classes. The distinction between the four L classes is clear from the figure. As in class T-2, the trajectories of classes L-1b and L-2b do not circumnavigate the moon, but always remain on the earth side of the moon. The reason that these two classes are not considered to be independent of L-1a and L-2a is because for three-dimensional trajectories, there is a continuous transfer of the a class into the b class. This, however, is not the case between classes T-1 and T-2 unless very low energies are included. Also, concerning three-dimensional trajectories, it is noted that in all cases the spacecraft must approach the western hemisphere of the moon and exit from the moon's eastern hemisphere. There are a few exceptions for class L trajectories that approach or exit the moon near the lunar poles.

Since the major emphasis in circumlunar trajectories is in a close approach to the moon, the loose classes are of the least concern. Also, there will most likely be little interest in class T-2 due to the lengthy flight times and greater midcourse sensitivities (which may be attributed to the long flight time). This leaves class T-1 trajectories[2] as the only set of trajectories which are of most interest. The remaining discussion is confined to this set.

---

[2]Very low energy trajectories cannot be computed, though they may be interesting, since the analytic model breaks down in this energy region.

A modification of the computer program discussed in Appendix B to allow a circumlunar trajectory search for fixed inclination in the return phase makes it possible to exhaustively study tight circumlunar trajectories of class T-1. This is possible since, by freeing the re-entry maneuver angle, trajectories may be chosen which return in a specified manner, in particular over the poles and in the earth-moon plane and not in such a manner as to satisfy certain terminal conditions. By choosing these extreme launch and return inclinations, one will obtain bounds on various parameters such as pericynthion altitude, re-entry velocity, time of flight, look angles, etc., as well as parametric relationships.

The motion of the spacecraft may then be characterized by the energy and inclination of the outward and return phases, where the perigee of the outward phase is the radial distance to the parking orbit and the return phase perigee is such that the re-entry angle at a specified re-entry altitude is satisfied. Figure 3 indicates the manner in which the trajectory may enter and exit the sphere of action (moon-centered). This would be for a fixed energy (or equivalently time of flight outward) and for a fixed position of the moon in its orbit. The inclinations are those of the earth phase conics, whereas the vectors converging on the moon approximate the direction of the moon-centered velocity at S.

Note, then, for a fixed position of the moon in its orbit, and a fixed time of flight in the outward phase, that fixing the inclinations of the outward and return phase conics will determine a specific circumlunar trajectory. These inclinations are taken with respect to the moon's plane in order to take advantage of the symmetry situation. That is, trajectories that have the outward and return inclination pair (i_{mo}, i_{mr}) (m indicating inclination with respect to the moon's plane) will have a reflection trajectory ($-i_{mo}$, $-i_{mr}$) with similar characteristics, where i_{mo} and i_{mr} are both measured from counterclockwise launch and approach toward the lunar north as shown in Fig. 3.

Having defined the manner in which inclinations will be measured, it is convenient to single out those inclinations of the outward phase conic which represent "extreme" trajectories. These extreme trajectories may be characterized in the following manner:

Case I. This applies to all trajectories in which the spacecraft is launched counterclockwise with respect to the earth and in such a manner that the outward phase conic lies in the moon's plane.

Case II. Here, the spacecraft is launched normal to the moon's plane. The symmetry relationship makes it immaterial as to whether it approaches the moon from above or below the moon's plane.

Case III. This applies to trajectories in which the vehicle is launched clockwise with respect to the earth and whose outward phase conic lies in the moon's plane.[#]

The next classification concerns the coasting orbit. Since the complete trajectory from launch to the moon may produce an in-plane angle greater than 360°, Mercator projections such as Fig. 4 are very convenient. This figure presents the two types of coasting orbits as defined in the computer program.

Type 1 is defined as that coasting orbit required to intercept the moon on the trajectory's downward swing. The moon's plane is shown by the dotted curve.

Type 2 is that coast trajectory, which will intercept the moon on the trajectory's upward swing. In the figure, this is the section labeled (2).

It is necessary to distinguish between these two types because of the two possible solutions for launching with a fixed launch azimuth. Figure 4 shows the moon displaced for the two possibilities; however, in actuality, the moon is fixed and earth rotates into position. The two types therefore result in very different times of launch. Also, the characteristics of the two trajectories will be distinctly different since the inclination of the trajectory plane with the moon's plane is different.

General Analysis

Fixed Conditions

This analysis will generally apply to circumlunar trajectories regardless of the manner of launch and the manner of re-entry; however, when using the Analytic Circumlunar Program, it is necessary to restrict the number of variables or the runs required for a parametric study will get completely out of hand. For this reason, these fixed parameters have been chosen to closely correspond to those desired in our

[#]These trajectories will not be considered in this paper.

present space program. Cape Kennedy has been chosen as the launch site, and since the time period of greatest interest is 1967 to 1970, the moon's orbital inclination has been chosen near its maximum value. This will permit launches from the Cape into the moon's orbit plane, which very conveniently supplies a realistic application of an extreme trajectory, i.e., case I. The powered flight angle and time, though not extremely important, were chosen as 30° and 10 min, respectively. The circular parking orbit altitude was fixed at 600,000 ft and the injection flight path angle at 90°. Also, anticipating that the greatest interest will be in manned flights, the re-entry angle was chosen as 96° (measured from the vertical) with a re-entry altitude of 400,000 ft.

Specifying the preceding launch and re-entry conditions is almost equivalent to fixing the outward and return phases perigee altitudes. The inclination of the outward phase is determined by the launch azimuth from the Cape and that of the return phase is an optional input. Since these are both inclinations with the equatorial plane, whereas the graphs presented here are with respect to the moon's plane, a conversion of one inclination to the other is desirable and graphs enabling one to do this have been generated, but are not included here.

Sphere of Action

There exists a simple method to add a considerable amount of intuitive knowledge to the behavior of free-flight circumlunar trajectories, and this is to consider the velocity translations required at entry and exit of the sphere of action. This is presented in Refs. 1 and 2, but is repeated here since it is directly applicable to results of this study.

First, considering the outward phase conic, it is clear that the magnitude of the earth-centered velocity ($\bar{u}$) of the spacecraft at S will depend primarily upon the time of flight from injection to pericynthion and the earth-moon distance at pericynthion arrival. The direction of $\bar{u}$ will depend upon the launch site latitude and launch azimuth, or, equivalently, the inclination of the outward phase conic. The magnitude and direction of the moon-centered velocity ($\bar{v}$) at S will in turn depend upon $\bar{u}$ and the moon's velocity $\bar{u}_m$, or

$$\bar{v} = \bar{u} - \bar{u}_m$$

If the magnitude of the earth-centered velocity $\bar{u}$ is held fixed, i.e., a fixed outward flight time and a fixed earth-moon distance, then the flight path angle β must remain fixed. The class of all moon-centered velocities $\bar{v}$ for varying outward inclinations must then originate at the tip of the moon's velocity vector and terminate on a spherical boundary whose radius is u and whose center is at the origin of $\bar{u}_m$. This is shown in Fig. 5. It should be remembered that the point of entry at S is not fixed for all cases, and for this reason this figure is not a complete representation of the problem.

The first observation one may make from this figure is that the moon-centered velocity magnitude increases as the inclination increases from counterclockwise to clockwise values, and that the direction points further away from the earth-moon line. The fact that the spacecraft's moon-centered energy increases implies that the return phase energy will increase. Also, for higher energies, the spacecraft must pass closer to the moon's center to turn through the same angle. However, since this angle to the earth-moon line will be greater for a higher moon-center velocity, no conclusion may be drawn concerning the effect of the inclination on the pericynthion altitude.

The analysis for the return phase velocity translation is similar. In this case, it is assumed that the outward phase conic is fixed in its time of flight and its inclination. Only the approach to the moon, such as the penetration point in the impact parameter plane, is allowed to vary and generate a set of circumlunar trajectories having any inclination on return. This is acceptable since small perturbations of the launch conditions will have considerable effect on the entry position at S but very little effect on the magnitude and orientation of the entry velocity. The approach to the moon, then, will vary the direction of the spacecraft's moon-centered velocity at exit of S but not its magnitude. The earth-centered velocity must then originate on a spherical boundary centered at the origin of the $\bar{u}_m$ vector, whose radius is v, and terminate at the tip of the $\bar{u}_m$ vector. This earth-centered velocity vector, $\bar{u} = \bar{v} + \bar{u}_m$, must be pointed in the general direction of the earth, and, from Fig. 5, it is clear that its magnitude will be smaller for clockwise re-entry than for counterclockwise re-entry. The conclusion is, then, that for a given outward phase conic as described previously, the flight time of the return phase will be greater for clockwise re-entry than for counterclockwise re-entry. Also, since the angle through which the moon-centered velocity vector must turn is smaller for clockwise re-entry than for counterclockwise re-entry, the pericynthion altitude will be greater for clockwise reentry. These characteristics are clearly present in the graphs that follow.

Another important observation that may be made from Figs. 1 and 5 involves the symmetry relation mentioned earlier, i.e., symmetry with respect to the moon's plane. This relationship may be extended if one notices that the velocity vector additions at the entry and exit points of S are independently symmetric with respect to the moon's plane. Thus, for a fixed outward phase conic, as defined previously, the return phase conic will be symmetric with respect to the moon's plane, except for variations occurring in the moon phase which may vary the time and position of exit from S. This implies that those parameters that are insensitive to the moon phase conic will be symmetric with respect to the earth-moon line. One such parameter is the time of flight. Thus, although (again for a fixed outward phase conic) the trajectories are not generally the same for return inclinations of i_{mr} and $-i_{mr}$, the return times of flight for these two trajectories are almost exactly the same. This has been considered in plotting the time graphs in Figs. 6 and 7.

In order to get a better feeling for the approach of circumlunar trajectories to the moon, it is worth looking at the impact parameter. The impact parameter plane is defined to be perpendicular to the incoming hyperbolic asymptote. The horizontal or $B \cdot T$ axis lies in the moon's plane and is directed positively opposite the moon's direction of motion. The vertical or $B \cdot R$ axis is directed positively below the moon's plane. The origin is fixed to the center of the moon. For all circumlunar trajectories considered here, the asymptote will be inclined only about $\pm 10^{\circ}$ to the moon's plane. Since the direction of the hyperbolic asymptote varies greatly with time of flight, outward inclination, and position of the moon, the impact parameter plane will be oriented differently with the earth-moon line for each point in Fig. 8.

Both graphs in Fig. 8 are drawn for the maximum distance of the moon. The outward flight time and the return inclination with the moon's plane are indicated. The larger magnitude of the impact parameter for a fixed time of flight implies a greater pericynthion distance. A greater time of flight results in a greater impact parameter since the incoming asymptote must be further from the moon for lower moon centered energies and the same turning of the spacecraft's velocity vector.

Time of Flight

The times of flight that are considered here are the times from injection to pericynthion, from pericynthion to re-entry, and the sum of these values, or what may be considered to be the total time of flight. For the fixed conditions discussed earlier, specifying the outward and return inclinations and the outward time of flight will determine a unique trajectory and in particular a specific return time of flight. Figs. 6 and 7 present this return time vs the outward time for constant return phase inclinations with the moon's plane. Each graph is drawn for an in-plane or normal firing with respect to the moon's plane and for a particular position of the moon in its orbit. Also plotted are constant total flight time lines.

In comparing these graphs, several observations may be made. The first is that launches counterclockwise and in the moon's plane (case I) results in equal or greater flight times in the return phase. Return flight times for launches normal to the moon's plane (case II) are nearly symmetric with 90° returns; that is, they are generally less than the outward flight time for return inclinations less than 90° and greater otherwise. Secondly, the difference in the outward and return times of flight may be considerable. In Fig. 7, for example, an outward flight time of 70 hr results in a return flight time of 92 hr if the return is clockwise in the moon's plane. The differences are even greater for Fig. 6.

As indicated on both time graphs, the dashed curves represent extrapolated values, i.e., values not obtainable with the present Analytic Circumlunar Program. There is no guarantee that these extrapolated trajectories exist. In fact it is possible to show that at some point they should not exist. Referring to Fig. 5 and the velocity vector addition at exit, if the moon-centered velocity $\bar{v}$ is smaller than the moon's velocity, then clockwise trajectories will not exist. Thus, in some cases, particularly those for which the spacecraft is launched in the moon's plane and returns in a clockwise manner, there will be a maximum allowable time of flight in the outward phase which will result in a circumlunar trajectory. No attempt is made in this study to determine these critical flight times, since the return flight time becomes large and this is the situation where the analytic model generates a poor approximation of the actual motion.

Pericynthion Distance

Some general conclusions concerning the behavior of the pericynthion altitude have been given. This behavior is very noticeable in the graphs presented here. Figures 9 and 10 show the variation of the pericynthion distance with the inclination of the return phase conic for fixed times of flight to the moon. Also plotted on each graph are two curves for fixed return times from the moon.

The first noticeable characteristic of these curves is that, for fixed outward flight times, the pericynthion altitude is greater for clockwise returns than for counterclockwise returns. This result was anticipated in the qualitative analysis of the previous section, and as shown in these figures, this distance may vary from below the surface of the moon to more than 1000 naut miles above the surface for a fixed outward phase flight time.**

A similar characteristic is the fact that, for fixed return flight times, the pericynthion altitudes are greater for counterclockwise return trajectories than for clockwise return trajectories. This might have been anticipated from Fig. 5 if the flight path angles at entry and exit of S were neglected. If this is done, then it is clear that for fixed return flight times (or energies), the moon phase velocity must be greater for cw trajectories than for ccw trajectories requiring a closer approach to the moon to turn through approximately the same angle.

Upon comparing the pericynthion altitudes for in-plane outward phase trajectories with out-of-plane outward phase trajectories, it is seen that for fixed outward phase flight times and return phase inclinations the former have lower altitudes than the latter. This indicates that the moon phase velocity change dominates the effect of the entry flight path angle.

Turning now to the effect of the moon's distance on the pericynthion altitude for a fixed flight time to the moon (see Appendix B) and fixed outward and return inclinations, since the moon phase velocity increases with increasing lunar distance, the pericynthion altitude must decrease. A similar case exists for a fixed return flight time.

A final characteristic of these curves is the symmetry condition. As mentioned before, circumlunar trajectories are

---

** In drawing these graphs, it was decided that altitudes less than 1200 naut miles would be of greater interest, although trajectories with pericynthion altitudes up to 5000 naut miles were obtained with the Analytic Circumlunar Program.

symmetric with respect to the moon's plane. Figure 9, for which the outward phase lies in the moon's plane, indicates this symmetry since values for i_r return phase inclinations are equal to $-i_r$ values. Figure 10, on the other hand, is launched normal to and below the moon's plane, i.e., the outward phase inclination is -90°. A comparison of pericynthion altitude for i_r return inclinations with $-i_r$ inclinations (where i_r is positive) indicates that negative return inclinations result in lower pericynthion distances for a given outward phase flight time. This may be explained by means of the general observation that the angle through which the moon phase velocity vector must turn at entrance and exit of the moon's sphere of action is greater for outward and return earth phases on the same side of the moon's plane than for these phases on opposite sides of the moon's plane. Thus, for a fixed outward time of flight, which implies an essentially fixed moon phase velocity magnitude, the pericynthion distance must be less to achieve the required greater turning of the velocity vector for (in this case) negative return phase inclinations.

Fixed Pericynthion Altitude

One of the more practical subclasses of free-flight circumlunar trajectories is the group that will approach very close to the moon. This subclass may be generated by cross-plotting figures similar to 9 and 10 for fixed pericynthion altitude. This was done for an altitude of 100 naut miles, and the results are shown in Figs. 11, 12, and 13. It is clear that one may choose specific outward and return inclinations and then obtain a specific pericynthion distance by controlling the outward time of flight. This time of flight will depend on the moon's distance from the earth as indicated in Figs. 11 and 12. It is interesting to note that the required outward time of flight decreases as the return inclinations go from 0° (ccw) to 180° (cw). The return time of flight, on the other hand, will increase. This behavior can be explained by means of the velocity vector diagrams in Fig. 5; however, it is more easily seen to be a reflection of the behavior observed previously, i.e., that the pericynthion altitude is greater for clockwise returns than for counterclockwise returns. Thus, to decrease pericynthion altitude for clockwise returns, holding the turning angle of the moon phase velocity nearly fixed, one must increase the velocity magnitude, which is equivalent to decreasing the outward flight time. A similar argument may be used to explain the increase in the return flight time when going from counterclockwise to clockwise returns. These variations in flight time are common to both Figs. 11 and 12 regardless of whether the spacecraft is launched in the moon's plain or perpendicular to the moon's plane.

Another observation may be made by considering Fig. 12, which has the spacecraft approaching the moon from below and normal to the moon's plane. The outward and return times of flight curves are not symmetric with the 180° return inclination. For example, these times of flight for position A and a return inclination of 90° are 62 and 64 hr, re-pectively, whereas the corresponding times of flight for a return inclination of -90° are 63° and 65°. This difference of one hr may be explained (as with pericynthion distance) by considering the angle that the moon phase velocity must go through. When approaching the moon from below, this angle is greater for returning below the moon's plane than returning above. For a fixed pericynthion distance at the moon, then, the velocity would have to be less for returning below than returning above and hence the flight times are greater.

It is interesting to note the variation between the outward and return times of flight in comparison with the total time of flight which is presented in Fig. 13. For trajectories launched in the moon's plane, the outward and return times of flight will be about equal for ccw returns and differ by almost 20 hr for cw returns (see Fig. 11). For trajectories launched normal to the moon's plane, the times of flight will be about equal for normal returns and differ by about 10 hr for ccw and cw returns (Fig. 12). Figure 13, on the other hand, indicates that the total time of circumlunar flight will vary only 2 or 3 hr for a specific outward inclination and any return inclination and it will vary only 3 or 4 hr for any outward and return inclination. The total time of flight is, however, strongly affected by the distance of the moon. This constancy in the total time of flight does not seem to be a basic property of free-flight circumlunar trajectories. That is, one can see from Figs. 9 and 10 that, for higher fixed pericynthion altitude, the variation in the total time of flight is greater than the 2 or 3 hr here. However, it is quite possible that one may be able to show why this should be approximately true.

Launch Windows

The analysis thus far has been carried out parametrically, choosing those variables, such as time of flight and inclinations which present the characteristics of circumlunar trajectories in a general manner. The development of the computer program, however, is of a more practical nature. In particular, it finds that trajectory that satisfies certain launch and re-entry constraints. This is important when performing a complete mission analysis.

The purpose here is not to perform such an analysis; however, one figure is shown to indicate what information might be necessary. Here the time period is early in January 1967, and the pericynthion altitude is fixed at 100 naut miles. The landing site is also fixed and in this case it is San Antonio, Texas. With these constraints, and also restricting the launch azimuths from 72° to 120° at Cape Kennedy, the Analytic Circumlunar Program finds only the trajectories shown in Fig. 14 to exist. They are of type I and range in outward flight times from 67 to 70 hr.

It is interesting to note that the re-entry maneuver angle, i. e., the great circle arc measured from the re-entry point to the landing site and in the plane of motion, must change with the liftoff time for a particular day. Other parameters, such as flight times, injection velocity, and approach to the moon are of interest but not presented here.

Comparison with the Exact Program

Before accepting the validity of the data generated by the Analytic Circumlunar Program, it is necessary to compare the results with those from a more exact model. The best model, of course, is that which considers all important forces (gravitational) acting on the spacecraft. This is accomplished with the Encke Program, which includes the effects of the oblateness of the earth and moon, the sun, and the nearby planets. This program integrates the equations of motion in time beginning with an initial position, velocity, and time of the spacecraft. To eliminate any discrepancies between the two programs due to the position of the moon, the analytic program uses the same ephemeris as the exact program.

The most direct method of comparing the two programs would be to input the injection out of parking orbit conditions directly into the exact program and see how the terminal at re-entry conditions compare. However, previous experience with an analytic earth-to-lunar impact program indicates that the comparison will be very poor. This may be stated without reservation since a similar comparison for direct lunar impact trajectories yield differences by as much as one earth radius in the impact parameter at the moon. This is for the longer flight times and near minimum distances of the moon. Because of the large discrepancies in the impact parameter at the moon, it cannot be expected that the trajectories generated by the exact program utilizing the analytic injection conditions will look anything like circumlunar trajectories.

The next possibility is to try to alter the analytic injection conditions so that some similarity in the moon phase will be obtained. This is most easily done by performing a linear search on the analytic impact parameter. The phrase "linear search" simply implies that the impact parameter will vary linearly with the injection parameters, allowing a linear interpolation in some chosen injection coordinates to satisfy specific impact parameter conditions. The required first derivatives are obtained from the variational equations that are solved in the exact program during the trajectory integration. It requires about four iterations to obtain the desired impact parameter within reasonable limits.

The revised injection conditions are then used to compute the exact trajectory around the moon and return to the earth. Table 1 presents a comparison of various parameters. It is interesting to note that the two programs now give reasonable agreement in all parameters. The maximum error in the spacecraft's return perigee distance is about 1000 naut miles.

Concerning the energy in the moon phase, it has been found that the hyperbolic excess velocity generated by the exact program is consistently 100 to 200 fps greater than the analytic value. This fact suggests that some empirical correction may be made. Or perhaps an alteration, presumably an improvement, in the model may be made, such as utilizing the Jacobian constant in the moon phase.[3] This adjustment may offer an improvement in the moon phase parameters and terminal conditions; however, it will not contribute greatly to the problem of obtaining exact circumlunar trajectories satisfying specific terminal conditions.

The problem of finding exact free-flight circumlunar trajectories may be attacked in several ways. The most direct method would be to attempt a linear search on definite terminal parameters at the earth such as re-entry angle, maneuver angle, and latitude and longitude at touchdown, with respect to some initial parameters at injection from a parking orbit.

However, although the Analytic Circumlunar Program together with a linear search on the impact parameter yields fair circumlunar trajectories (as shown), the terminal conditions are still not good enough to ignore nonlinear relationships with the initial conditions. This will cause the linear search to fail.

A more successful method has been presented by T. Gibson.[4] He has combined an analytic program similar to that discussed here with an exact integration program. In operation, the analytic program will find a first approximation to the desired circumlunar trajectory. The exact program will then take the injection conditions and integrate them out to the spacecraft's closest approach on return to the earth. The analytic program is then made to duplicate this incorrect trajectory. The differences in the analytic injection conditions between the desired trajectory and the incorrect trajectory is assumed to be the same as would be generated by the exact program, implying that this difference may be subtracted from the incorrect (exact program) trajectory injection conditions to yield the desired injection conditions for the exact program. Since this is not precisely the case, several iterations are required to converge on the desired trajectory. This method may profitably be combined with the impact parameter search procedure where this linear search would precede Gibson's search.

Appendix: Effect of the Moon's Orbit on the Results

The complex motion of the moon about the earth adds a considerable amount of latitude in the behavior of circumlunar trajectories.[5] Not only the moon's distance, which varies from 56 to 64 earth radii from the earth, but also the path angle of its velocity vector, greatly affects important parameters such as pericynthion distance, time of flight, etc. For this reason, this study had to include as a variable parameter the moon's distance together with some indication of the radial direction of the moon's motion.

A first approximation to the moon's orbit would be an ellipse with a mean semimajor axis and eccentricity of 60 earth-radii and 0.056, respectively. The inclination with the equator varies from 18.5° to 28.5°, and the node regresses, both with a period of approximately twenty years. In the year 1967, the inclination will be near its maximum value. Further, the line of apsides advances at a rate of approximately 40 deg/yr. All dependence on these angular variations is avoided if the circumlunar trajectories are referred to the moon's orbit plane. However, the mean ellipse is a poor approximation of the moon's in-plane motion. As may be seen in Ref. 5, the moon's perigee varies by about 2.5 earth-radii and its apogee varies about 0.5 earth-radius. A cycle of this variation occurs in approximately seven months, which is one half of the synodic period of the sun with the moon's line of apsides.

As examples of the effects of the lunar distance and position on various parameters, Figs. 15 and 16 present data generated for two-day intervals. Based on the classifications discussed earlier, the circumlunar trajectory chosen was I-70 hr - 0°, i.e., fired outward in the moon's plane with a 70-hr flight time to pericynthion and a 0° inclination (counterclockwise and in the moon's plane) for the return earth phase. The two months chosen were for maximum and minimum perigee distance to show the greatest variation. The arrows indicate increasing time.

As may be seen from the graphs, the injection velocity is highly dependent on the moon's distance but practically nondependent on the moon's velocity path angle, whereas the moon phase velocity v_{∞} is highly dependent on both factors. The large variation in v_{∞} for the same distance of the moon may be explained qualitatively on the basis of the moon's radial velocity component. It is clear that if the moon is moving away from the earth, then the relative velocity of the spacecraft with the moon, which is also moving away from the earth, will be lower than if the moon had a radial component towards the earth.

Figure 16 indicates the variation of the probe-moon-earth (PME) angle and the pericynthion distance with the position of the moon in its orbit. The PME angle is specifically defined as the angle between the spacecraft's incoming asymptote and the moon-earth line at the time of pericynthion passage. As indicated in this figure, both of these parameters decrease as the moon's distance increases. The explanation for this behavior has been presented.

References

[1]Egorov, V. A., "Certain problems of moon flight dynamics," Russian Literature of Satellites, Part I, pp. 107-174 (International Physical Index, Inc., New York, 1958).

[2]Penzo, P. A., "An analysis of moon-to-earth trajectories," ARS Preprint 2606-62 (November 1962).

[3]Brooks, J. E., "Formulation of the equations of motion, the Jacobian condition, and spheres of action for earth-moon-vehicle three-body systems," Space Technology Laboratories, Inc. Memo 9800.2-29 (January 2, 1963).

[4]Gibson, T., "Applications of matched conic and integrated circumlunar trajectories," ARS Preprint 2607-62 (November 1962).

[5]Woolston, D.S., "Declination, radial distance and phases of the moon for the years 1961 to 1971 for use in trajectory considerations," NASA TN D-911 (August 1961).

Table 1 Comparison of the analytic and integration programs after a linear search on B · T and B · R at the moon; the analytic value of x_p is about 0.2109×10^8 ft

Analytic Program

Case	T_o, hr	y_p, 10^7 ft	i_{mr}, deg	T_r, hr
1	60.0	0.757	270	72.1
2	70.0	0.707	0	74.2
3	70.0	0.668	90	80.5
4	70.0	1.127	180	101.7
5	70.0	0.825	90	75.4
6	75.0	0.705	90	75.0
7	75.0	0.658	270	74.8

Exact Program

Case	T_o, hr	y_p, 10^7 ft	i_{mr}, deg	T_r, hr	x_p, 10^8 ft
1	60.7	0.769	279	71.8	0.200
2	70.0	0.741	0	75.0	0.158
3	70.6	0.687	87	81.3	0.198
4	70.5	1.154	177	102.9	0.148
5	69.6	0.856	84	77.4	0.199
6	76.2	0.726	91	75.3	0.217
7	76.1	0.679	270	75.0	0.178

x_p = vacuum perigee of the return phase

T_o = outward time of flight

y_p = pericynthion radial distance

i_{mr} = return inclination with the moon's plane

T_r = return time of flight

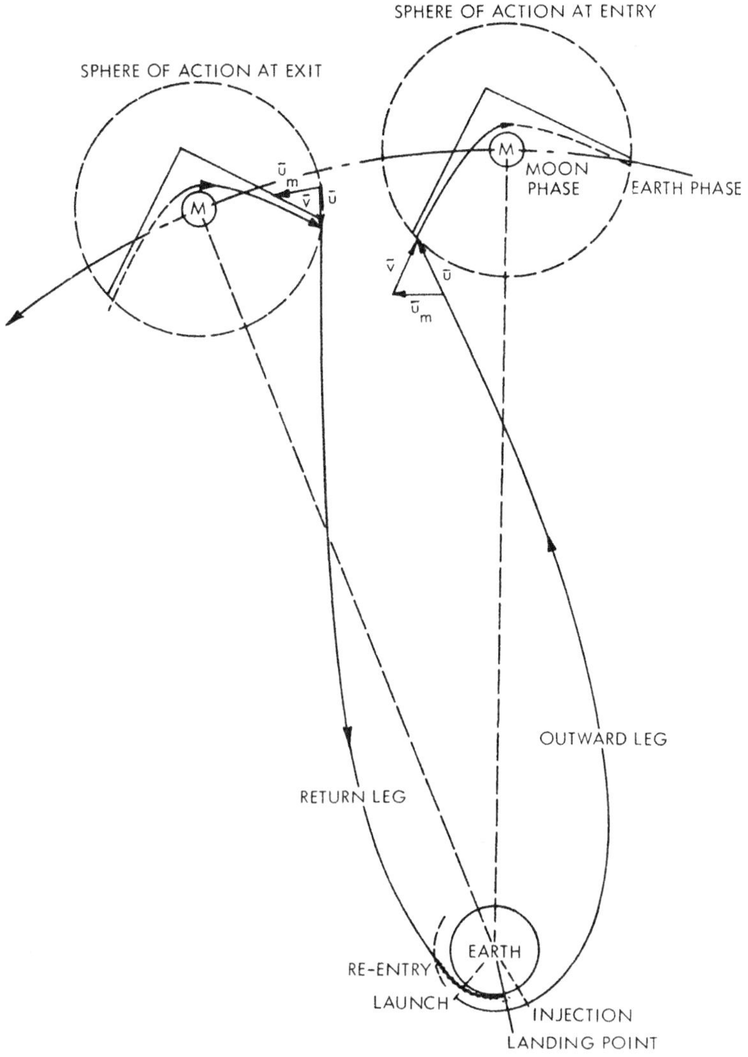

Fig. 1 Schematic of a free-flight circumlunar trajectory.

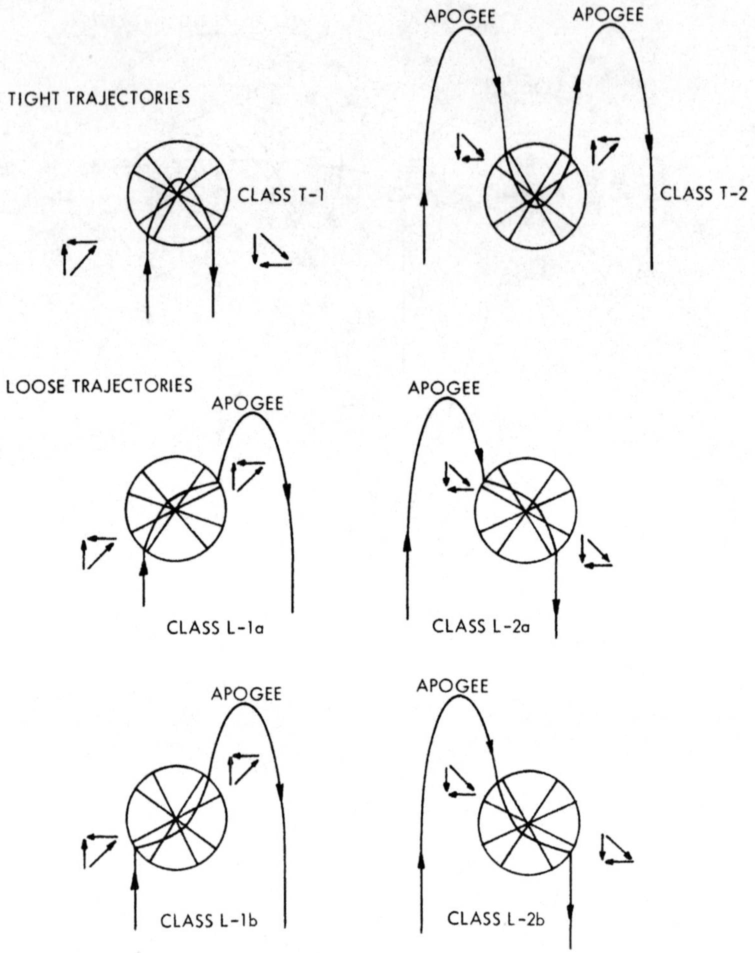

Fig. 2 Major classification of circumlunar trajectories.

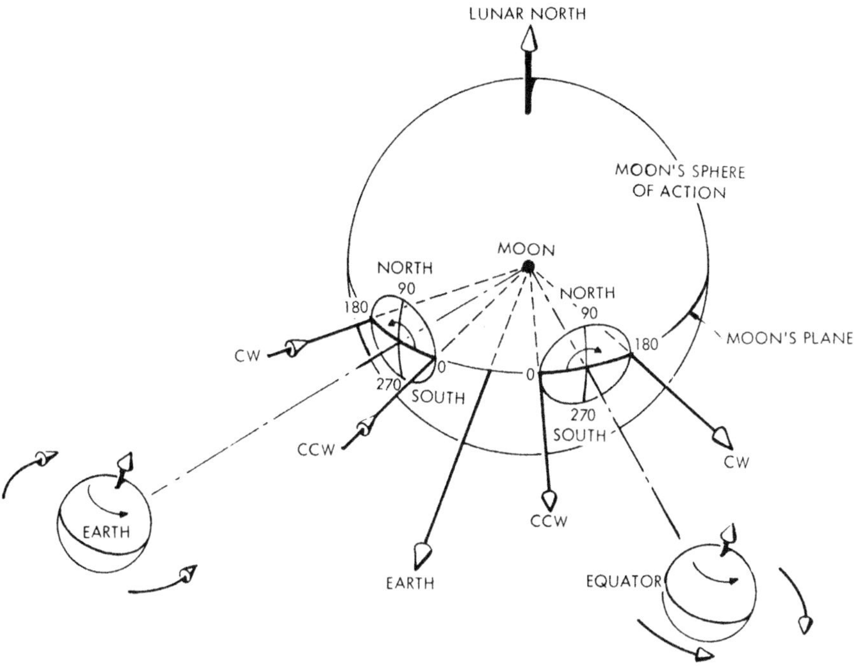

Fig. 3 Inclination definition diagram.

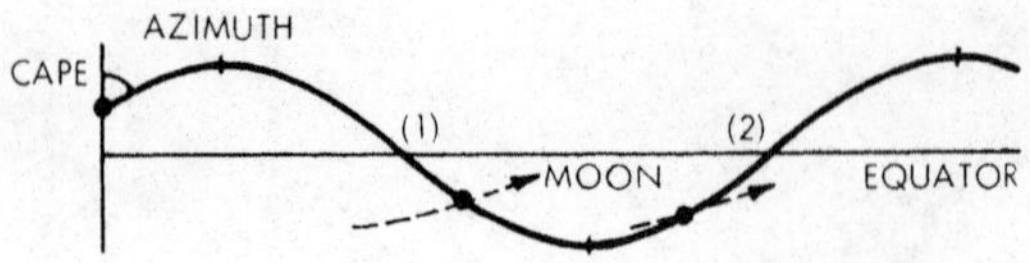

Fig. 4 Mercator projection of trajectory types.

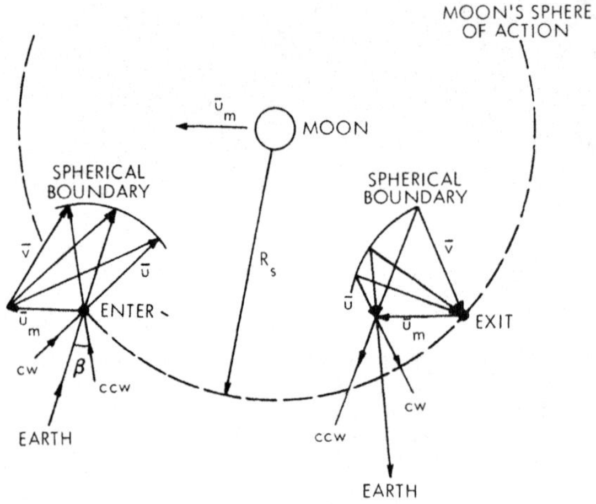

Fig. 5 Properties of the velocity vector addition at entry into and exit from S.

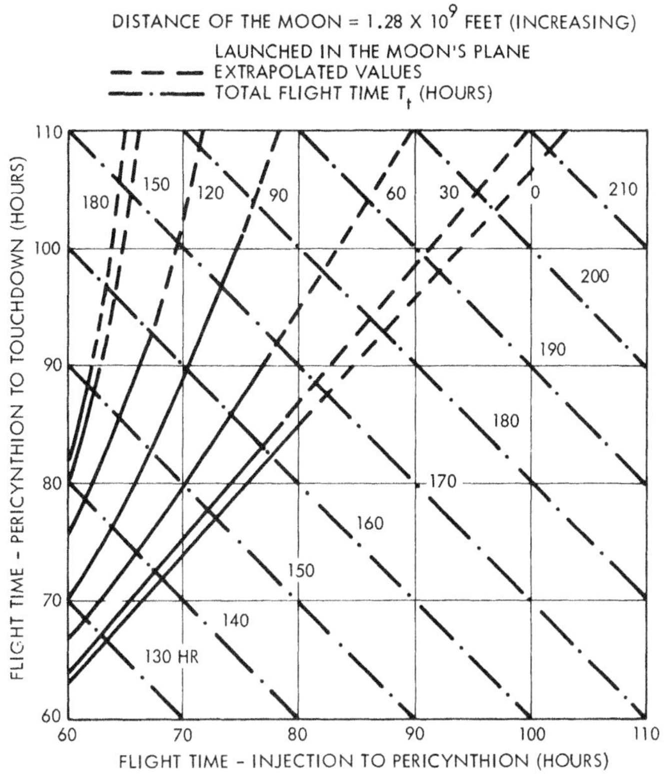

Fig. 6 The return flight time versus the outward flight time for constant return phase inclinations.

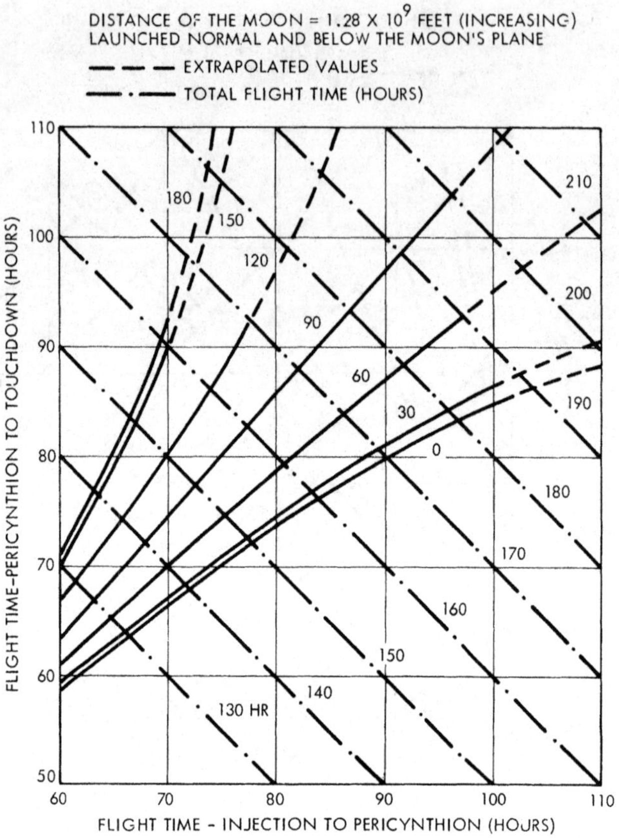

Fig. 7 The return flight time versus the outward flight time for constant return phase inclinations.

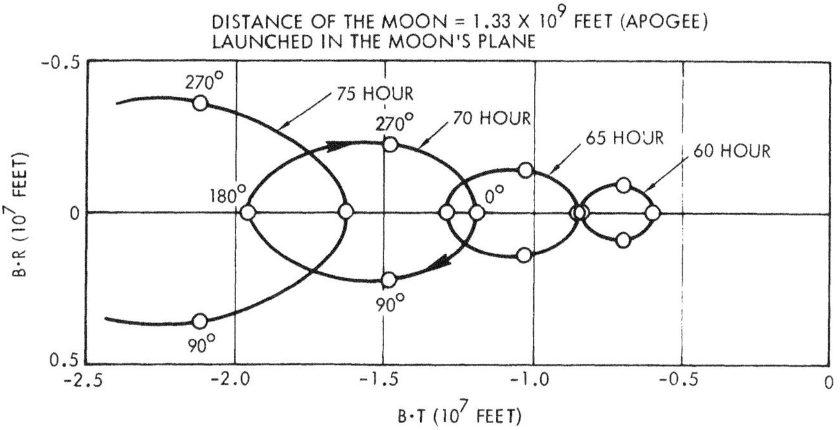

COMPONENTS OF THE IMPACT PARAMETER FOR CONSTANT OUTWARD TIMES OF FLIGHT

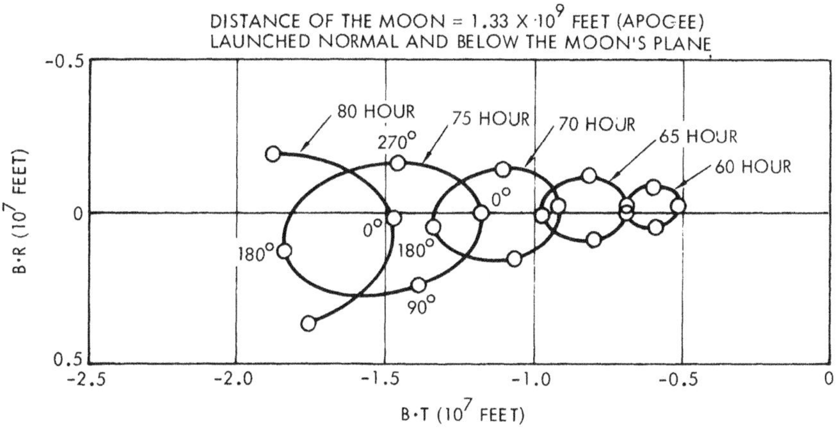

Fig. 8 Components of the impact parameter for constant outward times of flight.

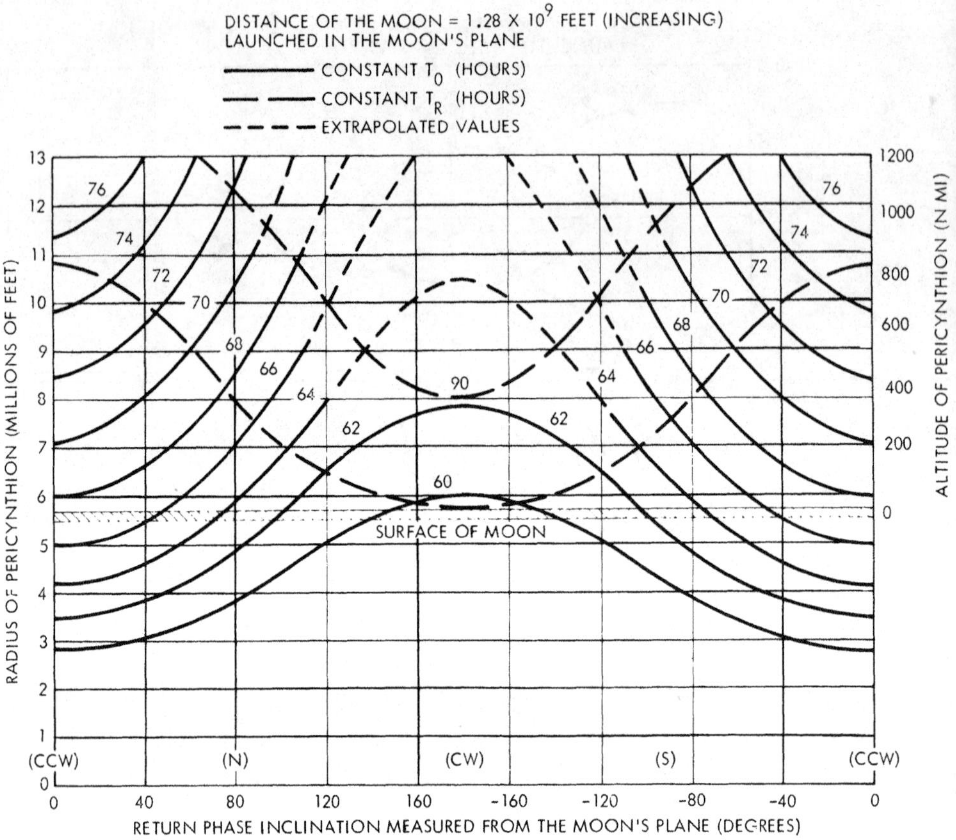

Fig. 9 The spacecraft's pericynthion distance versus the return phase inclination for constant outward phase flight times.

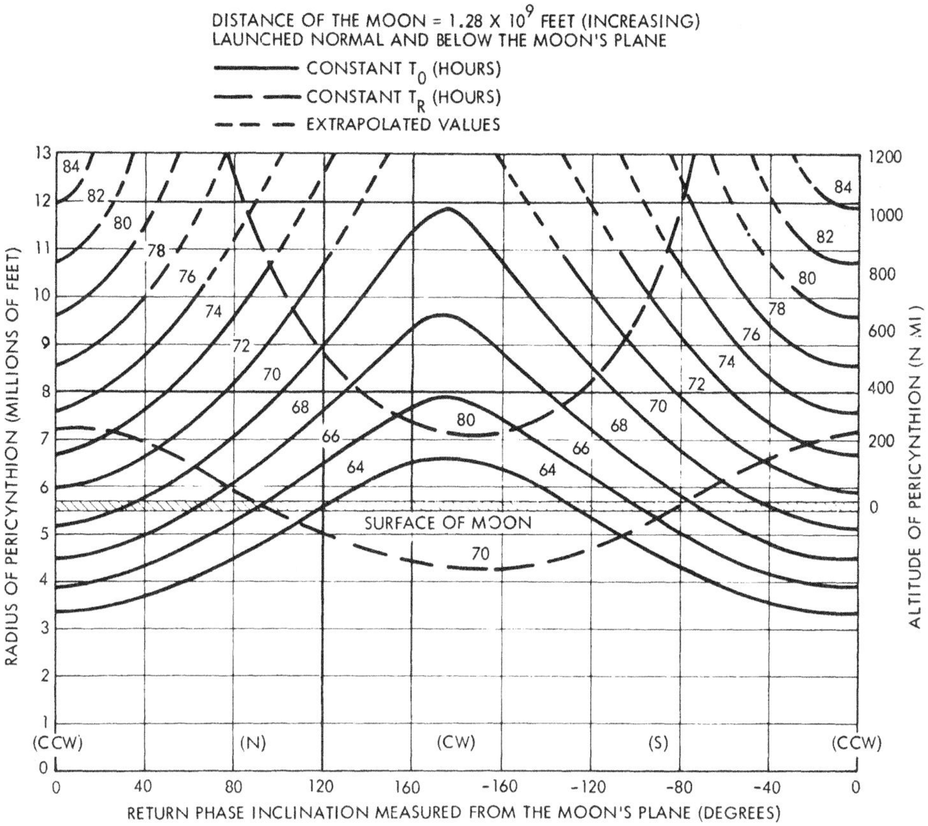

Fig. 10 The spacecraft's pericynthion distance versus the return phase inclination for constant outward phase flight times.

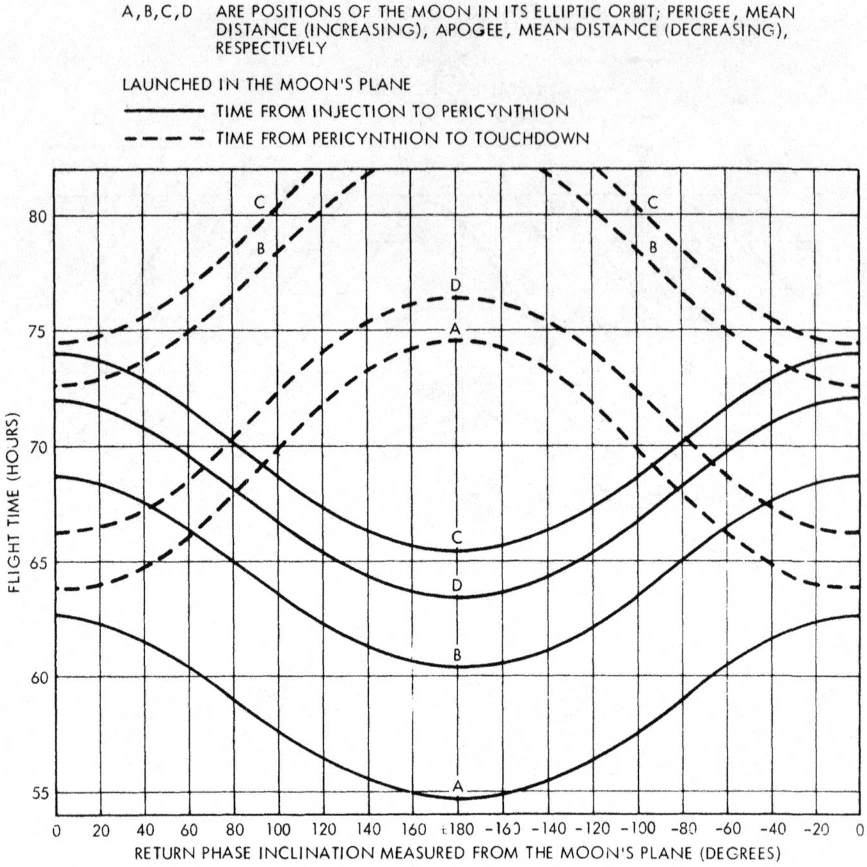

Fig. 11 Outward and return flight times versus the return phase inclinations for various distances of the moon. The pericynthion altitude is fixed at 100 naut miles.

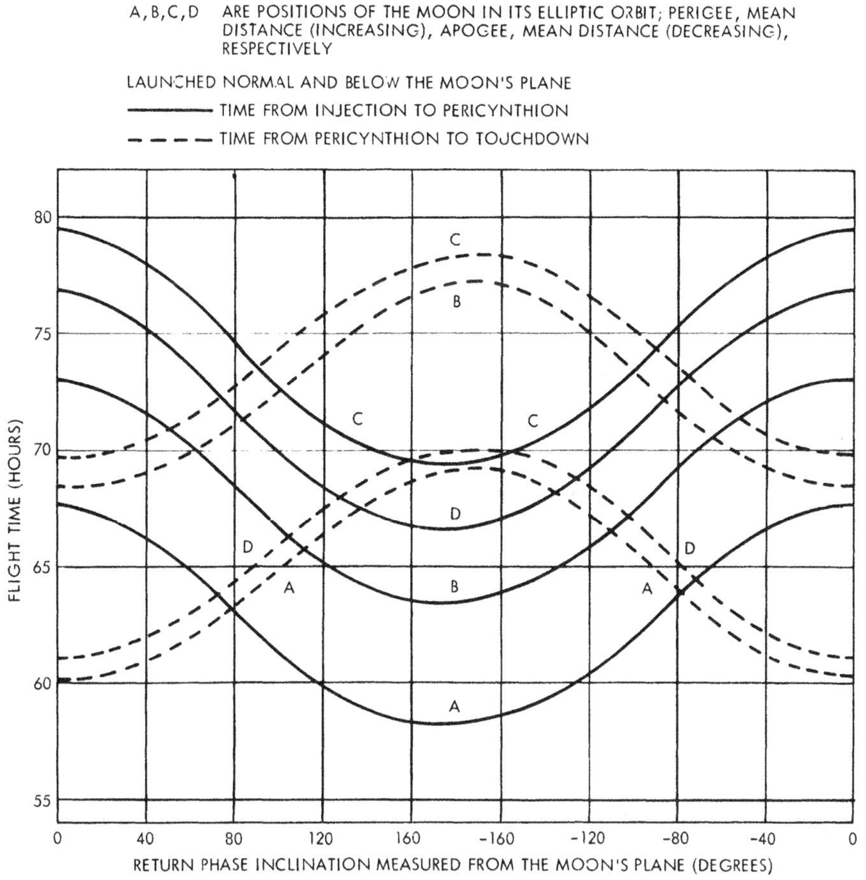

Fig. 12 Outward and return flight times versus the return phase inclinations for various distances of the moon. The pericynthion altitude is fixed at 100 naut miles.

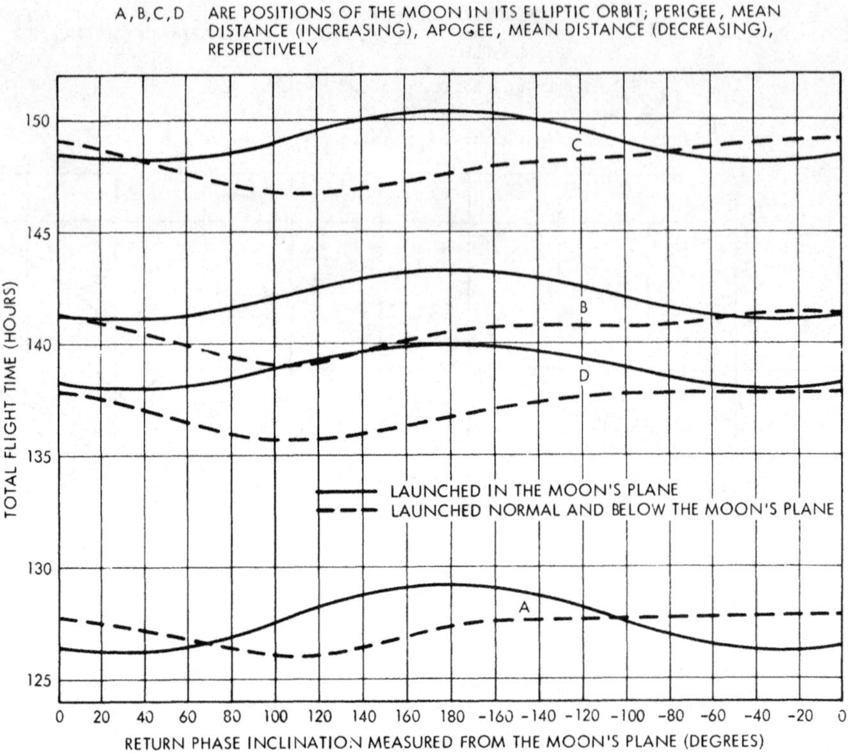

Fig. 13 The total flight time versus the return phase inclinations for various distances of the moon. The pericynthion altitude is fixed at 100 naut miles.

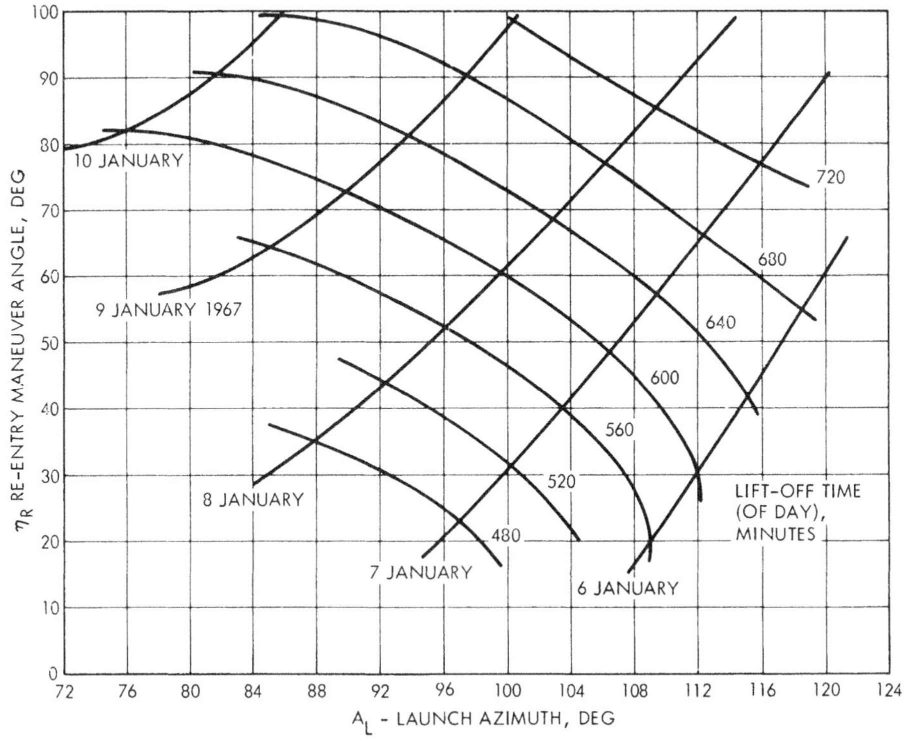

Fig. 14 The launch azimuth at Cape Kennedy versus the re-entry maneuver angle for landing at San Antonio, Texas. Shown are lines for constant days and lift-off times. The pericynthion altitude is fixed at 100 naut miles.

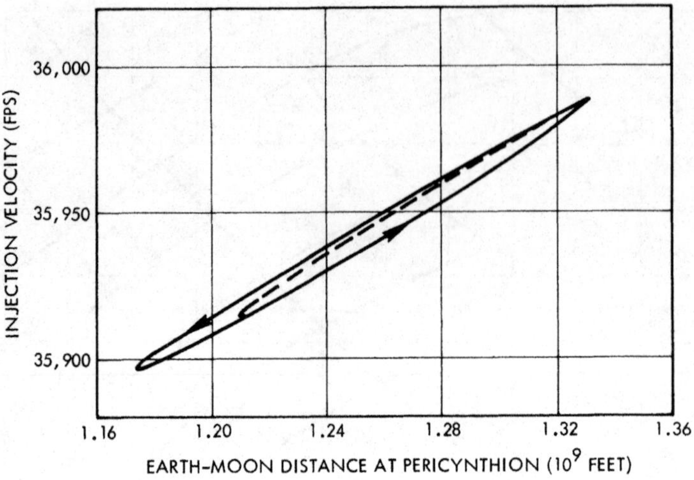

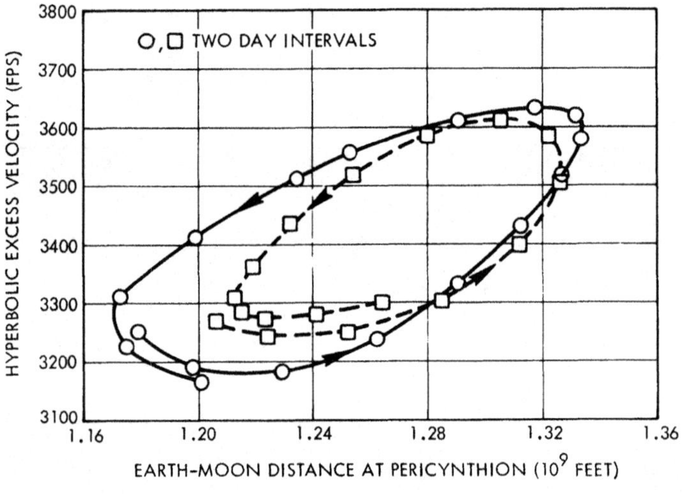

Fig. 15 Injection and hyperbolic excess velocity versus the earth-moon distance case I-70 hr, 0° return.

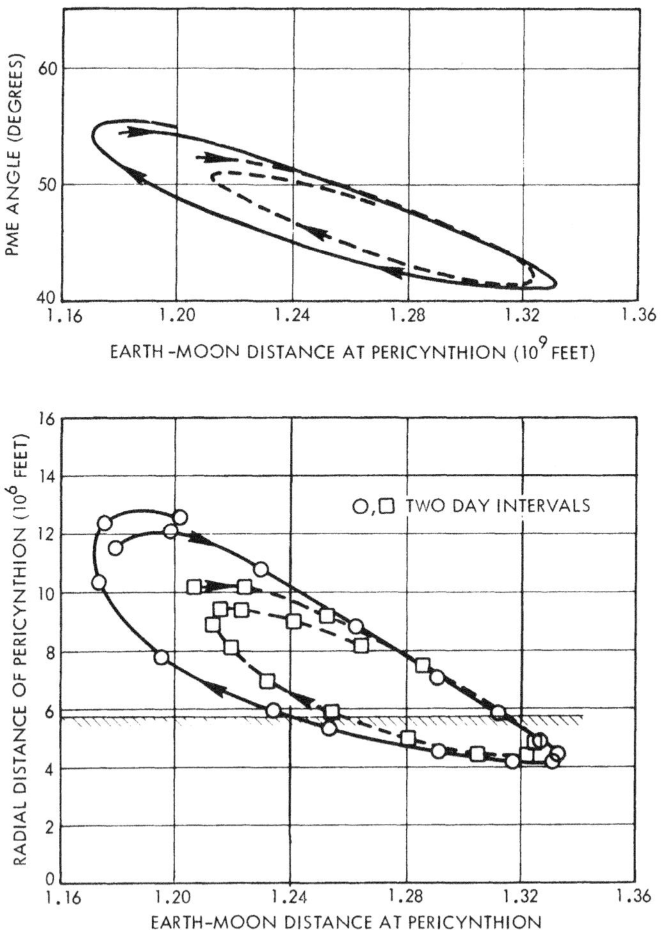

Fig. 16 Probe-moon-earth angle and pericynthion radial distance versus the earth-moon distance for case I-70 hr, 0° return.

FREE RETURN CIRCUMLUNAR TRAJECTORIES FROM LAUNCH WINDOWS WITH FIXED LAUNCH AZIMUTHS

Francis Johnson Jr.*

NASA Manned Spacecraft Center, Houston, Texas

Abstract

The term, fixed launch azimuth window "pane", is used to describe the time period within a variable launch azimuth window within which it is possible to launch into a specific geographically fixed orbit from which it is possible to attain free return circumlunar trajectories using coplanar injection. Within any such pane, trajectories having any return inclination within the total possible range of 360° can be attained. There is associated with any pane a closed, continuous locus of right ascension and latitude upon which, to a very accurate approximation, the vacuum perigees of all translunar trajectories must lie regardless of the coplanar injection thrust program employed. All trajectory characteristics are strongly tied to the position of trajectory perigee on this translunar perigee locus. As a result of this relationship, orbit launch time is found to be the most influencial factor in determining trajectory characteristics, and very little, if any, adjustment can be made at coplanar injection from an orbit having a specific launch time, to change the characteristics of the attainable trajectories. The factors determining the variation of different trajectory characteristics within panes and the length of panes are described. A translunar perigee locus presents a very comprehensive picture of all aspects of a specific pane, relating trajectory characteristics and injection requirements to orbit launch time. Translunar perigee loci are shown to be derived from the interaction of geographically fixed orbits with translunar perigee surfaces. These surfaces are much more general than the loci in that they present a comprehensive picture of the injection requirements and characteristics of all free return circumlunar trajectories existing in time and space. The concept of translunar perigee loci and surfaces can be applied to any lunar mission based on

---

Presented as Preprint 63-406 at the AIAA Astrodynamics Conference, New Haven, Conn., August 19-21, 1963.

*Aerospace Engineer.

a nominal free return circumlunar trajectory. It can provide significant insight into special problem areas such as non-coplanar injection, injection opportunities on succeeding orbits, and injection thrust programming. All trajectories involved in this study were calculated by numerical integration.

Nomenclature

translunar trajectory = first half of the complete free return circumlunar trajectory, going from the parking orbit around the earth to pericynthion at the moon

translunar perigee = vacuum perigee of the translunar trajectory

transearth trajectory = second half of the complete free return circumlunar trajectory, going from pericynthion at the moon back to earth

transearth perigee = vacuum perigee of the transearth trajectory

pericynthion nadir = dimensionless negative vector of the position vector of the moon at the time of trajectory pericynthion

i_{VTL} = inclination of the parking orbit and early translunar trajectory to the moon orbit plane (see Fig. 2)

i_{VTE} = inclination of the transearth trajectory to the moon orbit plane as defined at transearth perigee (see Fig. 2)

v_i = trajectory velocity at injection thrust cutoff

v_{lp} = trajectory velocity at translunar perigee

v_{ep} = trajectory velocity at transearth perigee

γ_i = trajectory flight path angle at injection thrust cutoff

t_{em} = translunar flight time, from translunar perigee to pericynthion

t_{me} = transearth flight time, from pericynthion to transearth perigee

t_{ee} = circumlunar flight time, from translunar perigee to transearth perigee ($t_{ee} = t_{em} + t_{me}$)

i_M = inclination of the trajectory to the moon orbit plane as defined at trajectory pericynthion (see Fig. 15)

θ_M = longitude of the ascending node of the trajectory plane as defined at trajectory pericynthion; θ_M is measured in the moon orbit plane about the moon's center from the earth-moon line (see Fig. 15)

ϕ_M = latitude of trajectory pericynthion relative to the moon orbit plane as measured at the moon's center (see Fig. 15)

λ_M = longitude of trajectory pericynthion, measured on the moon orbit plane about the moon's center from the earth-moon line (see Fig. 15)

Introduction

There are two requirements of Apollo missions as presently conceived, each of which by itself imposes considerable restraints on mission planning. The one requirement is that the nominal translunar trajectory have a free coast return to earth; the nominal mission trajectory must be of the free return circumlunar type. The other requirement is that the programming of the parking orbit launch azimuth be quantized, i.e., not varied continuously, but in discrete steps throughout an over-all launch azimuth window on a given launch day.

The quantity of information in both the onboard and ground-based computers associated with the programming of a specific parking orbit launch azimuth is such that it will take several minutes for a complete cycle of storage and checkout of this information prior to launch. Because of the finite length of this storage and checkout period, it will not be possible to program a continuous variation of launch azimuth throughout an over-all launch azimuth window. Launch azimuth programming will of necessity be quantized, taking the form of alternate go and no-go periods as shown in Fig. 1. During a no-go period, the information is stored and checked out for a launch at a specific azimuth during the immediately following go period. It seems only appropriate that these go periods, being discrete elements of the over-all launch window, be referred to as "window panes."

The required free return characteristic of a nominal trajectory is an obvious safety consideration. Because of the extreme accuracy requirements of injection conditions, it is not practical to expect to obtain a free return trajectory with safe re-entry characteristics without some midcourse corrections. By using a free return trajectory as a nominal, it is anticipated that the required midcourse corrections will be small in magnitude and can be achieved with the attitude control jets of the Service Module. The type of free return circumlunar trajectory with which this presentation is concerned is shown in Fig. 2.

The objective of this presentation is to show a comprehensive picture of the factors determining the length of panes in terms of parking orbit launch time, the variation of injection conditions within different panes, and the variation of the characteristics of all attainable free return circumlunar trajectories within different panes.

In calculating different free return circumlunar trajectories within any one pane during the course of this study, the only restrictions placed on the trajectories were that they all have the same altitudes of pericynthion and the same altitudes of transearth perigee; the trajectories were not restricted in their orientations at the moon or upon returning to earth. Injections into the translunar trajectories were always coplanar with the parking orbit. All of the trajectories involved in this study were calculated by the modified Encke method of numerical integration. It should be kept in mind throughout this presentation that these calculated free return circumlunar trajectories are highly sensitive nominals involving no midcourse corrections. The tolerances of injection velocity and times of launch and injection of these trajectories are in the order, respectively, of 0.01 to 0.001 fps and 1/30 sec.

Translunar Perigee Surface

Of all possible free return circumlunar trajectories of the type shown in Fig. 2, consider a restricted group of trajectories the members of which all have but three things in common: a specific i_{VTL}, a specific pericynthion altitude, and a specific transearth perigee altitude. No restriction is made of any other trajectory characteristic, such as an injection thrust cutoff condition, flight time, orientation at the moon or upon returning to earth, or time of performance, in defining this group of trajectories. Such a group of trajectories having these three characteristics in common are related by a very important mechanism; the vacuum perigees of all trajectories within the group lie on a moving closed surface similar to that shown in Fig. 3. Such a surface is referred to as a translunar perigee surface.

The position of a moving translunar perigee surface at any instant represents the inertial positions of the perigees of all trajectories of the group which are executed at that particular instant. The direction of any such trajectory at its vacuum perigee on the surface is defined by the requirements that the trajectory have the i_{VTL} represented by the surface and that the flight path angle be zero at perigee. The trajectory velocity at vacuum translunar perigee (v_{1p}) is in effect determined by the relative position of the perigee on the surface.

For a given set of values of i_{VTL}, pericynthion altitude, and transearth perigee altitude, there exists at any time a specific translunar perigee surface. The exact shape, dimensions, orientation, and motion of this surface are not only functions of these constant values of i_{VTL} and the two altitudes, but also of variables such as the position and velocity vectors of the moon, the position of the sun, earth oblateness, lunar triaxiality, and even the positions of the planets. Not a great deal is presently known regarding the collective effects of factors such as these on a translunar perigee surface. As shown in Fig. 4, the general shape of a translunar perigee surface in the region of low orbital altitudes is that of a very long cone having an elliptical-shaped cross section with its apex at the earth's center. The major axis of the cross section of a surface subtends about 2.7°, whereas the minor axis subtends about 0.2°. The orientation and displacement of a translunar perigee surface relative to the moon orbit plane are primarily determined by the i_{VTL} it represents. Because of this displacement from the moon orbit plane, the motion of a translunar perigee surface throughout a lunar month does not lie in a plane, but instead sweeps out a shallow cone, the axis of which passes through the earth's center and is perpendicular to the moon orbit plane. A translunar perigee surface rotates about this conical surface of motion throughout the lunar month roughly 140° behind the moon as it moves in its orbit plane.

Results indicate that the extreme values of v_{1p} at a given time and altitude on a translunar perigee surface occur at the extreme edges of the surface, i.e., at the poles of the elliptical-shaped cross section of the surface at that altitude. The lowest v_{1p} at a given altitude occurs at the pole facing the general direction of surface motion, the highest v_{1p} occurring at the other pole. The energies of trajectories having perigees at a given polar position on a surface (relative position between the two edges of the surface) are almost identical; this amounts to a rate of change of v_{1p} with perigee altitude of about -4.7 fps/mile at low orbital altitudes at a given polar position. Loci on a surface representing constant values of v_{1p} take the form of angled cross sections of the surface as shown in Fig. 4. These isovelocity loci are not geometrically fixed on a surface that represents any group of trajectories; they change with time, as do the shape, size, orientation, and motion of the surface.

Injection Conditions on a Parking Orbit Having a Specific Launch Time

Consider a parking orbit having a specific launch time which intercepts the translunar perigee surface representing its i_{VTL}. The launch time of the orbit determines both the inertial positions of the two points of interception and their relative positions on the surface. Coplanar injections into two different free return circumlunar trajectories can be performed mathematically but not realistically from the orbit by injecting instantaneously at the times and positions of either of these two interceptions with γ_i of 0° and v_i equal to v_{1p}, the value of the latter being determined by the relative position of an interception on the surface. However, in a realistic case, instantaneous injections are impossible, and it is necessary to incorporate a thrust program to achieve injection into a translunar trajectory.

In the vicinity of the time of either interception of the orbit with its associated translunar perigee surface, there exists at any instant a continuum of possible injection thrust cutoff conditions. This continuum of conditions is nothing more than the immediate conditions on those trajectories having perigees at different polar positions on the moving surface which were executed when they crossed the parking orbit plane. The only requirements of any one injection thrust cutoff condition is that extrapolating backwards in time both the motion of the vehicle on its translunar trajectory and the motion of the translunar perigee surface, the vehicle at vacuum perigee must lie on the surface, and v_{1p} must have the appropriate value as dictated by the relative position of the perigee on the surface. The continuum of possible injection cutoff conditions existing at any instant near the time of injection on an orbit having a specific launch time provides a great deal of freedom in injection thrust programming.

A proposed Apollo mission requirement is that, for a specific orbit launch time, there must be opportunities for injection on two succeeding orbits. Because of the very small difference between the i_{VTL}'s of any two such succeeding orbits due to orbit regression, the same moving translunar perigee surface can be considered as applying to both orbits. During an orbit period, the translunar perigee surface moves about 0.75° whereas the right ascension of the ascending node of the orbit regresses about 0.4° in the other direction. These opposing motions make the interceptions of the translunar perigee surface which are necessary for coplanar injection very difficult if not impossible on both of two succeeding orbits in most cases. The most favorable conditions for interceptions on succeeding orbits would be when the translunar perigee surface is near its extreme negative or positive latitude, and the $|i_{VTL}|$ involved is very small.

Translunar Perigee Locus of a Pane

Total Pane Length

In order for a specific translunar perigee surface to have application to a mission involving coplanar injection, the parking orbit must intercept the moving surface and the orbit must have the i_{VTL} represented by the surface. Consider a parking orbit having a specific launch azimuth and launch time such that it does intercept the translunar perigee surface representing its i_{VTL}. Varying only the launch time of this orbit, keeping launch azimuth and all other orbit parameters fixed, it is found that there is actually a range of launch time within which the orbits all intercept this surface. However, orbits having different launch times within this range do not have exactly the same i_{VTL}. The variation of i_{VTL} within this short range of launch time is usually in the order of 0.1°. The small size of this variation is because of the relatively small size of the cross section of the translunar perigee surface and the fact that the surface is moving at the rate of only about 0.5 deg/hr, whereas the plane of the intercepting orbit moves across the surface with changing launch time at the rate of 15° of right ascension per hour. A translunar perigee surface by previous definition represents a group of trajectories having one specific value of i_{VTL}. Thus, in the strictest sense of the word, each orbit having a different launch time within the range would be represented by a different translunar perigee surface; however, the differences between these surfaces, associated with the very small variation of i_{VTL}, are very small, and as a result these surfaces are nearly identical.

The interception of these orbits, having different launch times but the same launch azimuth and all other orbit parameters, with their nearly identical translunar perigee surfaces, defines an inertially fixed, closed, continuous locus of right ascension and latitude. These orbits constitute a fixed launch azimuth window pane. The inertial locus is referred to as the translunar perigee locus of the pane. An actual translunar perigee locus of a pane is shown in Fig. 5. A translunar perigee locus is actually a somewhat distorted picture of a cross section at orbit altitude of the translunar perigee surface of any one of the orbits within the pane, the distortion being due to the slight variation of i_{VTL} within the pane and the motions of the surface and the orbit plane. In view of the slight variation of i_{VTL} within a pane, in the strictest sense of the word, the i_{VTL} of a pane should be referred to as a mean value.

The total length of a pane is defined as that range of launch time within which all orbits launched at the same launch azimuth will intercept their essentially identical translunar perigee surfaces (and the translunar perigee locus of the pane). The boundaries of a pane are the launch times of orbits which are tangent to their essentially identical surfaces (and the translunar perigee locus). The total length of a pane is a function of the involved geometry among the moon orbit plane, equatorial plane, and the moving parking orbit plane and translunar perigee surface(s). Total pane lengths of from two to seven min have been defined. The translunar perigee surfaces involved in these panes were all located very near the equatorial plane. It would be expected that longer total pane lengths would be realized in panes wherein the surfaces were near their extreme negative or positive latitudes and values of $|i_{VTL}|$ were very small. The total length of any pane would not be considered operationally feasible because trajectory characteristics are essentially determined by orbit launch time within the pane, and within portions of every pane, the characteristics of the trajectories are operationally unsuitable.

Relationship between the Translunar Perigee Locus of a Pane and Trajectory Characteristics

Consider a parking orbit within a pane, i.e., an orbit that intercepts the translunar perigee surface representing its i_{VTL}. The vacuum translunar perigee of any free return circumlunar trajectory achieved from this orbit by coplanar injection must lie on this surface, the necessary v_{1p} being determined by the relative position of the perigee on the surface. If this translunar perigee surface were stationary, regardless of the injection thrust program used, the right ascension and latitude of the perigee would have to be equal to that of either of the orbit interceptions of the surface. In this hypothetical case, the right ascension and latitude of these two possible perigee positions are determined entirely by orbit launch time. However, in a realistic case, the translunar perigee surface is moving, but because of the relatively slow rate of surface motion (about 0.5 deg/hr) compared to the rate of vehicle motion in the process of injecting into a translunar trajectory from the parking orbit (240 to 340 deg/hr), the movement of the translunar perigee locus during the time period within which any conceivable injection process would be performed is insignificant. Consequently, to a very accurate approximation, the vacuum translunar perigees of all trajectories attainable from this orbit must lie at either of two fixed positions of right ascension and latitude regardless of the injection process used. These two possible positions correspond to the orbit intercepts of the translunar perigee surface which are determined solely by orbit launch time.

The translunar perigee locus of a pane is nothing more than the closed continuum of the right ascension and latitude of the intercepts of all possible orbits within the pane with their translunar perigee surface(s). Thus, to a very accurate approximation, the vacuum translunar perigees of all trajectories attainable within a pane, regardless of the injection process used, must lie on the translunar perigee locus of the pane; this relationship holds regardless of whether the trajectories are initiated by unrealistic but analytically convenient instantaneous injections at orbit altitude or by any realistic injection thrust program. This relationship is illustrated in Fig. 6, which shows a portion of the complete translunar perigee locus shown in Fig. 5. The perigees shown in Fig. 6 are at different altitudes, being of trajectories initiated by instantaneous injection at different γ_i at orbit altitude on orbits having different launch times within the pane.

All of the characteristics of a free coast return circumlunar trajectory are strongly related to the relative position of its vacuum perigee on its translunar perigee surface. As noted earlier, the necessary v_{1p} of the trajectory is determined by both the polar position of the perigee on the surface and its altitude. However, all other characteristics of the trajectory are primarily determined only by the polar position of the perigee; at a given polar position, variation of the altitude of translunar perigee (and v_{lp}) has very little effect upon the other characteristics of the resulting trajectory. Consequently, all of the characteristics of all trajectories attainable within a pane are strongly related to the relative position of their perigees on the translunar perigee locus of the pane. Since the position of a perigee on the locus is determined by orbit launch time, the launch time of an orbit within a pane essentially determines the characteristics of the trajectories attainable from that orbit by coplanar injection.

This relationship between the relative position of a perigee on a translunar perigee locus and several trajectory characteristics is illustrated in Fig. 7. Figure 7 shows the v_i, t_{em}, and i_{VTE} of the same trajectories represented in Fig. 6 as a function of the right ascension of their translunar perigees. The general continuity of the curves in Fig. 7 shows that trajectory characteristics are primarily determined by perigee position on the translunar perigee locus. The ineffectiveness of the injection process in determining trajectory characteristics is illustrated by the fact that these perigees are of trajectories initiated by instantaneous injections at different positions with different γ_i on orbits having different launch times. Note that the injection process has very little if any effect on t_{em} and i_{VTE}, whereas it does have a slight effect on v_i, as evidenced by the lack of over-all continuity of the data points on the v_i curve. The relationships between t_{em} and i_{VTE} and translunar perigee position shown in Fig. 7 are typical of that between all trajectory characteristics and perigee positions on the translunar perigee locus of any pane.

There is one trajectory characteristic, the total possible range of which is common to all panes. This characteristic is i_{VTE}. Figure 8 shows the 360° continuum of i_{VTE} attainable within any pane. The relative position of a perigee on a continuous translunar perigee locus in effect determines the i_{VTE} of the associated trajectory within this 360° continuum. It has been found convenient to divide the translunar perigee locus of a pane into four quadrants, each representing a 90° range of i_{VTE}; trajectories having translunar perigees within any one quadrant all have values of i_{VTE} within the 90° range represented by that quadrant. The boundaries of these quadrants are approximated by what would be considered the major and minor axes of the elliptical-shaped translunar perigee locus. These quadrants of a translunar perigee locus are shown in Fig. 9. As shown in Figs. 8 and 9, the terms used to identify the quadrants are the same as used to describe the orientation of the transearth trajectories relative to the moon orbit plane and the moon's orbital motion. The extreme ends of a translunar perigee locus, where the perigees of trajectories within the pane having i_{VTE} of 0° and 180° occur, are referred to, respectively, as the posigrade and retrograde poles of a locus. As shown in Fig. 9, the extremes of other trajectory characteristics are related to these same poles.

Since a translunar perigee locus closely resembles a cross section of the translunar perigee surface of any one of the orbits within the pane, the relationship between perigee positions on a locus and trajectory characteristics is very similar to the relationship between these same characteristics and the polar positions of perigees on the surface. A translunar perigee surface thus presents a comprehensive picture of the characteristics of a group of trajectories relative to the moon orbit plane. The surface represents the times, positions, directions, and velocities of the vacuum translunar perigees of trajectories within the group, all characteristics of these trajectories being related to the polar positions of their perigees on the surface. A translunar perigee locus, on the other hand, is much more specific in that it presents a comprehensive picture of every aspect of a specific fixed launch azimuth window pane, relating all trajectory characteristics and injection conditions to the launch times of orbits on a given date.

There are several different means of representing the variation of trajectory characteristics within panes. Parameter values can be represented as double-valued functions of the right ascension of the ascending node of the parking orbit on the equatorial plane. This form of representation, in effect, shows the variation of the parameters as a function of orbit launch time; it is very useful in defining the useful portion of a pane based on operational restrictions of trajectory characteristics. As shown in Fig. 7, it is also possible to represent the variation of trajectory parameters as a function of the right ascension (or latitude) of translunar perigee. However, the translunar perigee loci of different panes differ in shape, dimension, and position. Consequently, this form of representation does not facilitate the convenient correlation of characteristics of different panes.

It has been found very convenient to represent trajectory parameters as a function of i_{VTE} for several reasons. The complete 360° range of i_{VTE} is common to all panes. Over an abscissa range of 360°, any trajectory parameter can be expressed as a single-valued function. Extreme values of trajectory parameters within a pane usually occur on trajectories having extreme values of i_{VTE}.

As shown in the latter part of this section, variations of trajectory velocity at translunar perigee and transearth perigee (v_{1p} and v_{ep}) and the variations of translunar, transearth, and circumlunar flight times (t_{em}, t_{me}, and t_{ee}) within panes, can be readily expressed as functions of i_{VTE}. More involved trajectory characteristics, such as orientation and geometry at the moon and transearth perigee position, although just as strongly related to i_{VTE} and translunar perigee position as the forementioned velocities and flight times, are more comprehensively described as separate subjects in later sections.

As noted earlier, the total length of a pane would not be considered operationally acceptable because of the undesirable trajectory characteristics associated with portions of all panes. It has been found that what would be considered the useful portion of a pane, based on any given requirements of trajectory characteristics, is not proportional to the total length of a pane. For instance, of a pane having a total length of 126 sec, the portion yielding trajectories having $|i_{VTE}|$ less than 20° was found to be 39.6 sec long; the similarly defined portion of another pane having a total length of 377 sec was found to be 31.2 sec long.

Figure 10 shows v_{1p} at an orbital altitude of 650,000 ft in different panes as a function of i_{VTE}. As noted earlier, trajectory energy is almost invariant at any given polar position on a translunar perigee surface, resulting in a rate of change of v_{1p} with altitude of about -4.7 fps/mile. It can be seen that the range of v_{1p} at a given altitude is independent of i_{VTL} but is dependent upon the position of the moon. The results indicate that i_{VTL} has much less effect in determining v_{1p} (and v_i) than the position of the moon and the specific launch time of the parking orbit within the pane, the latter being represented indirectly in Fig. 10 by i_{VTE}. Notice that, within each pane, the range of v_{1p} of trajectories having $|i_{VTE}|$ less than 20° is only about 2 fps as compared to the total range of v_{1p} within the pane of between 65 and 75 fps.

Figure 11 shows the variation of v_{ep} within different panes as a function of i_{VTE}. At a given perigee position, v_{ep} has the same sensitivity to transearth perigee altitude as does v_{1p} to translunar perigee altitude; namely, trajectory energy is almost invariant, resulting in a rate of change of v_{ep} of about -4.7 fps/mile of altitude. There are 74 different trajectories represented in Fig. 11, the average absolute deviation of their transearth perigee altitudes from the nominal altitude of 30 miles being 2.21 miles. Because of the sensitivity of v_{ep} to transearth perigee altitude, it was necessary to take into consideration the deviations of the individual trajectories in deriving the curves shown in Fig. 11. The results indicate that, within a pane, a trajectory having an i_{VTE} of 0° will have the lowest v_{lp} and the highest v_{ep}, whereas a trajectory having an i_{VTE} of 180° will have the highest v_{lp} and the lowest v_{ep}.

The variations of the flight times within a typical single pane are shown in Fig. 12. Figure 13 shows the variation of translunar flight time within different panes. The longest translunar flight time, shortest transearth flight time, and shortest circumlunar flight time attainable within a pane occur on a trajectory having an i_{VTE} of 0°. The opposite extremes of these flight times within a pane occur on a trajectory having an i_{VTE} of 180°. The size of the range of each of these three flight times within a pane is independent of i_{VTL} but is dependent upon the position of the moon. For a given position of the moon, i_{VTL} has much less effect in determining flight times than the specific launch time of the parking orbit within the pane, the latter being represented indirectly in Fig. 13 by i_{VTE}. The circumlunar flight time of trajectories within a pane having $|i_{VTE}|$ less than 20° is essentially a constant value; this effect can be seen in Fig. 12. The results shown in Fig. 13 indicate that a translunar flight time of about 64 hr can be achieved at any time regardless of the position of the moon; however, the i_{VTE} of the 64-hr trajectory reaching the moon at apogee will be approximately 180°.

Trajectory Geometry near the Earth Relative to the Position of the Moon at Trajectory Pericynthion

A translunar or a transearth perigee locus by itself tells nothing about the geometric relationship between the planes and perigees of the trajectories, and the moon orbit plane and the position of the moon at the time of trajectory pericynthion. A detailed treatment of these very ordered geometric relationships within different panes is beyond the intended scope of this presentation; however, several of the more important relationships can be described briefly.

The angle between the translunar perigee and the pericynthion nadir of trajectories in panes having mean $|i_{VTL}|$ in the order of 8° or less was found to be $9.76^{\circ} \pm 0.05^{\circ}$ for trajectories having $|i_{VTE}|$ less than 20°, regardless of the position of the moon. The relative invariance of this parameter and the very ordered variation of other parameters defining the requirements of injection geometry, have proven very useful in initiating the iterative calculation of integrated trajectories from realistic parking orbits. Parking orbits which actually intercept the pericynthion nadir of their associated trajectories occur only in panes having mean $|i_{VTL}|$ of approximately 4° or less. The trajectories attainable from the parking orbit that intercepts the pericynthion nadir do not exhibit any unusual or extreme characteristics, nor do they have any unusual or extreme injection conditions. None of the parking orbits in panes having $|i_{VTL}|$ greater than about 4° intercept the pericynthion nadir.

In panes having higher values of $|i_{VTL}|$, the intersections of the orbits within the pane with the moon orbit plane tend to converge at a position approximately 1.5^o past the pericynthion nadir, i.e., the orbits tend to "aim" at the position of the moon two to four hr after pericynthion is actually achieved.

This same type of effect is observed in the transearth trajectory geometry in all panes. Except in cases when i_{VTE} is close to 0° or 180°, all transearth trajectories in a pane intersect the moon orbit plane between 1° and 2° behind the pericynthion nadir, i.e., the transearth trajectories tend to "aim" at the position of the moon two to four hr before pericynthion was actually achieved.

Variation of Transearth Perigee Position within a Pane

The transearth perigees of all trajectories attainable within a pane lie on a closed, continuous locus of right ascension and latitude just as do the translunar perigees. However, the shape and size of a transearth perigee locus and its orientation relative to the moon orbit plane are quite different from that of a translunar perigee locus. Figure 14 shows both the transearth and translunar perigee loci of a pane and the range of the pericynthion nadirs along the moon orbit plane; this is the same pane represented in Figs. 5-7. The position of the transearth perigee of a trajectory on the transearth perigee locus of a pane is, in effect, determined by the position of its translunar perigee on the translunar perigee locus. The perigees and pericynthion nadirs of several specific trajectories within the pane are identified alphabetically in Fig. 14.

The size, shape, and orientation of a transearth perigee locus have been found to be essentially independent of i_{VTL} and the position of the moon. It is possible that the declination of the moon, and consequently the latitude of the translunar perigee surface(s) and locus, have an effect on these properties of a transearth perigee locus. The general description to follow applies to panes wherein the translunar perigee surface(s) and locus are near the equator.

The shape of a transearth perigee locus is that of a distorted ellipse. What would be considered its minor axis subtends about 8.3° great circle arc and is coincident with the moon orbit plane. What would be considered its major axis subtends about 13.4° great circle arc. It is geometrically symmetric about its minor axis and the moon orbit plane but not about its major axis. Its geometric center is nearly coincident with the pericynthion nadir of the trajectory within the pane having an i_{VTE} of 180°.

It is not illustrated herein, but when the transearth perigee positions within a pane are normalized to their respective pericynthion nadirs, the resulting locus formed by the perigee positions is nearly a perfect circle. The center of this circular normalized transearth perigee locus lies on the moon orbit plane, but it is not coincident with either the pericynthion nadir or the intersection of the transearth trajectory planes with the moon orbit plane.

Trajectory Orientation in the Vicinity of the Moon

Within a pane, the orientation of the attainable trajectories in the vicinity of the moon follows a very ordered pattern, just as it does near translunar and transearth perigees. Parameters found convenient in describing the orientation of a trajectory at the moon are described in Fig. 15. The variation of these parameters within a pane is strongly related to the variation of the position of trajectory perigee on the translunar perigee locus of the pane. It is possible to represent these parameters as a function of i_{VTE}, but this form of representation, although facilitating convenient correlation of parameter variations in different panes and indirectly relating parameter values to the translunar perigee locus, does not present a complete physical picture of trajectory orientation.

The pericynthions of all trajectories attainable within a pane lie on a continuous, closed locus of ϕ_M and λ_M. This pericynthion locus, which is not inertially fixed as are the translunar and transearth perigee loci, is nearly a perfect circle. Figure 16 shows the pericynthion locus of the pane represented in Figs. 14 and 5. This locus has been found extremely useful in the iterative calculation of integrated free return circumlunar trajectories in that the position of the pericynthion of a trajectory has an extremely ordered relationship to both its i_{VTE} and transearth perigee radius. As with the translunar perigee surfaces and loci, the effects of all factors, such as earth oblateness, lunar traixiality, position of the sun, etc., on the size and position of pericynthion loci have not yet been completely determined. Available results indicate that the primary factors effecting the size and position of pericynthion loci are earth-moon distance, mean i_{VTL} of the pane, and the transearth perigee radius of the trajectories.

A most interesting and useful feature of a pericynthion locus is that the i_{VTE}'s of the trajectories within the pane are to a very accurate approximation, directly proportional to the angular position of their pericynthions around the locus as measured at its center. The two pericynthions having extreme values of λ_M on a locus are of trajectories having i_{VTE} of 0° and 180°, the pericynthion on the side of the locus facing the direction of lunar motion being of the trajectory having i_{VTE} of 180°. The two pericynthions having extreme values of ϕ_M on a locus are of trajectories having i_{VTE} of 90° and -90°, the pericynthion having the algebraically greater ϕ_M being of the trajectory having i_{VTE} of 90°. Thus, all trajectories having their pericynthions on the upper half of the locus of the pane have transearth trajectories which are below the moon orbit plane ($0 < i_{VTE} < 180$), whereas all trajectories having pericynthions on the lower half of the locus have transearth trajectories which are above the moon orbit plane ($-180^\circ < i_{VTE} < 0$).

A pericynthion locus representing a transearth perigee altitude of 30 miles has a diameter of about 13° when the moon is near apogee and a diameter of about 15° when the moon is near perigee. A pericynthion locus of a pane is actually but one of a continuous family of such circular loci concentric about a common center. To a very accurate approximation, the radius of any one of these circles is proportional to the square root of the transearth perigee radius of the trajectories it represents. The relationship between angular position of pericynthions and i_{VTE} applies to the complete family. The orientation of a family of pericynthion loci is such that a trajectory, having an i_{VTE} equal to its i_{VTL} and a transearth perigee radius equal to that of its translunar perigee, will have its pericynthion on the earth-moon axis (neglecting perturbations). This requirement of the pericynthion position of this type of trajectory is quite intuitive considering the over-all geometric symmetry of the trajectory. In realistic panes, transearth perigee radius is but a small percentage less than translunar perigee radius, and consequently, within any realistic pane the trajectory having i_{VTE} equal to its i_{VTL} comes very close to being the forementioned geometrically symmetric trajectory. Because of the "requirement" that the pericynthion of this type of trajectory lie on the earth-moon axis, the pericynthion loci of realistic panes pass through or very close to the earth-moon axis regardless of i_{VTL}. The center of a pericynthion locus is always on the opposite side of the moon orbit plane from the early translunar trajectories of the pane. In panes having mean $|i_{VTL}|$ of 20° or less, the ϕ_M of the center of the locus is approximately equal to $-i_{VTL}/8$ and the locus lies almost entirely on that side of the earth-moon axis facing the direction of lunar motion.

As noted earlier, the pericynthion locus is very useful in the iterative calculation of integrated free return circumlunar trajectories. The pericynthion locus of a pane can be quite accurately defined from a single trajectory having the desired pericynthion altitude but which has not yet been iterated to achieve the desired transearth perigee altitude; it is only necessary to know the i_{VTE}, i_{VTL}, and the λ_M and ϕ_M of the pericynthion of such a trajectory. From this information, the center of the locus is accurately defined on a λ_M, ϕ_M plot of the back side of the moon as the intersection of two lines, the one passing through the earth-moon axis and having a tangent equal to that of $-i_{VTL}$, the other line passing through the pericynthion position and having a tangent equal to that of $-i_{VTE}$. The size of the locus for the desired transearth perigee radius can be determined from the transearth perigee radius of the trajectory using the proportional relationship between locus radius and square root of transearth perigee radius. Having thus defined the pericynthion locus of a pane, the transearth perigee radii and i_{VTE}'s of subsequent trajectories, both in this same iteration process and throughout the pane, can be very accurately predicted without actually calculating the transearth half of the trajectory by integration, knowing only the λ_M and ϕ_M of pericynthions having the same desired altitude. The movement of a pericynthion during the iteration process of a trajectory is very ordered. The use of a pericynthion locus in the iteration process of a trajectory makes it possible to predict very early in the process whether the desired transearth perigee altitude can be reached from a specific geographic position of translunar perigee, and, if it can be reached, it is possible to predict quite accurately the i_{VTE}'s of the two attainable trajectories and the injection times and velocities necessary to achieve them.

Trajectories are far from being normal to a pericynthion locus. The trajectory azimuth at pericynthion is in the vicinity of -90°. The results indicate that a trajectory azimuth of exactly -90° occurs at a pericynthion position having a ϕ_M equal to the displacement of the center of the locus from the moon orbit plane. The implication is that if the center of the pericynthion locus is displaced from the moon orbit plane, a zero value of i_M does not occur in the pane. This in turn implies that the achievement of a zero value of i_M necessitates a zero value of i_{VTL}, in view of the factors governing the position of pericynthion loci.

Figure 17 shows the relationship between i_M and θ_M within the same pane represented in Fig. 16. The general shape of the curve expressing this relationship in Fig. 17 is the same in all panes defined in this study. In panes having higher mean $|i_{VTL}|$, the minimum value of i_M is approximately equal to 1/8 of the mean $|i_{VTL}|$ of the pane. In these same panes having higher mean $|i_{VTL}|$, the lower "corners" of the i_M-θ_M curve are more rounded than those of the curve shown in Fig. 17. No special effort was made in this study to define the extreme peaks of an i_M-θ_M curve of a pane; however, the results obtained indicate several definite trends.

All of the trajectories associated with one of the two "branches" of an i_M-θ_M curve have transearth trajectory orientations that are either all above or all below the moon orbit plane (see Fig. 8). Extreme values of i_M are associated with extreme values of i_{VTE}. The trajectory having the maximum value of i_M on either of the two branches of an i_M-θ_M curve has an i_{VTE} of either 90° or -90°. The two trajectories having the lowest values of i_M on an i_M-θ_M curve have i_{VTE}'s of 0° and 180°.

The magnitudes of the peak values of i_M within the panes defined in this study were found to be within 0.5° of 9.0° in all cases. The position of the moon has a much stronger influence than i_{VTL} in determining the magnitude of these peak i_M values; for a given i_{VTL}, their magnitude is greater when the moon is near perigee than when near apogee. The effect of the variation of i_{VTL} for a given position of the moon upon the magnitude of the peak values of i_M within a pane has not been completely defined. The results indicate that with increasing $|i_{VTL}|$, the two peak values of i_M within a pane become increasingly unequal in magnitude. This effect is related to the displacement of the center of the pericynthion locus from the moon orbit plane.

The maximum range of θ_M within a pane is but a few degrees more than the difference between the θ_M of the two i_M peaks; this difference decreases as $|i_{VTL}|$ increases. The position of the moon has very little effect upon this difference. The range over which θ_M varies within a pane tends to be centered at either 46° or -132°; the range of θ_M is centered about 46° if the early translunar trajectories are above the moon orbit plane, and about -132° if they are below. In the pane represented in Fig. 17, the early translunar trajectories are above the moon orbit plane. Regardless of whether the early translunar trajectories of a pane are above or below the moon orbit plane, the branch of its i_M-θ_M curve closest to a θ_M of -44° represents transearth trajectories below the moon orbit plane.

Considering the variation of transearth trajectory orientation, only a portion of the complete length of a pane is operationally feasible. The portion of a pane yielding trajectories having small values of $|i_{VTE}|$ is represented by one of the two continuous lines forming the lower 1/2 to 1/3 of the complete i_M-θ_M curve. Such a portion associated with $|i_{VTE}|$ less than 20° is shown in Fig. 17.

The general shape of the i_M-θ_M curves of all panes defined in this study implies that there is a small region relative to the moon orbit plane system which is common to the planes of all trajectories at pericynthion within a pane; the characteristics of this region have not been defined. If braking into the lunar orbit occurs at pericynthion and if it is coplanar with the trajectory, all possible orbits within a pane form a belt having a small common nodal region. In a mission involving no plane change at insertion into lunar orbit, in order for the orbit to pass over a specific selenographic point, this point must lie within the operationally feasible portion of the orbit belt of the pane, and a specific lunar orbit must be employed. The employment of a specific lunar orbit necessitates a specific position of the pericynthion of the circumlunar trajectory, which in turn necessitates a specific translunar perigee position. This in turn necessitates a specific parking orbit launch time within the pane, a very restrictive operational requirement. It would seem that some degree of plane change capability is necessary in any mission involving lunar orbits passing over a specific selenographic point.

In planning a lunar observation or landing mission involving a plane change at lunar orbit insertion, a proposed approach is to impose the range of potential plane change on the trajectory azimuth at pericynthion over the operationally feasible portion of the pericynthion locus of the pane. This defines an envelope of attainable lunar orbits within the pane relative to the moon orbit plane. Relating the selenographic system to this envelope of potential lunar orbits indicates the accessibility of selenographic points.

Conclusions

The characteristics of a free return circumlunar trajectory attained from a fixed launch azimuth window pane by coplanar injection are, to a very accurate approximation, determined solely by the launch time of the parking orbit. Every trajectory characteristic is essentially a double-valued function of orbit launch time within the pane. The relationships between orbit launch time and trajectory characteristics make it possible to use empirical models in planning missions involving fixed launch azimuth window panes and free return circumlunar trajectories.

Such an empirical model, having a specific application, is a translunar perigee locus. A translunar perigee locus is the continuum of all possible transearth perigee positions at orbital altitude of all such attainable trajectories within a pane. This locus can be derived by calculating about 20 or 30 integrated trajectories using instantaneous injections at γ_i of 0° at different "geographic" positions along the orbit (see Fig. 5); each such trajectory will be associated with a different orbit launch time. The iteration process of calculating these trajectories is greatly facilitated through the analyses of pericynthion position in relation to the readily derived pericynthion locus of the pane. This empirical model presents a very comprehensive picture of total pane length and the general relationship between all trajectory characteristics and orbit launch time within the pane. A very comprehensive method of determining the reliability of patched conic methods of calculating circumlunar free return trajectories is to derive the translunar perigee, pericynthion, and transearth perigee loci of a pane using this method, and then compare them with loci derived with integrated trajectories for the same pane. Injection conditions are only approximately represented by a translunar perigee locus, since it does not depict the variation of trajectory velocity at perigee as a function of altitude, nor does it define the time of an injection thrust cutoff condition. Translunar perigee loci are very specific empirical models in that each one represents a specific pane; they are actually produced by the interaction of fixed launch azimuth orbits with much more general empirical models, translunar perigee surfaces.

Consider the planning of a mission involving free return circumlunar trajectories from a parking orbit, the launch azimuth and all other parameters of which are fixed except for launch time. For a specific launch time, if the position of the translunar perigee surface that represents the desired trajectory altitudes at pericynthion and transearth perigee and the i_{VTL} of the orbit is known as a function of time following launch, it is possible to tell from this empirical model whether a nominal trajectory can be attained with that specific launch time. The condition necessary for coplanar injection is that the vehicle in orbit intercept this moving surface. If the relationship between trajectory velocity at perigee and translunar perigee position on the surface is known, this empirical model presents a picture of the continuum of suitable injection thrust cutoff conditions existing in any instant near the time of injection on the orbit which intercepts the surface. If the relationship between trajectory characteristics and perigee position on the translunar perigee surface are known, for this same orbit having a specific launch time, it is possible to determine from this empirical model the exact characteristics of the attainable trajectories, the general characteristics being determined primarily by orbit launch time.

The use of an empirical model of this type can greatly facilitate Apollo mission planning in that the accuracy of the calculation of trajectories by integration is achieved together with the speed of trajectory calculation by less accurate methods. The reliability of mission planning using an empirical model of this type is limited only by the accuracy of the integrated calculations from which the characteristics of the translunar perigee surface are defined and the accuracy of the empirical analytic representation of these characteristics.

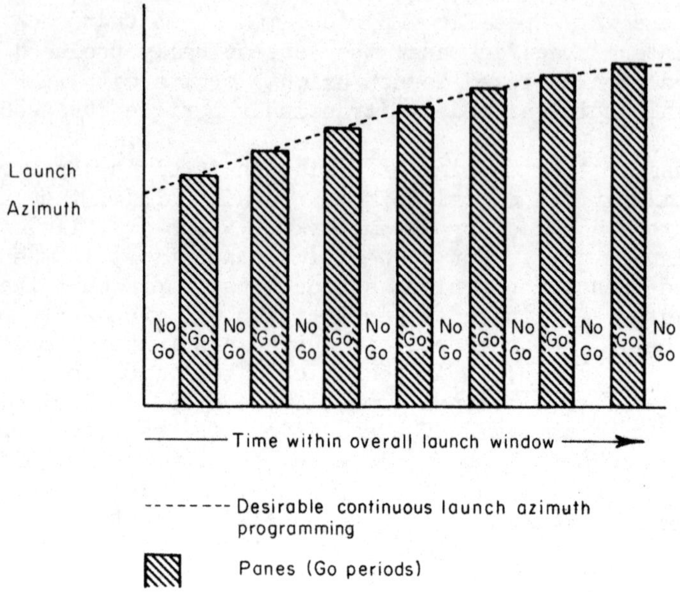

Fig. 1 Fixed launch azimuth window panes within an over-all launch window.

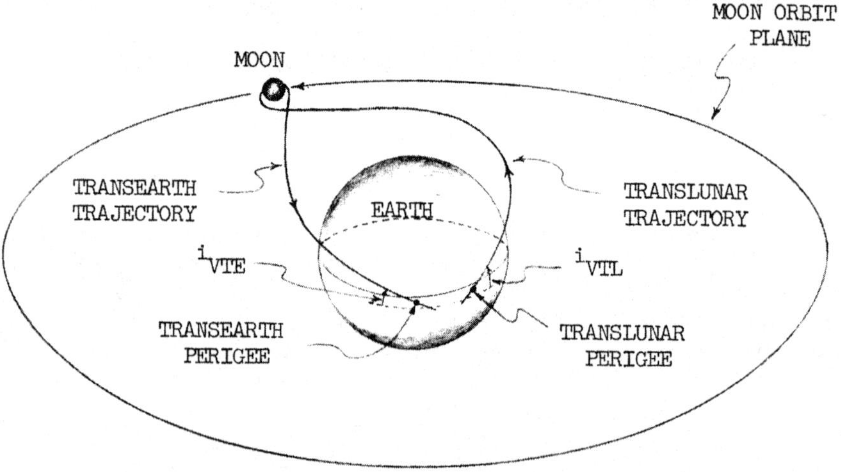

Fig. 2 Free return circumlunar trajectory.

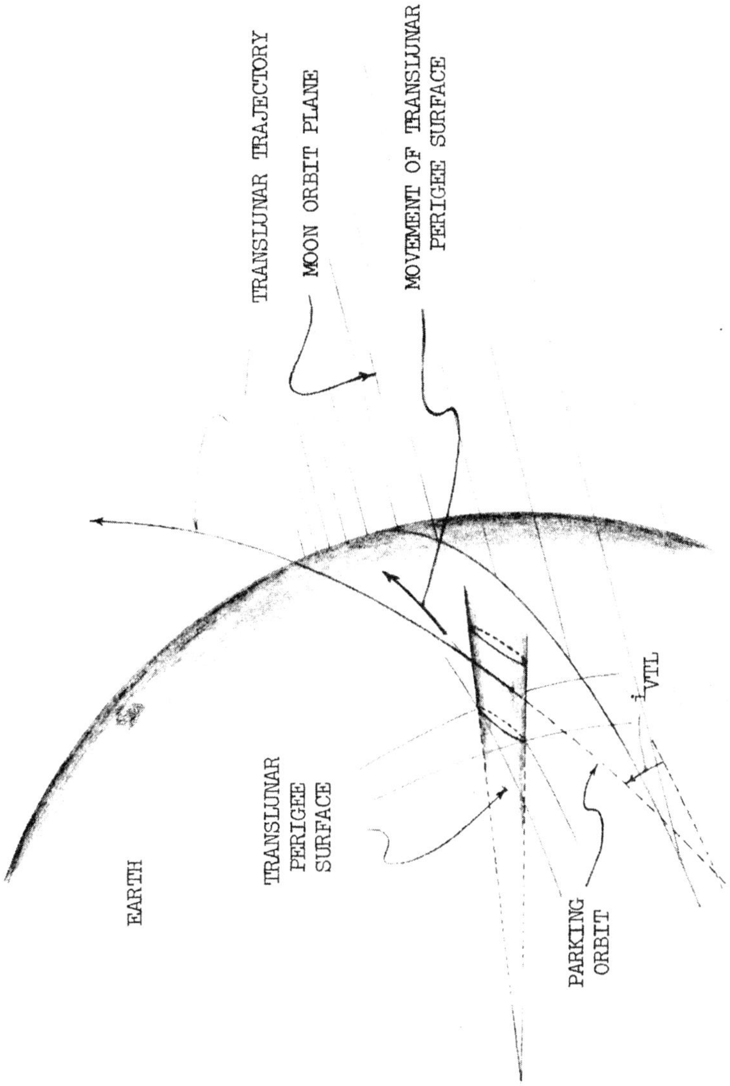

Fig. 3 A translunar perigee surface.

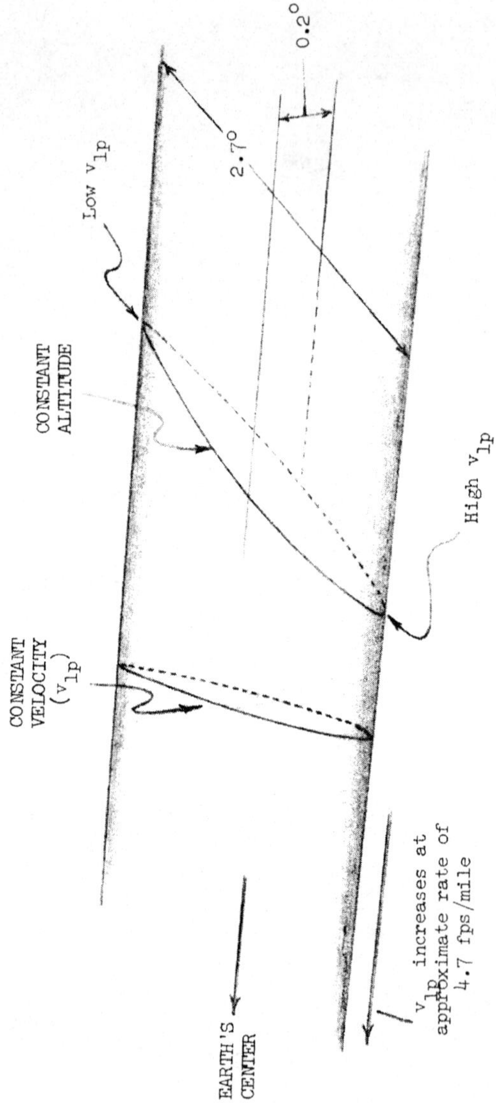

Fig. 4 Variation of trajectory velocity (v_{1p}) on a translunar perigee surface.

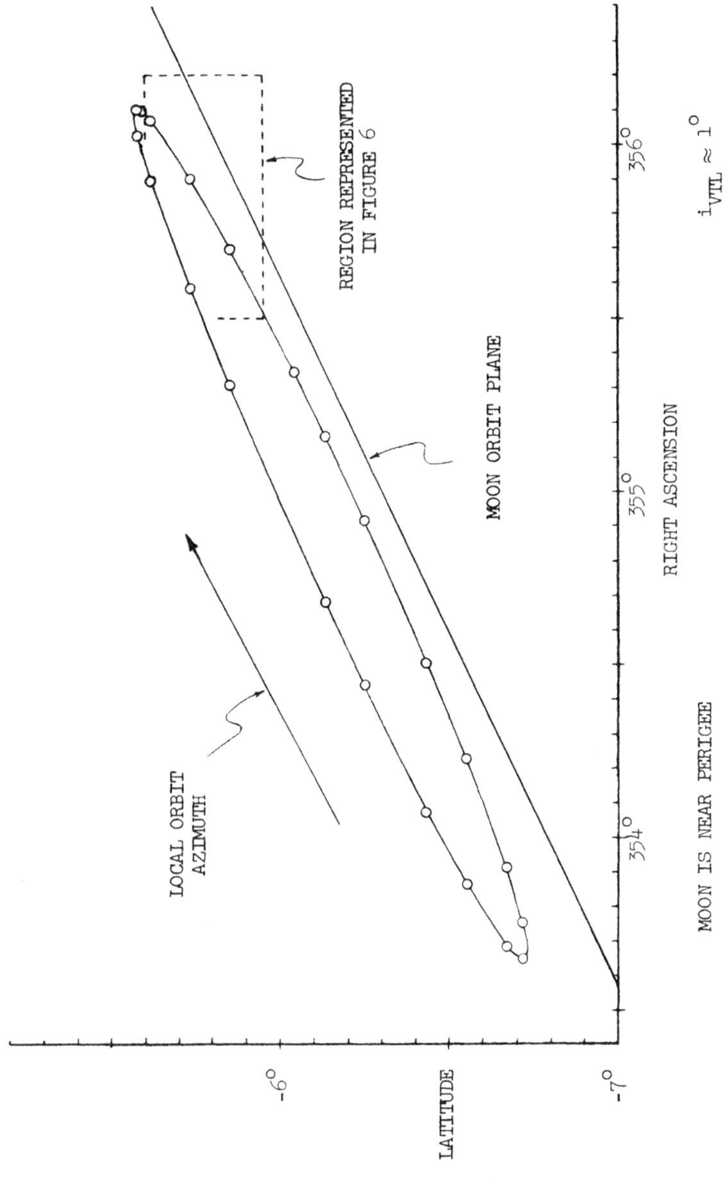

Fig. 5 Translunar perigee locus of a pane.

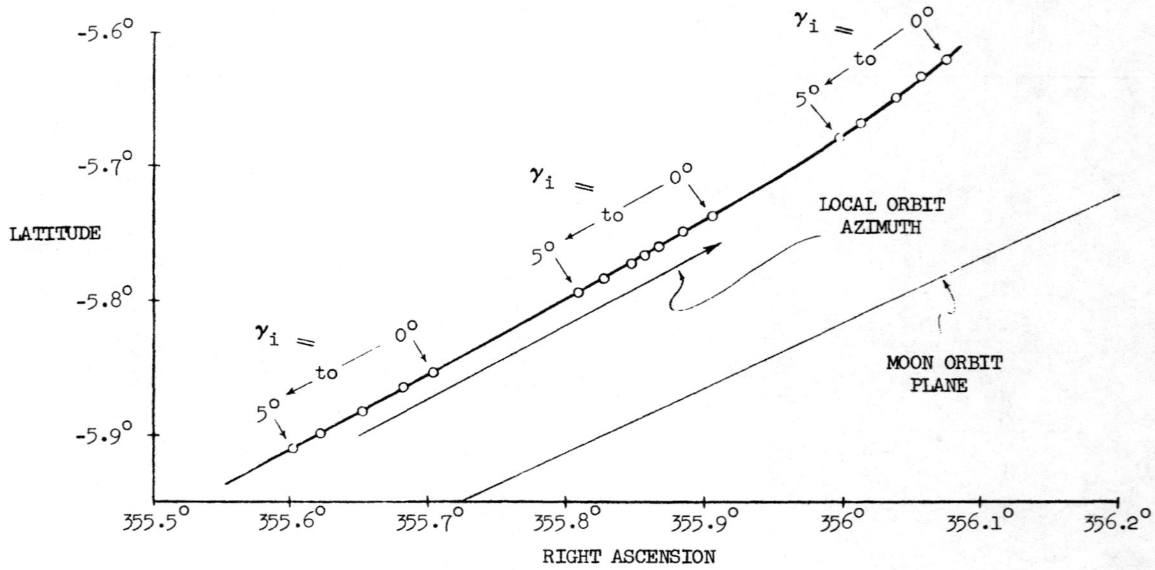

Fig. 6 Positions of perigees on part of a translunar perigee locus (see Fig. 5) of trajectories initiated at different flight path angles (γ_i) at orbital altitude of 650,000 ft.

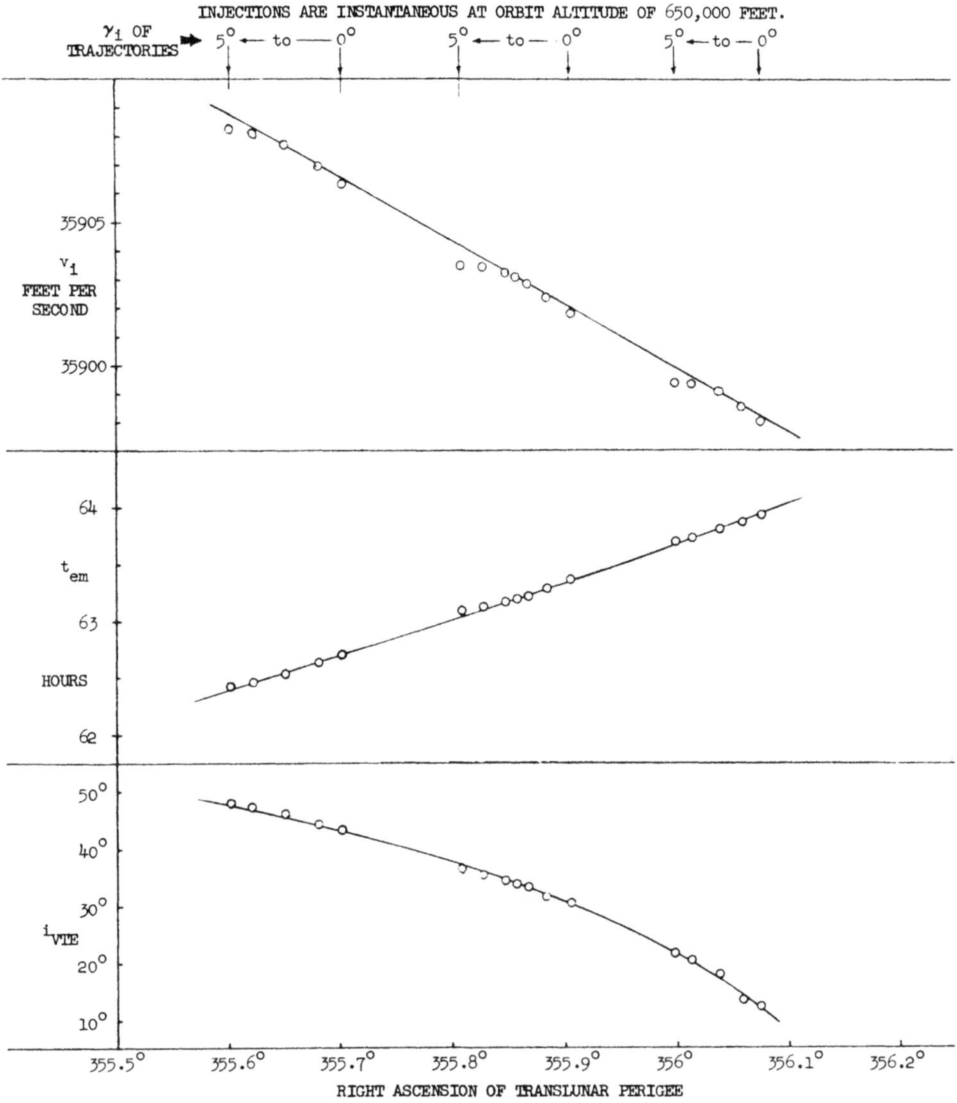

Fig. 7 Relationships of trajectory characteristics to inertial position of translunar perigee.

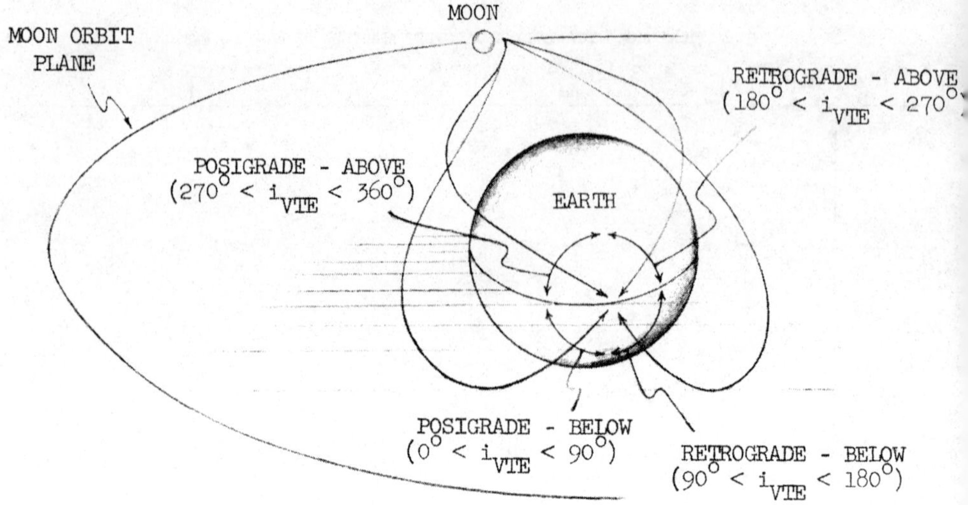

Fig. 8 Continuum of transearth trajectory orientations attainable within a pane.

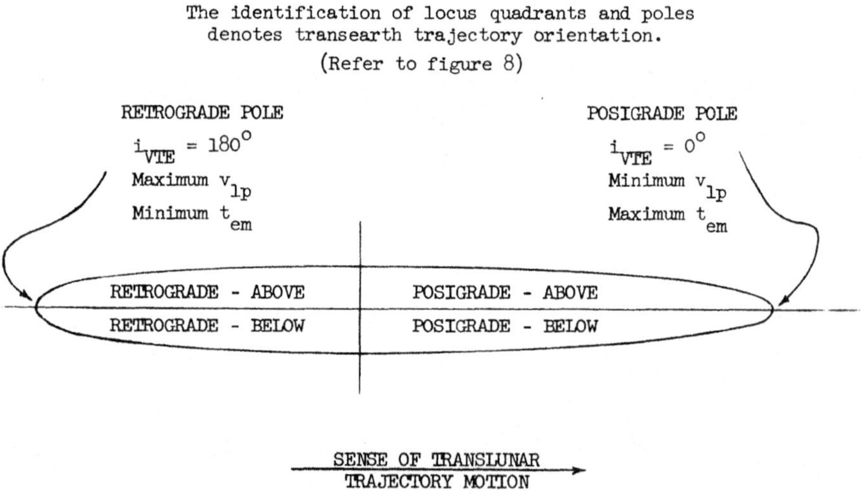

Fig. 9 The relation of trajectory characteristics to perigee position on a translunar perigee locus.

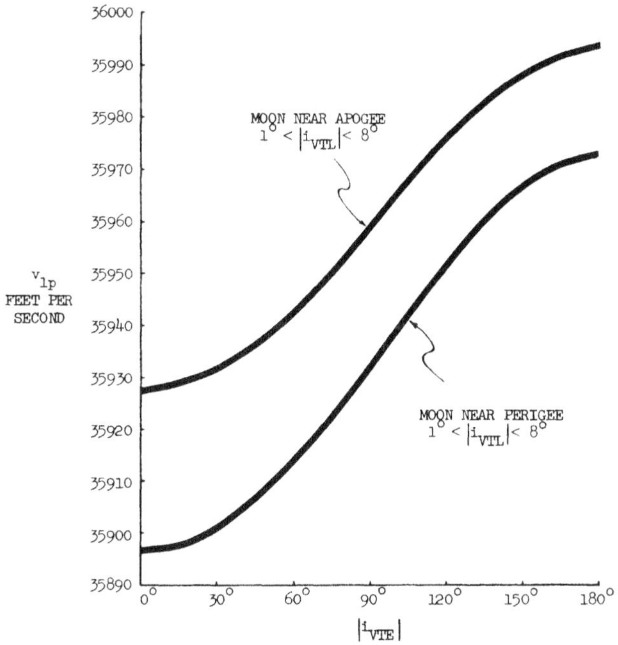

Fig. 10 Trajectory velocity (v_{1p}) at vacuum translunar perigee having altitude of 650,000 ft.

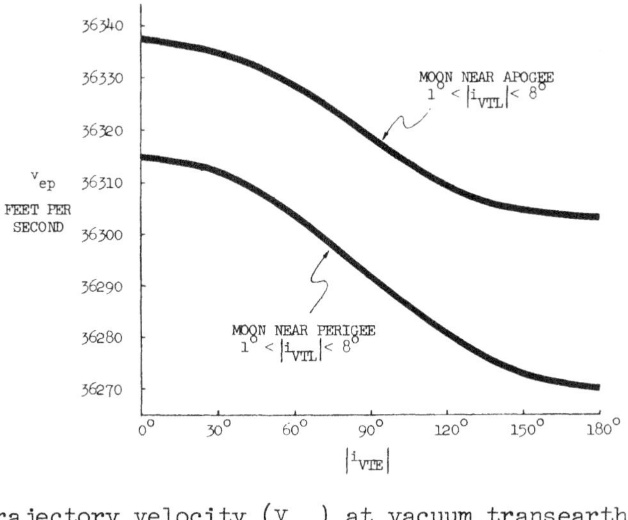

Fig. 11 Trajectory velocity (v_{ep}) at vacuum transearth perigee having altitude of 30 miles.

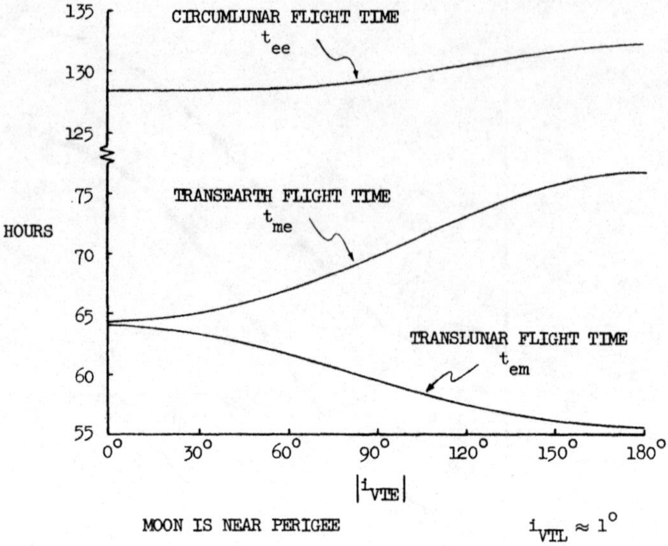

Fig. 12 Variation of flight times within a typical pane.

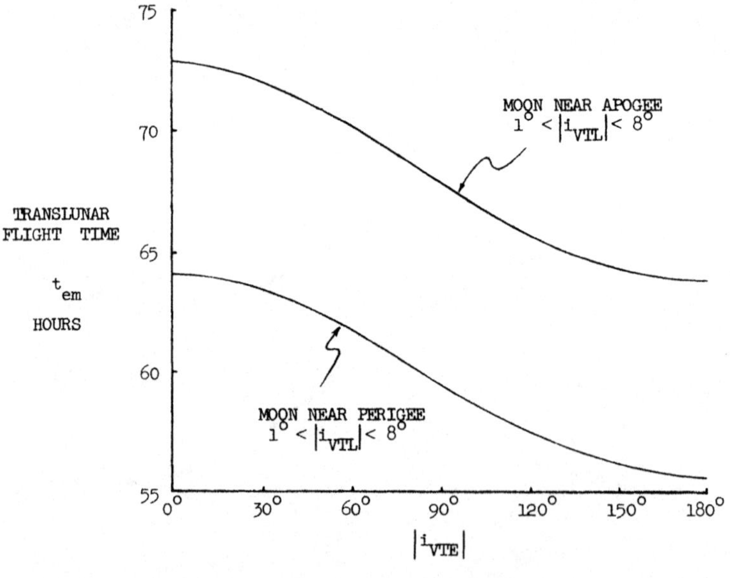

Fig. 13 Translunar flight time (t_{em}) within different panes.

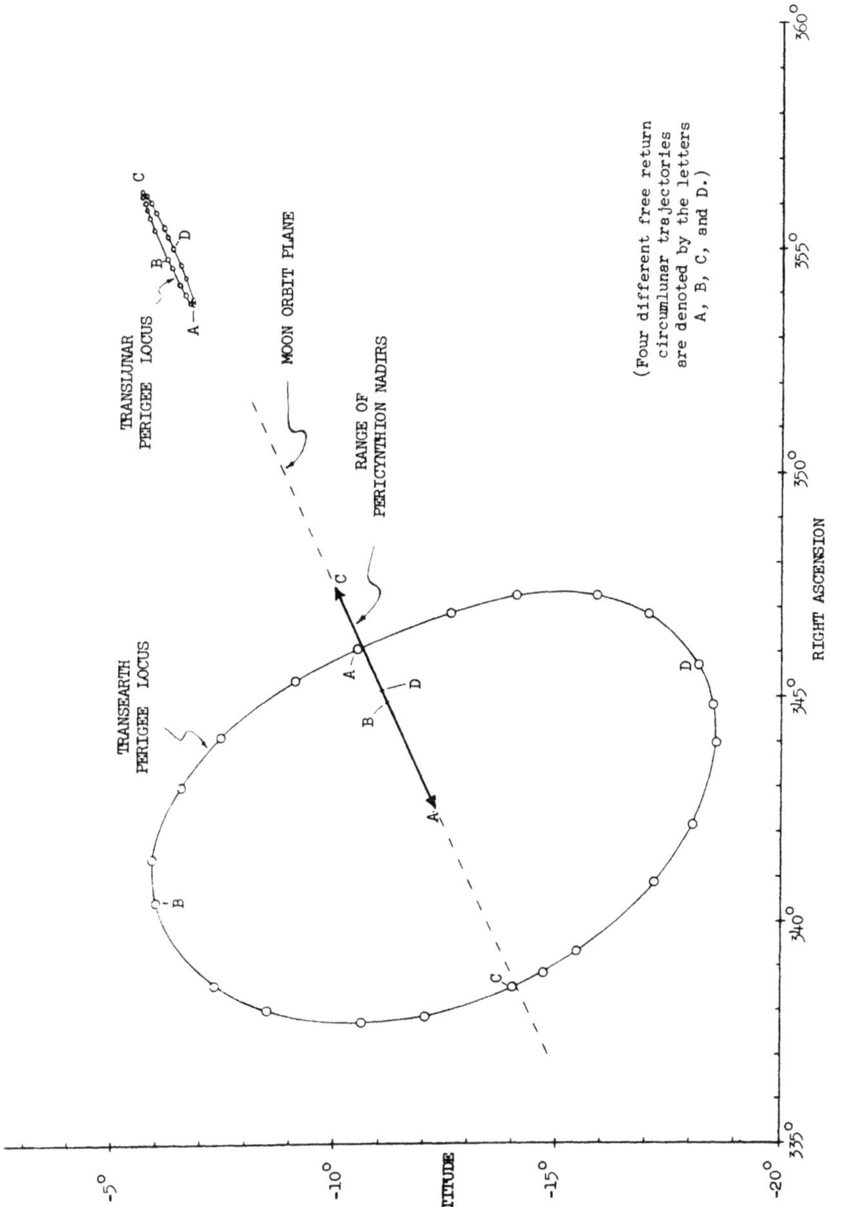

Fig. 14 Transearth and translunar perigee loci of a pane.

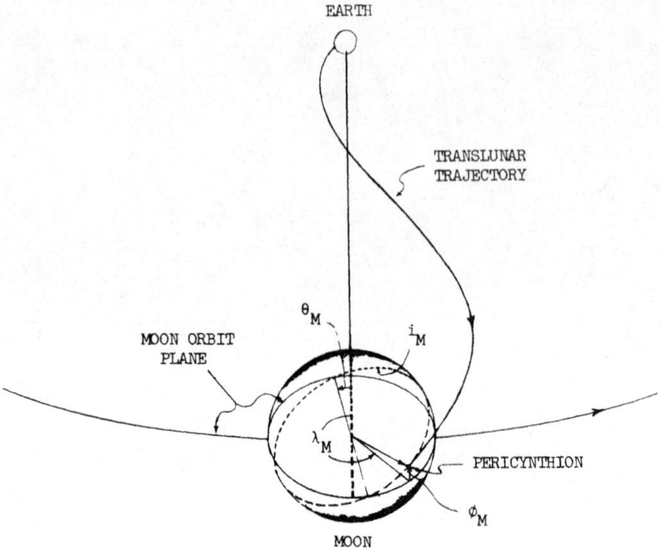

Fig. 15 Parameters describing trajectory orientation in the vicinity of the moon.

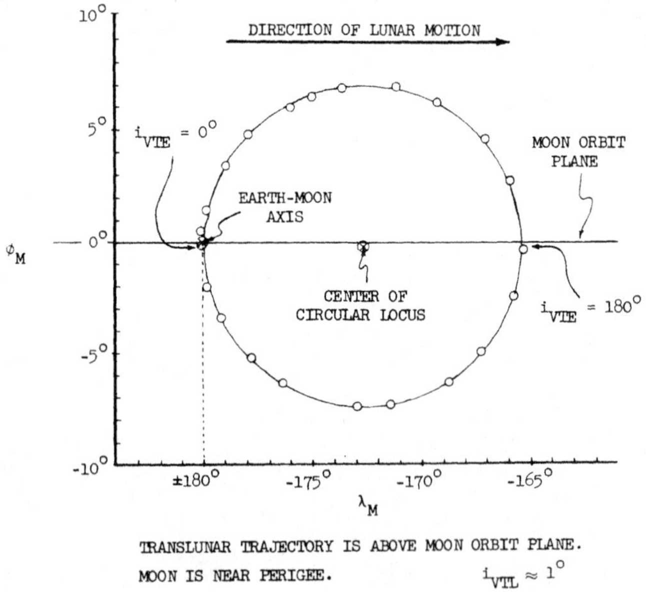

Fig. 16 Pericynthion locus of a pane.

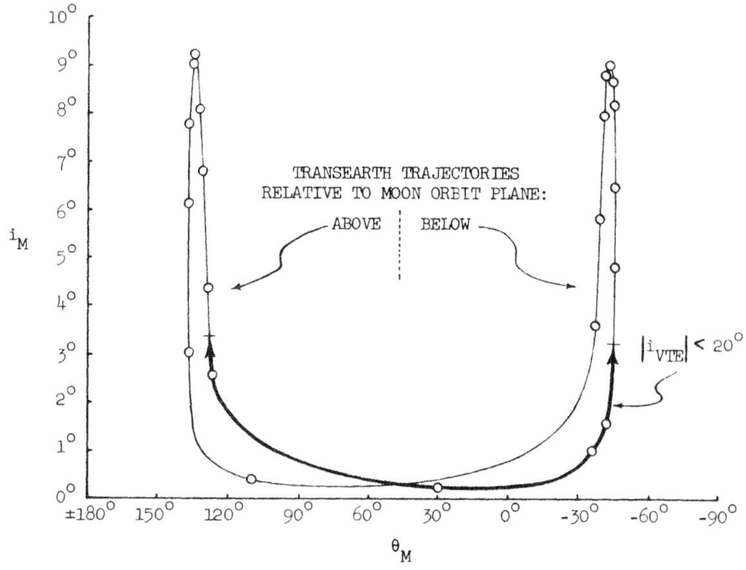

Fig. 17 Trajectory plane orientation at pericynthion.

IV.

ORBIT MECHANICS OF THE RANGER AND MARINER MISSIONS

TRAJECTORY DESIGN FOR RANGER AND MARINER MISSIONS

Victor C. Clarke Jr.*

Jet Propulsion Laboratory, Pasadena, Calif.

Abstract

Practical spaceflight trajectory design consists of a careful selection of a subset of trajectories. These trajectories are subject to several constraining factors, among which are the mission objectives and the limitations imposed by spacecraft hardware. After these factors have been fully evaluated, other characteristic parameters of the trajectory may be considered such as selection of flight time, launch dates, launch time, landing location (or other target coordinates), and retro velocity. It is the purpose of this paper to relate the trajectory design effort for the Ranger and Mariner missions at the Jet Propulsion Laboratory.

Presented as Preprint 63-408 at the AIAA Astrodynamics Conference, New Haven, Conn., August 19-21, 1963. This paper presents the results of one phase of research carried out at the Jet Propulsion Laboratory, California Institute of Technology, under Contract No. NAS 7-100, sponsored by NASA. The author wishes to acknowledge the excellent work of Saterios S. Dallas, who was trajectory design engineer for Rangers 1 and 2, William E. Kirhofer, for Rangers 3, 4, 5, and Willard E. Bollman for Mariner. In addition, the Trajectory Computation Group of the Jet Propulsion Laboratory, supervised by James F. Scott, deserves special credit for building the trajectory computer programs so vital to the trajectory design process. The services of Douglas B. Holdridge and Lee Laxdal of that group were especially outstanding.

The excellent cooperation of the trajectory engineers of the three participating organizations materially aided the successful completion of this rather formidable task. Particularly worthy of mention are Jack Buckingham, Jerry Reid, and William Lee of Space Technology Laboratories; Norman Schwalm, Robert English, and James Wilson of Lockheed Missiles and Space Company.

*Assistant Chief, Systems Analysis Section.

Introduction

The purpose of this paper (and its two companion papers,[1, 2]) is to relate, in general, the total trajectory engineering effort that was accomplished for the Ranger lunar, and Mariner interplanetary missions.

This effort was divided between three organizations: the Jet Propulsion Laboratory (JPL), Lockheed Missiles and Space Company (LMSC), and the Space Technology Laboratories (STL). As mission manager, JPL designed and built the Ranger and Mariner spacecrafts. The trajectory design performed at JPL was primarily concerned with the post-injection phase of flight, though sufficient cognizance of the preinjection phase was maintained so that over-all mission trajectories could be rationally determined. The techniques and methods used to accomplish the trajectory design task at JPL are the main topics of this paper.

LMSC, acting as Atlas/Agena boost vehicle system integrators, designed the boost vehicle trajectories and defined the guidance philosophy and technical requirements for the guidance equation synthesis. Finally, STL, under subcontract to LMSC, formulated the guidance equations and simulated the whole trajectory, with guidance equations, from launch-to-target, producing the firing tables which specify the guidance constants for actual flight.

General Description of the Trajectory Design Problem for Lunar and Interplanetary Missions

A striking aspect of lunar and interplanetary trajectory selection or design is the seemingly endless possibilities available. Apparently free to be chosen are a continuum of launch days, launch azimuths, and flight duration. The design process is concerned with the systematic selection of subsets of this mass of trajectories, the subsets being those trajectories that are best suited to satisfying mission objectives, launch vehicle and spacecraft restraints, and of imposing minimal constraints on the launch vehicle and spacecraft design.

The existence of such a large spectrum of possible trajectories requires trajectory computer programs[3] that can quickly and efficiently calculate the pertinent characteristics of all the possible trajectories. The analytical model of the flight path employed in such a program must necessarily be simplified, in order to permit speed and flexibility of operation. Thus, conic approximation models of the trajectories suffice to describe their major physical characteristics. Computer programs that evaluate such approximate analytical models are invaluable.

For example, with such a program, the interrelationships between such key trajectory parameters as launch date, flight time, injection energy requirements, communication distance at encounter, declination of the outgoing asymptote of the escape hyperbola (or declination of the moon at encounter), launch azimuth, launch time, and geographical injection location can be determined with sufficient accuracy to permit a good preliminary selection of the mission trajectories, in concert with the mission objectives and design restraints.

Although they are of considerable value, conic approximation models have shortcomings. It is inconvenient and impractical to calculate certain quantities by this technique. Furthermore, improved precision is needed as the trajectory design progresses, requiring a general n-body precision trajectory program.[4] Such a program must have sufficient flexibility of operation as an efficient design tool. In building a generalized, precision trajectory program, it is vital that many automatic features are employed which minimize input data requirements and eliminate unnecessary intermediate manual operations. Such a program has been developed at JPL for Ranger and Mariner trajectory design.[5]

Trajectory Selection for Lunar and Interplanetary Missions

Factors Affecting Trajectory Design and Selection

The selection of trajectories for a space flight mission is a complex engineering task which has definite interfaces with other engineering disciplines. To illustrate the depth of this problem and to describe the trajectory selection process, we shall take the design of the Ranger 3, 4, and 5 missions as an example. The trajectory selection process for an interplanetary mission is similar and will not be reported here, but it may be found in a paper by Bollman.[6]

The first step is to ascertain the mission objectives. For Rangers 3, 4, 5 (hereafter called Ranger Block II) these were: 1) to collect gamma-ray data in flight and in the vicinity of the moon, 2) to obtain photographs of the surface of the moon, 3) to transmit lunar seismic data after "rough" landing on the moon, 4) to perform a trajectory error correction,

5) to perform a terminal attitude maneuver, and 6) to continue development of the basic spacecraft technology.

To achieve these objectives, a 750 lb spacecraft system was developed with a basic bus consisting of (see Fig. 1) 1) a hexagonal-shaped structure, 2) solar panels and battery to provide power, 3) rotary high-gain and omnidirectional antennas, 4) monopropellant midcourse propulsion system; 5) instrument boxes for power conversion, transponder communication, central computer and sequencer, telemetry attitude control, etc., 6) vidicon and telescope system, 7) gamma-ray spectrometer system, 8) radar altimeter, 9) earth and sun sensors, and 10) cold gas bottles and attitude control jets.

The rough landing capsule system consisted of 1) solid propellant retrorocket with spin motors; 2) a 90-lb survival sphere contained in a balsa wood shell, and 3) a telemetry transmitter and lunar seismometer inside the survival sphere.

This configuration constituted a fully attitude-controlled spacecraft with a midcourse correction capability of 61 m/sec, and a retropropulsion system to deaccelerate the landing capsule by about 2690 m/sec. Further, for orbit determination purposes, a spacecraft-borne transponder permitted taking precision two-way doppler measurements at ground tracking stations of the NASA/JPL Deep Space Instrumentation Facility located at Goldstone, California, Woomera, Australia, and Johannesburg, South Africa.

Goals of Trajectory Design and Selection

Given the mission objectives and spacecraft configuration, the trajectory design and selection process consists of the following: 1) determination of permissible launch days (or period) during the month; 2) determination of permissible launch times (or windows) during the chosen launch periods; 3) determination of the landing location on the moon; 4) determination of the retro velocity increment for purposes of sizing the retro rocket; 5) determination of representative flight paths of the families of possible trajectories contained within the launch period window; and 6) presentation of the characteristics of these flight paths in parametric form to help determine such things as sequence of events and placement and utilization of mobile and ground-based tracking and telemetry instrumentation, plus any other pertinent trajectory-related data required to accomplish the mission.

Upon completion of these tasks, a "Lunar Targeting Criteria" specification is issued by JPL to LMSC which delineates the launch days, launch azimuths, and target conditions at the moon. Given this, LMSC and STL generate the firing tables and range safety data required to support the mission.

Constraining Factors

Theoretically, by use of the parking orbit[7] a spacecraft may be flown from Cape Kennedy to the moon on any day of the year and at any time of day, with any flight time, land at any place on the moon. However, practical considerations severely limit the choices of these various parameters. These shall be treated individually, and the major constraining factors on each will be cited.

Selection of flight time: The chief factor affecting flight time is the requirement that the terminal phase of flight and landing on the moon be visible from the Goldstone tracking station. This constraint automatically limits the flight time to nominal values of 42, 66, or 90 hr. The 42-hr time can be immediately eliminated because it requires excessive injection and retro velocities which greatly reduce the payload capability for the mission. For reasons that will be given later, the descent to the lunar surface should be as near vertical as possible and the landing location should be as near to the intersection of the earth-moon line and the lunar surface as possible. Since the vertical impact point of a lunar trajectory moves further from this point (in a westerly direction around the moon's leading edge) as flight time is increased, the 66-hr trajectory is preferred over the 90-hr time. This preference holds despite the fact that the energy requirements are greater for 66 than for 90 hr. The difference is small however, relative to the difference between 42 and 66 hr. For Ranger, the 66-90-hr difference amounts to about 25 lb of payload capability. The 66-hr trajectory is also slightly less sensitive to errors than the 90-hr, and the payload differential is reduced slightly by the requirement of less midcourse correction fuel.

Thus, for Ranger missions, the nominal 66-hr trajectory was chosen. Actual flight time varies from 63 to 71 hr, however, because of the position of the moon at encounter.

Selection of launch dates: The selection of launch dates is divided into two parts: 1) the choice of month, and 2) the choice of days within the month. The choice of month is dictated by project management and subject to many considerations such as manufacturing schedules, launch pad availability, etc. The choice of days within the month is controlled by applicable constraints, including 1) the attitude control system, 2) the vertical (or near vertical) descent requirement, and 3) a requirement that the spacecraft descend on a sunlit moon. The attitude control system affects launch date since the spacecraft achieves space stabilization by sensing the sun (along the roll axis) and the earth. Two vectors, along the probe-sun and probe-earth directions constitute a semi-space stabilized plane, which, with its normal, gives rise to a reference frame. The constraint arises because the angle (called the earth-probe-sun, or EPS angle) subtended at the probe between these two vectors must lie between 45° and 135°. Earth sensor field of view properties dictate this angular limitation.

The effect on launch date is illustrated in Fig. 2. Here it is seen that lunar encounters in symmetrical regions about first and third quarters of the moon provide acceptable EPS angles, while encounters in similar symmetrical regions about new and full moons result in unacceptable EPS angles. Thus, arrivals from the attitude control constraint are restricted to approximately 7-day intervals about first and third quarter.

The second constraint, to land vertically or nearly vertically, is dictated by the photographic properties of the vidicon experiment. The requirement is to provide proper nesting of the vidicon pictures during final descent.

The third constraint, that the spacecraft descend on a sunlit moon, is again dictated by the vidicon experiment. More precisely, the lighting angle (sun-moon-probe angle) at lunar impact should be between 30° and 60° to provide proper shadowing for the pictures. As seen from Fig. 2, this constraint, in combination with the vertical impact one, eliminates all of the trajectories that arrive in the first quarter interval and a portion of those that arrive during the third quarter interval. The net result is that a four to six day launch date interval remains, depending on the chosen month.

Selection of launch time: The selection of launch time window is chiefly dictated by the allowable launch azimuth interval as permitted by range safety considerations. For the Ranger program, waivers must be obtained from range safety to overfly land masses and launch over an azimuth interval. Thus far, this interval has been restricted to 93°-111° east of true north. Furthermore, restrictions on launch azimuth are generated by tracking and telemetry coverage limitations. Because of the properties of parking orbit type ascent trajectories, the second burn and injection locations vary widely over areas of the Atlantic Ocean (see Fig. 3), creating difficult tracking and telemetry instrumentation coverage problems. These can be alleviated somewhat by restricting the launch time window or azimuth, especially in the case of far downrange injections. In spite of the above mentioned restrictions, launch windows of from 50 min to 3 hr are possible.

Selection of landing location on the moon. The landing location is essentially determined by the requirement for vertical descent and lighting constraints as dictated by the vidicon experiment. However, there is another constraining factor because of the seismometer experiment that overlaps them. Within the survival sphere there is a transmitting antenna that erects itself to the lunar gravity vertical after impact. The antenna pattern is such that maximum antenna gain occurs along the lunar vertical. Thus, the ideal landing location for the sphere is at the point of intersection of the lunar surface and the earth-moon line. Such a landing location, however, would violate the vertical constraint of the vidicon experiment. Therefore, a loss in signal strength had to be taken, but the lower bound on the loss corresponded to a landing location no more than 45° in selenocentric angle from the earth-moon line. Furthermore, since the seismometer lifetime was to be three months, the 45° constraint had to be held despite the fact that this angle varies with lunar libration. Thus, the actual angle at impact had to be no more than about 31°. The net effect was that a compromise had to be made between the vidicon and seismometer experiments, which resulted in impact locations biased about 14° toward the earth-moon line.

Selection of retrovelocity increment. A further constraint resulting from the landing capsule was that the retrovelocity increment or unretarded impact speed must remain constant for all launch days during the launch period. This constraint was specified because a fixed impulse solid retrorocket was used. This constraint affects the flight time and injection energy also, in that these quantities are dependent on retrovelocity and thus remain relatively fixed on any launch day, but vary slightly from day to day to compensate for the changing position of the moon. Essentially, the retrovelocity must be selected so as to yield flight times that satisfy the Goldstone visibility constraint for all launch days of the launch period.

Conclusion

From the foregoing, it can be seen that a trajectory design does not start with a fully designed spacecraft and set of mission objectives. Rather, there is a constant iterative process, which occurs between the trajectory design and the spacecraft design.

References

[1]Schwalm, N., "Boost vehicle trajectories for Ranger and Mariner programs," AIAA Preprint 63-409 (August 1963).

[2]Reid, J. G., "Launch-to-mission completion targeting for Ranger and Mariner missions," AIAA Preprint 63-410 (August 1963).

[3]Clarke, V. C., Jr., Bollman, W. E., Roth, R. Y., and Scholey, W. J., "Design parameters for ballistic interplanetary trajectories -Part I. One way transfers to Mars and Venus," Jet Propulsion Laboratory, Tech. Rept. 32-77 (January 1963).

[4]Holdridge, D. B., "Space trajectories program for the IBM 7090 computer," Jet Propulsion Laboratory, Tech. Rept. 32-223 (Rev. 1) (September 1962).

[5]Laxdal, A. L. and Holdridge, D. B., "The JPL trajectory system," IAS Preprint 63-85 (January 1963).

[6]Bollman, W. E., "The engineering design of interplanetary ballistic trajectories," AIAA Preprint 63-414 (August 1963).

[7]Clarke, V. C., Jr., "Design of lunar and interplanetary ascent trajectories," Journal of the AIAA <u>1</u>, 1559-1567 (July 1963).

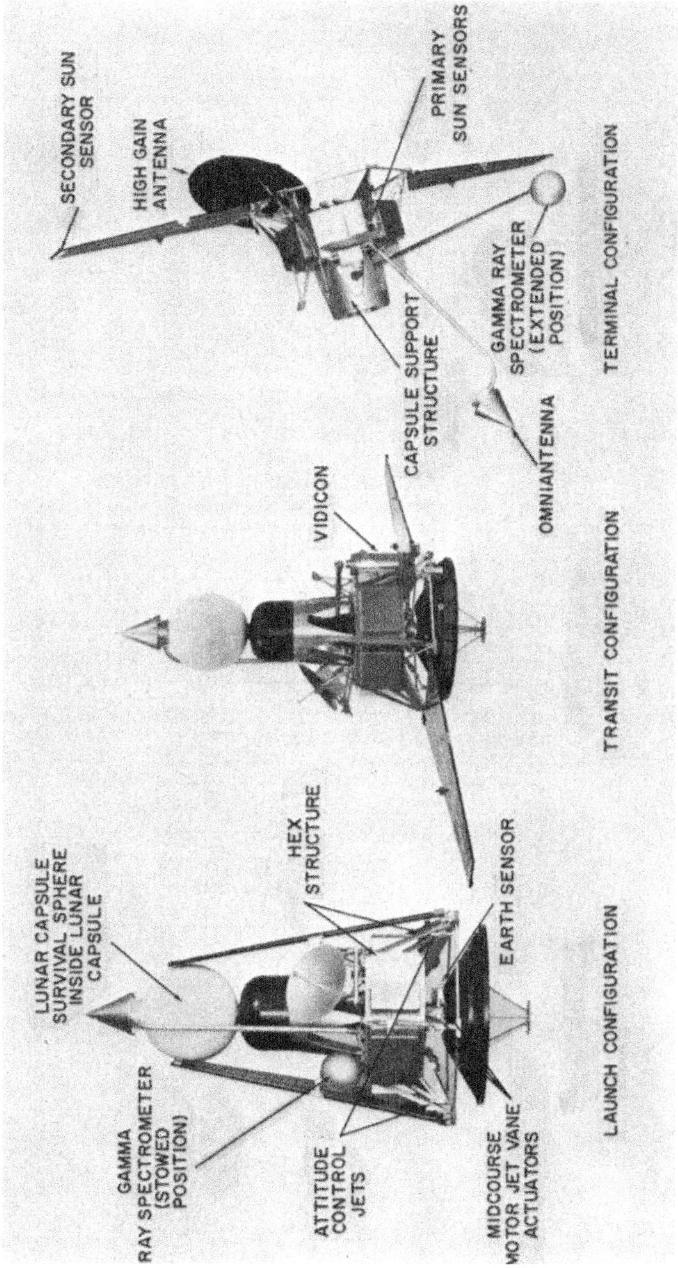

Fig. 1 Ranger spacecraft configuration.

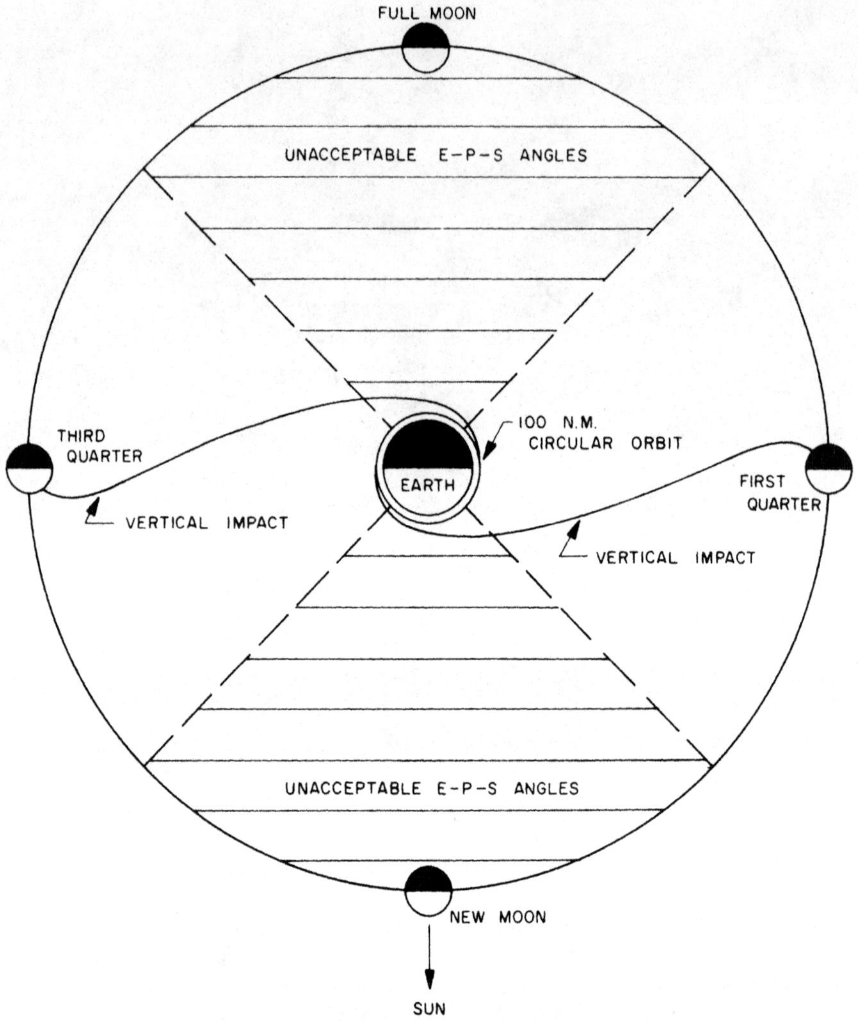

Fig. 2 Effect of lunar position on launch date.

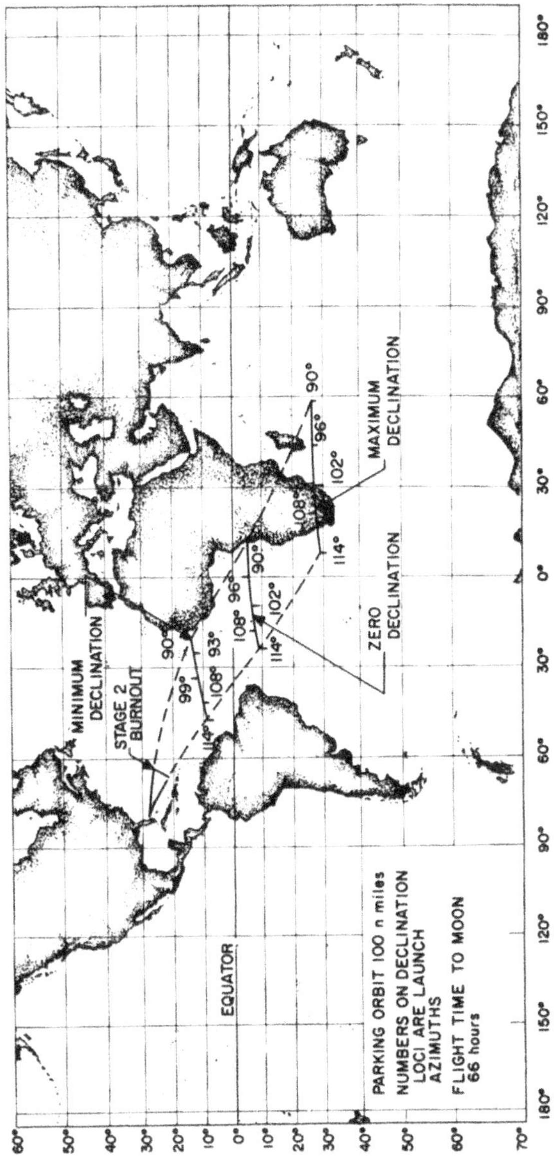

Fig. 3 Lunar injection loci,

BOOST VEHICLE TRAJECTORIES FOR RANGER AND MARINER PROGRAMS

Norman E. Schwalm*

Lockheed Missiles and Space Company, Sunnyvale, Calif.

Abstract

The Atlas/Agena trajectory profile for the Ranger and Mariner programs is presented. The boost guidance techniques for injecting spacecraft are discussed in relation to the variable launch azimuth "firing windows" of lunar and interplanetary missions. The Atlas and Agena vehicle and guidance systems, their objectives and limitations, are described only in terms of the development of the guidance philosophy, launch procedures, and pitch-plane trajectory shaping. The functions and methods of accomplishing boost trajectory dispersion analyses are also presented.

The techniques devised for boosting Ranger and Mariner spacecraft isolate the mission targeting from the ascent trajectory shaping. The determination of the Atlas/Agena performance characteristics thus become nonlaunch time dependent. The number of pre-launch settings that must be adjusted in the vehicle and guidance systems as a function of launch time are reduced, thereby simplifying imminent launch procedures. While these techniques were developed for current Atlas and Agena hardware, they may also apply to other booster vehicle configurations

---

Presented as Preprint 63-409 at the AIAA Astrodynamics Conference, New Haven, Conn., August 19-21,1963. The data presented in this paper related the boost guidance philosophy and the engineering and analysis techniques used to design boost-vehicle trajectories. As the booster-vehicle system analysis integrators, LMSC was responsible for generating this part of the Ranger and Mariner trajectory design. Many people with specialized backgrounds helped perform this task. The methods for solving the problems were generally developed in working meetings of all participating agencies. Within LMSC, engineering and guidance personnel from the Space Vehicle Guidance and Control Analysis department and Medium Space Vehicles Program took part in this effort

*Supervisor, Performance Analysis, Space Vehicle Guidance and Control Analysis Department.

Foreword

A team comprised of Jet Propulsion Laboratory (JPL), Lockheed Missiles and Space Company (LMSC), and Space Technology Laboratories (STL) personnel was established to design trajectories and firing tables for the Ranger lunar and Mariner interplanetary missions. A series of three papers describes the total effort and each paper relates the principal activities of the author's organization. This paper, second of the series, covers the ascent portion of the mission and the launch vehicle system, the Atlas D/Agena B space booster. The first and third papers of the series are listed as Refs. 1 and 2, respectively.

Introduction

The Atlas booster and General Electric (GE)/Burroughs radio guidance system were designed primarily for launching intercontinental ballistic missiles. Subsequently, the Atlas/Agena vehicle configuration was derived by fitting an upper-stage booster, the Agena vehicle, to the existing Atlas booster. Guidance equations for the GE/Burroughs system were then developed for earth-satellite missions utilizing the Atlas/Agena.

This booster system was also chosen to inject the Ranger and Mariner spacecrafts. An early criterion specified that existing vehicle and guidance hardware be used with only minor modifications, as required, for lunar and interplanetary variable launch azimuth "firing windows." It was then necessary to 1) develop a boost guidance philosophy which relates guidance capability requirements to mission criteria, and 2) in association with these requirements, devise methods for accomplishing the necessary launch-on-time procedures and the design of boost trajectories.

This paper describes the requirements and methods employed in support of the Ranger and Mariner missions. Because of the broad scope, each major task is presented as a section of the paper. The study is not intended to deal with hardware but with principles and techniques developed as trajectory analysis became better defined. To the best of the author's knowledge, no references exist for the material presented herein.

The contents of the paper are presented in the following order: The first section provides a general background and contains a description of the launch vehicle system objectives and the primary trajectory control components developed for earth-satellite missions. The next section introduces the time-variant characteristics for lunar and interplanetary missions and describes the resulting ascent guidance capability requirements. Launch-on-time procedures are then presented. A discussion of the STL-developed techniques for formulating guidance equations for the Atlas "pitch-plane" motion is followed by a description of the Agena "pitch-plane" guidance. A description of the boost trajectory shaping techniques and dispersion analyses completes the paper.

Flight Objectives and Primary Vehicle Control Parameters

Prior to the development of a boost guidance philosophy, it is necessary to select a basic trajectory profile whereby the booster provides the propulsive power and guidance to inject the spacecraft. For the Ranger and Mariner, a "parking orbit" approach was selected. For this type of trajectory, the vehicle is first placed in a low altitude, circular, earth-satellite orbit, where it coasts until the application of a final impulse injects the spacecraft into the proper lunar or interplanetary trajectory. The "parking orbit" approach, as compared to a direct ascent, yields several advantages. For example, greater payload weight can be achieved and the launch-time-window criteria are made less critical.

With current launch vehicles, no one booster vehicle, in itself, is able to accomplish the "parking orbit" approach, nor is there one capable of injecting another booster vehicle into a parking orbit with meaningful spacecraft weight. The Atlas D/Agena B space booster can achieve this approach with significant spacecraft weight because of the restart capability of the Agena B. Thus, the Atlas D and Agena first-burn combination can inject the Agena and spacecraft into a parking orbit and, after a prescribed coast time, the Agena second burn can inject the spacecraft. However, to insure successful spacecraft injection, each constituent system of the booster must achieve its flight objectives independently. These objectives are summarized as follows:

A. Atlas Vehicle System

1) Establish the desired trajectory heading after launch.

2) Follow a programmed pitch-down maneuver as it ascends through the atmosphere.

3) Transfer the required impulse to the Agena/spacecraft.

B. GE/Burroughs Radio Guidance System

1) Obtain a ballistic free-fall trajectory with preselected characteristics at Atlas thrust termination.

2) Orient the vehicle thrust attitude to a preselected reference prior to Atlas thrust termination.

3) Provide specified discrete commands to the Atlas/Agena vehicles.

C. Agena Vehicle System

1) Eject the shroud that protects the spacecraft against ascent aerodynamic heating and loads.

2) Achieve separation from the Atlas vehicle.

3) Ignite and accelerate itself and the spacecraft to achieve a circular, earth-satellite "parking orbit."

4) Reignite and accelerate itself and the spacecraft to achieve lunar or interplanetary injection conditions.

5) Separate the spacecraft.

The primary vehicle control components are listed in Table1.

Table 1

Vehicle Control Components and Their Functions

Component	Function
Atlas roll programmer	Establishes desired downrange azimuth after launch
Atlas pitch programmer	Determines vehicle pitch-attitude time history during ascent through the atmosphere
Atlas flight programmer	Sequences events within the vehicle as a function of flight time
General Electric/ Burroughs guidance	Provides discrete commands to the Atlas/ Agena and determines the vehicle pitch-and-yaw-attitude time histories during Atlas sustainer and vernier phases
Agena ascent sequence timer	Provides actuation time sequences during Agena flight to all equipment that must be started and stopped
Agena velocity meter	Provides Agena thrust-termination signals

Table 1 (Continued)

Component	Function
Agena attitude stabilization	Provides an attitude reference during the Agena flight, relative to the local horizontal in pitch and roll. (Initial reference is obtained at Atlas thrust termination)

Figure 1 illustrates the boost trajectory profile in the pitch plane and depicts the flight sequence resulting from the preceding primary vehicle control component functions.

Time-Variant Characteristics

For Ranger and Mariner, the pitch-plane characteristics of the parking orbit in inertial space (eccentricity, period, orbital altitude, etc.) were chosen to be independent of launch date or time of launch to eliminate one variable in the trajectory design. However, the traversed distance (parking orbit coast shown in Fig. 1) and spacecraft energy (length of Agena second burn shown in Fig. 1) are variable, and are a direct function of the launch date, time of launch, and the desired spacecraft target criteria. To illustrate, the traversed distance on a given launch date for Ranger is related to the moon's declination and the required earth-moon transfer time. For a constant transfer time (the injection occurring on the side of the earth away from the moon), the distance increases (injection moves downrange from the launch site) as the declination increases in a northerly direction; and, conversely, the distance descreases (injection moves uprange towards the launch site) as the declination decreases in a southerly direction.

During any given daily launch window, the required injection point remains almost stationary in inertial space. However, because of the earth's rotation, injection moves with respect to the earth's surface and traces a path from east to west. As the launch is delayed during any given day, the launch azimuth must be increased to maintain the injection point in the required trajectory plane, and the traversed distance in the parking orbit reduced because the launch site moves closer to the required injection point.

The launch vehicle system must therefore be capable of 1) accommodating dav-to-day changes in the parking orbit traversed distance and required spacecraft energy, and 2) compensating for the change in launch azimuth and traversed distance accompanying a delay in launch time. This capability is related to guidance requirements as follows:

1) As pertains to the pitch-plane motion, the non-launch-time dependent requirements for reaching a circular parking orbit at Agena first cutoff are identical to those of any earth-satellite mission .

2) The dependence upon liftoff time requires the GE/Burroughs guidance to continuously compute the required trajectory heading at Atlas termination, and the required parking orbit traversed distance in terms of coast time.

3) Since the GE/Burroughs guidance can command only the start of an Agena timer, and the runoff time between first and second burn is a variable, the requirement exists for two Agena sequence timers.

To meet Ranger or Mariner objectives, the present Atlas-Agena guidance system, developed for earth-satellite missions, will need only two additions. These are another Agena timer and a set of specific guidance equations for the General Electric/Burroughs guidance system. During the launch window, the settings of the guidance system must be adjusted to meet desired time-variant characteristics.

Launch Window Procedure

For a given launch date, the vehicle and guidance systems are set to establish the boost trajectory characteristics for the beginning of the launch window. Thus, day-to-day changes are made by gross adjustments to the settings. However, during each day, the trajectory heading and parking orbit coast time are continuous functions of the launch time. The settings that establish these parameters must also be adjusted on a continuous basis as a function of launch time.

Since all settings must be made prior to launch, they must be based on a predicted launch time. If the launch time is delayed after the settings are made, the settings must be changed to reflect the new predicted launch time. To change the settings, the countdown may have to be stopped (since time is required to make the change and obtain proper verification) and may even be recycled, causing a loss of available "firing" time from the allowable launch window. This sequence of events could prevent a launch unless appropriate procedures and techniques are established to 1) reduce the number of settings (as well as the number of systems which require settings) that are launch-time dependent, and 2) minimize the interplay between the countdown and the settings.

The number of launch-time dependent settings was reduced by the use of a nonlaunch time ascent pitch-plane trajectory, and the targeting technique, as described in Ref. 2, which determines, for each daily launch window, the Burroughs guidance target vector as a linear function of launch time. The remaining three systems that require setting adjustment as a function of launch time are as follows: 1) the Burroughs computer guidance equations, 2) the Atlas roll-program, and 3) the Agena timer that actuates the sequence for Agena second-burn ignition x sec after it is started.

To minimize the interplay between the countdown and adjusting settings, the simplest conceptual approach is to set the systems for a predicted launch time and to accept the resulting error if launch occurs at some other time. If this approach is followed, corrections must be made during the boost portion of the flight to prevent the expenditure of spacecraft midcourse propellants.

The practicality of allowing a setting to be in error at launch with subsequent correction of the trajectory during flight is determined by its effect on the ensuing trajectory and the available guidance correction capability. Each of the systems requiring adjustments will next be discussed as follows:

1) The correct time must be set into the Burroughs computer, when it is started, for the guidance equations to calculate continuously the required trajectory heading and parking orbit coast time as a function of subsequent time. The guidance system will steer to the wrong objectives if the start time is in error. The computer can be started prior to liftoff and thus is not dependent upon the exact launch time.

2) An Atlas roll-program error establishes the wrong downrange trajectory azimuth, but the ensuing trajectory characteristics are the same as if a correct azimuth were affected by a cross-wind blowing during ascent. In either case, the guidance equations during Atlas sustainer phase must correct this trajectory deviation prior to Atlas termination.

3) To insure the required parking orbit coast time, the guidance equations during Atlas sustainer phase must continuously predict the time to Agena second ignition. When the difference between this time and the Agena timer runoff time equals zero, a command is issued to start the timer. As the launch time is delayed, parking orbit time is reduced and, thus, for a constant Agena timer time, the start time would occur earlier in the Atlas sustainer phase.

Therefore, for the Atlas roll program and Agena timer settings, it becomes feasible to construct a "stair-type" curve as a function of launch time which is changed in discrete steps and maintained at a constant value between the steps. The guidance equations then can make the necessary corrections during flight by steering to the right or left and adjusting the start time of the Agena timer in flight. The selection of the discrete step intervals as a function of the launch time and the settings are "targeting" functions. Techniques for generating these data are described in Ref. 2. These settings are then presented to launch operations in the form of "firing tables."

The loss of available "firing time" resulting from the interplay between the countdown and the settings can thus be minimized. The launch controller can predict the time of launch, using the current countdown status and future estimated hold times, if a hold is expected. All corresponding settings can be made and verified. As the count proceeds, if unexpected delays occur, the settings remain unchanged unless the new launch time indicates the advancement to the next "stair." In this event, the process is repeated. For Ranger and Mariner missions, a discrete change occurs approximately every ten min during the "firing" window. Launch can occur at any instant during one of these ten-min periods and still achieve the same spacecraft injection accuracy.

In addition to setting the forementioned vehicle and guidance systems, range support agencies must be kept advised of these prelaunch changes in nominal trajectory characteristics. Subsequent to launch, radar tracking systems must acquire the vehicle, and Range Safety must possess the ability to distinguish between nominal and erratic vehicle behavior.

The guidance requirements are also used by STL in their development of equation formulation techniques for the GE/Burroughs radio guidance system. Since the determination of the ascent "pitch-plane" vehicle motion (nonlaunch time dependent) is LMSC's responsibility, the STL-developed guidance techniques are presented in the next section for purposes of continuity. The techniques for controlling the "yaw-plane" vehicle motion (launch time dependent) are presented in Ref. 2.

"Pitch-Plane" Guidance

"Pitch plane" control from launch through Agena first cutoff conforms to the requirements of an earth-satellite mission. This control is through the prelaunch settings of specific vehicle and guidance control components which are obtained from the trajectory-shaping studies discussed later in this paper. Since the functions of the control components have already been presented, this portion will deal with the GE/Burroughs and Agena guidance systems.

The GE/Burroughs radio guidance system consists of 1) a ground-based precision GE radar, providing position and velocity information and a command link from the ground to the vehicle, and 2) a ground-based general-purpose Burroughs digital computer to process the radar data and to initiate commands to the vehicle, as determined by the guidance equations contained within the computer. For Ranger and Mariner, the guidance equations are the responsibility of the Space Technology Laboratories.

A. Mathematical Definition of Mission Objectives

The explicit representation of actual objectives to be achieved at Atlas thrust termination (the "desired end" conditions) was the guidance method chosen for the Ranger and Mariner missions. This approach eliminates the need for a "stored" trajectory and can efficiently accommodate a wide range of initial conditions, since the interim path to be flown is determined by the equations rather than a return to the "stored" information.

The desired ballistic free-fall trajectory can be represented as a coast ellipse with a preselected semimajor axis, semiminor axis, and an orbital plane of motion (contains the target at the expected time of arrival). There are, therefore, an infinite number of combinations of Atlas terminal conditions (radius, inertial velocity, and flight-path angle) that satisfy entry into this desired coast ellipse. The guidance equations must command thrust termination of the Atlas when acceptable conditions are reached.

Earth oblateness and the gravitational attraction of other bodies must be considered for lunar and interplanetary flights. However, the problem to be solved would be unduly complicated if these effects were included explicitly. Thus, the earth is assumed to be a spherically symmetric body. The means of compensating for all other neglected effects is known as "target offset" or defining a "pseudo" coast ellipse that, when attained, will yield the correct solution.

In the pitch plane, the Keplerian relationships presented below can thus be employed for the solution of the required Atlas terminal conditions:

$$V = \left\{GM\left[\frac{2}{r} - \frac{1}{\alpha}\right]\right\}^{1/2} \tag{1}$$

$$\dot{r} = \left\{GM\left[\frac{2}{r} - \frac{1}{\alpha} - \frac{\beta^2}{\alpha r^2}\right]\right\}^{1/2} \tag{2}$$

where

V = inertial velocity of the vehicle

r = distance of the vehicle from the earth's center, geocentric radius

α = semimajor axis of the desired coast ellipse

β = semiminor axis of the desired coast ellipse

$\dot{r}$ = radial component of the inertial velocity

GM = gravitational constant of the earth

To terminate powered flight on a coast ellipse having a desired semimajor axis and semiminor axis, pitch steering and thrust cutoff must be arranged so that Eqs. (1) and (2) are satisfied simultaneously at cutoff.

Therefore, Keplerian relationships provide the means of establishing termination conditions. Methods must then be devised to predict these thrust termination conditions.

B. Time-To-Go

The first task is predicting accurately the remaining time of powered flight until thrust termination. A time-to-go function T_N must therefore be developed. It is desirable for this function to decrease linearly as a function of increasing real time and to equal zero when the vehicle's velocity equals the velocity required at cutoff. To accomplish this, an error signal V_N must be defined which is linear with increasing real time, and also equals zero at the required real time for time-to-go to equal zero. Thus the time-to-go can be expressed as

$$T_N = V_N/\dot{V}_N \qquad (3)$$

For Ranger and Mariner missions, a linear (in time) error signal was defined by the following equation:

$$V_N = 1 - m/m_f \qquad (4)$$

where m = mass at present and m_f = mass at termination. This error signal has all the aforementioned properties if the mass flow rate $\dot{m}$ is assumed to remain constant with remaining time. However, a knowledge of the mass time history is not available as input to the guidance equations. Therefore, it is necessary to associate V_N with the mass ratio through other parameters readily available. This is done by expressing the mass ratio as a function of the integral of thrust acceleration over the remainder of powered flight, or

$$V_N = 1 - \exp\left[\frac{1}{C^*}\int_0^{T_N} a_T \, dt\right] \tag{5}$$

where

a_T = thrust acceleration

C^* = effective rocket exhaust velocity

The integral can be evaluated from the difference in the present velocity and that required at predicted termination, corrected for the integrated effect of the angle-of-attack and the projection of gravity on the velocity vector.

C. Pitch Steering Philosophy

Trajectory shaping for optimum payload capability within vehicle constraints, the expenditure of additional propellants to correct dispersed trajectories, and ease of formulation are among the criteria for determining best pitch steering philosophy. The method for the Ranger and Mariner missions requires $(\ddot{r}/a_T)$ to be a preselected linear function of time, where $\ddot{r}$ represents the thrust, gravity, and centrifugal acceleration along the geocentric radius.

The steering method to be used must be known before predicting termination conditions, since these conditions are subject to any changes in thrust attitude that will occur after the present state is known. Two such conditions are the prediction of the geocentric radius r_f and the prediction of the radial component of the inertial velocity $\dot{r}_f$ found at termination. These predictions are derived from the following relationships:

$$r_f = r + \int_0^{T_N} \dot{r}\, dt + \int_0^{T_N}\int_0^{T_N} \left(\frac{\ddot{r}}{a_T}\right)_r a_T \, dt\, dt \tag{6}$$

$$\dot{r}_f = \dot{r} + \int_0^{T_N} \left(\frac{\ddot{r}}{a_T}\right) a_T \, dt \tag{7}$$

where r and $\dot{r}$ represents the present state, or initial conditions. $(\ddot{r}/a_T)$ represents the attitude if no corrective steering signals are sent, whereas $(\ddot{r}/a_T)_r$ represents a prediction of future steering required to correct any radial velocity errors.

Once r_f is known, the required velocity magnitude V_r and its required radial component $\dot{r}_r$ can be obtained from Eqs. (1) and (2) by substituting r_f for r. The difference between $\dot{r}_r$ and $\dot{r}_f$ represents the radial velocity error that will result at termination unless appropriate pitch steering is commanded to change the present thrust attitude. The steering will take the form of an initial attitude change followed by an approximately constant turning rate, as determined by the preselected linear function of $(\ddot{r}/a_T)$ with remaining time. Nominal trajectories $(\dot{r}_r = \dot{r}_f)$ will have only the approximate constant turning rate.

The initial thrust attitude change for making $\dot{r}_r$ and $\dot{r}_f$ equal at termination is found from Eq. (7). This equation can also be written as

$$\dot{r}_r = \dot{r} + \int_0^{T_N} \left(\frac{\ddot{r}}{a_T}\right)_r a_T \, dt \tag{8}$$

From Eqs. (7) and (8), the difference between $(\ddot{r}/a_T)_r$ and the present state $(\ddot{r}/a_T)$ directly represents the present attitude error $\delta\psi$ that must be corrected to achieve $\dot{r}_r$ at termination. The one assumption made is that $\sin\psi_r - \sin\psi = \delta\psi$.

D. Altitude Steering Control and Agena First-Burn Timer Start

For Ranger and Mariner missions, the point of Atlas thrust termination on the required coast ellipse is limited to a narrow segment by controlling the geocentric radius and hence the altitude at termination. Early in the sustainer phase, r_f as determined from Eq. (6) is checked to see if it lies outside the required segment. If it does, the vehicle is turned accordingly up or down in an open-loop fashion and allowed to remain at this attitude until r_f lies within the limits. Pitch steering during the remainder of the sustainer phase is based only upon the $\delta\psi$ necessary for the correct velocity vector orientation at termination.

Altitude steering control insures that the Agena first-burn timer will be started by the guidance system during the Atlas vernier solo phase at which time the vehicle is essentially on the coast ellipse. The timer is started by the guidance equations at a preselected time prior to the coast ellipse apogee time.

E. Agena Guidance

Agena guidance is an inertial system incapable of determining the best path to be flown and associated cutoff velocity and possessing no knowledge of its initial conditions. The objective, therefore, is to reproduce the flight path of a precomputed nominal trajectory.

The system maintains its thrust-attitude time history relative to the local vertical. Since a predetermined thrust attitude time history is maintained, the system can command thrust termination when a preset amount of velocity has been gained along the thrust axis. The velocity gained added vectorially to the initial velocity at ignition therefore will result in a predetermined velocity vector at cutoff. To insure the preceding, the vehicle must always be ignited at a predetermined point, relative to coast apogee, along its coast trajectory.

This concludes the guidance philosophy portion of this paper. The remaining sections pertain to the design of boost trajectories and to dispersion analyses.

Trajectory Design

A digital computer simulation model of the launch vehicle system serves as a primary tool in the boost trajectory design. The simulation model uses six-degrees-of-freedom and does not include vehicle dynamics, as it is assumed that the rocket engines are instantaneously positioned to balance any aerodynamic or center-of-gravity offset moments. The simulation consists of a series of separately coded subroutines linked together by a driver routine. It is therefore easily possible to obtain simulations with varying degrees of complexity (e.g., preliminary studies or the determination of flight settings), and to update the model to represent the latest experience gained from all Atlas/Agena flights.

By detailed trajectory and guidance analyses, the expected vehicle performance characteristics are found through the simulation model. These analyses insure that nominal as well as dispersed flight trajectories deliver the maximum payload to final burnout. For an established booster configuration such as Atlas/Agena, the optimum design boost trajectory is determined by analyses conducted within constraints imposed by the vehicle and guidance systems. This part of the paper explains this design process.

The boost guidance philosophy for the Ranger and Mariner missions permits the independent design of the Atlas and Agena portions of the trajectory. That is, once the Atlas guidance coast ellipse is specified, matching of the Agena first- and second-burn phases at the appropriate points results in the desired parking orbit and spacecraft injection criteria.

The "pitch-plane" characteristics of the specified Atlas guidance coast ellipse, like those of the parking orbit, are chosen to be non-launch time dependent. This choice recognizes that the rotation of the earth imparts velocity to the Atlas vehicle, thereby contributing to its energy, and this imparted velocity varies with launch azimuth. As launch azimuth increases, the guidance system causes additional Atlas propellants to be expended to achieve the same "pitch-plane" coast ellipse. The elimination of this variable in the trajectory design thus results in a slight degradation of the optimum payload capability.

The pitch-plane profile determined by LMSC, together with the JPL mission target criteria, is then integrated by STL to complete the boost trajectory design. This is accomplished by determining, for a given launch time, 1) the required trajectory azimuth, 2) the coast time in the parking orbit, and 3) the magnitude of the final impulse. Since Ref. 2 relates the targeting activities, this paper will discuss only the design of the ascent pitch-plane profile.

Hardware design limitations are found throughout the ascent trajectory. For the Atlas, the following must be considered: 1) aerodynamic heating of skins and equipment, as well as structural bending-moment limitations, influence both the shape and magnitude of the booster-stage pitch-program time history. After a program has been designed, analyses have shown that a constant percentage change can be applied throughout the booster phase to "loft" or "depress" the trajectory. The percentage change is, however, constrained to insure that design limitations are not exceeded; 2) staging the booster engines above the maximum axial G-load may cause structural failure, whereas staging too soon may cause a propellant pump cavitation; and 3) steering maneuvers are carefully considered to insure that the radar tracker can always "see" the vehicle antenna and maintain enough signal strength. A nominal sustainer pitch program is selected to insure that dispersed trajectory excursions are within limitations. From the guidance pitch steering philosophy employed, it is known that the response for a nominal flight will be an approximate constant turning rate. Thus, the selection of the program involves only the selection of the rate magnitude.

There must be enough time between Atlas/Agena separation and Agena first burn to allow such necessary events as ejecting the nose shroud. The Agena attitude control system cannot exceed a maximum above the local horizontal (nose up). The altitude at Agena first cutoff (injection into the parking orbit) may be limited by free molecular heating of the exposed payload, since the protective nose shroud was ejected at Atlas/Agena separation.

Through an iterative process between preliminary trajectories and hardware design groups, the foregoing limitations are defined, consistent with the applicable mission. A definitive payload optimization study can then be undertaken. Assuming that the weight, propellant loadings, and propulsion characteristics of the vehicles are known, the pitch-plane trajectory control parameters to be optimized are shown in Table 2.

Table 2

Pitch Plane Trajectory Control Parameters

Atlas	Agena
Booster-stage pitch program	Agena coast time prior to first burn
Staging time of the booster engines	Agena first-burn pitch attitude
Sustainer-stage pitch program	Agena first-burn velocity gained

The Atlas portion of the flight will be considered first, based upon a given coast apogee altitude. For a constant expenditure of Atlas propellants, the optimized trajectory will result in the maximum coast apogee velocity. To determine the maximum Atlas coast ellipse apogee velocity, for a given apogee altitude, we select in combination 1) the appropriate constant percentage change to be applied to the booster-stage pitch program, 2) the booster staging time (or axial G-load), and 3) the rate magnitude for the sustainer-stage pitch program. The steps followed are:

1) For a given booster-staging time and expenditure of propellants (constant Atlas burnout weight), the sustainer pitch-program rate magnitude that yields the desired radar antenna-angle time history is determined. The booster-stage pitch program is then varied, all other parameters remaining fixed, to determine the resulting coast apogee velocity as a function of coast apogee altitude.

2) The process is repeated, varying the booster-staging time. For a given coast apogee altitude, the associated velocity thus can be determined as a function of the staging time, and this result examined to determine the optimum. A final trajectory is then generated.

3) If desired, the preceding process can be repeated to determine optimized trajectories as a function of coast apogee altitudes.

From the coast apogee conditions, altitude and velocity, the semimajor and semiminor axes of the desired guidance coast ellipse can be specified and the guidance equations incorporated into the simulation. Dispersed trajectories are then computed to insure that no hardware design limitations are violated.

For the desired parking orbit injection conditions (inertial velocity and radius vector), the Agena first-burn phase is matched at the appropriate point along the Atlas coast ellipse. This is done by varying the Agena ignition time, pitch attitude, and impulse gained. For low-altitude parking orbits, the optimum ignition time is normally dictated by the minimum allowable coast time from Atlas separation. The problem therefore reduces to two unknowns: 1) the pitch attitude reference with respect to the local vertical, and 2) thrust duration time. In reality, pitch attitude primarily affects the flight-path angle and thrust duration primarily affects the velocity magnitude.

The techniques for determing the optimum ascent pitch-plane profile apply to the design of preliminary trajectories, as well as to the design of preflight reference trajectories that determine the actual flight hardware and guidance settings.

After receipt of the STL preflight reference "targeted" trajectories, which are based upon nominal vehicle performance, vehicle and system component tolerances are investigated for the effects of trajectory dispersions.

Dispersion Analysis

Nominal as well as dispersed flight trajectories must result in spacecraft injection within specified tolerances. The effect of deviations in vehicle and guidance performance characteristics, therefore, must be extensively evaluated to find if the resulting trajectories result in mission success. The simulation of non-nominal but probable performance deviations thus accomplishes two basic functions:

1) The first function insures that expected trajectory dispersions lie within the boost hardware design limitations. Also, it insures that enough propellant margins are maintained to meet specified guidance cutoff probabilities.

2) The second function provides the expected trajectory errors at spacecraft injection. These errors must be subsequently corrected by a spacecraft midcourse to meet the desired target criteria. The required midcourse maneuver also serves as a basis for evaluating the actual flight performance.

Deviations in the vehicle and guidance performance characteristics must be computed in view of the standard deviation of each control and performance tolerance to establish a specified confidence level (e.g., If 10,000 vehicles were flown, 9974 would not exceed the specified tolerance). These component tolerances are generally available from one or more of the following sources: 1) hardware specification and ground tests, 2) analytical studies, and 3) flight test results. In general, these tolerances exhibit two important characteristics: 1) the magnitude is a random quantity having Gaussian distribution, and 2) the resulting dispersed trajectories differ only slightly from the nominal, so that the assumption of linear perturbations is justified. In addition, the tolerances are considered as independent random variables, e.g., gyro drift is independent of rocket-thrust tailoff uncertainties.

The effect of each tolerance on the trajectory is determined by assuming all other inputs to be nominal. The component is varied in accordance with the tolerance specified. The resulting trajectory from launch to mission completion is obtained from the launch vehicle system-digital computer-simulation model, linked to an interplanetary program. Also, an interative procedure is available for the determination of the spacecraft midcourse maneuver needed for the desired target criteria. Simulating the trajectories from liftoff to mission completion eliminates any inaccuracies introduced by relating trajectory errors from a point in the boost phase to target dispersions by partial derivatives determined from a reference case.

To complete a detailed launch-vehicle dispersion analysis as mentioned above, 56 component tolerances must be individually simulated with both positive and negative variations. Analytical techniques then must be derived, combining the individual results to establish the overall launch vehicle probable deviations.

Since positive and negative variations do not always yield symmetric trajectory deviations, the variation that gives the largest deviation is conservatively selected. The overall probable deviation in any parameter (such as propellants expended) resulting from the boost system tolerances is then taken as the RSS (root-sum-square) of all the individual parameters.

The specified accuracy with which the spacecraft must be injected is established in terms of the amount of midcourse maneuver required to correct injection errors. This is determined by a unit of variance analysis (UOV) devised by JPL. The actual variance of the midcourse velocity, $\sigma^2_{V_j}$, due to each component tolerance e_j, is derived from the following relationship:

$$\sigma^2_{V_j} = P^2_j \, e^2_j \tag{9}$$

where P_j is the partial derivative of midcourse velocity with respect to the jth tolerance considered, and can be expressed as

$$P_j = \partial V_M / \partial e_j \tag{10}$$

where the P_j's are determined directly from the midcourse requirement to correct the trajectory deviations resulting from the jth tolerance.

To obtain the percent variance of the midcourse requirement due to each tolerance, a figure-of-merit (FOM) can be defined to represent a basis for comparison. The number of UOV for each tolerance is then the ratio of the actual variance $\sigma^2_{V_j}$ to the $(FOM)^2$ expressed in percent, or

$$UOV_j = [100/(FOM)^2] \, \sigma^2_{V_j} \tag{11}$$

Two important results are obtained from the UOV analysis. First, the RSS of the midcourse velocity requirements is obtained from the actual variance due to each error source, or

$$\sigma_V \, (RSS) = \left[\sum_{j=1}^{j=n} P^2_j \, e^2_j \right]^{1/2} \tag{12}$$

and secondly, a list of the UOV contributions reveals the relative importance of each error source. The latter allows the most significant tolerances to be singled out as well as the elimination of those tolerances that produce a negligible effect.

Another result that can be obtained from the dispersion analysis is a mapping of the probable errors at the spacecraft target without midcourse correction. The method generally accepted to accomplish this task involves the determination of an ellipse of target errors due to boost-trajectory perturbations.

For each launch-vehicle system tolerance considered, Δe, let M_1 and M_2 represent the parameters employed to measure errors at the target. Assuming small perturbations,

$$\Delta M_1 = (\partial M_1/\partial e)\,\Delta e \tag{13}$$

$$\Delta M_2 = (\partial M_2/\partial e)\,\Delta e \tag{14}$$

where the partial derivatives are determined by simulating the perturbed trajectory and observing the effect on M_1 and M_2.

To extend the method to more than one error source, let the n error sources be Δe_j $(j = 1, 2, 3 \ldots n)$. Computing n trajectories, one for each of the n sources, the following $(n \times 2)$ matrix can be built:

$$\Delta M = \begin{bmatrix} \frac{\partial M_1}{\partial e_1}\Delta e_1 & \frac{\partial M_2}{\partial e_1}\Delta e_1 \\ \frac{\partial M_1}{\partial e_2}\Delta e_2 & \frac{\partial M_2}{\partial e_2}\Delta e_2 \\ \frac{\partial M_1}{\partial e_n}\Delta e_n & \frac{\partial M_2}{\partial e_n}\Delta e_n \end{bmatrix} \tag{15}$$

The moment matrix of target errors is then

$$M = \Delta M^T \Delta M \tag{16}$$

or

$$M = \begin{bmatrix} \delta M_1^2 & \rho \\ \rho & \delta M_2^2 \end{bmatrix} \tag{17}$$

where δM_1^2 equals the sum of the squares of the n ΔM_1's. δM_2^2 equals the sum of the squares of the n ΔM_2's and ρ represents the cross correlation of ΔM_1 and ΔM_2. By process of "matrix diagonalization" the (M_1, M_2) coordinate frame can be rotated through an angle θ into the axes of the ellipse so that the off-diagonal terms are zero. This is accomplished by introducing the rotation matrix R, where

$$R = \begin{bmatrix} \cos\theta & \sin\theta \\ -\sin\theta & \cos\theta \end{bmatrix} \tag{18}$$

and it can be shown that

$$\theta = 1/2 \tan^{-1}\left(\frac{2\rho}{\delta M_1^2 - \delta M_2^2}\right) \tag{19}$$

thus,

$$RMR^T = \begin{bmatrix} \lambda_1 & 0 \\ 0 & \lambda_2 \end{bmatrix} \tag{20}$$

where λ_1 and λ_2 are the squares of the semimajor and semiminor axes, and θ is the inclination of the axis of the ellipse to the M axis. It can also be shown that

$$\lambda_1 = 1/2 \left[\left(\delta M_1^2 + \delta M_2^2\right) + \sqrt{\left(\delta M_1^2 - \delta M_2^2\right)^2 + 4\rho^2}\right] \tag{21}$$

$$\lambda_2 = 1/2 \left[\left(\delta M_1^2 + \delta M_2^2\right) - \sqrt{\left(\delta M_1^2 - \delta M_2^2\right)^2 + 4\rho^2}\right] \tag{22}$$

An ellipse of the target errors can then be formed. If the n error sources considered were each of three-sigma deviation, then the probability that the miss for any trajectory will be within this ellipse is also three-sigma.

References

[1] Clarke, V. C., Jr., "Trajectory design for Ranger and Mariner missions," AIAA Preprint 63-408 (August 19–21, 1963).

[2] Reid, J. G. and Lee, W. R., "Launch-to-mission completion targeting for Ranger and Mariner missions," AIAA Preprint 63-410 (August 19–21, 1963).

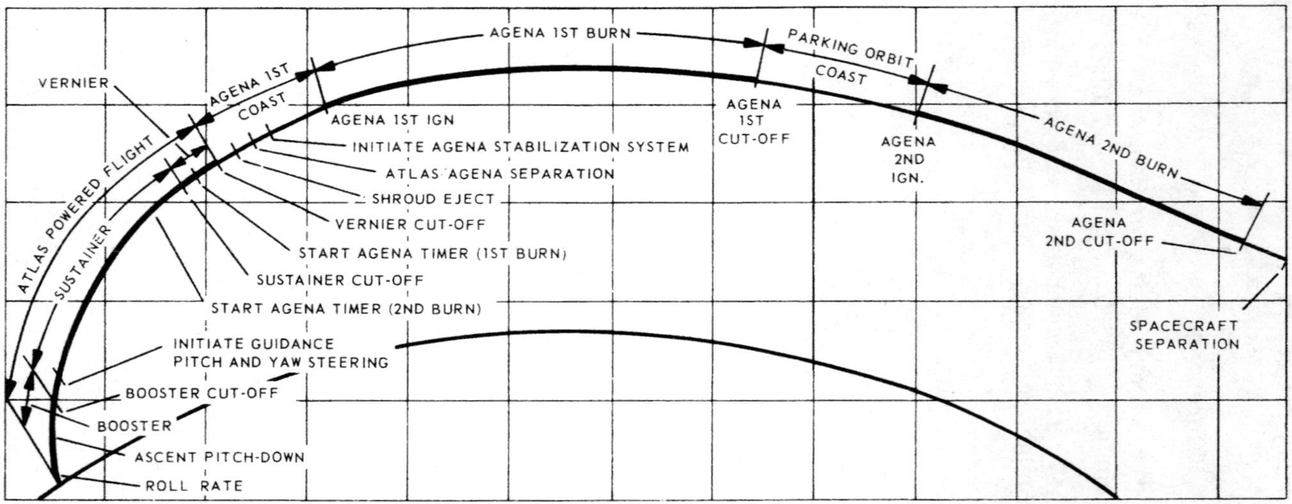

ATLAS EVENT	CONTROL COMPONENT	AGENA EVENT	CONTROL COMPONENT
ROLL RATE	ATLAS ROLL PROGRAMMER	INITIATE AGENA STABILIZATION SYSTEM	TIMER
ASCENT PITCH-DOWN	ATLAS PITCH PROGRAMMER	AGENA PITCH AND ROLL REFERENCE	STABILIZATION SYSTEM
BOOSTER CUT-OFF	GE/BURROUGHS GUIDANCE	AGENA 1ST IGNITION	TIMER
SUSTAINER STEERING	GE/BURROUGHS GUIDANCE	AGENA 1ST CUT-OFF	VELOCITY METER
START AGENA TIMER (2ND BURN)	GE/BURROUGHS GUIDANCE	AGENA 2ND IGNITION	TIMER
SUSTAINER CUT-OFF	GE/BURROUGHS GUIDANCE	AGENA 2ND CUT-OFF	VELOCITY METER
START AGENA TIMER (1ST BURN)	GE/BURROUGHS GUIDANCE	SPACECRAFT SEPARATION	TIMER
VERNIER CUT-OFF	GE/BURROUGHS GUIDANCE		
SHROUD EJECT	GE/BURROUGHS GUIDANCE		
ATLAS AGENA SEPARATION	GE/BURROUGHS GUIDANCE		

Fig. 1 Trajectory pitch-plane profile.

LAUNCH-TO-MISSION COMPLETION TARGETING FOR RANGER AND MARINER MISSIONS

Jerry G. Reid* and William R. Lee†

TRW Space Technology Laboratories, Redondo Beach, Calif.

Abstract

Launch-to-mission completion targeting techniques, computer program descriptions, and related topics are presented. These techniques are employed in support of the Ranger/Mariner Lunar and Interplanetary Program. Requirements for an effective and highly automated targeting and firing table generation program are discussed in relation to the high density launch schedule and precise trajectory information required. Ascent guidance philosophy, guidance constants mechanization, and preflight/post-flight trajectory efforts are described only as related to the launch-to-mission completion targeting procedures. Analytic and variational methods may be utilized effectively to make targeting of lunar and interplanetary missions a relatively inexpensive operation. Also the techniques used are very general and are easily applicable to both simpler and more complicated missions.

Introduction

Support of the Ranger/Mariner Lunar and Interplanetary Program includes preflight mission analyses in the area of guidance equations, targeting, firing table generation, range safety and range support trajectory information, and post-flight trajectory, guidance and performance analyses. The area of targeting and firing table generation is the prime subject of this paper. The other areas will be discussed only as they relate to the prime topic.

The task of supporting a high density launch schedule of approximately 1 lunar vehicle every 2 or 3 months with inter-

---

Presented as Preprint 63-410 at the AIAA Astrodynamics Conference, New Haven, Conn., August 19-21, 1963.

*Ranger/Mariner Project Engineer, Mission Analysis and Simulation Laboratory.

†Member of the Technical Staff, Space Navigation Department.

planetary attempts as they occur (2 or 3 launch attempts during a planetary opportunity) requires effective and highly automated techniques of obtaining trajectory and target dependent parameters. This launch schedule for the Ranger/Mariner Program required that 10 lunar opportunities and 1 planetary opportunity be targeted during the 15-month period from October 1961 through December 1962. This task amounted to targeting approximately 130 daily windows. Each launch day has a launch window of 2 to 4 hr during which the vehicle guidance and control system must be capable of properly guiding the vehicle to its objective for any instant of launch during the launch period. Therefore, some number of trajectories must be specifically targeted on a given day and the required parameters fitted as a launch time dependent function for the daily window in order to adequately specify the targeting parameters to the guidance system. This is generally 3 to 5 evenly time spaced trajectories per launch date or a total of 390 to 650 targeted trajectories during this 15-month period. In addition to the targeting of this large number of discrete trajectories, firing tables, firing table check cases, range safety and range support information, and post-flight trajectory and performance analyses are required. This emphasizes the need for computer programs that are automated to the extent that a small group of engineers and support personnel can handle the task.

Targeting

Targeting is the computational procedure by which a required set of ascent guidance and trajectory parameters are determined with sufficient precision to meet a set of mission and target constraints. Accuracy requirements dictate the use of high precision trajectory programs that adequately simulate all factors affecting the trajectory. Also, because of the amount of effort involved in a launch time dependent targeting scheme, an operational targeting program must be efficient. Therefore, use must be made of analytic solutions or approximate rapid computational techniques in obtaining initial estimates of trajectory or target parameters. Also, any applicable analytic or rapid integration techniques for determining partial derivatives should be utilized to decrease computation time.

Guidance Philosophy

The guidance equations used in the Ranger/Mariner Program are a complex set of relationships used to cause Atlas burnout on a desired coast ellipse. This coast ellipse is defined such that the first Agena burn phase will result in a circular parking orbit with the proper altitude and inclination. Note that only the Atlas has complete, closed-loop guidance; the Agena, using a body-mounted accelerometer for measurement, adds two fixed, predetermined velocity increments, the first increment for entry into parking orbit, and the second for final injection. For the purposes of this discussion, only those portions of the equations applicable to the targeting will be considered. With this major simplification, the guidance equations may be treated as a system whose only requirements are the maintenance of the parking orbit inclination and the termination of the parking orbit at the proper time. These two functions are accomplished through the use of a "target" vector. The "target" vector is, approximately, the hyperbolic excess velocity vector for interplanetary trajectories and the earth-moon position vector for lunar trajectories; more precisely, it is that quantity that, when substituted into the equations below, yields a trajectory that hits the target. The determination of the "target" vector represents the major portion of the targeting effort. The inclination of the Atlas coast ellipse may be defined at Atlas burnout as

$$i = \cos^{-1} \frac{\bar{r} \times \bar{v}}{|\bar{r} \times v|} \cdot \bar{Z}$$

where r and v are missile position and velocity, respectively, and $\bar{Z}$ defines the polar direction.

The desired inclination is

$$i_D = \cos^{-1} \frac{\bar{r} \times \bar{r}_T}{|\bar{r} \times \bar{r}_T|} \cdot \bar{Z}$$

where $\bar{r}_T$ is the target vector.

From these expressions, an error signal may be generated:

$$\dot{\epsilon}_y = \frac{\bar{r} \times \bar{r}_T}{|\bar{r} \times \bar{r}_T|} \cdot \bar{v}$$

where $\dot{\epsilon}_y$ is the velocity normal to the desired orbit plane and must be driven to zero at burnout.

The time for parking orbit termination is determined from

$$t = t_{poi} + \left(\cos^{-1} \frac{\bar{r} \cdot \bar{r}_T}{|r| \; |\bar{r}_T|} - \varphi \right) \frac{1}{\omega}$$

where t_{poi} is the time of parking orbit injection, φ is the sum of all angles (exclusive of the parking orbit angle) between the current position vector and the target vector, and ω is the parking orbit angular rate.

Following Atlas burnout, Agena attitude is controlled by horizon scanner referenced gyros, and injection velocity is determined by integrating accelerometers. This injection velocity, with the two components of the target vector, completes the set of variables for targeting.

The guidance target is expressed in the guidance equations in polar, geocentric, nonrotating coordinates. The target vector varies with time during each daily launch window; this variation results from both the motion of the approximate target vector (the hyperbolic excess vector or the earth-moon vector) and from the changing character of the geocentric part of the trajectory. The target vector time variation is accounted for by a linear function in each of the two target components.

Targeting Procedures

The primary purpose of targeting for the Ranger/Mariner lunar and interplanetary missions is to determine the Atlas and Agena guidance and control settings for a launch during the available time period. After these settings are determined, they are formulated into a set of firing tables as a function of liftoff time.

The lunar and interplanetary targeting procedure, as applied to the Ranger/Mariner Program, may be divided into a set of chronological steps as follows:

1) Ascent trajectory pitch plane shaping parameters and a time sequence of events to parking orbit injection are received from Lockheed Missiles and Space Company as input to the targeting exercise.[1]

2) Launch dates and post-injection trajectory specifications are received from the Jet Propulsion Laboratory. This con-

tains the results of the mission design and includes lunar or interplanetary arrival conditions that are used as inputs to the targeting program.[2]

3) A determination is made of initial estimates of lift-off time and coast times for a selected number (3 to 5) of trajectories for each launch date. This is accomplished by use of an Analytic Lunar Program in which a patched conic technique is utilized or, in the case of interplanetary trajectories, an Analytic Interplanetary Program. The resulting output of these search programs are best estimates of ascent guidance and control parameters to be used in the launch-to-mission completion targeting program.

4) Launch-to-mission completion search trajectories are performed utilizing inputs from the foregoing 1, 2, and 3. The outputs of this program are precise guidance and control parameters for the complete launch-to-mission completion trajectory. Three to five of these trajectories are targeted for each launch date.

5) Generation of a firing table for each launch day from the output of the foregoing item 4. This consists of fitting each time dependent parameter and formulating them into a table of these parameters which is compatible with the ascent guidance and control systems. A selected number (at least 3/day) of trajectory check cases are then simulated using the firing table data as input. The adequacy and accuracy of the firing table data are partially checked in this fashion. An additional check, which is similar, is performed by Lockheed Missiles and Space Company using the firing table data as input to a similar but independently derived launch-to-mission completion program. The Jet Propulsion Laboratory also serves a check function by simulating a number of earth injection to mission completion trajectories using TRW Space Technology Laboratories' injection points as origin. This multitude of checks on independent programs insures the adequacy and accuracy of the complicated procedure of firing table constants generation.

The foregoing set of procedures (except 1 and 2, which were discussed in Refs. 1 and 2) are discussed in greater detail in the following paragraphs and are shown in the following flow diagram of procedures (Fig. 1).

Initial Estimates of Targeting Parameters

Analytic search programs are used in obtaining initial best estimates of the guidance and control parameters required to sufficiently define the launch-to-mission completion trajectory. The output of these programs are best estimates of liftoff time, parking orbit coast time, the guidance target, inplane angles, and guidance and control parameters which represent these quantities.

Launch-to-Mission Completion Targeting Program

The lunar and interplanetary targeting program consists of a simulation of the guided ascent trajectory, the lunar or interplanetary transfer trajectory, and associated program controls and options to allow automatic convergence to desired end conditions in an optimum fashion. The ascent trajectory portion of the targeting program includes simulation of all vehicle parameters and external forces affecting the trajectory. The transfer trajectory accounts for all significant perturbations from the bodies of the solar system. A primary feature of the targeting program is a convergence routine that allows selected vehicle and guidance parameters to be changed automatically in such a way that the trajectory converges to a required set of terminal conditions. At the end of this convergence, the program computes and prints (or stores on tape) the necessary vehicle, guidance, and control parameters required to duplicate the converged trajectory. A sufficient number, usually three to five, of these targeted trajectories are made per launch day in order to determine and make a linear fit of the launch time dependent parameters. Then, the day dependent and launch time dependent parameters are combined into a firing table (or a targeting tape) for that day. This table or tape, properly mechanized, will allow launch at any time during a daily launch window. Each day of the launch period is targeted in the same fashion.

Convergence Routine

It is necessary to determine that "target" vector that, when used in the guidance equations, yields acceptable terminal conditions. The approximate target vector may be used to begin an iterative differential correction procedure as follows:

$$\delta c = \mathbf{A}^{-1} \delta e$$

where

δc = column matrix of corrections to the target vector

δe = column matrix of terminal errors

$\mathbf{A}$ = transformation matrix of partial derivatives relating errors to corrections

The matrix $\mathbf{A}$ may be obtained by several means. Suppose $\mathbf{A}$ to be a square matrix with elements such as the partial derivatives of terminal conditions with respect to the independent search parameters. The most obvious technique of obtaining the matrix $\mathbf{A}$ is to make computer runs perturbing the necessary quantities. However, another technique of obtaining the $\mathbf{A}$ matrix has been used. This technique may be described as

$$\mathbf{A} = \mathbf{TVP}$$

where $\mathbf{T}$ relates the desired terminal parameters to position and velocity at the target body, $\mathbf{V}$ relates terminal position and velocity to earth injection conditions, and the $\mathbf{P}$ matrix relates earth injection conditions to the independent search parameters. This technique is useful in that the $\mathbf{P}$ matrix may be computed analytically, the $\mathbf{V}$ matrix may be obtained from variational equations, and the $\mathbf{T}$ matrix may be obtained analytically. The $\mathbf{V}$ matrix may also be obtained analytically, but much of the program flexibility is lost, e.g., midcourse velocity and error analyses options. This technique is superior to the perturbed run technique in obtaining partial derivatives in that each computer run (after the first) is a convergence pass. The latter technique reduces the required computer time per targeted trajectory to approximately one-third that required by the more laborious method.

In the initial phase, the independent variables are, in general, the guidance parameters of interest. Thus, it is necessary to describe conditions at earth injection into the transfer phase as functions of the guidance parameters of interest. The terminal phase may be a two-body representation of the desired terminal conditions as a function of position and velocity at the terminal condition. Individual and properly sized perturbations are then made in the independent parameters for the initial and terminal analytic representations in order to form average rates that may be formulated into the required sensitivity matrices $\mathbf{P}$ and $\mathbf{T}$. The sensitivity matrix $\mathbf{V}$ of position and velocity at the target body, with respect to earth injection position and velocity, is obtained from the variational equations. Thus, the three required sensitivity matrices are generated and combined to form the matrix $\mathbf{A}$ as just described.

Alternate Technique

The launch-to-mission completion targeting technique just discussed requires repetition of the ascent powered flight phase during the iteration procedure. The constraints placed on the ascent phase of the trajectory make this phase very nearly identical (for a given launch azimuth) from day to day. This suggests the possibility of eliminating the necessity for repeating this phase of the trajectory after it has once been defined for all launch azimuths of interest for one launch day. The technique of targeting one launch day in the normal manner, and initiating all other days at the parking orbit injection conditions for the given launch azimuths has been tested and proven feasible. This new approach eliminated the need for searching on the target vector and substituted the simpler search on liftoff time, parking orbit coast time, and injection energy. These quantities are then transformed into the guidance parameters representing the "target" vector.

Firing Tables

For each Ranger or Mariner mission, the primary purpose of targeting is to determine the ascent guidance computer target dependent constants, the parking orbit coast time and associated guidance and Agena timer constants, the Atlas booster autopilot roll program for an optimum pitch plane azimuth, and the velocity meter setting required for the final Agena burn. These parameters must be determined and mechanized for launch at any time during the available launch periods. These parameters are derived from the targeting procedure and presented in a set of firing tables which are actually a listing of trajectory and day dependent constants and a series of launch plans of approximately 10 min in length which contain time dependent constants.

Constants Mechanization

The targeting data, which represents almost exact targeted trajectories for discrete liftoff times, are appropriately linearized and quantized for each daily launch window. The guidance constants are, in general, represented as a "wired" (constants appropriately chosen and usually not changed for a given launch window or mission) constant part and a manual and/or a launch time dependent part. The Agena timer runoff time is chosen to be compatible with constants chosen for the ascent guidance computer to properly begin the Agena first and second burn, which controls the parking orbit altitude and coast time. The Agena second burn velocity meter setting is chosen such that, with a nominal second burn shutdown impulse and spacecraft separation impulse, the spacecraft will be injected into the proper transfer trajectory. The velocity meter setting is not a time dependent function. However, it is day dependent. The planned time of liftoff is an input to the guidance computer in the form of a Greenwich Mean Time just prior to liftoff, at which time the computer is turned on and counts down to the exact time of liftoff. If a long hold occurs, a new time to restart the computer can be input and the sequence repeated for the new planned liftoff time.

The initial pitch plane for the trajectory is set by the Atlas roll programmer. This launch azimuth is generally set at approximately 10-min intervals and corresponds to the optimum trajectory for the middle time of the 10-min launch plan. The ascent guidance corrects for the out-of-plane error by yaw steering for any other launch time during the particular launch plan. This approach is acceptable from a performance and guidance viewpoint.

The coast time is mechanized by input guidance constants that control the start time of an Agena timer during the Atlas phase and an Agena timer runoff time (also input). After the constants are set at the beginning of a launch interval, the ascent guidance equations control the timer initiation time and, thus, the parking orbit coast time. After about 10 min of window has been penetrated and the parking orbit coast time decreased by approximately 30 sec, a 30-sec increment is induced into the Agena timer runoff time, a similar (but oppositely directed) increment is input to the guidance computer, and the sequence is repeated. This procedure is continued until the vehicle is launched or until the window has elapsed. The firing table launch plan lengths of 10 min each are dictated by discrete timer runoff times in the Agena timer and a comparable period of the Atlas ascent trajectory during which the guidance functions are accurate enough to make good predictions of time-to-go, thereby, parking orbit coast time. The launch plans normally have a small overlap to allow for momentary holds during the countdown. However, if the hold is of sufficient length to place the predicted liftoff time in the next or some later plan, the manual settings must be made for that respective launch plan. In special situations, the launch conductor may call a hold to allow the proper launch plan settings to be made.

Thus, the liftoff time, launch azimuth, guidance target, and the parking orbit coast time are mechanized into the ascent guidance control systems. Additionally, velocity meter settings are made in the Agena stage termination control system. Integrating accelerometers then terminate the Agena first burn on the predetermined velocity meter setting required to place the vehicle into the required parking orbit. The vehicle then coasts the required amount and reignites and burns a second time until a second predetermined velocity meter setting terminates the powered flight when the vehicle is placed on the proper lunar transfer trajectory.

Accuracy

The targeting accuracy must be such (for any given guidance system) that the final firing tables are as accurate as required by the allotment of midcourse velocity to targeting. For the current system, lunar targeting is constrained to approximately 2% of the midcourse correction capability. This tolerance was relaxed to approximately 7% of the midcourse correction capability for the Mariner targeting because of the large window of more than 60 days (compared to approximately 6 days for lunar missions). This relaxation allowed the use of a constant target vector over the daily launch window and lowered the number of required discrete targeted trajectories to a reasonable number (1/day).

Concluding Remarks

Trajectory support of an operational lunar and interplanetary program such as Ranger/Mariner requires special computer programs and operational techniques in order to produce the vast quantities of precise trajectory data needed to adequately describe the missions. Analytic and variational methods may be utilized effectively to make targeting of lunar and interplanetary missions a relatively inexpensive operation. The complexity of the computer program logic required to mechanize the targeting procedures made the guidance equations an important subroutine in the complete targeting program rather than the major consideration, as has been the case in ICBM and earth satellite missions. This is best illustrated by the parking orbit targeting technique in which only one day utilizes the guidance equations directly, but, rather, external transfer functions are used to compute the required guidance constants to mechanize a launch-to-mission completion trajectory.

The techniques used in targeting Ranger/Mariner missions are general in concept and have been successfully applied to other missions such as Apollo and Gemini for mission design purposes. The targeting program search mechanism is quite general in the choice of dependent and independent variables, such that, the program may be applied easily, by input alone, to a new mission. In fact, trajectory shaping, injected weight optimization, and targeting to any computed set of terminal conditions (or any number of intermediate conditions) are possible in a single computer simulation by proper input to the program.

References

[1]Schwalm, N. E., "Boost vehicle trajectories for Ranger and Mariner programs," AIAA Preprint 63-409 (August 19-21, 1963).

[2]Clarke, V. C., Jr., "Trajectory design for Ranger and Mariner missions," AIAA Preprint 63-408 (August 19-21, 1963).

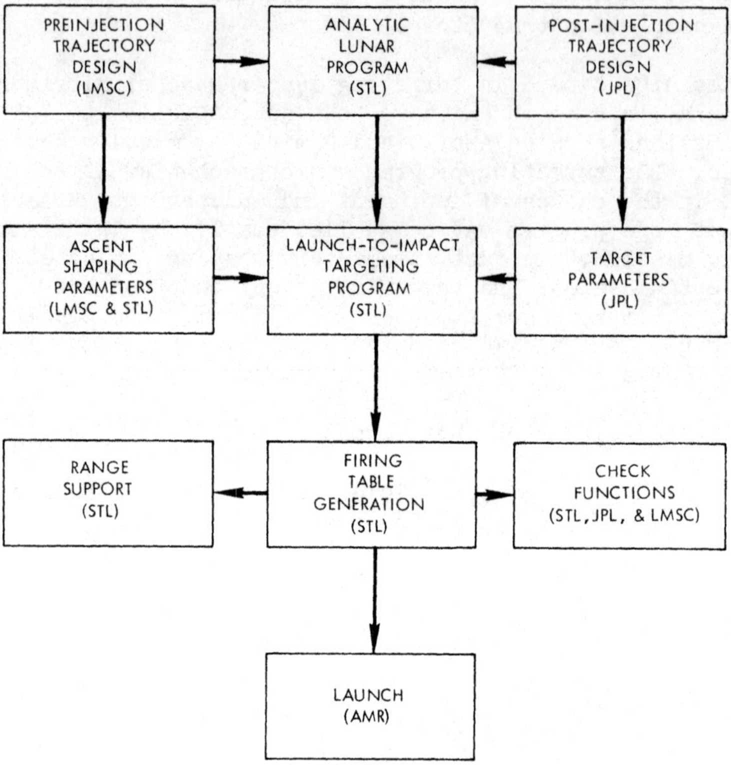

Fig. 1 Mission design, targeting, and launch support.

V.

ORBIT TRANSFER AND TRAJECTORY OPTIMIZATION

A TRAJECTORY OPTIMIZATION TECHNIQUE BASED UPON THE THEORY OF THE SECOND VARIATION

Henry J. Kelley*

Analytical Mechanics Associates, Inc., Uniondale, N.Y.

and

Richard E. Kopp† and H. Gardner Moyer‡

Grumman Aircraft Engineering Corporation, Bethpage, N.Y.

Abstract

A successive approximation method based upon the theory of the second variation is developed. In the early stage of computation, the process behaves much like the gradient/penalty function process with boundary conditions met only approximately. In the terminal stage, convergence more rapid than that of a gradient method is achieved with "exact" satisfaction of boundary conditions an integral part of the process. Since the equations of variation of the Euler-Lagrange equations are employed in the computational scheme, only slight additional effort is required to perform a check of the generalized Jacobi (Mayer) condition.

Introduction

Research in methods for numerically determining optimal trajectories has taken, in the main, two directions: study of steep descent processes in various versions,[1-8] and development of iterative solution schemes for the Euler-Lagrange equations.[9-19]

The strong points of steep descent processes are that convergence does not depend upon availability of a good initial estimate of the optimal trajectory as a starting point, and that they seek out weak relative minima as distinct from points at which the functional is merely stationary. The main weakness in practical applications is that convergence slows in the terminal phase of the process as the optimal trajectory is approached. As

---

Presented as Preprint 63-415 at the AIAA Astrodynamics Conference, New Haven, Conn., August 19-21, 1963. This research was sponsored by U.S. Air Force Office of Scientific Research, Applied Mathematics Division, under Contract AF49(638)-1207, and by NASA Manned Space Flight Center under Contract NAS9-1147.

*Vice-President.

†Section Head - Systems Research.

‡Systems Analyst.

with other methods, there is difficulty if either the Legendre-Clebsch condition or the generalized Jacobi condition is met only marginally, i.e., if the solution exhibits either singular subarcs or conjugate endpoints, and this appears in the form of convergence so poor that, practically speaking, the method fails. Except in these cases, an a posteriori check of the Weierstrass condition will establish the weak relative minimum obtained by the process as a strong relative minimum if the strengthened form of the condition is met.

Iterative solution of the Euler equations requires a good first estimate of multiplier initial conditions in order to converge at all. Convergence of the process is assured, theoretically, if the trial initial conditions are sufficiently close and if the solution is nonsingular with nonconjugate endpoints. The establishment of minimality requires separate checks of the Weierstrass and generalized Jacobi conditions. The latter is somewhat complex computationally and has been only rarely performed in practice. Convergence as a practical matter is troublesome, particularly so in the case of atmospheric flight of lifting vehicles.[9] Some success has been realized in the initial value iteration approach in the computation of optimal rocket trajectories in vacuum.[11-13] Even in this class of comparatively well-behaved Euler solutions, the main practical difficulty is in obtaining a trial solution whose end conditions approximate those desired. An attractive feature of the method is fast convergence in the terminal phase of the computation. A refinement of the method is the use of a linearized version of the Euler equations to obtain the elements of the transition matrix needed in the iteration.[14,15] In the work of Ref. 15, a separate computation via gradient method was employed to obtain the first estimate of the multiplier initial values. A different sort of iteration scheme tailored to "bang-bang" control problems is reported in Ref. 16. A method based upon the Euler-Lagrange equations and a generalization of Newton's method has been investigated in Refs. 17, 18, and 19, but has received little attention in trajectory applications. The relationship of this method to that which we discuss herein will be examined in the latter portion of the paper.

In the present paper, we present a successive approximation technique based upon the theory of the second variation. As with gradient methods, the initial trajectory estimate is required to be neither optimal nor necessarily a good approximation. In the initial phase of computation, the penalty function treatment of terminal conditions is employed and the behavior of the process strongly resembles that of a gradient/penalty function process as a result of step size constraints being operative which limit the amount of improvement sought during each cycle. These constraints are progressively relaxed, finally dropped, and the terminal penalty scheme discarded in favor of "exact" terminal conditions that are ultimately satisfied if a solution exists, in the sense that the specified conditions are attainable. This second phase of the process is computationally similar to iteration on the

Euler equations and shares the feature of fast terminal convergence.

Problem Formulation and Penalty Function Approximation

We begin with a statement of the trajectory optimization problem in the usual Mayer format. Given a system of first-order differential equations

$$\dot{x}_i = g_i(x_1, \ldots, x_n, y_1, \ldots, y_\ell, t) \tag{1}$$

it is required to find a solution of this system satisfying certain specified initial and terminal conditions and providing a minimum of some function $P = P(x_{1_f}, \ldots, x_{n_f}, t_f)$ of the terminal values of the variables x_i and the terminal time. The variables x_i, $i = 1, \ldots, n$, are state variables and the y_k, $k = 1, \ldots, \ell$, control variables. The latter may be subject to inequality constraints of the form

$$y_{k1} \leq y_k \leq y_{k2} \qquad k = 1, \ldots, \ell \tag{2}$$

as subsidiary conditions of the problem. We will deal primarily with the relatively simple case in which such inequality constraints are absent, adding some comments in the latter portion of the paper on the treatment of inequalities.

For simplicity of presentation, we will assume that all of the initial values of the x_i are fixed at a specified initial time t_o:

$$x_i(t_o) = \tilde{x}_{i_o} \qquad i = 1, \ldots, n \tag{3}$$

The terminal time t_f will be regarded as unspecified, which is more often the case than not in applications. The terminal values of the first m of the x_i will be taken as fixed:

$$x_i(t_f) = \tilde{x}_{i_f} \qquad i = 1, \ldots, m \tag{4}$$

and those of the remaining ones unspecified. Some or all of the terminal values of the x_i, $i = m + 1, \ldots, n$ may appear as arguments of the function P whose minimum is sought. The particular form of boundary conditions chosen here for definiteness is reasonably typical of problems arising in applications, and, in any case, modification of the ensuing analysis to accommodate other types of boundary conditions will present no essential difficulty.

An alternate formulation of the problem is given in terms of an approximation employing an augmented function of the terminal values:

$$P'(x_{1_f},\ldots,x_{n_f},t_f) = P(x_{m+1_f},\ldots,x_{n_f},t_f) + \frac{1}{2}\sum_{j=1}^{m} K_j(x_{j_f} - \tilde{x}_{j_f})^2 \tag{5}$$

A minimum of the function P' is to be sought without specification on the terminal values of the x_i. With $K_j > 0$, $j = 1, \ldots, m$, the second member of (5), which may be termed a "penalty function," will be positive if there are deviations from the desired terminal values $\tilde{x}_{j_f}$. If the K_j are chosen to be numerically large, it may be anticipated that a trajectory, optimal in the sense of minimizing P', will come close to meeting the desired terminal conditions, provided, of course, that these are attainable. One advantage of a penalty function treatment of terminal conditions is that a solution of the problem may be computed even though the desired terminal conditions are unattainable, i.e., even if no solution exists for the corresponding problem stated in terms of fixed terminal conditions. In such cases, the resulting solution, which fails to closely approximate the desired terminal conditions, may be of considerable value to the analyst in establishing physically reasonable terminal specifications for families of solutions, information which is available a priori only rarely. The basis, genesis, and application of the penalty function technique are discussed in Ref. 1.

An Expansion about a Reference Trajectory

In the classical theory of the Mayer problem, the constraints given by the differential equations (1) are adjoined to the functional P' by means of Lagrange multipliers:

$$J = P'(x_{1_f},\ldots,x_{n_f},t_f) + \int_{t_o}^{t_f} \sum_{i=1}^{n} \lambda_i(-\dot{x}_i + g_i)dt \tag{6}$$

and an expansion of J is performed in the neighborhood of a reference trajectory $\bar{x}_i(t)$, $\bar{y}_k(t)$:

$$J = J_o + J_1 + \tfrac{1}{2} J_2 + \ldots \tag{7}$$

Here, $J_o = P' (\bar{x}_{1_f}, \ldots, \bar{x}_{n_f}, \bar{t}_f)$, since the integrand in (6) vanishes along the reference trajectory, which is presumed to satisfy the system (1). J_1 and J_2 are, respectively, the collections of first- and second-order terms in the variations δx_i, δy_k of the state and control variables from those of the reference trajectory; they are known as the first and second variations. Since the analytical form of the second variation J_2 appearing in the classical literature corresponds to the case of a reference trajectory that satisfies the Euler-Lagrange equations of the problem as well as the system (1), the following derivation of the slightly more general form corresponding to a reference trajectory that satisfies only (1) is needed.

An expansion of the function P' in the neighborhood of the terminal point of the reference trajectory is given by

$$\begin{aligned} P' = P' (\bar{x}_{1_f}, \ldots, \bar{x}_{n_f}, \bar{t}_f) &+ \\ &\sum_{i=1}^{n} \frac{\partial P'}{\partial x_{i_f}} \Delta x_{i_f} + \frac{\partial P'}{\partial t_f} \delta t_f + \\ &\tfrac{1}{2} \sum_{i=1}^{n} \sum_{j=1}^{n} \frac{\partial^2 P'}{\partial x_{i_f} \partial x_{j_f}} \Delta x_{i_f} \Delta x_{j_f} + \\ &\sum_{i=1}^{n} \frac{\partial^2 P'}{\partial x_{i_f} \partial t_f} \Delta x_{i_f} \delta t_f + \\ &\tfrac{1}{2} \frac{\partial^2 P'}{\partial t_f^{\,2}} \delta t_f^{\,2} + \ldots \end{aligned} \tag{8}$$

Since we desire ultimately to obtain an approximation from (8) valid to second order in control variations $\delta y_k(t)$, evidently we must employ estimates of the state variable terminal increments Δx_{i_f} which are correct to second order in the control variations, at least in the first-order terms of (8). A first-order estimate of the variations in the state variables is given by the solution of a linearized version of (1):

$$\delta \dot{x}_i = \sum_{j=1}^{n} \frac{\partial g_i}{\partial x_j} \delta x_j + \sum_{k=1}^{\ell} \frac{\partial g_i}{\partial y_k} \delta y_k \tag{9}$$

$$i = 1, \ldots, n$$

$$\delta x_i(t_o) = 0 \tag{10}$$

in which the partial derivatives of the functions g_i are evaluated along the reference trajectory. An alternate and equivalent first-order estimate of δx_i at the terminal time of the reference trajectory $\bar{t}_f$ is given by

$$\delta x_i(\bar{t}_f) = \int_{t_o}^{\bar{t}_f} \sum_{k=1}^{\ell} \frac{\partial H_i}{\partial y_k} \delta y_k \, dt \tag{11}$$

The function $H_i \equiv \sum_{j=1}^{n} \lambda_j^{(i)} g_j$ is defined in terms of that solution $\lambda_j^{(i)}$ of the adjoint system

$$\dot{\lambda}_j = - \sum_{i=1}^{n} \lambda_i \frac{\partial g_i}{\partial x_j} \qquad j = 1, \ldots, n \tag{12}$$

which corresponds to the special boundary conditions

$$\begin{aligned} \lambda_j(\bar{t}_f) &= 1 \qquad & j = i \\ &= 0 \qquad & j \neq i \end{aligned} \tag{13}$$

A second-order estimate of the increment in x_i at $t = \bar{t}_f$ is given by the integral

$$\xi_i(\bar{t}_f) = \int_{t_o}^{\bar{t}_f} \left(\sum_{k=1}^{\ell} \frac{\partial H_i}{\partial y_k} \delta y_k + \omega_i \right) dt \tag{14}$$

where

$$2\omega_i \equiv \sum_{p=1}^{n} \sum_{q=1}^{n} \frac{\partial^2 H_i}{\partial x_p \, \partial x_q} \delta x_p \, \delta x_q + 2 \sum_{p=1}^{n} \sum_{k=1}^{\ell} \frac{\partial^2 H_i}{\partial x_p \, \partial y_k} \delta x_p \, \delta y_k + \sum_{k=1}^{\ell} \sum_{s=1}^{\ell} \frac{\partial^2 H_i}{\partial y_k \, \partial y_s} \delta y_k \, \delta y_s \tag{15}$$

The second member of the integrand of (14) utilizes the influence functions of (12) and (13) to obtain an estimate of the effects of the second-order terms in the g_i that were omitted in (9).

A corresponding estimate for the increment in x_i at a variable terminal time $t_f = \bar{t}_f + \delta t_f$ is given by

$$\Delta x_{i_f} = \xi_{i_f} + \left(\bar{g}_{i_f} + \sum_{j=1}^{n} \frac{\partial g_{i_f}}{\partial x_{j_f}} \delta x_{j_f} + \sum_{k=1}^{\ell} \frac{\partial g_{i_f}}{\partial y_{k_f}} \delta y_{k_f} \right) \delta t_f + \frac{1}{2} \left(\sum_{j=1}^{n} \frac{\partial g_{i_f}}{\partial x_{j_f}} \bar{g}_{j_f} + \sum_{k=1}^{\ell} \frac{\partial g_{i_f}}{\partial y_{k_f}} \dot{y}_{k_f} + \frac{\partial g_{i_f}}{\partial t_f} \right) \delta t_f^2 \tag{16}$$

in which the abbreviations $\delta x_{i_f} \equiv \delta x_i(\bar{t}_f)$, $\xi_{i_f} \equiv \xi_i(\bar{t}_f)$, $\bar{g}_{i_f} \equiv g_i(\bar{x}_{1_f}, \ldots, \bar{x}_{n_f}, \bar{y}_{1_f}, \ldots, \bar{y}_{\ell_f}, \bar{t}_f)$ are employed.

Substituting (16) into (8), and discarding terms of order higher than second, we then obtain the desired second-order approximation to P':

$$J_o + J_1 + \tfrac{1}{2} J_2 = P'(\bar{x}_{1_f},\ldots,\bar{x}_{n_f},\bar{t}_f) +$$

$$\sum_{i=1}^{n} \frac{\partial P'}{\partial x_{i_f}} \left[\xi_{i_f} + \left(g_{i_f} + \sum_{j=1}^{n} \frac{\partial g_{i_f}}{\partial x_{j_f}} \delta x_{j_f} + \right.\right.$$

$$\left. \sum_{k=1}^{\ell} \frac{\partial g_{i_f}}{\partial y_{k_f}} \delta y_{k_f} \right) \delta t_f \; \tfrac{1}{2} \left(\sum_{j=1}^{n} \frac{\partial g_{i_f}}{\partial x_{j_f}} \bar{g}_{j_f} + \right.$$

$$\left.\left. \sum_{k=1}^{\ell} \frac{\partial g_{i_f}}{\partial y_{k_f}} \dot{\bar{y}}_{k_f} + \frac{\partial g_{i_f}}{\partial t_f} \right) \delta t_f^{\,2} \right] + \frac{\partial P'}{\partial t_f} \delta t_f +$$

$$\tfrac{1}{2} \sum_{i=1}^{n} \sum_{j=1}^{n} \frac{\partial^2 P'}{\partial x_{i_f} \partial x_{j_f}} (\delta x_{i_f} + \bar{g}_{i_f} \delta t_f) \cdot (\delta x_{j_f} + \bar{g}_{j_f} \delta t_f) + \tag{17}$$

$$\sum_{i=1}^{n} \frac{\partial^2 P'}{\partial x_{i_f} \partial t_f} (\delta x_{i_f} + \bar{g}_{i_f} \delta t_f) \delta t_f +$$

$$\tfrac{1}{2} \frac{\partial^2 P'}{\partial t_f^{\,2}} \delta t_f^{\,2}$$

Here, δx_{i_f} and δt_f have been regarded as first-order quantities and the difference between ξ_{i_f} and δx_{i_f} as of second order.

It is of interest to relate this expression to the classical development in which the reference trajectory is an Euler solution, for the case of open terminal conditions on the state variables and the terminal time. The functions H and ω appearing in the classical development are

$$H = \sum_{i=1}^{n} \frac{\partial P'}{\partial x_{i_f}} H_i \tag{18}$$

$$\omega = \sum_{i=1}^{n} \frac{\partial P'}{\partial x_{i_f}} \omega_i \tag{19}$$

and the multiplier functions are corresponding linear combinations of the fundamental solutions $\lambda_j^{(i)}$ of the system (12) defined by the unit terminal conditions (13). If inequality constraints on the control variables y_k are absent, the Euler-Lagrange equations for the control variables are

$$\frac{\partial H}{\partial y_k} = 0 \qquad k = 1, \ldots, \ell \tag{20}$$

and certain terms in (17) which contain $\delta y_k(t)$, δy_{k_f}, and $\dot{y}_{k_f}$ disappear.

Successive Approximation Process

We consider the possibility of determining control increments $\delta y_k(t)$ that minimize the second-order approximation to P' given by (17). This variational problem is of Bolza form owing to the appearance of the integral

$$\sum_{i=1}^{n} \frac{\partial P'}{\partial x_{i_f}} \xi_{i_f} = \int_{t_o}^{\bar{t}_f} \left(\frac{\partial H}{\partial y_k} \delta y_k + \omega\right) dt \tag{21}$$

in the expression (17). As subsidiary conditions of the variational problem, we have the system of differential equations (9) which defines the $\delta x_i(t)$. If the reference trajectory satisfies the specified initial conditions, which we shall assume to be the case, the appropriate initial conditions on the δx_i are $\delta x_i(t_o) = 0$. Terminal conditions on the δx_i are unspecified, as is the increment in terminal time δt_f. The quadratic/linear format of this Bolza variational problem is computationally attractive, and this provides a primary motivation for the approach to the successive approximation process presently under consideration.

Adjoining the differential constraints (9) with Lagrange multipliers $\delta\lambda_i$, $i = 1, \ldots, n$, we proceed to write the Euler-Lagrange equations, the Weierstrass necessary condition, and the transversality conditions for the problem. The Euler-Lagrange equations corresponding to the state variables are:

$$\delta\dot{\lambda}_i = - \sum_{j=1}^{n} \delta\lambda_j \frac{\partial g_j}{\partial x_i} - \frac{\partial}{\partial \delta x_i} \omega \tag{22}$$

$$i = 1, \ldots, n$$

The Weierstrass necessary condition takes the form

$$h(\delta y_1^*, \ldots, \delta y_\ell^*) \geq h(\delta y_1, \ldots, \delta y_\ell) \tag{23}$$

in which the δy_k^*, $k = 1, \ldots, \ell$, are arbitrary. The function h is given by

$$h = \sum_{i=1}^{n} \delta\lambda_i \left(\sum_{j=1}^{n} \frac{\partial g_i}{\partial x_j} \delta x_j + \sum_{k=1}^{\ell} \frac{\partial g_i}{\partial y_k} \delta y_k \right) + \sum_{k=1}^{\ell} \frac{\partial H}{\partial y_k} \delta y_k + \omega \tag{24}$$

Owing to the linear/quadratic form taken by h, the Weierstrass necessary condition is equivalent to the Legendre-Clebsch necessary condition and the Euler equations for the δy_k. This would not be the case if considerations included inequality constraints on the control variables.

The transversality conditions corresponding to open δx_{j_f} are

$$\sum_{i=1}^{n} \frac{\partial P'}{\partial x_{i_f}} \frac{\partial g_{i_f}}{\partial x_{j_f}} \delta t_f + \sum_{i=1}^{n} \frac{\partial^2 P'}{\partial x_{i_f} \partial x_{j_f}} (\delta x_{i_f} + \bar{g}_{i_f} \delta t_f) + \frac{\partial^2 P'}{\partial x_{j_f} \partial t_f} \delta t_f - \delta\lambda_{j_f} = 0 \qquad j = 1, \ldots, n \tag{25}$$

and to open δt_f

$$\sum_{i=1}^{n} \frac{\partial P'}{\partial x_{i_f}} \bar{g}_{i_f} + \frac{\partial P'}{\partial t_f} + \sum_{i=1}^{n} \frac{\partial P'}{\partial x_{i_f}} \left[\left(\sum_{j=1}^{n} \frac{\partial g_{i_f}}{\partial x_{j_f}} \delta x_{j_f} + \sum_{k=1}^{\ell} \frac{\partial g_{i_f}}{\partial y_{k_f}} \delta y_{k_f} \right) + \right.$$

$$\left(\sum_{j=1}^{n} \frac{\partial g_{i_f}}{\partial x_{j_f}} \bar{g}_{j_f} + \sum_{k=1}^{\ell} \frac{\partial g_{i_f}}{\partial y_{k_f}} \dot{\bar{y}}_{k_f} + \frac{\partial g_{i_f}}{\partial t_f}\right) \delta t_f \Big] +$$

$$\sum_{i=1}^{n} \sum_{j=1}^{n} \frac{\partial^2 P'}{\partial x_{i_f} \partial x_{j_f}} \bar{g}_{j_f} (\delta x_{i_f} + \bar{g}_{i_f} \delta t_f) + \tag{26}$$

$$\sum_{i=1}^{n} \frac{\partial^2 P'}{\partial x_{i_f} \partial t_f} (\delta x_{i_f} + 2\bar{g}_{i_f} \delta t_f) + \frac{\partial^2 P'}{\partial t_f^{\,2}} \delta t_f = 0$$

The analysis of our successive approximation process bears a resemblance to that of the classical accessory minimum problem for the second variation. In the classical analysis, the reference trajectory satisfies the Euler-Lagrange equations and the transversality conditions for the problem of minimizing P'; the requirement of positive semidefiniteness for the second variation suggests the problem of minimizing the second variation, the so-called accessory minimum problem. The analysis leads to the Legendre-Clebsch and generalized Jacobi (Mayer) necessary conditions. A main feature of the analysis, in the absence of inequality constraints on the control variables, is that the Euler-Lagrange equations and transversality conditions of the accessory problem are precisely linearized versions of those for the original problem of minimizing P'. Such is also the case in the present analysis, with the slight but important difference that certain zero-order terms remain in the linearized expressions due to the nonoptimality of the reference trajectory. A somewhat analogous approach, employing an optimal reference trajectory, has been taken in Ref. 20 in connection with an optimal guidance approximation scheme.

In the present application - the determination of an optimal trajectory through successive improvements on a nonoptimal reference trajectory - there is a question concerning the existence of a minimum of (17) and a related question concerning the convergence of the process. If the reference trajectory were close to the optimal trajectory sought, existence and convergence arguments of sorts could be built around this fact. Such a requirement on the reference trajectory chosen as a starting point for the computational process would, however, obviously represent an undesirable restriction. On the other hand, if the reference trajectory satisfies only the basic system (1) and the initial conditions, but is otherwise arbitrary, there is no assurance that a minimum of (17) exists, and, in fact, it will commonly be the case that (17) is unbounded below. If, for example, the function h given by (24) has no minimum in the δy_k, the question of existence of a minimum of the approximation is settled in the negative. Thus, at least in the early phase of the computational process, it appears necessary to introduce restrictions on the "step size" as measured in terms of the norms of the functions $\delta y_k(t)$. We therefore alter the problem by the introduction of additional subsidiary conditions given by the equations

$$\delta \dot{x}_{n+k} = \tfrac{1}{2}\,\delta y_k^{\,2} \qquad \delta x_{n+k}(t_o) = 0 \qquad k = 1, \ldots, \ell \tag{27}$$

defining variables δx_{n+k} whose terminal values $\delta x_{n+k}(\bar{t}_f)$ are integral square measures of the magnitudes of the control variable increments $\delta y_k(t)$.

With the constraints (27) adjoined by means of additional multipliers $\delta\lambda_i$, $i = n+1, \ldots, n+\ell$, the analysis proceeds as before, and the Euler-Lagrange equations (22) and the transversality conditions (25) and (26) are unchanged. The Euler-Lagrange equations corresponding to the variables δx_{n+k}, $k = 1, \ldots, \ell$ are

$$\delta\dot{\lambda}_{n+k} = 0 \qquad\qquad k = 1, \ldots, \ell \tag{28}$$

indicating the constancy of these multipliers. The Weierstrass necessary condition is given by

$$\hat{h}(\delta y_1^{\,*}, \ldots, \delta y_\ell^{\,*}) \geq \hat{h}(\delta y_1, \ldots, \delta y_\ell) \tag{29}$$

in which the $\delta y_k^{\,*}$ are arbitrary, and the function $\hat{h}$ is given by

$$\hat{h} = \sum_{i=1}^{n} \delta\lambda_i \left(\sum_{j=1}^{n} \frac{\partial g_i}{\partial x_j}\,\delta x_j + \sum_{k=1}^{\ell} \frac{\partial g_i}{\partial y_k}\,\delta y_k \right) + \sum_{k=1}^{\ell} \frac{\partial H}{\partial y_k}\,\delta y_k + \omega + \tfrac{1}{2} \sum_{k=1}^{\ell} \delta\lambda_{n+k}\,\delta y_k^{\,2} \tag{30}$$

the last member arising from the additional constraints (27).

The Weierstrass necessary condition (29) provides information of value in the choice of the constraint multipliers $\delta\lambda_i$, $i = n+1, \ldots, n+\ell$, this choice being equivalent to the establishment of the step size parameters $\delta x_i(t_f)$, $i = n+1, \ldots, n+\ell$. In the case of an unbounded control variable y_s, for example, a requirement for the function $\hat{h}$ to possess a minimum is

$$\frac{\partial^2 \hat{h}}{\partial \delta y_s^{\,2}} = \frac{\partial^2 H}{\partial y_s^{\,2}} + \delta\lambda_{n+s} \geq 0 \tag{31}$$

and if $\partial^2 H/\partial y_s^{\,2}$ takes on negative values along the reference trajectory, a $\delta\lambda_{n+s} > 0$ at least large enough in magnitude to satisfy (31) will be required to satisfy (29). More generally, the $\delta\lambda_i$, $i = n+1, \ldots, n+\ell$, must be chosen at least large enough in magnitude to insure that the function $\hat{h}$ possesses a minimum.

If the multipliers $\delta\lambda_i$, $i = n+1, \ldots, n+\ell$, are assigned large positive values, corresponding to a restriction to very small step size, the successive approximation process described by our analysis becomes a gradient process. In this case, the process would ultimately approach a weak relative minimum of P', provided that a minimum exists, since its nature is not such as to seek out stationary points of nonminimal character, and the generalized Jacobi necessary condition would automatically be satisfied.

As a practical matter, it seems appropriate to choose values for these multipliers somewhat larger than necessary to satisfy the Weierstrass condition (29), but not so large as to adversely affect the speed of convergence of the process. A conservative, but computationally expensive procedure would be to perform the generalized Jacobi test for the problem of minimizing the approximation (17) at each step of the process, thus insuring that the step-size multipliers have been chosen large enough to exclude generalized conjugate points from the interval $t_o < t \leq \bar{t}_f$. A more practical procedure, having an element of a gamble, would be merely to check at each step whether or not a decrease in P' has been realized, and to perform the generalized Jacobi test only on the specimen finally obtained after the process has converged. We will discuss the Jacobi test procedure subsequently.

A point neglected in the preceding analysis is the determination of control increments at the terminal point of the reference trajectory $\delta y_{k_f} = \delta y_k(\bar{t}_f)$ which enter the expression (17) whose minimum is sought and which appear consequently in the transversality condition (26). If continuity were required of the control variables y_k, and hence of the δy_k, the control increments at $t = \bar{t}_f$ would be determined by the operation $\min\limits_{\delta y_k} \hat{h}$ just as at interior points of the interval $t_o \leq t \leq \bar{t}_f$. The introduction of such a continuity requirement at the terminal point is a feasible, if rather arbitrary, means of handling the matter. If, on the other hand, the δy_{k_f} are regarded as free of choice, it follows that they must be chosen so as to provide a minimum of (17). Since the δy_{k_f} appear linearly in (17) with coefficients

$$\sum_{i=1}^{n} \frac{\partial P'}{\partial x_{i_f}} \frac{\partial g_{i_f}}{\partial y_{k_f}} \delta t_f = \frac{\partial H_f}{\partial y_{k_f}} \delta t_f$$

it is possible that no such minimum exists. This situation will definitely arise in the case of a control variable y_s, which is not subject to an inequality constraint of the form (2) if

$$\frac{\partial H_f}{\partial y_{s_f}} \neq 0 \qquad \text{and} \qquad \delta t_f \neq 0.$$

Such considerations suggest the possibility that the control variables chosen for the reference trajectory should not be completely arbitrary but rather should be taken such as to minimize

$$\sum_{i=1}^{n} \frac{\partial P'}{\partial x_i} g_i$$

in the vicinity of the terminal point. The course of action adopted is probably not of key importance computationally, since the effect is local, and

$$H_f = \sum_{i=1}^{n} \frac{\partial P'}{\partial x_{i_f}} g_{i_f}$$

will rapidly approach a minimum in the y_{k_f} in any case.

Computational Procedure for the Penalty Function Process

A possible sequence of calculations is the following:

1) Integrate numerically the system (1) employing the given initial conditions and stored first estimates of the control variables $\bar{y}_k(t)$.

2) Terminate the trajectory at a time $\bar{t}_f$ determined so that P' regarded as a function of t_f along the trajectory attains a minimum, and hence that $\frac{d}{dt_f} P' = 0$. This technique has previously been employed in the gradient/penalty function process.[1,2]

3) Integrate the adjoint system (12) backwards in time employing terminal values $\lambda_i(\bar{t}_f) = \frac{\partial P'}{\partial x_{i_f}}$. Store the initial values $\lambda_i(t_o)$. Calculate the coefficients appearing in the function h (24) during this integration, and select the $\delta\lambda_i$, $i = n+1, \ldots, n+\ell$.

4) Generate by numerical integration a matrix solution of the combined system (9) and (22) with δy_k determined by the operation $\min_{\delta y_k} \hat{h}$, i.e., perform n integrations concurrently with unit matrix initial values of the $\delta\lambda_i$, $i = 1, \ldots, n$, and all δx_i initial values zero. Regenerate the reference trajectory and the adjoint solution concurrently for the purpose of calculating the coefficients of the combined system.

5) By linear algebraic operations, determine the initial values of the $\delta\lambda_i$, $i = 1, \ldots, n$, and the value of δt_f that satisfy the transversality conditions (25) and (26).

6) Employ the $\delta\lambda_i$ initial values so determined for another integration of the combined system (9) and (22). Add the δy_k generated in this solution to the stored $\bar{y}_k$. This furnishes the control functions for a new reference solution.

7) Repeat step 1, starting a new cycle. Compare the value of P' obtained in step 2 of the new cycle with the previous value of P'. The process repeats until decrements in P' become small.

Refinement Process

On account of the penalty function approximation, the process described will converge to a solution whose terminal state variable values differ from those prescribed. For large positive values of the penalty constants K_j, the differences will be small unless the prescribed terminal values are unattainable. The penalty function formulation of the problem serves its purpose in permitting a determination of whether or not this is the case and by providing a scalar measure of convergence - the decrement in P'.

In examining the refinement process described in the following, we assume that the penalty function process has converged closely enough to a minimum that the step size constraints (27) are no longer necessary and that the adjoint variables approximate the multiplier functions of the fixed endpoint problem.

We adopt an expansion for the function P similar to that given by (8) for the function P':

$$P = P(\bar{x}_{m+1_f}, \ldots, \bar{x}_{n_f}, t_f) + \sum_{i=m+1}^{n} \frac{\partial P}{\partial x_{i_f}} \Delta x_{i_f} + \frac{\partial P}{\partial t_f} \delta t_f +$$

$$\frac{1}{2} \sum_{i=m+1}^{n} \sum_{j=m+1}^{n} \frac{\partial^2 P}{\partial x_{i_f} \partial x_{j_f}} \Delta x_{i_f} \Delta x_{j_f} + \sum_{i=m+1}^{n} \frac{\partial^2 P}{\partial x_{i_f} \partial t_f} \Delta x_{i_f} \delta t_f + \tag{32}$$

$$\frac{1}{2} \frac{\partial^2 P}{\partial t_f^2} \delta t_f^2 + \ldots$$

The fixed state variable terminal conditions will be

$$\bar{x}_{i_f} + \Delta x_{i_f} - \tilde{x}_{i_f} = 0 \qquad i = 1, \ldots, m \tag{33}$$

Here the $\tilde{x}_{i_f}$ are specified values, the $\bar{x}_{i_f}$ those of the reference trajectory obtained via the penalty function version of the process, and the Δx_{i_f} are the second-order approximations to terminal value increments given by (16). The constraints (33) may be adjoined to (32) by means of additional multipliers μ_i, $i = 1, \ldots, m$, and an approximation sought to the augmented expression which is valid to second order. Approximating the μ_i to zero and first order terms as

$$\mu_i = \bar{\mu}_i + \delta\mu_i \qquad i = 1, \ldots, m \tag{34}$$

and taking μ_i for the first refinement cycle as the terminal value of the adjoint variable λ_i obtained in the penalty function approximation

$$\bar{\mu}_i = \frac{\partial P'}{\partial x_{i_f}} \qquad i = 1, \ldots, m \tag{35}$$

we obtain the desired second-order approximation to P as

$$J_o + J_1 + \tfrac{1}{2} J_2 = P(\bar{x}_{1_f}, \ldots, \bar{x}_{n_f}, \bar{t}_f) + \sum_{i=m+1}^{n} \frac{\partial P}{\partial x_{i_f}} \Big[\xi_{i_f} + \big(g_{i_f} +$$

$$\sum_{j=1}^{n} \frac{\partial g_{i_f}}{\partial x_{j_f}} \delta x_{j_f} + \sum_{k=1}^{\ell} \frac{\partial g_{i_f}}{\partial y_{k_f}} \delta y_{k_f}\Big) \delta t_f + \tfrac{1}{2}\Big(\sum_{j=1}^{n} \frac{\partial g_{i_f}}{\partial x_{j_f}} \bar{g}_{j_f} +$$

$$\sum_{k=1}^{\ell} \frac{\partial g_{i_f}}{\partial y_{k_f}} \dot{y}_{k_f} + \frac{\partial g_{i_f}}{\partial t_f}\Big) \delta t_f^{2}\Big] + \frac{\partial P}{\partial t_f} \delta t_f + \tfrac{1}{2} \sum_{i=m+1}^{n} \sum_{j=m+1}^{n} \frac{\partial^2 P}{\partial x_{i_f} \partial x_{j_f}} \times$$

$$\left(\delta x_{i_f} + \bar{g}_{i_f} \delta t_f\right) \left(\delta x_{j_f} + \bar{g}_{j_f} \delta t_f\right) +$$

$$\sum_{i=m+1}^{n} \frac{\partial^2 P}{\partial x_{i_f} \partial t_f} \left(\delta x_{i_f} + \bar{g}_{i_f} \delta t_f\right) \delta t_f + \tfrac{1}{2} \frac{\partial^2 P}{\partial t_f^{2}} \delta t_f^{2} +$$

$$\sum_{i=1}^{m} \bar{\mu}_i \Big[\xi_{i_f} + \Big(\bar{g}_{i_f} + \sum_{j=1}^{n} \frac{\partial g_{i_f}}{\partial x_{j_f}} \delta x_{j_f} + \sum_{k=1}^{\ell} \frac{\partial g_{i_f}}{\partial y_{k_f}} \delta y_{k_f}\Big) \delta t_f + \qquad (36)$$

$$\tfrac{1}{2}\Big(\sum_{j=1}^{n} \frac{\partial g_{i_f}}{\partial x_{j_f}} \bar{g}_{j_f} + \sum_{k=1}^{\ell} \frac{\partial g_{i_f}}{\partial y_{k_f}} \bar{y}_{k_f} + \frac{\partial g_{i_f}}{\partial t_f}\Big) \delta t_f^{2} + \bar{x}_{i_f} - \tilde{x}_{i_f}\Big] +$$

$$\sum_{i=1}^{m} \delta\mu_i \left(\delta x_{i_f} + \bar{g}_{i_f} \delta t_f + \bar{x}_{i_f} - \tilde{x}_{i_f}\right)$$

In this second-order approximation, δx_{i_f} and δt_f have been regarded as of first order and the difference between ξ_{i_f} and δx_{i_f} as of second order.

The transversality conditions corresponding to δx_{j_f} are

$$\sum_{i=m+1}^{n} \frac{\partial P}{\partial x_{i_f}} \frac{\partial g_{i_f}}{\partial x_{j_f}} \delta t_f + \sum_{i=1}^{m} \bar{\mu}_i \frac{\partial g_{i_f}}{\partial x_{j_f}} \delta t_f + \delta\mu_j - \delta\lambda_{j_f} = 0$$

$$j = 1, \ldots, m \qquad (37)$$

$$\sum_{i=m+1}^{n} \frac{\partial P}{\partial x_{i_f}} \frac{\partial g_{i_f}}{\partial x_{j_f}} \delta t_f + \sum_{i=1}^{m} \bar{\mu}_i \frac{\partial g_{i_f}}{\partial x_{j_f}} \delta t_f +$$

$$\sum_{i=m+1}^{n} \frac{\partial^2 P}{\partial x_{i_f} \partial x_{j_f}} (\delta x_{i_f} + \bar{g}_{i_f} \delta t_f) +$$

$$\frac{\partial^2 P}{\partial x_{j_f} \partial t_f} \delta t_f - \delta \lambda_{j_f} = 0 \qquad j = m+1, \ldots, n \tag{38}$$

to δt_f

$$\sum_{i=m+1}^{n} \frac{\partial P}{\partial x_{i_f}} \left[\bar{g}_{i_f} + \sum_{j=1}^{n} \frac{\partial g_{i_f}}{\partial x_{j_f}} \delta x_{j_f} + \sum_{k=1}^{\ell} \frac{\partial g_{i_f}}{\partial y_{k_f}} \delta y_{k_f} + \left(\sum_{j=1}^{n} \frac{\partial g_{i_f}}{\partial x_{j_f}} \bar{g}_{j_f} + \right.\right.$$

$$\left.\left. \sum_{k=1}^{\ell} \frac{\partial g_{i_f}}{\partial y_{k_f}} \dot{\bar{y}}_{k_f} + \frac{\partial g_{i_f}}{\partial t_f} \right) \delta t_f \right] + \frac{\partial P}{\partial t_f} + \sum_{i=m+1}^{n} \sum_{j=m+1}^{n} \frac{\partial^2 P}{\partial x_{i_f} \partial x_{j_f}} (\delta x_{i_f} + \bar{g}_{i_f} \delta t_f) \bar{g}_{j_f} +$$

$$\sum_{i=m+1}^{n} \frac{\partial^2 P}{\partial x_{i_f} \partial t_f} (\delta x_{i_f} + 2\bar{g}_{i_f} \delta t_f) + \frac{\partial^2 P}{\partial t_f^2} \delta t_f +$$

$$\sum_{i=1}^{m} \bar{\mu}_i \left[\bar{g}_{i_f} + \sum_{j=1}^{n} \frac{\partial g_{i_f}}{\partial x_{j_f}} \delta x_{j_f} + \right. \tag{39}$$

$$\left. \sum_{k=1}^{\ell} \frac{\partial g_{i_f}}{\partial y_{k_f}} \delta y_{k_f} + \left(\sum_{j=1}^{n} \frac{\partial g_{i_f}}{\partial x_{j_f}} \bar{g}_{j_f} + \sum_{k=1}^{\ell} \frac{\partial g_{i_f}}{\partial y_{k_f}} \dot{\bar{y}}_{k_f} + \frac{\partial g_{i_f}}{\partial t_f} \right) \delta t_f \right] +$$

$$\sum_{i=1}^{m} \delta \mu_i \bar{g}_{i_f} = 0$$

and to $\delta \mu_j$

$$\delta x_{j_f} + \bar{g}_{j_f} \delta t_f + \bar{x}_{j_f} - \tilde{x}_{j_f} = 0 \qquad j = 1, \ldots, m \tag{40}$$

Examination of these relationships establishes that they are linearized versions of the transversality conditions for the original problem. In the case of (39) it is convenient, computationally, to eliminate the $\delta\mu_j$ employing (37), and introduce the $\delta\lambda_{j_f}$ from (38) where appropriate. Since the $\delta\mu_j$ do not appear elsewhere, (37) can then be dropped.

Computational Procedure for the Refinement Process

The procedure for refinement process calculations is generally quite similar to that sketched earlier for the penalty function process, with the following exceptions. Trajectory termination time t_f is taken to be $t_f + \delta t_f$ of the preceding cycle. The adjoint system solution is integrated numerically forward in time, with initial values $\bar{\lambda}_{i_o} + \delta\lambda_{i_o}$, $i = 1, \ldots, n,$ of the preceding cycle. Terminal conditions of the equations of variation are (37-40).

Generalized Jacobi Test

The generalized Jacobi test may be applied to the solution obtained after convergence of the process with only slight additional computational effort, since it requires the equations of variation of the Euler-Lagrange equations and the same transversality conditions employed in the successive approximation process. The version of the test dealt with here is applicable only to normal nonsingular extremals. It should be mentioned that a similar restriction applies to the successive approximation process itself. The process fails in the case of a singular subarc appearing in the solution due to the indeterminacy of the Weierstrass condition, and it requires modification in the case of abnormality phenomena.

The generalized Jacobi test matrix is the matrix whose inverse is required in the computation of the $\delta\lambda_{i_o}$, $i = 1, \ldots, n,$ and δt_f satisfying, through the equations of variation of the Euler-Lagrange equations (22) and the equations of variation of the basic system (9), the terminal conditions (37-40). Values of the independent variable $t^* > t_o$ at which the matrix becomes singular determine generalized conjugate points. The generalized Jacobi necessary condition is the requirement that there exist no such points in the interior of the interval $t_o < t < t_f$.

For computational test purposes, a succession of times coinciding with values employed in numerical integration are regarded, each in turn, as terminal points, the elements of the test matrix evaluated just as in the successive approximation process, and the determinant computed.

The vanishing of the determinant along the trajectory

indicates the existence of a nontrivial solution of the combined equations of variation that satisfies the linearized version of the terminal conditions. According to the usual argument, the nontrivial solution exhibits a corner at the point t^*, at which point the Weierstrass-Erdmann corner conditions are not met, and hence the zero value of the second variation given by the nontrivial solution cannot be the minimum value. The possibility of a negative second variation if $t_o < t^* < t_f$ disqualifies the test extremal as a candidate for a minimizing arc. The generalized Jacobi condition, called the condition of Mayer in the stronger form appropriate to sufficiency proofs, is treated in Refs. 21 and 22.

Treatment of Inequality Constraints on Control Variables

If inequality constraints of the form (2) are operative, an analysis similar to that preceding may be carried out. The difference is that the minimum operation on the function $\hat{h}$ given by (30) is subject to inequality constraints on the δy_k derived directly from (2) and the control functions $\bar{y}_k(t)$ of the reference trajectory. In this case, the solution of the two-point boundary problem for the minimum of the approximation (17) or (36) cannot be carried out by linear operations, and an important advantage is lost.

In flight performance applications, an inequality constraint on a control variable is usually associated with the appearance of the control variable in the basic system (1) linearly. If the optimal control is bang-bang, it will usually be advantageous to deal with switching times as control parameters, in which case linear methods may then be employed for treatment of the two-point boundary problem arising in the successive approximation process. If not, the occurrence of a singular subarc in the solution is implied and existing numerical schemes fail.

Relationship to Other Computational Techniques

The present scheme has similarities to two existing techniques based upon the Euler-Lagrange equations. The first of these is the generalization of Newton's method studied in Refs. 17-19 but never applied numerically to trajectory problems, to the writers' knowledge. The main difference is in the penalty function approximation and in the use of the step-size constraints employed in the present method to insure the satisfaction of the Legendre-Clebsch necessary condition. It should be noted that the straightforward use of Newton's method may yield a process in which the function h of (24) is maximized rather than minimized over certain intervals, and that the trajectory obtained by the converged process may consequently fail to satisfy the Legendre-Clebsch condition, and hence not furnish a minimum.

The refinement process presently described would be similar to the methods of Refs. 14 and 15 if each new reference trajectory were obtained via the Euler-Lagrange equations, with the solution of the two-point boundary value problem employed only for computation of new multiplier initial values. The limitation of such a technique is that a good first estimate of multiplier initial values is required for convergence. In the present method, the equivalent of this is obtained via the penalty function version of the process which produces the first approximation for the refinement process.

Low Thrust Example

To illustrate this second-order computational technique and compare it with the first-order gradient method, a constant low-thrust transfer between coplanar circular orbits has been calculated. This particular problem, an Earth to Mars transfer, was treated by the present authors in Ref. 2 using first-order steep descent theories. The system of equations governing the motion was given by:

Radial Acceleration

$$\dot{u} = g_1 = \frac{v^2}{R} - A_o\left(\frac{R_o}{R}\right)^2 + \frac{T \sin \theta}{m_o - T/V_e \, t} \tag{41}$$

Circumferential Acceleration

$$\dot{v} = g_2 = -\frac{uv}{R} + \frac{T \cos \theta}{m_o - T/V_e \, t} \tag{42}$$

Radial Velocity

$$\dot{R} = g_3 = u \tag{43}$$

All the initial and final values of the state variables were specified and the transfer time was to be minimized, i.e.,

$$P = t_f \tag{44}$$

The second-order descent process was coded for the IBM 7094 computer. A modified Adams numerical integration scheme was used. The integration step was fixed at two days, dividing most of the trajectories into approximately one hundred intervals.

For the purpose of coding the first-stage penalty function process (but not the second-stage refinement process) the complexity of the numerical calculation was greatly reduced by setting δt_f to zero. Thus, (17) simplified to

$$J_o + J_1 + \tfrac{1}{2} J_2 = P' + \sum_{i=1}^{n} \frac{\partial P'}{\partial x_{i_f}} \xi_{i_f} + \tag{45}$$

$$\frac{1}{2}\sum_{i=1}^{n}\sum_{j=1}^{n}\frac{\partial^2 P'}{\partial x_{i_f}\,\partial x_{j_f}}\,\delta x_{i_f}\,\delta x_{j_f}$$

The control variable increment $\delta\theta(t)$ was then calculated to obtain the maximum reduction in P' at time t_f. Of course when the trajectory with control variable $\theta + \delta\theta$ was computed, it was not terminated at time t_f but rather at the point of the trajectory with minimum P'. With respect to over-all computational time, this technique represents a compromise between a true second variation calculation and additional programming complexity. For this particular problem, it was advantageous to treat the problem as a fixed time problem when computing $\delta\theta$ using penalty functions.

The initial $\theta(t)$ function corresponded to constant circumferential thrust. This resulted in terminal boundary value errors that averaged 20%. After 6 descent cycles, using the penalty function procedure, the terminal errors averaged 3% with the transfer time at 180 days. After 5 additional cycles of the refinement process, the average boundary value error was reduced to 0.05% and the transfer time had reached its minimum of 193 days. The over-all computer time was two min., thus representing half the computer time required by the first-order gradient program.

Conclusions

The second variation trajectory optimization method described in this paper is appreciably more complicated than the first-order gradient theory. It appears, however, to be economical in the sense of computing time when many optimal trajectories are to be computed. In addition, the generalized Jacobi test may be applied with only slight additional computational effort. As with any computational approach to the solution of optimal trajectories, the effectiveness of the method rests with its judicious application.

References

[1]Kelley, H.J., "Method of gradients," Optimization Techniques, edited by G. Leitmann (Academic Press Inc., New York, 1962), Chap. 6.

[2]Kelley, H.J., Kopp, R.E., and Moyer, H.G., "Successive approximation techniques for trajectory optimization," 1961 IAS Vehicle Systems Optimization Symposium (Institute of Aerospace Sciences, New York, 1961), pp. 10-25.

[3]Bryson, A.E., Denham, W.F., Carroll, F.J., and Mikami, K., "Determination of lift or drag programs to minimize re-entry heating," . Aerospace Sci. 29, 420-430 (March 1962).

[4]Ho, Y.C., "A successive approximation technique for optimal control systems subject to input constraints," 1961 Joint Automatic Control Conference sponsored by the Instrument Society of America and others, Boulder, Colo. (June 28-30, 1961).

[5]Ho, Y.C., "A computational technique for optimal control problems with state variable constraints," J. Math. Anal. Appl. 5, 216-224 (October 1962).

[6]Bryson, A.E. and Denham, W., "A steepest-ascent method for solving optimum programming problems," Trans. ASME - J. Appl. Mech. 84(3), 247-257 (June 1962).

[7]Dreyfus, S., "Variational problems with inequality constraints," J. Math. Anal. Appl. 4, 297-308 (April 1962).

[8]Dreyfus, S., "Numerical solution of variational problems," J. Math. Anal. Appl. 5, 30-45 (August 1962).

[9]Mengel, A.S., "Optimum trajectories," Proceedings of Project Cyclone Symposium 1 on REAC Techniques, Reeves Instrument Corp. and U.S. Navy Special Devices Center, New York (March 15-16, 1951), pp. 7-13.

[10]Breakwell, J.V., "The optimization of trajectories," Soc. Indust. Appl. Math. 7, 215-248 (June 1959).

[11]MacKay, J.S., Rossa, L.G., and Zimmerman, A.V., "Optimum low-acceleration trajectories for Earth-Mars transfer," 1961 IAS Vehicle Systems Optimization Symposium (Institute of Aerospace Sciences, New York, 1961) pp. 157-167.

[12]Melbourne, W.G., Sauer, C.G., Jr., and Richardson, D.E., "Interplanetary trajectory optimization with power-limited propulsion systems," 1961 IAS Vehicle Systems Optimization Symposium (Institute of Aerospace Sciences, New York, 1961), pp. 138-150.

[13]Miner, W.E. and Silber, R., "Launch window studies employing variational calculus to determine optimum three dimensional trajectories leading to injection into a space fixed circular orbit," ARS Preprint 2626-62 (November 13-18, 1962).

[14]Jurovics, S.A. and McIntyre, J.E., "The adjoint method and its applications to trajectory optimization," ARS J. 32, 1354-1358 (September 1962).

[15]Speyer, J.L., "Optimization and control using perturbation theory to find neighboring optimum paths," Fall 1962 Meeting of the Society of Industrial and Applied Mathematics, Cambridge, Mass. (November 1-3, 1962).

[16]Neustadt, L.W., "Synthesizing time optimal control systems," J. Math. Anal. Appl. 1, 484-493, no. 4 (1960).

[17]Hestenes, M.R., "Numerical methods of obtaining solutions of fixed end-point problems in the calculus of variations," Rand Rept. RM 102 (August 1949).

[18]Stein, M.L., "On methods for obtaining solutions of fixed end-point problems in the calculus of variations," J. Res. Nat. Bur. St. 50, 277-297 (May 1953).

[19]Kalaba, R., "On nonlinear differential equations, the maximum operation, and monotone convergence," J. Math. Mech. 8, 519-574 (1959).

[20]Kelley, H.J., "Guidance theory and extremal fields," Inst. Radio Engrs. Trans. Auto. Control AC-7, 75-82 (October 1962).

[21]Hestenes, M.R., "Sufficient conditions for the general problem of Mayer with variable end-points," Trans. Am. Math. Soc. 35, 479-90 (1933); also Contributions to the Calculus of Variations 1931-32 (University of Chicago Press, Chicago, Ill.), pp. 339-359.

[22]Bliss, G.A. and Hestenes, M.R., Sufficient conditions for a problem of Mayer in the calculus of variations," Trans. Am. Math. Soc. 35, 305-326 (1933); also Contributions to the Calculus of Variations 1931-32 (University of Chicago Press, Chicago, Ill.), pp. 295-337.

MINIMUM IMPULSE TRANSFER

John V. Breakwell*

Lockheed Missiles & Space Company, Sunnyvale, Calif.

Abstract

Minimum impulse transfer between coplanar ellipses is treated by variation-of-parameters with imparted velocity as the independent variable. The resulting necessary conditions for optimal control are satisfied by n-impulse transfers, as well as by Lawden's continuous-thrust spiral.

1. Introduction

It has been shown by Lawden[1] that the necessary conditions for minimum impulse transfer between coplanar orbits in an inverse-square central field are satisfied by a particular type of spiral solution, although, if the coplanar orbits have low eccentricity, it is more advantageous to make a two-impulse transfer.

The use by Lawden of time as an independent variable is nevertheless awkward, not only because the optimization of thrust magnitude is "degenerate,"[2] in the sense that the magnitude is not directly obtainable from a maximum principle, but more especially because impulsive transfers have to be treated in a special way.

In this paper, a variation-of-parameters formulation of the transfer, with imparted velocity as independent variable,[3] is applied to transfer between coplanar ellipses. The state variables, then, are the three parameters specifying an ellipse (in two dimensions); the control variables are (1) the position on the ellipse at which the next velocity increment is to be imparted, and (2) the direction of this velocity increment. The degeneracy of the problem is thereby removed, and continuous-thrust arcs such as Lawden's spiral and n-impulse transfers are treated in the same way.

Presented as Preprint 63-416 at the AIAA Astrodynamics Conference, New Haven, Conn., August 19–21, 1963.

*Senior Scientist.

2. The Equations of Motion

Let us define any ellipse by values of three parameters, p_i $(i = 1, 2, 3)$, as follows: $p_1 = L = \ln h$, h being the angular momentum, $p_2 = \epsilon$, the eccentricity, $p_3 = \beta$, the "argument of perigee," so that the polar equation of the ellipse is

$$r = (1/\mu)\, e^{2L}/[1 + \epsilon \cos (\theta - \beta)]$$

μ being the product of the central attracting mass by the universal gravitational constant.

The rates of change of the parameters p_i in the presence of a thrust acceleration T/m in the direction ψ have the form

$$\dot{P}_i = (1/\mu)\,(T/m)\, f_i(P_j, \theta^*, \psi) \tag{2.1}$$

where $\theta^* = \theta - \beta$, the "true anomaly." Explicitly,

$$f_1 = \frac{e^L \cos \psi}{1 + \epsilon \cos \theta^*}$$

$$f_2 = e^L \left\{ \sin \theta^* \sin \psi + \frac{2 \cos \theta^* + \epsilon(1 + \cos^2 \theta^*)}{1 + \epsilon \cos \theta^*} \cos \psi \right\} \tag{2.2}$$

$$f_3 = \frac{e^L}{\epsilon} \left\{ - \cos \theta^* \sin \psi + \frac{(2 + \epsilon \cos \theta^*)}{1 + \epsilon \cos \theta^*} \sin \theta^* \cos \psi \right\}$$

Also, the velocity V, imparted by the engine, increases at a rate $\dot{V} = T/m$. Introducing a new independent variable $u = V/\mu$, we may write

$$dp_i/du = f_i(p_j, \theta^*, \psi) \tag{2.3}$$

In this basic set of equations, both θ^* and ψ are to be regarded as control variables, namely, the position and direction of the next "push" on the ellipse. If we can determine an optimum control history $\theta^*(u)$, $\psi(u)$, the time history of the push is literally "whenever we get around to it."

3. Necessary Conditions for Optimum Control

The determination of the optimum control, in order to transfer from an ellipse $p^{(1)}$ to ellipse $p^{(2)}$ with minimum u, is now a nondegenerate problem in the calculus of variations, reducing to the maximization of a Hamiltonian which is nonlinear in the control

$$F(\lambda_j, p_j; \theta^*, \psi) = \sum_{i=1}^{3} \lambda_i f_i(p_j, \theta^*, \psi) \tag{3.1}$$

where the adjoint "sensitivities" λ_i $(i = 1, 2, 3)$ satisfy

$$d\lambda_i/du = -(\partial F/\partial p_i) \tag{3.2}$$

yielding two integrals

$$F = \text{const} = H \tag{3.3}$$

$$\lambda_3 = \text{const} \tag{3.4}$$

The rate of change of λ_1 has, furthermore, a simple form:

$$d\lambda_1/du = -H \tag{3.5}$$

The maximization of F, with respect to the control ψ, gives us no trouble since F is expressible in the form $A(\lambda_j, p_j, \theta^*) \sin \psi + B(\lambda_j, p_j, \theta^*) \cos \psi$. Thus, say,

$$\max_{\psi} F = \sqrt{A^2 + B^2} = F_1(\lambda_j, p_j, \theta^*)$$

The maximization of F_1^2 (and thus F_1) with respect to θ^* is troublesome, however; this function may have up to three (maybe even four) local maxima.

4. Lawden's Spiral and its "Mirror Image"

The above optimal control conditions are satisfied in particular, and with two coincident maxima of F_1^2 as a function of θ^*, by Lawden's spiral, which may be expressed as

$$\begin{aligned} L &= \ln \frac{s^3(3 - 4s^2)}{3 - 5s^2} \\ \epsilon \cos \theta^* &= \frac{-21s^2 + 63s^4 - 48s^6}{(3 - 5s^2)^2} \\ \epsilon \sin \theta^* &= \frac{6s\sqrt{1 - s^2}\,(1 - 2s^2)(3 - 4s^2)}{(3 - 5s^2)^2} \end{aligned} \tag{4.1}$$

where $s = \sin\psi$, with the λ's equal to

$$\lambda_1 = \frac{3\sqrt{1-s^2}}{s^3(3-5s^2)}(1-7s^2+10s^4)$$

$$\lambda_2 = \frac{3\sqrt{1-s^2}}{\epsilon}\,\frac{18-91s^2+149s^4-80s^6}{5(3-4s^2)^2(3-5s^2)} \qquad (4.2)$$

$$\lambda_3 = 3$$

normalized so that $H = 1$.

However, computation shows that as soon as $\sin\psi$ exceeds 0.439 and, hence, ϵ exceeds 0.9245, F_1^2 has another (simple) maximum greater than that corresponding to θ^* on the spiral, so that the spiral can no longer be locally optimal.

There is a "mirror image" trajectory obtained by changing the sign of λ_3, $d\beta/du$, $d\theta/du$, θ^*, and ψ, along which fuel expenditure is the same as on the spiral. This transfer can only be approximated in finite time by a large number of small thrusts separated by coasting arcs through almost 360°.

5. n-Impulse Transfer

Except for these singular cases where F has two coincident maxima, however, the optimum control is such that

$$\frac{d\theta}{du} = \left[\frac{d(\theta^* + \beta)}{du}\right] = 0$$

and

$$d\psi/du = 0$$

indicating an <u>impulse</u> for which, of course, we can obtain simple analytic expressions for p_i as functions of u and $p_i^{(1)}$, and also, rather easily, expressions for λ_1 and λ_2 using (3.3) and (3.5). Thus, given initial parameters L_1, ϵ_1, β_1 at $u = u_1$ and constant control variables ψ, θ, we may obtain, corresponding to any $u > u_1$,

$$\epsilon \cos(\theta - \beta) = \left\{ \frac{e^{L_1}(u - u_1)\cos\psi}{\sqrt{1 + \epsilon_1 \cos(\theta - \beta_1)}} + \sqrt{1 + \epsilon_1 \cos(\theta - \beta_1)} \right\}^2 - 1$$

$$\epsilon \sin(\theta - \beta) = \frac{\epsilon_1 \sin(\theta - \beta_1)}{\sqrt{1 + \epsilon_1 \cos(\theta - \beta_1)}} - \tan\psi \sqrt{1 + \epsilon_1 \cos(\theta - \beta_1)}$$

$$+ \tan\psi \{1 + \epsilon \cos(\theta - \beta)\} \tag{5.1}$$

$$L = L_1 + \frac{1}{2} \ln \frac{1 + \epsilon \cos(\theta - \beta)}{1 + \epsilon_1 \cos(\theta - \beta_1)}$$

If, furthermore, we normalize the λ's so that $H = 1$, we have

$$\lambda_1(u) = \lambda_1(u_1) - (u - u_1)$$

$$\lambda_3(u) = \lambda_3(u_1) \tag{5.2}$$

$$\lambda_2(u) = \frac{1 - \lambda_1(dL/du) - \lambda_3(d\beta/du)}{(d\epsilon/du)}$$

A program for generating the trajectories in the three-dimensional parameter space operates as follows:

(1) Input $p_i^{(1)}$ $(i = 1, 2, 3)$ and the two initial ratios $\lambda_1^{(1)} : \lambda_2^{(1)} : \lambda_3^{(1)}$.

(2) Maximize F_1^2 by a search procedure.

(3) Set $\sin\psi = A/F_1$, $\cos\psi = B/F_1$.

(4) Integrate (analytically), with respect to u, with θ and ψ constant until, using a search procedure, F_1 is found to have a new maximum equal to the one associated with the previous θ^*.

(5) Transfer by "coast" to the θ associated with the new maximum and compute the corresponding new direction ψ.

(6) Integrate forward with respect to u until another maximum takes over.

By different choices of the two initial ratios $\lambda_1^{(1)} : \lambda_2^{(1)} : \lambda_3^{(1)}$, we must obtain a two-parameter family of trajectories in the three-dimensional "phase space," which must include any optimum n-impulse transfer from the initial ellipse $p^{(1)}$ to any second ellipse $p^{(2)}$.

The intersection of these phase trajectories with planes β = const may be examined, since β is necessarily monotonic throughout the n-impulse transfer, in order to compare n-impulse transfers to the same final ellipse with different n . It will suffice to consider planes between $\beta = \beta_1$ and $\beta = \beta_1 + \pi$; transfer to $\beta < \beta_1$ can be obtained by "mirror image," and transfer of β by more than π is superfluous (this is presumably the reason why the low eccentricity comparison near the end of Ref. 1 is so unfavorable to the spiral).

Figure 1 shows a plot corresponding to $\beta = 20°$ for the case of an initial eccentricity $\epsilon_1 = 0.5$. (The scale and orientation of the ellipse is chosen so that $L_1 = \beta_1 = 0$.)

In no case was transfer by more than two impulses (one coast) cheaper than a two-impulse transfer to the same final ellipse.

References

[1] Lawden, D. F., "Optimal intermediate-thrust arcs," Astronaut. Acta VIII, 106–123 (1962).

[2] Breakwell, J. V., "The optimization of trajectories," J. Soc. Ind. Appl. Math. 7, 215–247 (June 1959).

[3] Contensou, P., "Etude theorique des trajectoires optimales dans un champ de gravitation. Application au cas d'un centre d'attraction unique," Astronaut. Acta VIII, 134–150 (1962).

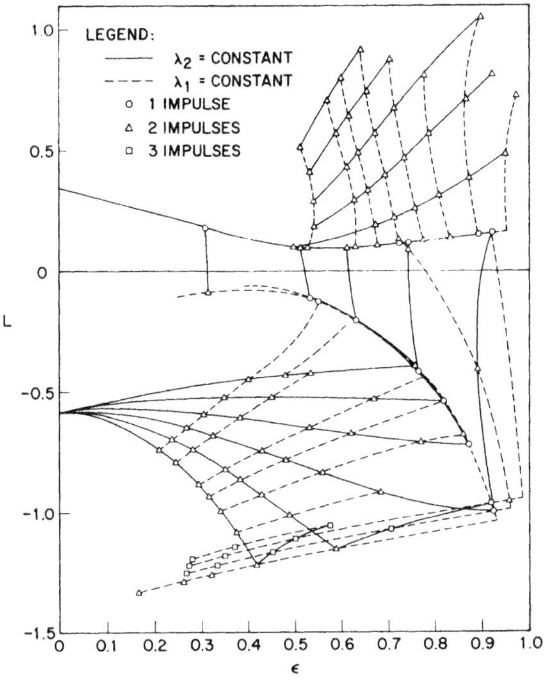

Fig. 1 Cross section of parameter space: λ_1 and λ_2 curves at $\beta = 20°$; $\epsilon_0 = 0.5$.

ON SOME SINGLE-IMPULSE TRANSFER PROBLEMS

Maurice L. Anthony* and Frank T. Sasaki†

Martin Company, Denver, Colo.

Abstract

Several problems pertaining to optimum single-impulse maneuvers in a Newtonian field are discussed in this paper. They include: 1) transfers between conics with two intersections, 2) circularization of trajectories, 3) directional escape from nearly circular orbits, 4) intersection and non-intersection with a prescribed region, and 5) an application to a two-impulse transfer.

Nomenclature

A, A_I, A_{II} = Flight path angle

c = Non-dimensional speed, V_1/V_{c_1}

r, R = Radial coordinate

v, V = Vehicle speed

V_c, V_e, V_∞ = Circular speed, escape speed, and speed at infinity

α = Flight path angle

μ = Gravitation constant

Subscripts

1(2) = Conditions at the nearer (farther) of two intersections to the center of force

a(p) = Conditions at apoapsis (periapsis)

Presented as Preprint 63-421 at the AIAA Astrodynamics Conference, New Haven, Conn., August 19-21, 1963.

*Manager, Astrodynamics Staff.

†Associate Research Scientist.

I. Introduction

This paper investigates certain optimum single-impulse maneuvers of a space vehicle in a Newtonian field. Four transfer maneuvers that can be accomplished by a single pulse are discussed and application is made to a two-pulse problem.

In Sec. II transfers are investigated between conics with two intersections. At each intersection a single-pulse transfer may be effected. The smaller of the two transfer pulse magnitudes is determined for both coplanar and non-coplanar conics.

The pulse required to transfer from a given trajectory to a circular orbit is studied in Sec. III. The minimum magnitude transfer pulse is obtained when the maneuver is accomplished at one or the other extremes of the radius. A condition involving the eccentricity of the initial trajectory is obtained by which the absolute minimum transfer pulse is determined.

In Sec. IV, the minimum pulse is determined in order to escape from a nearly circular orbit so that as the vehicle recedes its velocity vector asymptotically approaches a limit vector coplanar with the initial orbit.

In Sec. V, a space vehicle with specified initial velocity and position is required to either intersect or evade a prescribed region. The problem of determining the minimum pulse for this maneuver, either with or without constraints, can be geometrically interpreted as determining the minimum distance from a given point to a portion of a conic section. Although not discussed in this paper, the geometric approach may be applied to central force fields more general than Newtonian[1].

In Sec. VI, using results of single-pulse problems, the optimum two-pulse transfer between coplanar circular orbits with the same circulation is easily shown to be the cotangential transfer of Hohmann.

II. Transfers Between Conics with Two Intersections

A single impulsive change in velocity to transfer from one conic trajectory to another is discussed in this section. The trajectories are assumed to have two points in common, but they need not be closed or coplanar. The goal is to determine which intersection point requires the smaller pulse to effect transfer.

1. Coplanar case

The geometry of the problem of transfer between coplanar conics is shown in Fig. 1. Initially a vehicle is moving along the first conic, identified by the parameters e and r_p. At one of the intersection points, an instantaneous velocity pulse is provided so that thereafter the vehicle moves along the second conic identified by the parameters E and R_p. The central angle η is measured from the periapsis of the first conic counterclockwise to periapsis of the second conic, and θ is measured counterclockwise from the periapsis of the second conic. The radii of the two intersections are designated by r_1 and r_2, where $r_1 \leq r_2$.

a) Same circulation

By using the above nomenclature, the radius, speed, and flight path angle at any point on the first conic can be written as follows:

$$\left.\begin{aligned} r &= \frac{r_p\ (1+e)}{1+e\ \cos\ (\theta+\eta)} \qquad v = \left\{\frac{\mu\ \left[1+e^2 + 2e\ \cos\ (\theta+\eta)\right]}{r_p\ (1+e)}\right\}^{1/2} \\ \sin\alpha &= \frac{e\ \sin\ (\theta+\eta)}{\sqrt{1+e^2 + 2e\ \cos\ (\theta+\eta)}} \qquad \cos\alpha = \frac{1+e\ \cos\ (\theta+\eta)}{\sqrt{1+e^2 + 2e\ \cos\ (\theta+\eta)}} \end{aligned}\right\} \quad (1)$$

For the second conic, the corresponding quantities are given by

$$\left.\begin{aligned} R &= \frac{R_p\ (1+E)}{1+E\ \cos\ \theta} \qquad V = \left[\frac{\mu\ \left(1+E^2 + 2E\ \cos\ \theta\right)}{R_p\ (1+E)}\right]^{1/2} \\ \sin A &= \frac{E\ \sin\ \theta}{\sqrt{1+E^2 + 2E\ \cos\ \theta}} \qquad \cos A = \frac{1+E\ \cos\ \theta}{\sqrt{1+E^2 + 2E\ \cos\ \theta}} \end{aligned}\right\} \quad (2)$$

In these expressions for the flight path angles, the conics are assumed to have the same circulation, i.e., the orbital paths are both traced out in counterclockwise sense around the common focus in Fig. 1.

The values of θ at the two points of intersection are obtained by setting $r = R$ from their expressions in Eqs. (1) and (2). This leads to an expression for the anomalies at the two intersections:

$$\cos\theta = \frac{\left(H^2 - h^2\right)\left(Eh^2 - eH^2 \cos\eta\right) \pm eH^2\sqrt{\Gamma(\eta)}\,|\sin\eta|}{\left(Eh^2 - eH^2\right)^2 + 2eEh^2H^2\left(1 - \cos\eta\right)} \tag{3}$$

$$\Gamma(\eta) = \left(Eh^2 - eH^2\right)^2 + 2eEh^2H^2\left(1 - \cos\eta\right) - \left(h^2 - H^2\right)^2 \tag{4}$$

where the angular momenta of the motions are given by

$$h = \sqrt{\mu r_p (1 + e)} \qquad H = \sqrt{\mu R_p (1 + E)} \;. \tag{5}$$

In Eq. (3), if $\Gamma(\eta) = 0$, $\cos\theta$ is restricted to a single value. This condition, together with $r = R$, defines a unique value of θ, where the two conics are tangent. Since no comparison is possible in this case, it is excluded from the analysis. Also in Eq. (3) the denominator can only be positive or zero. The latter condition may also be excluded, for vanishing of the denominator implies both $Eh^2 = eH^2$ and $\cos\eta = 1$, simultaneously. The path equations of Eqs. (1) and (2) indicate that either no intersection occurs or the paths are identical, and hence no transfer problem exists. In view of the foregoing considerations, together with the condition that $r_1 \leq r_2$, the plus sign in Eq. (3) is associated with θ_1 and the minus sign with θ_2.

At the two points of intersection, the magnitudes of the pulses required to transfer may be written:

$$F_i = \sqrt{V_i^2 + v_i^2 - 2V_i v_i \cos\left(A_i - \alpha_i\right)}, \quad i = 1,2 \tag{6}$$

where subscripts 1 and 2 refer to the properties at θ_1 and θ_2 as given by Eq. (3). The difference of the squares of the pulses

$$G = F_1^{\,2} - F_2^{\,2} \tag{7}$$

is a function of the single variable η if the conic properties are considered to be fixed. By expanding each trigonometric function in Eq. (6), and by using Eqs. (1) and (2) for the speeds and the flight path angles, coupled with the use of Eq. (3) for θ_1 and θ_2, an equivalent expression may be obtained for the function G after considerable but straightforward algebraic manipulation:

$$G = \frac{-4\mu^2 eE\sqrt{\Gamma(\eta)}\,(h - H)^2\,|\sin\eta|}{hH\left[\left(Eh^2 - eH^2\right)^2 + 2eEh^2H^2\left(1 - \cos\eta\right)\right]} \tag{8}$$

The requirement that the trajectories have two distinct intersections implies that both $\Gamma(\eta)$ and the bracketed term of the denominator are positive. Hence G is always negative or zero; thus, for conics with two points of intersection and the same circulation, transfer at the intersection farther from the focus never requires the smaller transfer pulse. The special cases in which $G = 0$ are now examined.

Case i: $eE = 0$: Either $e = 0$ or $E = 0$, i.e., one of the conics is a circle.

Case ii: $\sin \eta = 0$: Either $\eta = 0$ or π, and hence the major axes of the conics are aligned. In cases (i) and (ii), the intersections are equidistant from the focus. Conversely, if $r_1 = r_2$, then either $eE = 0$ or $\sin \eta = 0$. Thus, if the two intersections are equidistant from the focus, the transfer pulses are of the same magnitude.

Case iii. $h-H = 0$: (Alternatively stated, the semi-parameters are equal.) The equal magnitude pulses required for transfer are applied radially. If there are two intersections, they occur 180° apart. This can be shown by considering the differential equations of two paths:

$$(1/r)'' + 1/r = \mu/h^2 \qquad (1/R)'' + 1/R = \mu/H^2 \tag{9}$$

where prime $\equiv \frac{d}{d\theta}$. It is seen that $W = 1/r - 1/R$ satisfies

$$W'' + W = \mu\left(h^{-2} - H^{-2}\right) \tag{10}$$

If the trajectories cross at $\theta = \gamma$, then

$$W = \mu\left(h^{-2} - H^{-2}\right)\left[1 - \cos(\theta - \gamma)\right] + B \sin(\theta - \gamma) \tag{11}$$

where $B \neq 0$. If $h = H$, then $W = B \sin(\theta - \gamma)$, and the two intersections occur 180° apart. Conversely, if two intersections occur 180° apart, the angular momenta have the same magnitude.

b) Opposite circulation

In the case of motions with opposite circulation, the flight path angle relationship of Eq. (1) must be replaced by

$$\sin \alpha = \frac{-e \sin(\theta + \eta)}{\sqrt{1 + e^2 + 2e \cos(\theta + \eta)}}$$
$$\cos \alpha = -\frac{1 + e \cos(\theta + \eta)}{\sqrt{1 + e^2 + 2e \cos(\theta + \eta)}} \tag{12}$$

By using Eqs. (1) and (2) except for the modification suggested by Eq. (12), the following expression is obtained for G:

$$G = \frac{4\mu^2 eE \sqrt{\Gamma(\eta)}\left(h + H\right)^2 |\sin \eta|}{hH\left[\left(Eh^2 - eH^2\right)^2 + 2eEh^2H^2 (1 - \cos \eta)\right]} \tag{13}$$

which is non-negative. Thus for orbits with opposite circulation and two points of intersection, transfer at the intersection point nearer to the focus never requires the smaller transfer pulse. The exceptional cases in which the transfer pulses are equal at both intersections occur when the intersections are the same distance from the focus.

2. Noncoplanar case

If two noncoplanar conics intersect at all, they must do so on the line of intersection of the two planes of motion which passes through the common focus of the conics. If they intersect twice, they do so on opposite sides of the focus, so that the intersections occur 180° apart, as measured from the focus. But this requires that their angular momenta be of the same magnitude. Using this fact, it can be shown that transfer at the larger radius of intersection is the more economical. The vectors $\overline{v}$ and $\overline{V}$ of Fig. 2 represent the velocity vectors of the two trajectories that lie in the planes I and II, respectively. The pulse required to transfer is given by $\overline{F} = \overline{V} - \overline{v} = \overline{m} + \overline{n}$, where $\overline{m}$ is the radial component of $\overline{F}$. From Fig. 2 the magnitudes of $\overline{m}$ and $\overline{n}$ satisfy

$$m^2 = (v \sin \alpha - V \sin A)^2 \qquad n^2 = 4V^2 \cos^2 A \sin^2 (i/2) \tag{14}$$

Since the pulse magnitude is given by $F = (m^2 + n^2)^{1/2}$, where m is the same for both intersections, and n^2 decreases as r increases, it follows that transfer at the larger radius requires the smaller transfer pulse.

The conclusions of this section are summarized in Table 1.

III. Circularization of Trajectories

Initially, a vehicle moves along a conic path. At any point on this path, a single impulsive change in velocity can be used to "circularize" the path, i.e., to effect a transfer to a circular orbit. The magnitude of the pulse required for circularization is determined as a function of the radius of the circular orbit and is shown to be a local minimum at both periapsis and apoapsis. The absolute minimum pulse is shown to occur at the maximum radius for $0 < e < 5/4$, and at periapsis for $e > 5/4$. Also, the smallest radius at which circularization is less costly than that at periapsis is determined.

In terms of r_p and e, v and α at any radius are related by the conservation of angular momentum and energy:

$$rv \cos \alpha = \sqrt{\mu r_p (1 + e)} \qquad v^2 = \mu \left[2/r - (e - 1)/r_p \right] \tag{15}$$

The magnitude of the pulse required to transfer to a circular orbit is given by

$$\begin{aligned} F &= \sqrt{v^2 + V_c^2 - 2vV_c \cos \alpha} \\ &= \sqrt{\frac{\mu}{r_p}} \left[3\left(\frac{r_p}{r}\right) - 2\sqrt{(1 + e)\left(\frac{r_p}{r}\right)^3} + e - 1 \right]^{\frac{1}{2}} \end{aligned} \tag{16}$$

through the use of Eq. (15). It is easily shown that as r increases from r_p, F begins at

$$F_p = \sqrt{\frac{\mu}{r_p}} \left(\sqrt{1 + e} - 1 \right) \tag{17}$$

increases to its maximum value of

$$F_{max} = \frac{e}{\sqrt{1 + e}} \sqrt{\frac{\mu}{r_p}} \tag{18}$$

(which occurs at $r = r_p (1+e)$, i.e., $\theta = \pm 90°$), and then decreases to the value

$$F_a = \sqrt{\frac{\mu}{r_p} \left(\frac{1 - e}{1 + e} \right)} \left(1 - \sqrt{1 - e} \right) \tag{19}$$

for bound orbits $(0 < e < 1)$, and to the value

$$F_\infty = \sqrt{\frac{\mu}{r_p}} \sqrt{e - 1} \tag{20}$$

for escape trajectories $(e \geq 1)$. Thus, the magnitude of the pulse required for circularization is a minimum at either extreme radius. This is verified by the results shown in Fig. 3.

For all bound orbits, it is clear from the relation

$$F_p - F_a = \sqrt{\frac{\mu}{r_p(1 + e)}}\left[2 - \left(2 + 2\sqrt{1 - e^2}\right)^{\frac{1}{2}}\right] > 0 \qquad (21)$$

that apoapsis requires a smaller pulse than periapsis. For unbound orbits

$$F_p - F_\infty = \sqrt{\frac{\mu}{r_p}}\left[\sqrt{1 + e} - 1 - \sqrt{e - 1}\right] \qquad (22)$$

where the function in brackets decreases monotonically as eccentricity increases and is zero for $e = 5/4$. Thus, the absolute minimum pulse required for circularization occurs at the largest radius for $0 < e < 5/4$, and at the smallest radius for $e > 5/4$. Fig. 4 also shows this by directly comparing the relative minima. For convenient reference, the formulas for F_{min} are collected here.

$$\left.\begin{aligned} F_{min} &= \sqrt{\frac{\mu}{r_p}}\sqrt{\frac{1 - e}{1 + e}}\left[1 - \sqrt{1 - e}\right] && \text{for } 0 < e \leq 1 \\ &= \sqrt{\frac{\mu}{r_p}}\sqrt{e - 1} && \text{for } 1 \leq e \leq \frac{5}{4} \\ &= \sqrt{\frac{\mu}{r_p}}\left[\sqrt{e + 1} - 1\right] && \text{for } e \geq \frac{5}{4} \end{aligned}\right\} \qquad (23)$$

In addition to illustrating the analytical results just obtained, Fig. 3 also shows that it is possible to circularize in the vicinity of the absolute minimum pulse in a non-optimum manner, and still require a smaller pulse than the other local minimum. For example, in the case $e = 0.6$, circularization at any radius larger than $3.5\ r_p$ requires a smaller pulse than that at periapsis. More generally, for eccentricities in the range $0 < e < 5/4$, the smallest radius for which circularization does not exceed periapsis circularization is given by

$$\frac{r}{r_p} = \frac{1}{4}\left[\sqrt{1 + \frac{8\sqrt{1 + e}}{3 - 2\sqrt{1 + e}}} - 1\right]^2 \qquad (24)$$

The radial position at which the two pulses are equal is shown in the lower left portion of Fig. 3 as a dotted line. Similarly, for $e > 5/4$, the pulse required for circularization in a vicinity of periapsis, specifically in the range

$$1 \leq r/r_p < \frac{4}{9}(1 + e) \tag{25}$$

is smaller than that required at $r = \infty$. The radial position at which the two pulses are equal is shown as a dotted line in the upper right portion of Fig. 3.

IV. Directional Escape from Nearly Circular Orbits

This section investigates the minimum magnitude velocity pulse required for a vehicle to escape from a nearly circular orbit so that, as it recedes from the center of the field, its velocity vector asymptotically approaches a prescribed limiting velocity that lies in the plane of the initial orbit and is denoted by $\bar{V}_\infty$.

As indicated in Fig. 5, the direction of $\bar{V}_\infty$ is given by the angle ω measured from periapsis of the initial orbit which is characterized by r_p and e. The departure point P is located by the angle ψ measured from the ultimate direction of the escape trajectory. The magnitude of the velocity pulse is

$$F = \sqrt{v_1^2 + V_1^2 - 2v_1V_1 \cos(\alpha_1 - A_1)} \tag{26}$$

Before the pulse:

$$v_1 = \sqrt{\frac{\mu\left[1 + e^2 + 2e\cos(\omega - \psi)\right]}{r_p(1 + e)}} \tag{27}$$

$$\tan\alpha_1 = \frac{e\sin(\omega - \psi)}{1 + e\cos(\omega - \psi)} \qquad R_1 = \frac{r_p(1 + e)}{1 + e\cos(\omega - \psi)} \tag{28}$$

The required speed at the departure point can be expressed in terms of V_∞ and R_1 and, with the aid of Eq. (28), can be written

$$V_1 = \sqrt{V_\infty^2 + \frac{2\mu[1+e\cos(\omega - \psi)]}{r_p(1+e)}} \tag{29}$$

An additional relation is required to assure that the velocity approaches the direction of $\bar{V}_\infty$ as the vehicle recedes from the force field. This relation is obtained by considering the path equation which can be written[2]

$$\left.\begin{aligned} \frac{R}{R_1} &= \frac{\beta}{1 + (\beta-1)\cos\Theta - \beta\delta\sin\Theta} \\ \beta = c^2\cos^2 A_1 \qquad \delta &= \tan A_1 \qquad c^2 = 2 + \frac{V_\infty^2 r_p (1+e)}{\mu[1+e\cos(\omega-\psi)]} \end{aligned}\right\} \tag{30}$$

and Θ is measured from point of departure from the original orbit. As Θ approaches ψ, the radius grows indefinitely. This condition requires that

$$1 + (\beta - 1)\cos\psi - \beta\delta\sin\psi = 0 \tag{31}$$

which may be solved for $\tan A_1$:

$$\tan A_1 = \frac{c^2\sin\psi \pm \sqrt{(c^2-2)^2\sin^2\psi + 4(c^2-2)(1-\cos\psi)}}{2(1-\cos\psi)} \tag{32}$$

It can readily be shown[2] that the upper sign leads to an escape trajectory of circulation opposite to the original trajectory, and the lower sign leads to an escape trajectory of the same circulation. Hence, for each departure point of the initial orbit, there are two admissible escape trajectories, and consequently F, as given by Eq. (26), is a double valued function of ψ. The magnitudes of the transfer pulses associated with the two escape trajectories, with and opposite the circulation of the original orbit, are denoted by F_I and F_{II}, respectively. The minimum value of F_I is determined by setting the derivative of F_I (using Eq. (32) with the minus sign) with respect to Θ equal to zero. This leads to an equation which cannot be explicitly solved for the critical value of ψ. As an alternative, the equation is solved approximately by expanding the expression to first order terms in the eccentricity of the initial orbit. This leads to the approximate solution

$$\psi = \psi_o + \psi_1 e \tag{33}$$

$$\psi_o = \cos^{-1}\left(\frac{-1}{1+K}\right) \qquad K = \frac{V_\infty^2 r_p}{\mu} \tag{34}$$

$$\begin{aligned} \psi_1 = {} & \left\{\frac{-1}{1+K}\sqrt{\frac{K}{2+K}} + \frac{1}{(1+K)^3}\left[(1+K)(1+2K) + (2+K)\left(\sqrt{2+K} - 1\right)^2\right]\right\}\sin\omega \\ & - \frac{1}{(1+K)^3}\sqrt{\frac{K}{2+K}}\left\{(1+K)\left[1-(1+K)(2+K)\right] - (2+K)^2\left(\sqrt{2+K} - 1\right)^2\right\}\cos\omega \end{aligned} \tag{35}$$

In Eq. (34), ψ_o represents the transfer angle when the initial orbit is circular and is therefore in the interval $90° < \psi_o < 180°$.

Since, for the optimal departure, it can be shown that $\cos(\alpha_1 - A_1) = 1 + O(e^2)$, the optimal pulse, correct to the first order in e, is:

$$F_I^* = \min\left(V_1 - v_1\right) \tag{36}$$

where terms of the first order in e are retained. This can be written

$$F_I^* = \sqrt{\mu/r_p}\left(F_o + eF_1\right) \qquad F_o = \sqrt{2 + K} - 1$$
$$F_1 = \left[\frac{\sqrt{2 + K} - 1}{(1+K)\sqrt{2+K}}\right]\left[1 + K + \cos\omega - \sqrt{K(2+K)}\,\sin\omega\right] - \frac{1}{2} \tag{37}$$

It can be shown[3] that $F_{II}^* - F_I^* > 0$ and hence F_I^* is the absolute minimum admissible F.

The nondimensional pulses, $F_I\sqrt{r_p/\mu}$ and $F_{II}\sqrt{r_p/\mu}$, are shown in Fig. 6 which also indicates the approximate analytical values of ψ^* and $F_I^*\sqrt{r_p/\mu}$ as given by Eqs. (33) and (37). Additional results are shown in Fig. 7. Further calculations indicate that the accuracy of the analytical solution is still in good agreement for larger values of eccentricities.

In summary, an approximate analytical solution has been obtained that determines the minimum pulse maneuver required to escape from a nearly circular orbit so that, as it recedes, the vehicle's velocity approaches a prescribed limit velocity that is coplanar with the initial orbit. Analytical approximate results for the minimum required pulse magnitude and the departure point are determined, which agree well with numerical results.

V. A Geometric Approach for Single-Impulse Transfer Problems

This section presents a geometric method which pertains to certain minimum single-impulse maneuvers of a space vehicle. Initially, the vehicle is assumed to be at a radius R_I with speed v_I and flight path angle α_I. The boundary of a region is given by $r = R_{II}$. We discuss the following questions: If $R_I < R_{II}$, what is the minimum pulse required for the vehicle to intersect or evade the region $r > R_{II}$? Alternatively, if

$R_I > R_{II}$, what is the minimum pulse required to intersect or evade the region $r < R_{II}$? The same questions are discussed with restrictions placed on the central transfer angle, or the terminal flight path angle. For convenience four problems are treated separately.

1. $R_I < R_{II}$; no constraints on A_{II}, θ_T

The minimum pulse required is to be determined to 1) intersect the region $r \geq R_{II}$ when the original trajectory is a non-intersecting one, or 2) evade the region when the original orbit is destined to intersect. A schematic illustrating the geometry and nomenclature is shown in Fig. 8.

The energy and angular momentum equations of the transfer trajectory are

$$V_I^2 = V_a^2 + 2\mu\left(1/R_I - 1/R_a\right) \qquad R_I V_I \cos A_I = R_a V_a \tag{38}$$

where the subscript a refers to apoapsis conditions. Elimination of V_a yields

$$V_I^{\,2}\left[1 - \left(\frac{R_I \cos A_I}{R_a}\right)^2\right] = 2\mu\left(\frac{R_a - R_I}{R_I\, R_a}\right) \tag{39}$$

This equation may be considered as a relationship that V_I and A_I must satisfy for a fixed apoapsis. By introducing

$$X_I = V_I \cos A_I \qquad Y_I = V_I \sin A_I \tag{40}$$

so that X_I and Y_I represent the horizontal and vertical components of the velocity just after the pulse, Eq. (39) transforms to the ellipse:

$$\left(X_I/\hat{\sigma}\right)^2 + \left(Y_I/\hat{\tau}\right)^2 = 1 \tag{41}$$

where

$$\hat{\sigma} = \sqrt{\frac{2\mu R_a}{R_I\left(R_I + R_a\right)}} \qquad \hat{\tau} = \sqrt{\frac{2\mu\left(R_a - R_I\right)}{R_I\, R_a}} \tag{42}$$

Hence, $\overline{V}_I$ must terminate on the ellipse of Eq. (41) for a given R_a. Since both $\hat{\sigma}$ and $\hat{\tau}$ increase with R_a, points in region D (Fig. 9), on or exterior to the ellipse for which

$R_a = R_{II}$, are associated with velocities which ensure that $R_a \geq R_{II}$. Alternatively, points in region $\tilde{D}$ on or interior to the ellipse, are associated with velocities for which $R_a \leq R_{II}$. Thus, if $\bar{v}_I$ lies interior to the ellipse for which $R_a = R_{II}$, all admissible pulses compatible with the requirement that the vehicle intersect the region $r \leq R_{II}$, must terminate in the region D. Similarly, if $\bar{v}_I$ terminates exterior to the ellipse, all admissible pulses compatible with the requirement that the vehicle evade the region $r \geq R_{II}$ must terminate in the region $\tilde{D}$. In either case, the minimum pulse must terminate on the ellipse for which $R_a = R_{II}$, i.e., on the ellipse common to both regions:

$$\left(X_I/\sigma\right)^2 + \left(Y_I/\tau\right)^2 = 1 \qquad \sigma = V_{e_I}\Big/\sqrt{1+\epsilon} \qquad \tau = V_{e_I}\sqrt{1-\epsilon}$$
$$\epsilon = R_I/R_{II} \qquad V_{e_I} = \sqrt{2\mu/R_I} \tag{43}$$

Hence, the problem of determining the minimum pulse is equivalent to that of determining the minimum distance from a given point to an ellipse. Since the direction of the minimum pulse must be orthogonal to the ellipse, it will not, in general, be directed collinearly with $\bar{v}_I$. Furthermore, the trajectory associated with the minimum pulse intersects tangentially the circle $r = R_{II}$.

The problem of minimizing the distance from a given point to a conic section generally involves finding the solution to a quartic equation. However, when either the horizontal component (x_I) or the radial component (y_I) of $\bar{v}_I$ is zero, the solution is obvious. For example, when $x_I = 0$ (i.e., the motion is initially radial) the minimum pulse vector $\bar{F}_{min}$ is also directed radially. For the case $y_I = 0$ (i.e., pulse injection is applied initially at an apse) it can be easily shown that the minimum pulse is directed collinearly with $\bar{v}_I$ only if $x_I \geq x_c = (\sigma^2 - \tau^2)/\sigma$. Hence even when injection occurs at an apse, minimum injection pulses are not necessarily collinear with the initial velocity. Geometrically, x_c represents the abscissa of the center of curvature of the ellipse at the point $(\sigma, 0)$. The heavy solid lines in Fig. 9 represent the region in which $\bar{v}_I$ must terminate to minimize collinear pulse injection. Otherwise, the minimum magnitude velocity pulse is not collinear with $\bar{v}_I$. It is of interest to observe that since $x_c < V_{c_I} = \sqrt{\mu/R_I}$, if the vehicle is initially in a circular orbit, the minimum pulse is directed horizontally. Table 2 lists the expressions for F_{min} and the corresponding A_I when injection occurs at an apse.

2. $R_I > R_{II}$; no constraints on A_{II}, θ_T

By referring to periapsis rather than apoapsis conditions, an equation corresponding to Eq. (41) is obtained:

$$\left(X_I/\hat{S}\right)^2 - \left(Y_I/\hat{T}\right)^2 = 1 \tag{44}$$

where

$$\hat{S} = \sqrt{\frac{2\mu R_p}{R_I(R_I + R_p)}} \qquad \hat{T} = \sqrt{\frac{2\mu(R_I - R_p)}{R_I\, R_p}} \tag{45}$$

Hence in this case, $\overline{V}_I$ must terminate on a hyperbola that is determined when R_p is prescribed. The region $\tilde{L}$, on or to the left of the hyperbola for which $R_p = R_{II}$ (Fig. 10), is associated with $R_p \leq R_{II}$, and in region L, $R_p \geq R_{II}$. Each of the regions $\tilde{L}$ and L is divided by the circle with a radius equal to the escape speed at R_I. Points in region $\tilde{L}$ below the circle are associated with finite trajectories that intersect the region $r \leq R_{II}$. Points in region $\tilde{L}$ above the circle are associated with trajectories which, although possessing periapsis radii less than R_{II}, are on the outward bound branch of an escape trajectory, and hence do not intersect. Points in L exterior to the shaded region are associated with escape trajectories that do not intersect the region $r \leq R_{II}$, and hence may by considered as "fly-by" trajectories. Points in the shaded region correspond to trajectories that neither intersect nor escape and thus represent satellite orbits.

By reasoning similar to that used before, the minimum pulse required to either intersect or evade the region, $r \leq R_{II}$, corresponds to the minimum distance from $\overline{v}_I$ to the hyperbola given by

$$\left(X_I/S\right)^2 - \left(Y_I/T\right)^2 = 1 \quad S = V_{e_I}\Big/\sqrt{1 + \epsilon} \quad T = V_{e_I}\sqrt{\epsilon - 1} \tag{46}$$

As in the case $R_I < R_{II}$, the determination of the minimum pulse requires the solution of a quartic equation. However, the solution is obvious when either component of $\overline{v}_I$ is zero. It can be easily shown that the minimum pulse is directed collinearly with $\overline{v}_I$ only if $y_I = 0$ and $0 < x_I \leq X_c = (S^2 + T^2)/S$. The quantity X_c represents the abscissa of the center of curvature of the hyperbola at its vertex. Since $V_{c_I} > X_c$, if the vehicle is initially in a circular orbit, the minimum pulse required to intersect the region $r \leq R_{II}$ is directed horizontally. Table 3 shows F_{min} and A_I when injection occurs at an apse.

3. Flight path angle A_{II} constrained

The questions discussed earlier are now reconsidered with the flight path angle at intersection being either specified or restricted to be within specified limits.

Using the energy and angular momentum relationships of the transfer trajectory the following equation is obtained:

$$\frac{X_I^2}{(1 - \epsilon) \cos^2 A_{II} \left(\cos^2 A_{II} - \epsilon^2\right)^{-1}} + \frac{Y_I^2}{1 - \epsilon} = V_{e_I}^2 \qquad (47)$$

Hence $\bar{V}_I$ must terminate on a conic section that depends on the parameter A_{II} and the radii ratio ϵ. For convenience, the cases $R_I > R_{II}$ and $R_I < R_{II}$ are treated separately.

a) For $\epsilon > 1$, Eq. (47) represents a hyperbola for all A_{II}. The curves of constant A_{II} are indicated in Fig. 11. For $A_{II} > 0$ the curves are truncated by the circle with radius equal to escape speed. The minimum pulse corresponds to the minimum distance from $\bar{V}_I$ to the A_{II}-curve, or, if the interval on A_{II} is prescribed, to a region bounded by two A_{II}-curves.

b) If $\epsilon < 1$, then for a constant A_{II}, Eq. (47) represents an ellipse for $\cos A_{II} > \epsilon$, a hyperbola for $\cos A_{II} < \epsilon$, and a straight line for $\cos A_{II} = \epsilon$ (see Fig. 12). The minimum pulse terminates orthogonal to an A_{II}-curve. If $A_I < 0$, the transfer trajectories intersect $r = R_{II}$ only after periapsis is attained. Thus, if $r = \rho < R_I$ represents the surface of a planetary body, an additional constraint must be imposed, namely $R_p > \rho$. This constraint defines the region shown shaded in Fig. 12. In this case, the admissible A_{II}-curves are terminated by their intersection with the shaded region.

4. Central transfer angle θ_T constrained

A geometric interpretation is now presented for the case where the central transfer angle θ_T, is prescribed. From the path relationship given by Eq. (30) the following equation, which meets the condition that $\Theta = \theta_T$ when $r = R_{II}$, holds:

$$(\epsilon - \cos \theta_T) X_I^2 + \sin \theta_T X_I Y_I = (1 - \cos \theta_T) V_{c_I}^2 \qquad (48)$$

where the subscript 1 in Eq. (30) is replaced by I to be in accordance with the notation of this section. For a fixed θ_T, the above equation describes a hyperbola (or its degenerate case) on which $\overline{V}_I$ must terminate. The asymptotes are given by

$$X_I = 0 \qquad Y_I = -\left(\frac{\epsilon - \cos \theta_T}{\sin \theta_T}\right) X_I \qquad (49)$$

a) $R_I > R_{II}$: The family of hyperbolas given by Eq. (49) is indicated in Fig. 13. The envelope of the family is the hyperbola given by Eq. (47), i.e., the hyperbola for which $A_{II} = 0$. At the tangency point the transfer trajectory is a grazing one with speed V_G and flight path angle A_G. Points on a θ_T-curve lying to the side of the tangency point for which $A_I > A_G$, are associated with transfer trajectories that reach the radius R_{II} with a negative flight path angle. Since $A_{II} < 0$ implies that apoapsis of the transfer trajectory is attained before intersection, the portion of a θ_T - curve for which $A_{II} < 0$ is terminated at its intersection with the circle representing escape speed. If $A_I < A_G$, then $A_{II} > 0$. Thus if R_{II} represents the radius of a planetary body, positive values of A_{II} are not admissible and hence the A_{II} - curves are terminated at their point of contact with the envelope.

b) $R_I < R_{II}$: Constant θ_T - curves obtained from Eq. (48) are shown in Fig. 14. The envelope of these is the ellipse for which $A_{II} = 0$ as given by Eq. (47). The point of tangency of a θ_T - curve with its envelope yields grazing conditions. For other velocity vectors that terminate on the same θ_T - curve, the corresponding transfer trajectory reaches R_{II} with $A_{II} > 0$ if $A_I < A_G$, and $A_{II} < 0$ if $A_I > A_G$. Constraints on the trajectories which are imposed by escape conditions or a planetary body (discussed in Secs. 3.b and 4.a) are also indicated.

For both cases a) and b), the magnitude of the minimum pulse corresponds to the minimum distance from $\overline{v}_I$ to a portion of a hyperbola. If θ_T is restricted to be within certain limits, the minimum pulse corresponds to the minimum distance from a point to the region enclosed by the two curves corresponding to $\theta_{T\,min}$ and $\theta_{T\,max}$. Except in certain isolated cases, a quartic equation must be solved to determine the minimum pulse. However, when $\overline{v}_I = 0$ and A_{II} is not constrained, the minimum pulse can be easily determined.[3,4]

In summary a geometric interpretation has been presented to understand more clearly certain optimum single-impulse maneuvers. The determination of the minimum pulse required to intersect or evade a region with a boundary given by $r = R_{II}$, either with or without constraints on the central transfer angle or the flight path angle at intersection, is equivalent to the determination of the minimum distance from a point to a portion of a conic section, which generally is further equivalent to solving a quartic equation. If the maneuver is accomplished with no constraints on A_{II} or θ_T, however, and if the initial velocity is radial or circumferential, the minimum pulse is easily determined and the results are tabulated. Geometrically, it is evident that the minimum pulse is not generally collinear with the initial velocity vector even when pulse injection occurs at an apse.

VI. Application to a Two-Pulse Problem

This section presents an application of the geometrical method of Sec. V to the problem of determining the minimum two-pulse maneuver that can effect transfer between two coplanar circular orbits of the same circulation, i.e., to the Hohmann problem. A pulse $\overline{F}_i$ is introduced so that the vehicle departs from the first orbit of radius r_i and, at arrival at radius r_f, a second pulse $\overline{F}_f$ is used to establish a circular orbit. The solution can be easily obtained by resorting to the velocity-flight path angle diagrams discussed in Sec. V. To simply reach the radius r_f from r_i, the initial velocity on the transfer trajectory $\overline{V}_i$ must terminate either on or exterior to the ellipse:

$$(X_i/\sigma)^2 + (Y_i/\tau)^2 = 1 \tag{50}$$

where

$$X_i = V_i \cos A_i \qquad Y_i = V_i \sin A_i$$

$$\sigma = \sqrt{\frac{2\mu R_f}{R_i(R_i + R_f)}} \qquad \tau = \sqrt{\frac{2\mu\left(R_f - R_i\right)}{R_i R_f}} \tag{51}$$

The expressions in Eq. (51) are the same as those in Eq. (43) if the subscripts i and f are associated with the subscripts I and II, respectively. All admissible pulses begin at P_i and terminate in region D as indicated in Fig. 15.

At the terminal end of the transfer trajectory:

$$\frac{X_f^{\,2}}{\dfrac{2\mu\left(R_i - R_f\right)R_i\cos^2 A_i}{R_f\left[R_i^{\,2}\cos^2 A_i - R_f^{\,2}\right]}} + \frac{Y_f^{\,2}}{\dfrac{2\mu\left(R_i - R_f\right)}{R_i R_f}} = 1 \tag{52}$$

where

$$X_f = V_f \cos A_f \qquad Y_f = V_f \sin A_f \tag{53}$$

Equation (52) is identical to Eq. (47) if the subscripts i and f are associated with the subscripts II and I respectively, since in Sec. V the subscript I always refers to conditions at the initial point of the transfer trajectory. Using the results of Sec. V, it follows that the terminal velocity vector $\overline{V}_f$ must lie in region $\tilde{L}$, i.e., either on or to the left of the hyperbola ($\cos A_i = 1$) of Fig. 15. The second pulse is then represented by a vector that begins at the end of $\overline{V}_f$ and terminates at the point P_f.

Since $V_{c_i} > x_c$, the smallest distance from P_i to region D is that directed along the X_i axis to the point (σ , 0). This distance corresponds to the minimum value of F_i given by $(F_i)_{min} = \sigma - V_{c_i}$. Similarly, since $V_{c_f} < X_c$, the smallest distance from P_f to region $\tilde{L}$ is that along the X_f axis to the point (S,0), where S is given by Eq. (45) with the appropriate change between the subscripts:

$$S = \sqrt{\frac{2\mu R_i}{R_f\left(R_i + R_f\right)}} \tag{54}$$

The minimum value of F_f is given by $(F_f)_{min} = V_{c_f} - S$. A transfer is possible using these two pulses; the transfer path is cotangential with the circular paths, and the pulses are both tangential to the circular paths.

The solution of the two pulse problem must be at least as large as the sum of the minima found in the two single-pulse problems just discussed. Usually, the conditions pertaining to simultaneous minimization of the two quantities F_i and F_f treated as single-pulse problems are in conflict, so that the solution of a two-pulse problem is usually not the sum of two minima but is a compromise between gains and losses at the two ends of the transfer arc. However, the two-pulse problem of this section is an exception, since the conditions required to minimize F_i lead to precisely the conditions required to minimize F_f also. Since the simultaneous minimization of these two pulses is possible, the minimum two-pulse requirement for this problem $F_{min} = (F_i)_{min} + (F_f)_{min}$, i.e.

$$F_{min} = \sqrt{\frac{2\mu R_f}{R_i(R_i + R_f)}} - \sqrt{\frac{\mu}{R_i}} + \sqrt{\frac{\mu}{R_f}} - \sqrt{\frac{2\mu R_i}{R_f(R_i + R_f)}} \qquad (55)$$

which is the pulse associated with the classical Hohmann maneuver.

The optimum two-impulse transfer between coplanar circular orbits with the smae circulation has been found geometrically to be the cotangential transfer ellipse of Hohmann.

References

[1] Anthony, M.L. and Sasaki, F.T., "Geometric Aspects of Single-Pulse Transfers in Central Force Fields," to be published in the Proceedings of the Fifth International Symposium on Space Technology and Science, Tokyo, Japan (September 1963).

[2] Anthony, M.L., "Free Flight Missile Trajectories," pp. 219-236, Vistas of Astronautics, Vol. II (New York Pergamon Press, 1959).

[3] Anthony, M.L. and Sasaki, F.T., "On Some Single-Impulse Transfer Problems," R-63-9, The Martin Company, Denver, Colorado (May 1963).

[4] Anthony, M.L. and Sasaki, F.T., "A Property of Optimum Paths Common to Newtonian and Uniform Force Fields," AIAA J. Vol. I, pp. 1897-1899 (August 1963).

Table 1 Summary of pulse magnitude relations

$r_1 = r_2$				$F_1 = F_2$
$r_1 < r_2$	Coplanar	Same circulation	$h = H$	$F_1 = F_2$
			$h \neq H$	$F_1 < F_2$
		Opposite circulation		$F_1 > F_2$
	Noncoplanar			$F_1 > F_2$

Table 2 F_{min} and A_I for apsidal injection and $\epsilon < 1$

	$0 \le x_I \le x_c = \frac{\epsilon^2}{\sqrt{1+\epsilon}} V_{e_I}$	$x_I \ge x_c$
F_{min}	$(1-\epsilon^2)^{\frac{1}{2}}\left[\frac{V_{e_I}^2}{1+\epsilon} - \left(\frac{x_I}{\epsilon}\right)^2\right]^{\frac{1}{2}}$	$\left\lvert\frac{V_{e_I}}{\sqrt{1+\epsilon}} - x_I\right\rvert$
A_I	$\tan^{-1}\left\{\pm(1-\epsilon^2)^{\frac{1}{2}}\left[\left(\frac{x_c}{x_I}\right)^2 - 1\right]^{\frac{1}{2}}\right\}$	0

Table 3 F_{min} and A_I for apsidal injection and $\epsilon > 1$

	$0 \le x_I \le X_c$	$x_I \ge X_c = \frac{\epsilon^2}{\sqrt{1+\epsilon}} V_{e_I}$
F_{min}	$\left\lvert\frac{V_{e_I}}{\sqrt{1+\epsilon}} - x_I\right\rvert$	$(\epsilon^2-1)^{\frac{1}{2}}\left[\left(\frac{x_I}{\epsilon}\right)^2 - \frac{V_{e_I}^2}{1+\epsilon}\right]^{\frac{1}{2}}$
A_I	0	$\tan^{-1}\left\{\pm(\epsilon^2-1)^{\frac{1}{2}}\left[1 - \left(\frac{X_c}{x_I}\right)^2\right]^{\frac{1}{2}}\right\}$

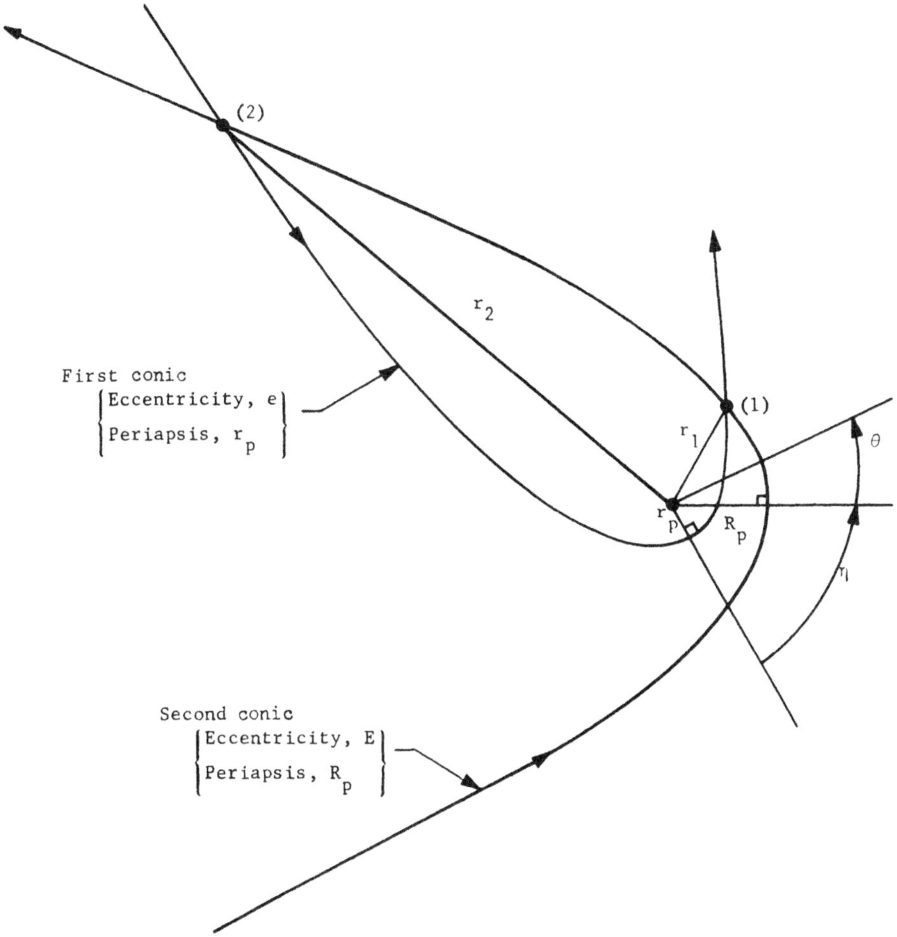

Fig. 1 Geometry of two intersecting conics.

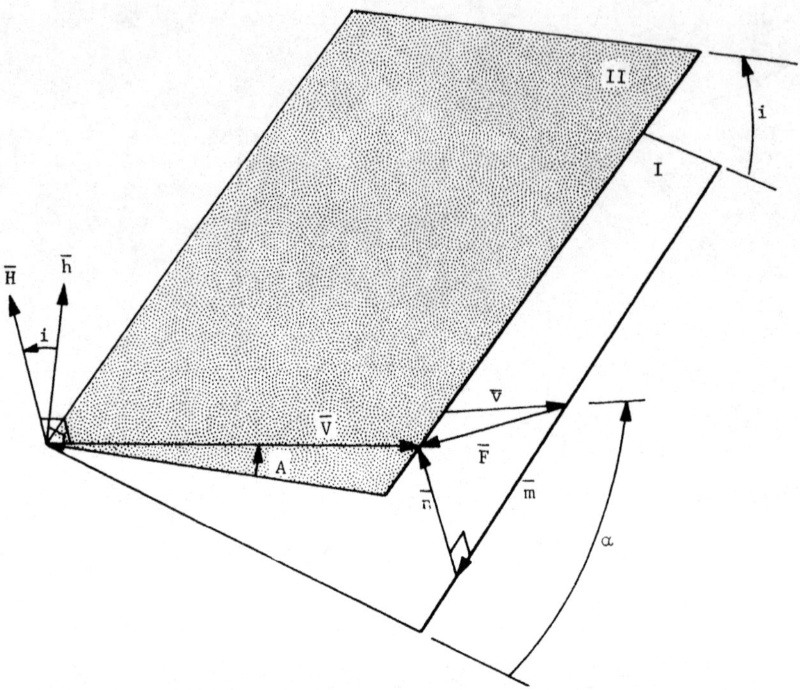

Fig. 2 Pulse geometry for non-coplanar transfer.

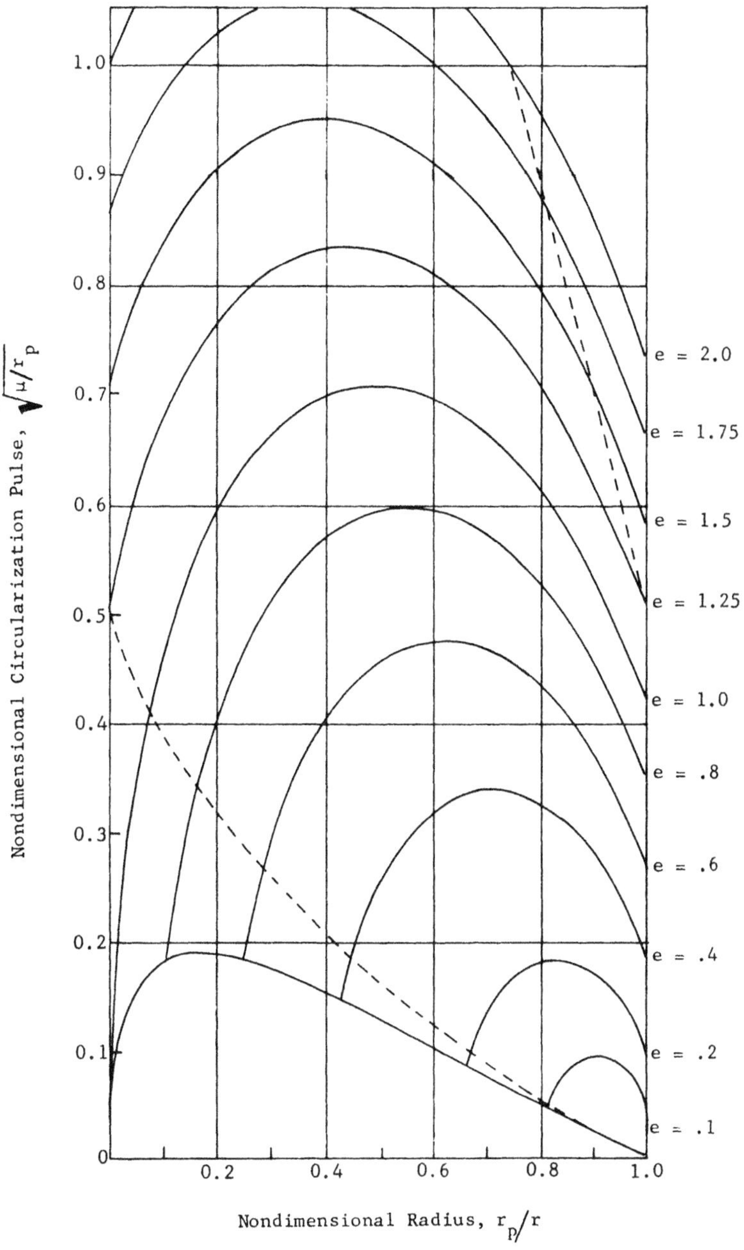

Fig. 3 Nondimensional circularization pulse as a function of r_p/r for several values of eccentricity.

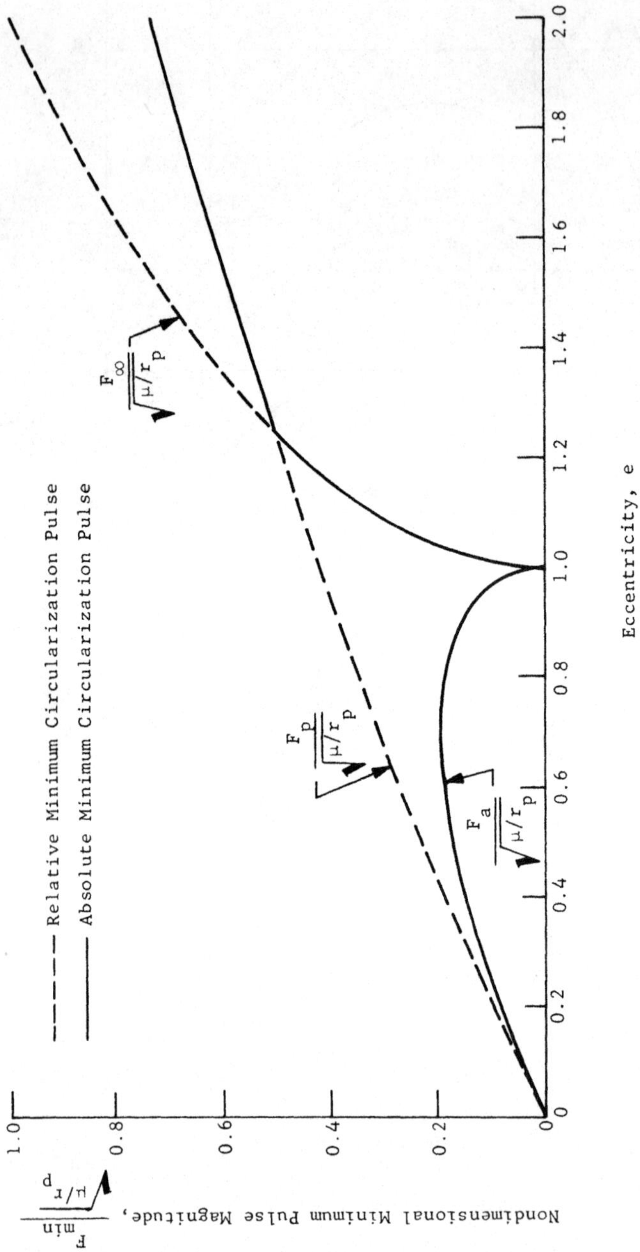

Fig. 4 Minimum magnitude circularization pulse as a function of eccentricity.

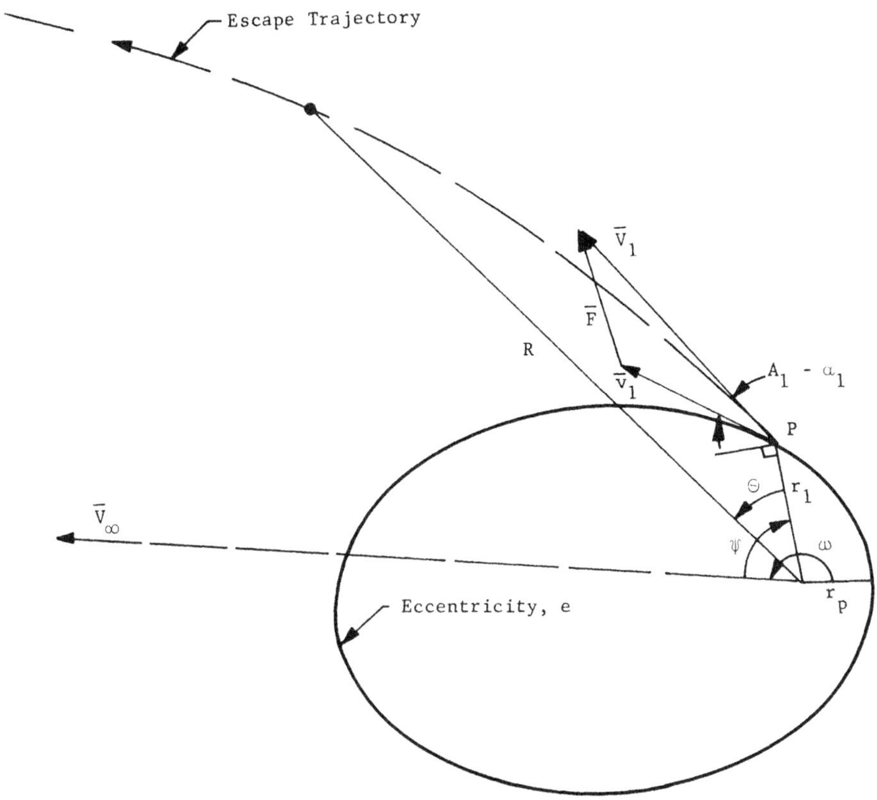

Fig. 5 Geometry of paths and escape pulse.

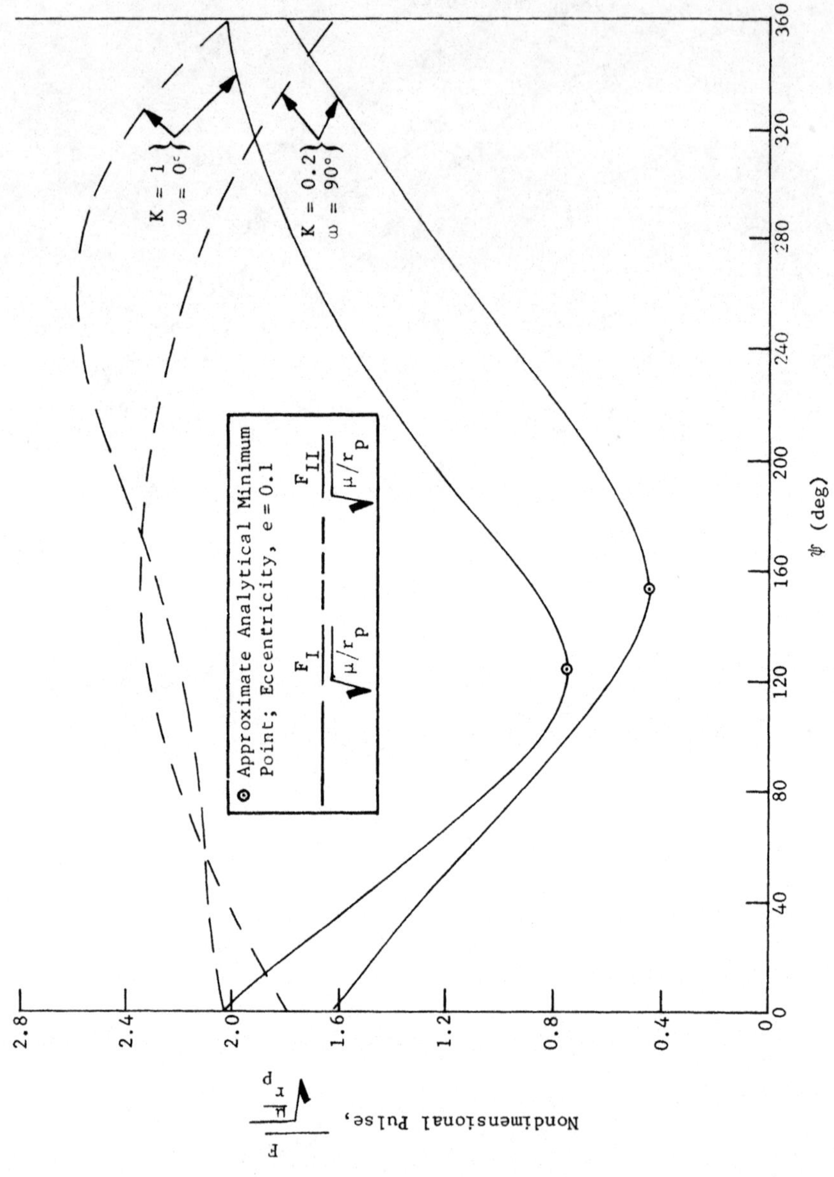

Fig. 6 Nondimensional escape pulse magnitude as a function of ψ for two values of K and ω.

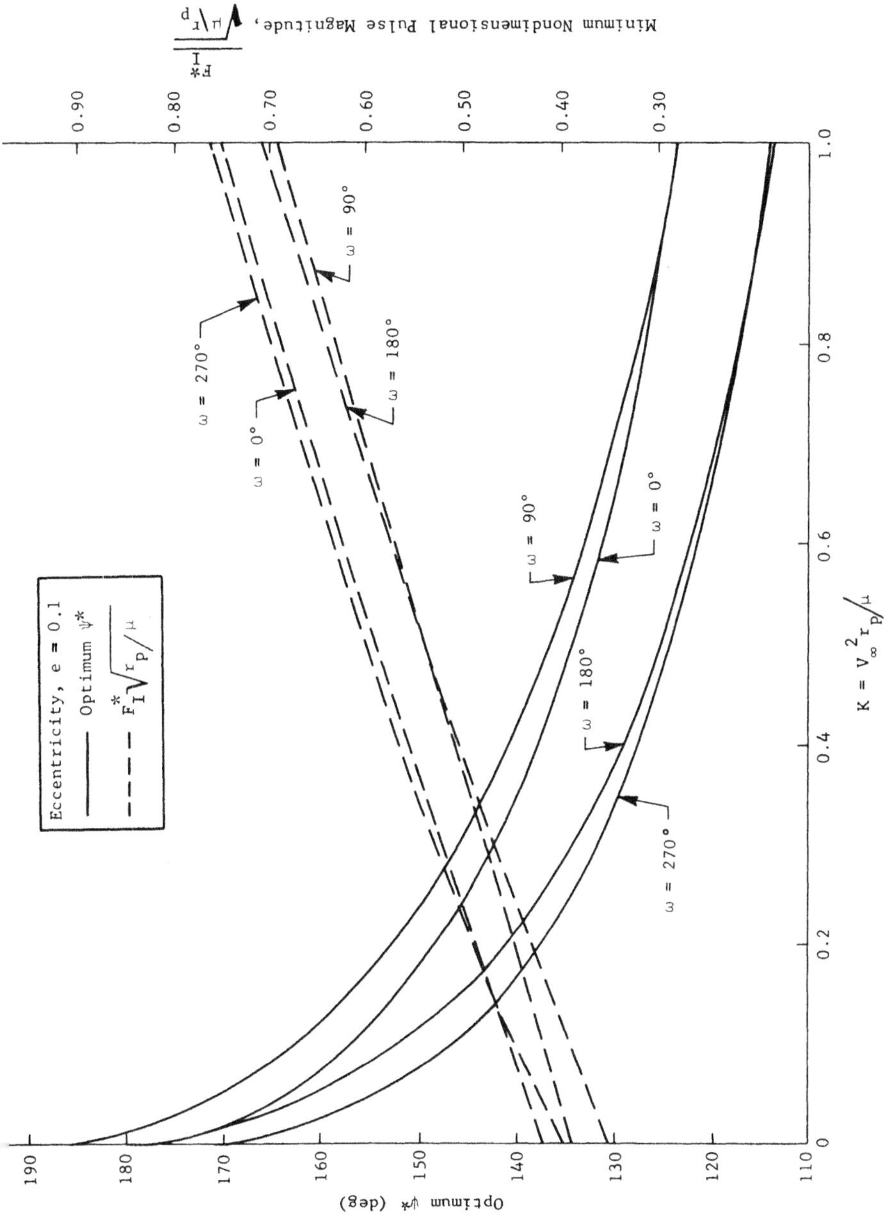

Fig. 7 Minimum nondimensional pulse magnitude and ψ* as functions of residual speed.

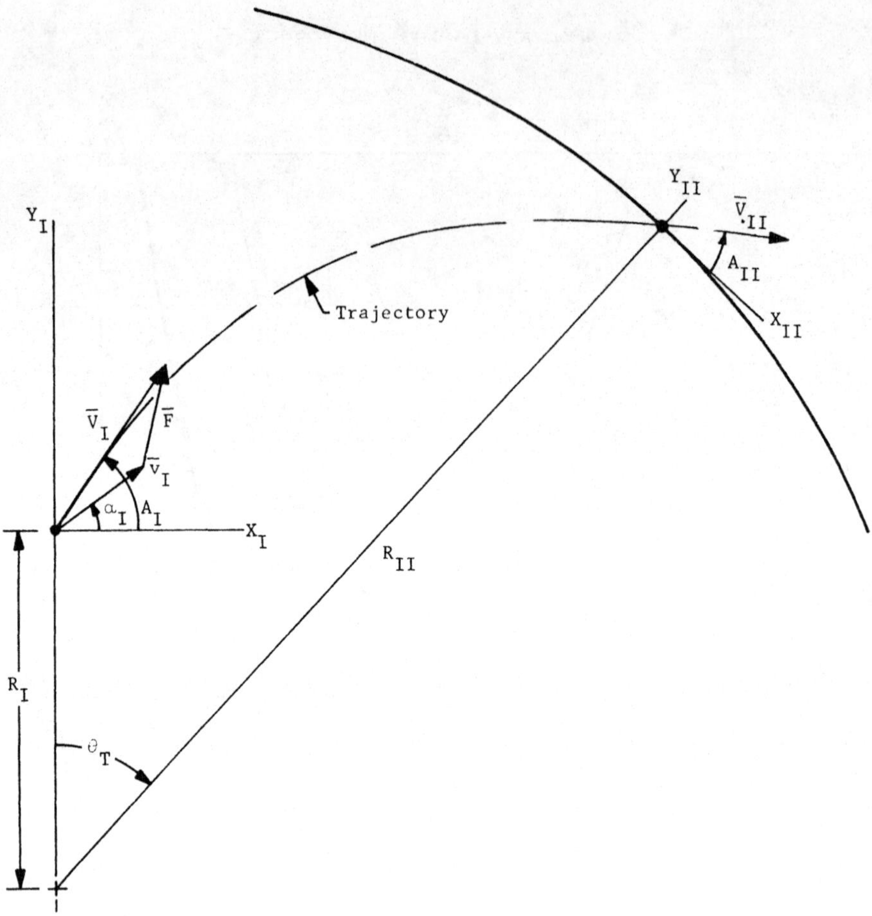

Fig. 8 Schematic illustrating geometry and nomenclature.

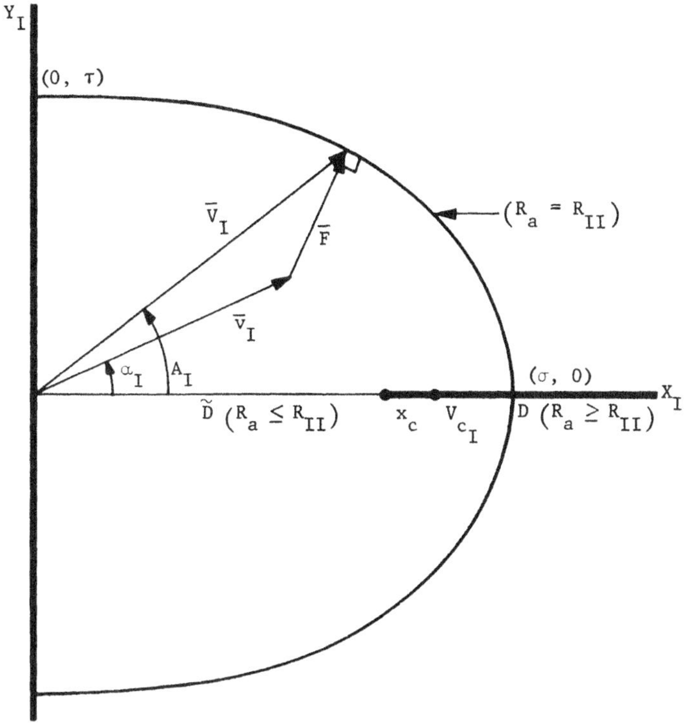

Fig. 9 Minimum pulse geometry for $R_I < R_{II}$.

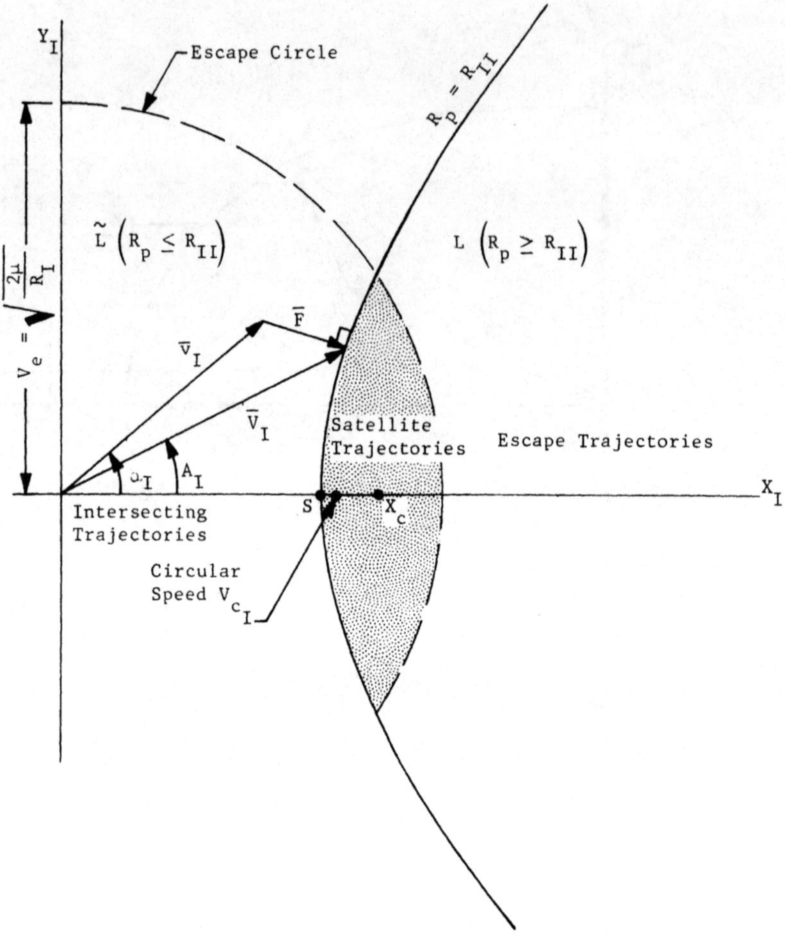

Fig. 10 Minimum pulse geometry for $R_I > R_{II}$.

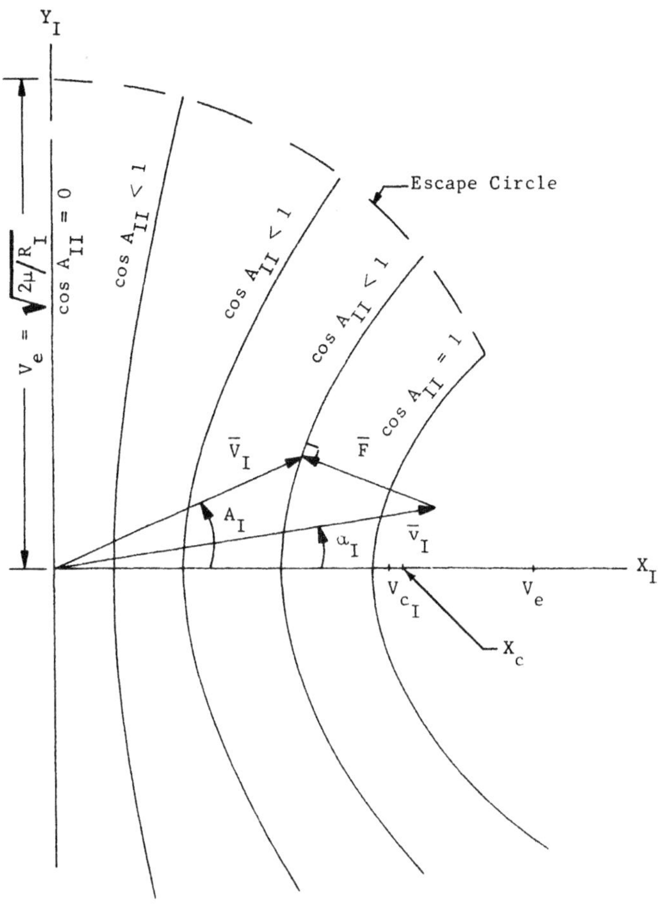

Fig. 11 Schematic illustrating curves of constant A_{II} for $R_I > R_{II}$.

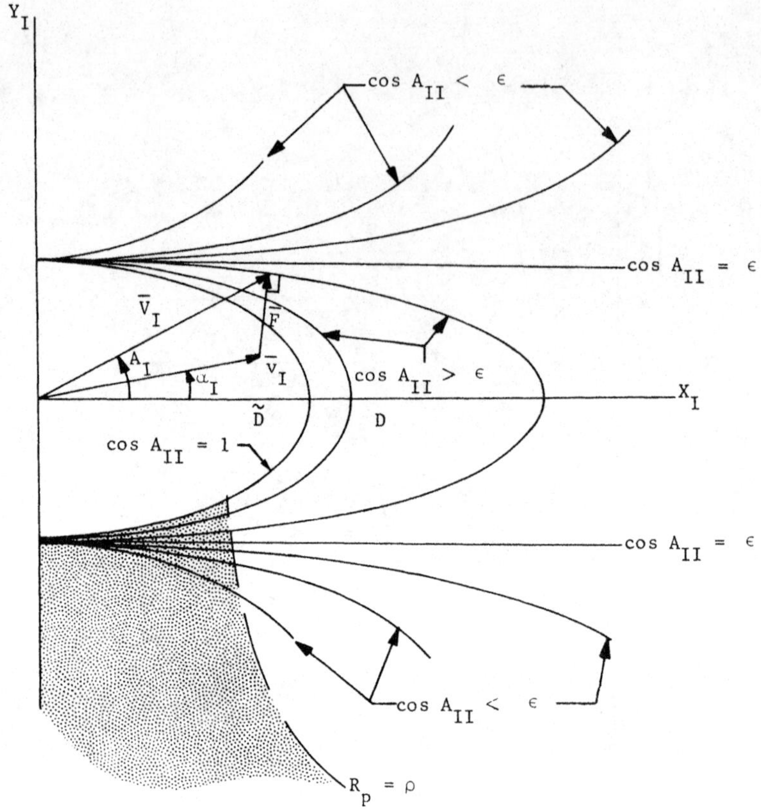

Fig. 12 Schematic illustrating curves of constant A_{II} for $R_I < R_{II}$.

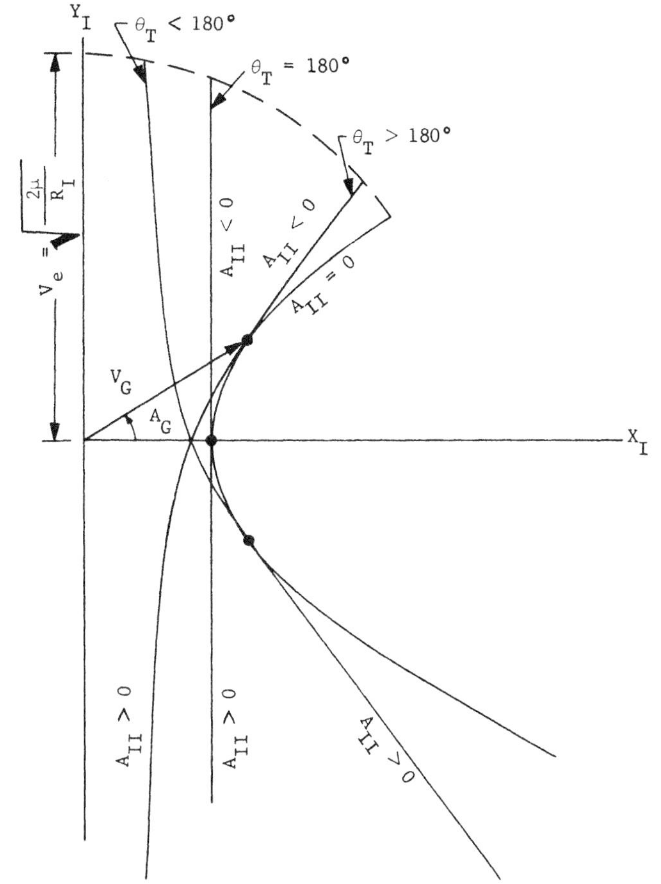

Fig. 13 Schematic illustrating θ_T-curves for $R_I > R_{II}$.

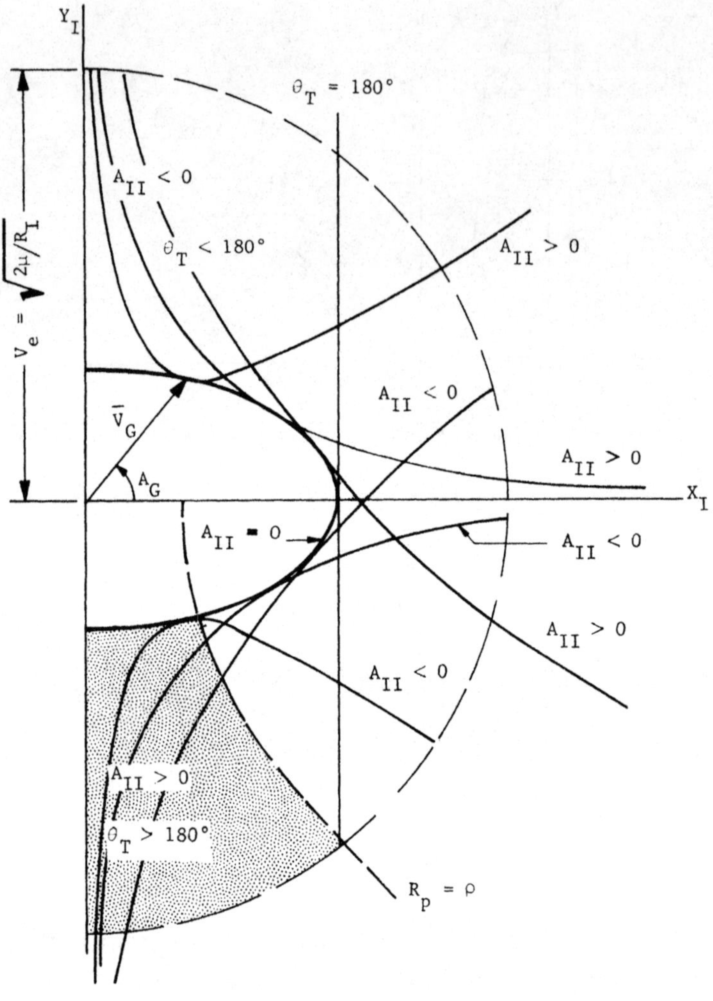

Fig. 14 Schematic illustrating θ_T-curves for $R_I < R_{II}$.

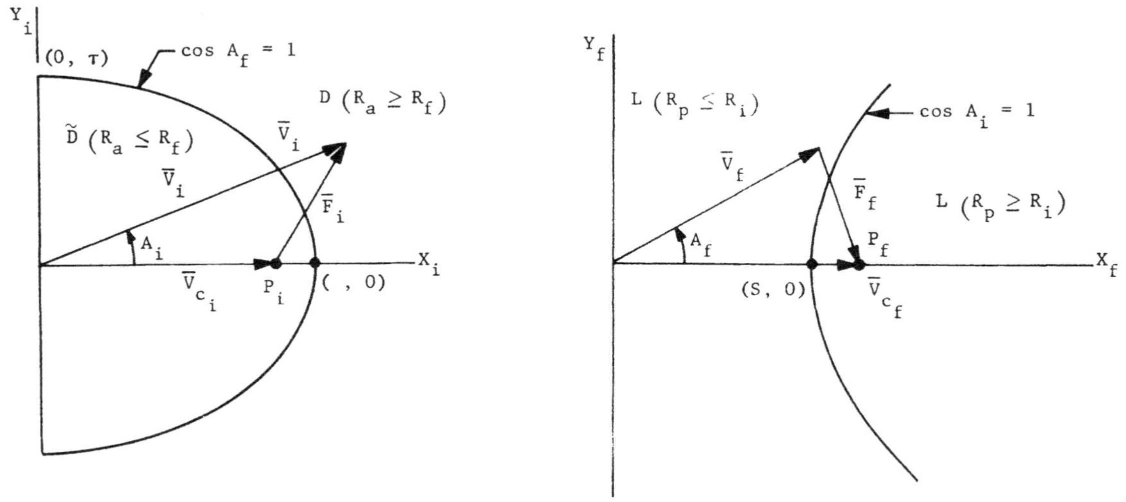

Fig. 15 Pulse geometry associated with typical admissible pulses $\overline{F}_i$ and $\overline{F}_f$.

ANALYSIS OF THE ORBITAL TRANSFER PROBLEM IN THREE-DIMENSIONAL SPACE

Samuel P. Altman* and Josef S. Pistiner+

United Aircraft Corporate Systems Center, Farmington, Conn.

Abstract

This paper extends previous analysis of two-dimensional (or coplanar) orbital transfer to three-dimensional space, by use of the orbital hodograph. The transfer orbit is obtained by the use of two impulsive changes of the vehicle velocity vector at two specified endpoints. The gravitational field of the attracting celestial body is assumed as a simple spherical harmonic (i.e., of zero order). The basic equations for the three-dimensional transfer are formulated in terms of the three hodograph parameters (of which one is independent) and the angles of intersection between the transfer orbit plane and the initial and final orbit planes. The transfer orbit solution requiring minimum total velocity increment is shown to be the real root of an eighth-order polynomial equation with constant coefficients. The basic equations and polynomial coefficients reduce to those for the previous case of coplanar transfer. The possible provision of the oblateness effects upon the nodal and apsidal lines of the transfer orbit are discussed.

Nomenclature

A,B,D,E,F,G,H,J = terms defining the polynomial coefficients K_i as functions of the endpoint conditions

A',B',D',E',F',G',H',J' = terms defining the nondimensional

Presented as Preprint 63-411 at the AIAA Astrodynamics Conference, New Haven, Conn., August 19-21, 1963.

*Supervisor, Space Sciences; presently Manager, Systems Analysis Engineering, The General Electric Missile and Space Division, Valley Forge, Pa.

+Supervisor, Guidance and Control.

		polynomial coefficients as functions of the end-point conditions
a,b,c,d,e	=	coefficients of the velocity increment equation
C,R	=	velocity parameters of the orbit
E	=	total point-mass potential and kinetic energy of orbital motion
h	=	total point-mass angular momentum of orbital motion
$\bar{i}, \bar{j}, \bar{k}$	=	orthogonal unit vectors of a rotating polar coordinate system
i	=	incidence angle between the orbital plane and the equatorial plane of the attracting celestial body
J	=	first coupling constant of the spherical harmonic expansion for the gravitational potential of the celestial body, described as an oblate spheroid
a'	=	semimajor axis
K_i	=	coefficients of the general octic equation, where i = 0, 1, 2,..., 8
m	=	mass of the orbital vehicle
O	=	center of the celestial body, described as a sphere
P	=	point in space defining the terminal point of a position vector $\bar{r}$ originating from the celestial body center 0
$\bar{r}$	=	position vector originating from the celestial body center 0
r_e	=	equatorial radius of the terrestrial sphere

r	=	with subscripts 1 or 2, radial distance between the gravitational centers of the orbital and celestial bodies; with all other subscripts, nondimensional ratio
r_o	=	normalized circular orbit velocity at the final transfer point $(= v_{c2}/v_{c1})$
r_A	=	normalized initial orbit parameter $C\ (= C_A/v_{c1})$
r_B	=	normalized final orbit parameter $C\ (= C_B/v_{c1})$
$\bar{s}$	=	unit vector normal to the initial orbit plane
$\bar{t}$	=	unit vector normal to the transfer orbit plane
u	=	normalized initial velocity increment $(= \Delta v_1/v_{c1})$
$\bar{v}$	=	velocity vector
v	=	scalar magnitude of the velocity vector $\bar{v}$
Δv	=	scalar magnitude of the vector difference $\Delta\bar{v}$ between two velocity vectors
w	=	normalized final velocity increment $(= \Delta v_2/v_{c1})$
X	=	rectilinear coordinate along the apsidal line, fixed in inertial space
x	=	velocity parameter of the transfer orbit normalized with respect to the circular orbit velocity at the initial transfer point $(= C_T/v_{c1})$
Y	=	rectilinear coordinate normal to the apsidal line in the orbital

		plane, fixed in inertial space
y	=	normalized velocity parameter of the transfer orbit $(=R_T/v_{c1})$
Z	=	rectilinear coordinate perpendicular to the orbital plane, fixed in inertial space
z	=	linear velocity normalized with respect to the circular orbit velocity at the initial transfer point (e.g., $z_{T2} = v_{T2}/v_{c1}$)
α	=	angular difference between the apsidal lines of the initial orbit and the transfer orbit at perigee with sense established by that order
β	=	angle between the hodograph v_ϕ axis and the hodograph total velocity
γ	=	angle between the hodograph v_ϕ axis and the hodograph radius R to the body point on the hodograph
δ, ρ, η	=	direction angles for the orthogonal inertial coordinate axes X, Y, Z, respectively
ϵ	=	eccentricity of the conic section
ζ	=	angle of intersection between the plane of the initial orbit and the plane of the transfer orbit
η	=	angular difference between the apsidal lines of the transfer orbit and final orbit at perigee with sense established by that order
θ	=	angle between the total velocity vector and the local horizon in the orbital plane

K_i = nondimensional coefficients of the general nondimensional octic equation, where i = 0, 1, 2, 3..., 8

λ = angle of intersection between the plane of the final orbit and the plane of the transfer orbit

μ = gravitational constant of the celestial body

σ = angular difference between the transfer orbit phases for the two transfer points

Φ = direction of the radial line between centers of the orbital and celestial bodies

ϕ = true anomaly of the vehicle in orbit

Ψ = apsidal line of the orbit at perigee

Ω = reference angle of the line of nodes for the given orbital plane

ψ = angle between the line of nodes and the apsidal line of the orbit at perigee

$(\sim)$ = specific value, i.e., per unit-mass of the (   )-term

$(\cdot)$ = first time derivative of the (   )-term

Subscripts

A = initial orbit

B = final orbit

c1 = circular orbit of radius = r_1

c2 = circular orbit of radius = r_2

H	=	projection upon the local horizon plane
i,j,k	=	projection upon the $\bar{i}$, $\bar{j}$, $\bar{k}$ axes, respectively
J	=	due to the coupling constant J of the harmonic expansion of the potential field
r	=	radial
T	=	transfer orbit
1	=	point of transfer from the initial orbit to the transfer orbit
2	=	point of transfer from the transfer orbit to the final orbit
ϕ	=	normal to the radius, in the orbital plane

Introduction

Orbital transfer of a space vehicle subject to the gravitational attraction of a major celestial body constitutes a classical problem of astronautics. The most rudimentary class of orbital transfer is provided by the use of two impulsive thrusts causing instantaneous velocity vector changes at each of two transfer points. In this case, the transfer trajectory is then an orbital segment that is defined by the radius and velocity vectors at each of the two end-points.

Although this transfer problem is basic in nature, much still remains unknown about it and its characteristics. In general, most previous work has been based upon determination of various optimum solutions. Of course, many different optimization criteria and problem formulations are possible. At present, one of the most urgent optimization criteria requires the determination of that one transfer trajectory requiring the least total change in velocity; that is, the least total fuel consumption for the transfer maneuver. An equally vital problem that does not involve an optimization is the determination of the transfer trajectory that requires a specified time interval for the transfer.

The best known optimization for coplanar transfer has been

the Hohmann transfer ellipse.[1] The Hohmann transfer ellipse requires the least total velocity change for transfer between circular orbits or between elliptical orbits with collinear apsidal lines. The two transfer points must be 180° displaced in space (i.e., the central angle = 180°). However, it should be noted that, for certain end-point conditions, the use of three transfer points (or bi-elliptical transfer between coplanar circular orbits) has been shown to require even less total velocity change[2] than Hohmann transfer.

Various studies have been conducted on the two-impulse orbital transfer problem.[3-10] Most of such work required optimization of the total velocity change for transfer between two given orbits, with the end-points free for suitable selection. However, in some cases, realistic mission requirements will probably specify desired end-points. For example, the transfer point may be specified over a ground tracking (or command and control) site. Also, specific trajectory characteristics, not necessarily based upon an optimization solution, may be desired. Consequently, the two-impulse orbital transfer discussed in this paper will provide a general formulation for transfer between arbitrarily specified end-points on given orbits.

In this paper, the orbital transfer analysis is based upon the use of the orbital hodograph parameters. These parameters provide unique transfer relations between the various velocity vectors in very basic mathematical form. Earlier work developed the hodograph equations for coplanar transfer[6] and some optimization solutions.[10-12]

The present paper presents 1) the complete equations for two-impulse orbital transfer in three-dimensional space, 2) the optimization solution for minimum total velocity change of three-dimensional orbital transfer, and 3) various suggested applications to possible mission requirements. It is recognized that the astrodynamics specialist will probably define the orbits and end-points by sets of orbital parameters or by latitude - longitude data, for reference to a fixed rectilinear coordinate system with origin at the gravitational center of the celestial body. Consequently, the angle of intersection (between orbital planes) and the transfer central angle (between end-points) are defined (in the Appendix) by vector relations that are functions of these parameters.

It should be noted that the transfer equations and optimization solutions presented in this paper are not restricted to elliptical orbits but are valid for parabolic or hyperbolic orbits as well. Consequently, transfer between orbits of any form of conic section (e.g., from elliptic to hyperbolic) can be analyzed by the one general formulation presented herein. However, as presently stated, this formulation is only valid for the potential field of a regular, spherically symmetric celestial body. If analytic provision for the effects of higher-order harmonic terms such as those due to oblateness or triaxiality of the celestial body is desired, these requirements may be incorporated as suggested in the paper.

Velocity Hodograph of an Orbit

In the 19th century, Hamilton and Möbius independently developed the concept of the velocity hodograph for a planar orbit.[13-14] As an orbital vehicle proceeds along a planar path, the locus of the velocity vector terminus with its origin fixed in space, describes a planar figure defined as the hodograph for the trajectory. ‡ The hodograph of the velocity vector for an orbit within the gravitational field of a regular spherical celestial body is a circle.

In planar nonorthogonal coordinates, the orbital velocity is defined (as shown in Fig. 1) by the vector addition of the constant magnitude vector $\overline{C}$ directed normal to the radius vector $\overline{r}$, and the constant magnitude vector $\overline{R}$ directed perpendicular to the major axis (or line of apses) for the orbital conic.

By the use of rotating reference axes with origin at the vehicle center of mass, the velocity hodograph can be presented in planar orthogonal coordinates. Let the reference coordinate system be defined by the unit vector $\overline{i}$, and the unit vector $\overline{j}$ normal to $\overline{i}$, positive in the sense shown in Fig. 1. Then the energy equation for the planar orbit (in vacuum) of a vehicle within the gravitational field of a regular spherical celestial body is as follows:[14]

$$E = m(v_{\phi}^2 + v_r^2)/2 - \mu m/r = \text{const} \quad (1)$$

Since the total angular momentum is

$$h = mrv_{\phi} = \text{const} \quad (2)$$

‡ Hodograph of a variable vector: The locus of one end of a radius vector (the other end being fixed) that represents in magnitude and direction the rate of change of the variable vector.[21]

Equation (1) may now be rewritten as

$$R^2 = v_r^{\,2} + (v_\phi - C)^2 \tag{3}$$

where

$$C = \mu m/h = \mu/rv_\phi \tag{4}$$

$$R = \sqrt{2E/m + (\mu m/h)^2} = \sqrt{2\tilde{E} + C^2} \tag{5}$$

In rectilinear coordinates fixed in inertial space, it can be shown that

$$\dot{X}^2 + (\dot{Y} - R)^2 = C^2 \tag{6}$$

It is seen that Eq. (3) represents a circle (as shown in Fig. 2) which is the polar velocity hodograph of the orbit in space (as shown in Fig. 2). In general, all conic sections (ellipse, parabola, and hyperbola) in space are transformed into circles in the velocity hodograph plane.[15]

For the impulsive transfer problem, two orbits differing in one or more orbital parameters must be related analytically. The necessary relations are provided by the condition (imposed by the instantaneous orbital change) that the position vector $\bar{r}$ at a transfer point must be defined identically by either orbit. That is,

$$r_{A1} = r_{T1} \tag{7}$$

$$\Phi_{A1} = \Phi_{T1} \tag{8}$$

where the direction of the position vector $\bar{r}$ in inertial space is defined by Φ. Previous two-dimensional analysis[6] has shown that these relations are most amenable to mathematical treatment when the definitive equations for the respective orbits are suitably normalized. The resultant velocity hodographs for orbits A and T are circles whose components in the rotating coordinate system are

$$Cv_r = CR \sin \phi \tag{9}$$

$$Cv_\phi = C^2 + CR \cos \phi \tag{10}$$

This hodograph scale change enables the transfer problem to be formulated by use of simple trigonometric relations and conformal mapping of all space angles.

General Equations for Three-Dimensional Orbital Transfer

The three-dimensional orbital transfer is shown schematically in Fig. 3. The vehicle will proceed in the initial orbit A until the initial transfer point P_1 (defined by the position vector $\bar{r}_1$) is attained. An instantaneous change in the velocity vector will cause the orbital vehicle to depart from point P_1 along the segment $P_1 - P_2$ of the transfer orbit T. Upon arrival at the final transfer point P_2 (defined by the position vector $\bar{r}_2$), a final velocity vector change will cause the orbital vehicle to proceed along the final orbit B.

The initial and final orbits are defined by the given sets of orbital parameters. The arbitrarily selectable transfer points may be defined by the true anomaly ϕ from perigee passage Ψ if the position vectors $\bar{r}_1$ and $\bar{r}_2$ are not known explicitly. As an alternative, the latitude and longitude coordinates of the intersections of the celestial body surface by the position vectors may define the transfer points.

Having selected the transfer points P_1 and P_2 (and thereby assigning the directions Φ of the position vectors), the space orientation of the transfer orbit T is defined. That is, the plane of the transfer orbit is determined by three points in space (P_1, P_2, 0) or by two vectors with one common origin ($\bar{r}_1, \bar{r}_2$). Consequently, the angles of intersection ζ and λ between the orbital planes are determined by the initial and final orbits, respectively, and the transfer points. Also, the central angle σ , which lies between $\bar{r}_1$ and $\bar{r}_2$ in the transfer orbit plane, is defined by these conditions. The analytic definitions of the central angle σ and the intersection angle ζ (or λ) are presented in the Appendix.

The equations for transfer at point P are obtained by reference to Fig. 4. A required velocity vector increment (not shown in the figure) must be added to the initial velocity vector $\bar{v}_A$ lying in the initial orbit plane, so that the transfer velocity vector $\bar{v}_{T1}$ lying in the transfer orbit plane is obtained. Noting that the intersection angle lies in the plane of the local horizons for P_1 (i.e., $\bar{r}_1$ is the line of intersection between the initial and transfer orbit planes), the orthogonal components of the velocity vectors are as follows:

$$v_{A/i} = v_A \sin \theta_A \tag{11}$$

$$v_{A/H} = v_A \cos \theta_A \tag{12}$$

$$v_{A/j} = v_{A/H} \cos \zeta = v_A \cos \theta_A \cos \zeta \tag{13}$$

$$v_{A/k} = v_{A/H} \sin \zeta = v_A \cos \theta_A \sin \zeta \tag{14}$$

$$v_{T1/i} = v_{T1} \sin \theta_{T1} \tag{15}$$

$$v_{T1/j} = v_{T1} \cos \theta_{T1} \tag{16}$$

$$v_{T1/k} = 0 \tag{17}$$

Also, $\Delta v_1^{\,2} = (v_{T1/i} - v_{A/i})^2 + (v_{T1/j} - v_{A/j})^2 + (v_{T1/k} - v_{A/k})^2$ (18)

so that $\Delta v_1^{\,2} = (v_{T1} \sin \theta_{T1} - v_A \sin \theta_A)^2 + (v_T \cos \theta_{T1} - v_A \cos \theta_A \cos \zeta)^2 +$
$(v_A \cos \theta_A \sin \zeta)^2$ (18a)

In the three-dimensional problem, the hodograph for orbit A must be rotated about the Cv_ϕ axis through the intersection angle ζ. With this orientation, the hodograph for orbit A would then lie in the plane of the transfer orbit. The hodograph relations described in the previous section are directly applicable. Consequently,

$$v_A^{\,2} = v_{A/i}^{\,2} + v_{A/H}^{\,2} \tag{19}$$

where

$$v_{A/i} = (v_{c1}^{\,2}/C_A - C_A) \tan \phi_A \tag{20}$$

$$v_{A/H} = v_{c1}^{\,2}/C_A \tag{21}$$

and

$$v_{T1}^{\,2} = v_{T1/i}^{\,2} + v_{T1/j}^{\,2} \tag{22}$$

where

$$v_{T1/i} = (v_{c1}^{\,2}/C_T - C_T) \tan \phi_{T1} \tag{23}$$

$$v_{T1/j} = v_{c1}^{\,2}/C_T \tag{24}$$

That is, the radial components ($v_{A/i}$ and $v_{T1/i}$) of the respective velocity vectors ($\bar{v}_A$ and $\bar{v}_{T1}$) are comprised of the abscissa or (Cv_r/C) components of the respective CR radii to the intersections with the $(Cv_\phi)_1$ line. The horizontal components ($v_{A/H}$ and $v_{T1/j}$) of the respective velocity vectors ($\bar{v}_A$ and $\bar{v}_{T1}$) are comprised of the ordinate or (Cv_ϕ/C) components of the respective CR radii. Note that

$$v_{cm}^{2} = (Cv_{\phi})_{m} = \mu/r_{m} \tag{25}$$

where m = 1 or 2; also, that the true anomaly ϕ_A lies in the initial orbit plane, whereas the true anomaly ϕ_T lies in the transfer orbit plane.

The equations for transfer at point P_2 are obtained similarly, upon reference to Fig. 4. The complete set of transfer equations are as follows:

$$v_{nm}^{2} = (v_{cm}/C_{n})^{2} + (v_{cm}/C_{n} - C_{n})^{2} \tan^{2} \phi_{nm} \tag{26-29}$$

where

m = 1 and n = A or T

m = 2 and n = T or B

$$\Delta v = \left|\Delta v_1\right| + \left|\Delta v_2\right| \tag{30}$$

$$\Delta v_1^{2} = \left[a_1(C_A - C_T \cos\zeta)^2 + (b_1 C_T^{2} + c_1 C_T + d_1)^2\right] / C_A^2 C_T^2 \sin^2\sigma + e_1^2 \tag{18b}$$

$$\Delta v_2^{2} = \left[a_2(C_B - C_T \cos\lambda)^2 + (b_2 C_T^{2} + c_2 C_T + d_2)^2\right] / C_B^2 C_T^2 \sin^2\sigma + e_2^2 \tag{31}$$

where

$$\sigma = (\phi_B - \phi_A) + (\Psi_B - \Psi_A) \tag{32}$$

$$a_m = v_{cm}^{4} \sin^2\sigma \tag{33, 34}$$

$$b_1 = C_A (1 - \cos\sigma) \tag{35}$$

$$c_1 = (C_A^{2} - v_{c1}^{2}) \tan\phi_A \sin\sigma \tag{36}$$

$$d_1 = C_A (v_{c1}^{2} \cos\sigma - v_{c2}^{2}) \tag{37}$$

$$e_1 = v_{c1}^{2} \sin\zeta / C_A \tag{38}$$

$$b_2 = C_B (\cos\sigma - 1) \tag{39}$$

$$c_2 = (C_B^{2} - v_{c2}^{2}) \tan\phi_B \sin\sigma \tag{40}$$

$$d_2 = C_B (v_{c1}^{2} - v_{c2}^{2} \cos\sigma) \tag{41}$$

$$e_2 = v_{c2}^{2} \sin\lambda / C_B \tag{42}$$

and

$$\phi_{T1} = \arctan\left\{\left[\cos\sigma - (C_T^{2} - v_{c2}^{2})/(C_T^{2} - v_{c1}^{2})\right]/\sin\sigma\right\} \tag{43}$$

$$\phi_{T2} = \text{arc tan}\left\{\left[(C_T^{\,2} - v_{c1}^{\,2})/(C_T^{\,2} - v_{c2}^{\,2}) - \cos\sigma\right] / \sin\sigma\right\} \quad (44)$$

$$C_T R_T = (v_{c1}^{\,2} - C_T^{\,2})/\cos\phi_{T1} = (v_{c2}^{\,2} - C_T^{\,2})/\cos\phi_{T2} \quad (45)$$

or

$$C_T R_T = (v_{c2}^{\,2} - v_{c1}^{\,2})/(\cos\phi_{T2} - \cos\phi_{T1}) \quad (45a)$$

A nondimensional form of the transfer equations is as follows:

$$z_{A1}^{\,2} = 1/r_A^{\,2} + (1/r_A - r_A)^2 \tan^2\phi_{A1} \quad (26a)$$

$$z_{T1}^{\,2} = 1/x^2 + (1/x - x)^2 \tan^2\phi_{T1} \quad (27a)$$

$$z_{T2}^{\,2} = (r_o^{\,2}/x)^2 + (r_o^{\,2}/x - x)^2 \tan^2\phi_{T2} \quad (28a)$$

$$z_{B2}^{\,2} = (r_o^{\,2}/r_B)^2 + (r_o^{\,2}/r_B - r_B)^2 \tan^2\phi_B \quad (29a)$$

$$u^2 = \left[(r_A - x\cos\zeta)/r_A x\right]^2 + \left[x(1 - \cos\sigma) + (r_A^{\,2} - 1)\tan\phi_{A1}\sin\sigma/r_A + (\cos\sigma - r_o^{\,2})/x\right]^2 + (\sin\zeta/r_A)^2 \quad (18c)$$

$$w^2 = \left[(r_B - x\cos\lambda)r_o^{\,2}/r_B x\right]^2 + \left[x(\cos\sigma - 1) + (r_B^{\,2} - r_o^{\,2})\tan\phi_{B2}\sin\sigma/r_B + (1 - r_o^{\,2}\cos\sigma)/x\right]^2 + (\sin\lambda/r_B)^2 \quad (31a)$$

$$\phi_{T1} = \text{arc tan}\left\{\left[\cos\sigma - (x^2 - r_o^{\,2})/(x^2 - 1)\right]/\sin\sigma\right\} \quad (43a)$$

$$\phi_{T2} = \text{arc tan}\left\{\left[(x^2 - 1)/(x^2 - r_o^{\,2}) - \cos\sigma\right]/\sin\sigma\right\} \quad (44a)$$

$$xy = (1 - x^2)/\cos\phi_{T1} = (r_o^{\,2} - x^2)/\cos\phi_{T2} \quad (45b)$$

or

$$xy = (r_o^{\,2} - 1)/(\cos\phi_{T2} - \cos\phi_{T1}) \quad (45c)$$

Some of the equations presented above are undefined for $\sigma = \pi$. Consequently, the corresponding transfer equations that are defined, for this case, are as follows:

$$C_T^{\,2} = (v_{c1}^{\,2} + v_{c2}^{\,2})/2 \quad (46)$$

$$\Delta v_1^2 = \left[v_{c1}^2 (C_A - C_T)/C_A C_T \right]^2 + \left[(v_{c1}^2 - C_T^2) \tan \phi_{T1}/C_T - (v_{c1}^2 - C_A^2) \tan \phi_{A1}/C_A \right]^2 \qquad (47)$$

$$\Delta v_2^2 = \left[v_{c2}^2 (C_B - C_T)/C_B C_T \right]^2 + \left[(v_{c2}^2 - C_T^2) \tan \phi_{T2}/C_T - (v_{c2}^2 - C_B^2) \tan \phi_{B2}/C_B \right]^2 \qquad (48)$$

$$\phi_{T1} = \phi_{T2} - \pi \qquad (49)$$

The equations were derived from the basic velocity vector components expressed by hodograph parameters. It is apparent that here, $\phi_{T1} = 0$ or π for the required total velocity change to be a minimum. Note that all orbits must then lie in one plane.

Optimization Solutions for Minimum Total Velocity Increment

The general equations for two impulse orbital transfer in three-dimensional space define all possible transfer trajectories and the attendant velocity vector changes in terms of the end-point position and velocity conditions and one independent variable, the hodograph parameter C_T. Given the end-point conditions, a single infinity of such orbits is possible. If one trajectory characteristic or additional condition is imposed (such as due to the space mission requirements), then one orbit, out of the single infinity of solutions, will comprise a unique solution.

One such additional constraint, of great import to the vehicle stage design, is the optimization of the on-board fuel consumption. In terms of the orbital transfer dynamics, the one orbit that requires the least total velocity change [see Eq. (30)] will require the least fuel consumption. Note that the fuel mass is not explicitly present in the transfer equations, because these equations are valid for unit mass in orbit.

Since the total velocity change Δv_T is expressed in terms of one independent variable, the interior maxima and minima for Δv_T over the C_T range of solution are provided by the positive real roots of

$$d(\Delta v)/d C_T = 0 \qquad (50)$$

Of these roots, one will provide the least Δv_T interior to the C_T range. When the C_T range of solution is finite (or bounded), then the end-points must be examined to assure that the absolute least minimum does not occur at such points. In Ref. 6, the end-points of the C_T range for elliptical orbits was defined analytically so that this consideration could be examined. However, the transfer equations and Eq. (50) are continuous and valid over the entire C_T range of solutions, elliptic, parabolic or hyperbolic, for C_T from zero to values approaching infinity. But examination of Eqs. (30, 18b and 31) shows that the velocity increment $\Delta v_T \to \infty$ for $C_T \to 0$ or $C_T \to \infty$. Consequently, the absolute minimum will be an interior point given by Eq. (50).

The use of Eqs. (18, 30 and 31) with Eq. (50) results, upon reduction, in the eighth-order polynomial (or octic) equation (51) that follows:

$$K_8 C_T^8 + K_7 C_T^7 + K_6 C_T^6 + K_5 C_T^5 + K_4 C_T^4 + K_3 C_T^3 + K_2 C_T^2 + K_1 C_T + K_0 = 0 \tag{51}$$

where

$$K_8 = A\left[(G^2 - B^2) + 4A\,(D - H)\right] \tag{51-1}$$

$$K_7 = 4A\left[2A(J-E) + (DG-BH)\right] + \left[BG(G-B)\right] \tag{51-2}$$

$$K_6 = 2A\left[4(BJ-EG) + (GJ-BE)\right] + \left[DG^2 - B^2H\right] \tag{51-3}$$

$$K_5 = 4A\left[2F(G-B) + (DJ-EH)\right] + \left[2BG(J-E) + B^2J - EG^2\right] \tag{51-4}$$

$$K_4 = A\left[(J^2-E^2) - 8F(D-H)\right] + \left[F(G^2-B^2)\right] + \left[2(DGJ - BEH)\right] \tag{51-5}$$

$$K_3 = -4F\left[2A(J-E) + (DG-BH)\right] - \left[2EJ(G-B)\right] + \left[BJ^2 - E^2G\right] \tag{51-6}$$

$$K_2 = 2F\left[-4(BJ-EG) + (GJ-BE)\right] + \left[DJ^2 - E^2H\right] \tag{51-7}$$

$$K_1 = -4F\left[2F(G-B) + (DJ-EH)\right] - \left[EJ(J-E)\right] \tag{51-8}$$

$$K_0 = F\left[(J^2-E^2) + 4F(D-H)\right] \tag{51-9}$$

and

$$A = b_1^2/C_A^2 \tag{51a}$$

$$B = 2b_1c_1/C_A^2 \tag{51b}$$

$$D = (a_1^2 + 2b_1d_1 + c_1^2)/C_A^2 \tag{51c}$$

$$E = 2(C_A a_1 \cos \zeta - c_1 d_1)/C_A^2 \tag{51d}$$

$$F = v_{c1}^4 - 2v_{c1}^2 v_{c2}^2 \cos \sigma + v_{c2}^4 \tag{51e}$$

$$G = 2b_2c_2/C_B^2 \tag{51f}$$

$$H = (a_2^2 + 2b_2 d_2 + c_2^2)/C_B^2 \tag{51g}$$

$$J = 2(C_B a_2 \cos \lambda - c_2 d_2)/C_B^2 \tag{51h}$$

Note that the terms a, b, c and d are defined in Eqs. (33-41). The constant coefficients K_i (i=0, 1, 2,..8) are defined by the boundary conditions of the problem, i.e., the initial transfer point 1 on the initial orbit A and the final transfer point 2 on the final orbit B. The one positive real root of the polynomial equation, which results in the minimum total velocity increment Δv, provides the absolute minimum velocity increment solution. If both the initial and final orbits are circular, then $c_1 = c_2 = 0$ so that B = G = 0 and $K_6 = 0$. If the entire transfer lies in one plane, then $\zeta = \lambda = 0$ so that Eq. (51) would reduce to the two-dimensional form of Ref. 6.

The nondimensional form of the optimizing polynomial will be identical to Eq. (51). In this form, the nondimensional variable x replaces the orbital hodograph variable C_T, with the nondimensional coefficients $\bar{K}_i$ of identical algebraic form as the dimensional coefficients K_i. Then the corresponding forms of Eqs. (51-10) to (51-17) will be as follows:

$$A' = (1 - \cos \sigma)^2 \tag{52a}$$

$$B' = 2(r_A^2 - 1)(1 - \cos \sigma) \tan \phi_A \sin \sigma / r_A \tag{52b}$$

$$D' = \left[\sin^2 \sigma + 2r_A^2 (1 - \cos \sigma)(\cos \sigma - r_o^2) + (r_A^2 - 1)^2 \tan^2 \phi_A \sin^2 \sigma\right] / r_A^2 \tag{52c}$$

$$E' = 2\left[\sin^2 \sigma \cos \zeta - (\cos \sigma - r_o^2)(r_A^2 - 1) \tan \phi_A \sin \sigma\right] / r_A \tag{52d}$$

$$F' = 1 - 2r_o^2 \cos\sigma + r_o^4 \tag{52e}$$

$$G' = 2(r_B^2 - r_o^2)(\cos\sigma - 1)\tan\phi_B \sin\sigma / r_B \tag{52f}$$

$$H' = \left[r_o^4 \sin^2\sigma + 2r_B^2(\cos\sigma - 1)(1 - r_o^2\cos\sigma) + (r_B^2 - r_o^2)^2 \tan^2\phi_B \sin^2\sigma \right] / r_B^2 \tag{52g}$$

$$J' = 2\left[r_o^4 \sin^2\sigma \cos\lambda - (1 - r_o^2\cos\sigma)(r_B^2 - r_o^2) \tan\phi_B \sin\sigma \right] / r_B \tag{52h}$$

Study of the terms of the coefficients [Eqs. (52a to 52h)] reveals an interesting generalization of the transfer characteristics. Suppose transfer occurs between circular orbits, so that $c_1 = c_2 = 0$. Then the central angle σ occurs only as $\sin^{2n}\sigma$ and $\cos^n\sigma$, so that the equations are identical for σ and $(2\pi - \sigma)$ since

$$\sin^{2n}(2\pi - \sigma) = \sin^{2n}\sigma$$

$$\cos^n(2\pi - \sigma) = \cos^n\sigma$$

That is, the polynomial solutions are symmetrical about the $0 - \pi$ axis. Consequently, solutions over the σ -range from 0 to π comprise all solutions as a function of the σ end-point conditions. However, such symmetry is not present in the general case, since $\sin^n\sigma$ occurs in the coefficient terms B, E, G and J.

The polynomial equation vanishes for $\sigma = 0$ or π, since then all coefficients become zero. However, only $\sigma = \pi$ is a realizable solution. For this central angle value, all orbits must be coplanar. Consequently, such solutions must be obtained from Eqs. (46-49).

Potential Mission Applications

The transfer equations and optimization presented in the preceding section enable the general analysis of various mission applications that are amenable to the impulsive thrust approximation. Recognizing that realizable thrust must be active over a finite interval rather than instantaneous, it is nevertheless useful to study the transfer characteristics due to impulsive thrust. Either the actual thrust program is closely approximated by impulsive thrust, or the transfer characteristics of impulsive thrust may be used as idealized (or asymptotic) performance objectives.

Although various potential mission applications of the three-dimensional transfer analysis are easily visualized, a few basic applications will be noted and briefly discussed. The planetary orbital transfer (such as for terrestrial or lunar satellites) is of immediate interest. In this mission class, the basic equations are most tractable to the selection of transfer orbits that will meet arbitrarily specified conditions. For example, the end-points may be selected in the immediate vertical region (or zenith) above ground tracking stations. Moreover, it may be desirable to minimize the radial velocity change at the end-points, because of higher tracking accuracy of normal velocity components. The hodograph equations enable the desired transfer orbits to be identified by orbital parameters. Subject to the concurrent fulfillment of the time constraint for rendezvous, the minimum velocity change solution would be provided by Eq. (52) or its nondimensional form.

With arbitrarily specified end-point conditions, the minimal velocity change solution will provide a specific transfer time which, in general, will differ from the desired transfer time. However, the rendezvous time constraint may be treated by further extension of the basic hodograph theory. Five basic approaches are possible. First, the analytic solution for minimum velocity change with specified transfer time may be attempted, by use of the analytic relations of Eqs. (26-45), and the time equations[12]. Second, the neighborhood of the specified end-point conditions may be mapped, in respect to the velocity change and transfer time. That is, a range of variation for velocity change and transfer time will be provided by proximate end-point conditions. Third, with the given end-point conditions, the region of transfer orbit solutions in the neighborhood of the absolute optimum transfer orbit could be used. That is, a transfer orbit with hodograph parameter C_T of value close to that provided by Eq. (51) will require almost minimal velocity change but with a somewhat different transfer time. Fourth, the use of two transfer orbits (such as used in bielliptical transfer) would provide an additional degree of freedom in order to enable optimization subject to a specified time constraint. Fifth, the required transfer time, rather than the position and velocity end-point conditions, may be selected by use of multiple period rendezvous. That is, if the mission permits the free use of more than one orbital period of elapsed time for the rendezvous, then the relative phase (and consequently the transfer time) of target and space vehicle may be selected so that suitable values of the central angle σ are applicable. In this case, however, the oblateness effects upon the orbits must then be considered. Before proceeding to discussion of these effects, it should be noted that these five basic approaches to the transfer time constraint could be investigated in analytic form.

For short-period rendezvous, the central angle σ between the end-points will be less than 2π (i.e., less than one orbital period), so that oblateness effects should be negligible. If such effects are not considered negligible, they may be provided principally upon consideration of the J term of the spherical harmonic expansion for the gravitational potential field of an oblate spheroid:

$$V = -(\mu m/r_e)\left[r_e/r + J(r_e/r)^3(1/3 - \cos^2\eta) + \ldots\right] \tag{53}$$

As shown by Blitzer, Weisfeld and Wheelon [16-18] the apsidal and nodal lines of an orbit are rotating due to the oblateness terms, through the following angles of rotation respectively in each orbital revolution:

$$\Delta\Psi_J = 2\pi J\left[r_e/a'(1-e^2)\right]^2 (2 - 5\sin^2 i/2) \tag{54}$$

$$\Delta\Omega_J = 2\pi J\left[r_e/a'(1-e^2)\right]^2 \cos i \tag{55}$$

By use of the hodograph equations for the semilatus rectum[15] and a time average for the nodal and apsidal line rotations,

$$\Delta\Psi_{JT} = \Delta\Psi_J(\sigma/2\pi) = J\sigma\left[C^2 r_e/\mu\right]^2(2 - 5\sin^2 i/2) \tag{56}$$

$$\Delta\Omega_{JT} = \Delta\Omega_J(\sigma/2\pi) = J\sigma\left[C^2 r_e/\mu\right]^2 \cos i \tag{57}$$

Consequently, the apparent final transfer point P_2' (shown in Fig. 5) will be different from the actual point P_2 in space. That is, the apparent transfer is analyzed for transfer from P_1 to P_2'; then the orbital vehicle will actually travel so that it arrives at point P_2 in space. As a result of these oblateness effects, the angles of intersection (ζ, λ) between orbital planes and the central angle σ are no longer constants, but are functions of the orbital parameter C. Although analytic solution with the resulting set of transfer equations is possible, the complexity indicates resort to machine computation as most desirable.

Interplanetary flight can be analyzed by the three-dimensional orbital transfer as a first approximation, in accordance with current analysis practice. That is, the interplanetary trajectory is best approximated by a restricted four-body problem: the vehicle, sun, departure planet (or moon) and arrival planet (or moon). However, the planetary influences might be neglected so as to consider the vehicle as a solar satellite transferring between points in the neighborhood of the planetary bodies.

Rendezvous between a satellite orbiting about a celestial body, such as the earth, and a propelled space vehicle might be provided by a program of successive impulses of thrust. As the ballistic body and space vehicle approach the terminal state, the range will close so that the space vehicle trajectory can be determined as a function of the satellite orbit and the relative range and bearing (or time derivatives thereof) between the satellite and space vehicle. At suitable time of thrust application, the required thrust vector for minimum fuel consumption can be applied to achieve rendezvous. This process can be repeated as desired. The time instants of thrust application and the total number of such impulses will be determined by guidance considerations such as trajectory updating and sensor data accuracy. However, the thrust vector itself can be explicitly specified in any such case by a reduced form of the optimization equation. In this terminal phase, the central angle σ will be small, so that

$$\cos\sigma \approx 1 \quad \text{and} \quad \sin\sigma \approx \sigma$$

Then A = B = G = 0, and K_3 through K_8 reduce to zero, so that the polynomial [Eq. (51)] becomes a quadratic whose constant coefficients are

$$K_2 = DJ^2 = E^2 H \tag{58}$$

$$K_1 = E\,(4\,FH + EJ) - J\,(4FD + EJ) \tag{59}$$

$$K_0 = F\left[(J^2 - E^2) + 4F\,(D - H)\right] \tag{60}$$

Consequently, the C_T value for minimum velocity change transfer, and the velocity increment vector (Δv) itself can be obtained analytically. The orbital parameters can be expressed as functions of the relative range and bearing of the line of sight data such as obtained from rendezvous guidance systems. As a result, the guidance functions to command the required thrust magnitude and direction can be finally obtained in a form suitable for study and/or subsystem mechanization.

The hodograph equations and the optimization polynomial can be extremely valuable in the computation of alternative orbital transfer trajectories subject to the inflight selection of the initial transfer point, final transfer point, and the transfer orbit by a human operator. This selection would be feasible by use of an integrated man-machine system employing an orbital transfer situation display such as proposed in Ref. 19.

Hoelker and Silber[2] showed for transfer between coplanar circular orbits that, under certain end-point conditions, three impulses could require less total velocity change than two impulses. This interesting and previously unsuspected result leads naturally to speculation about the possible existence of a comparable optimization in the general three-dimensional orbital transfer problem. That is, given a set of end-points in specified initial and final orbits (or two sets of position and velocity vectors), is there a third point or points at which a third impulse may be applied such that the sum of the three velocity changes will be less than the sum of the two velocity changes for the two-impulse optimization?

It appears that the hodograph analysis is suitable for a study of this problem. In this approach, the intermediate point would be free, without definition of an orbital plane; however, the two transfer planes are defined by the point sets (O, P_1, P_i) and O, P_i, P_2). The problem could be formulated so that C_{T1} and C_{T2} appear in the velocity increment equation as apparent independent variables, but with auxiliary equations stating the constraints due to velocity change at the intermediate transfer point. That is, the terminal velocity vector for the first transfer orbit at the intermediate transfer point becomes the initial velocity vector for the second transfer orbit, so that the C_{T1} and C_{T2} are not truly independent. With this formulation, the variational calculus could then be used, with machine computation for solution, to determine the intermediate transfer point requiring the least total velocity change. This technique has already been developed for analysis of the two-impulse orbital transfer formulation as stated by Vargo.[3]

In conclusion, it is noted that a one-impulse optimization,[11] such as for atmospheric re-entry from orbit, provides a fourth order polynomial for solution, whereas the two-impulse optimization provides an eighth order polynomial for solution. If a formulation of the three-impulse problem were available in suitable form (i.e., as a function of one independent variable directly), then a twelfth order polynomial probably would result.

Appendix: Transfer and Orientation Geometry

The transfer central angle σ is defined by the position vectors $\bar{r}_1$ and $\bar{r}_2$ as follows:

$$\cos \sigma = \bar{r}_1 \cdot \bar{r}_2 / r_1 r_2 \tag{A1}$$

The intersection angles ζ and λ are defined by the initial and final orbital planes and the transfer orbit plane that is defined by the position vectors $\bar{r}_1$ and $\bar{r}_2$. Consider the derivation of the intersection angle ζ between the initial orbit plane and the transfer orbit plane. Since the initial orbit plane is known, the unit vector $\bar{s}$ normal to it is defined. The unit vector $\bar{t}$ of the transfer orbit plane is

$$\bar{t} = \bar{r}_1 \times \bar{r}_2 / \left| \bar{r}_1 \times \bar{r}_2 \right| \tag{A2}$$

Then the intersection angle ζ is the angle between the normal unit vectors $\bar{s}$, $\bar{t}$ as defined by

$$\cos \zeta = \bar{s} \cdot \bar{t} \tag{A3}$$

Reference 20 presents a detailed description of the transfer and orientation geometry in terms of the orbital parameters (and consequently the direction angles) or planetocentric surface coordinates.

References

[1]Hohmann, W., "Die Erreichbarkeit der Himmelskörper" (Oldenbourg, Munich, 1925).

[2]Hoelker, R. F. and Silber, R., "The bi-elliptical transfer between coplanar circular orbits," Proceedings of the 4th AFBMD/STL Symposium 1959 (Pergamon Press, London 1961), Vol. III, pp. 164-175.

[3]Vargo, L. G., "Optimal transfer between coplanar terminals in a gravitational field," Advances in Astronautical Sciences (Plenum Press, N. Y., 1958), Vol. 3, pp. 20-1 to 20-9.

[4]Horner, J. M., "Optimum impulse orbital transfers between

coplanar orbits," ARS J. 32, 1082-1089 (1962).

[5]Rider, L., "Characteristic velocity requirements for impulsive thrust transfers between non-coplanar circular orbits," ARS J. 31, 345-351 (1961).

[6]Altman, S. P. and Pistiner, J. S., "Hodograph analysis of the orbital transfer problem for coplanar nonaligned elliptical orbits," ARS J. 31, 1217-1225 (1961).

[7]Barrar, R. B., "Two-impulse transfer vs. one-impulse transfer: analytic theory," AIAA J. 1, 65-68 (1963).

[8]Bender, D. F., "Optimum coplanar two-impulse transfers between elliptical orbits," IAS Paper 62-4 (January 1962).

[9]Eckel, K., "Optimum transfer between non-coplanar elliptical orbits," Astronaut. Acta. VIII, Fasc. 4, 177-192 (1962).

[10]Altman, S. P. and Pistiner, J. S., "Minimum velocity increment solution for two-impulse coplanar orbital transfer," AIAA J. 1, 435-442 (1963).

[11]Altman, S. P. and Pistiner, J. S., "Comment on the correlation of Stark's two-terminal trajectory optimization with an orbital hodograph analysis," ARS J. 32, 983 (1962).

[12]Altman, S. P. and Pistiner, J. S., "Minimum velocity increment solution for two-impulse coplanar orbital transfer," UA SCTM 124 (January 1962).

[13]Whittaker, E. T., A Treatise of the Analytical Dynamics of Particles and Rigid Bodies (Cambridge University Press, England, 1959), p.89.

[14]MacMillan, W. D., Statics and the Dynamics of a Particle (Dover Publications, Inc., N. Y., 1958), p. 286.

[15]Altman, S. P. and Pistiner, J. S., "Hodograph transformations and mapping of the orbital conics," ARS J. 32, 1109-1111 (1962).

[16]Blitzer, L., Weisfeld, M., and Wheelon, A. D., "Perturbations of a satellite's orbit due to the earth's oblateness," J. Appl. Phys. 27, 1141-1149 (1956).

[17]Blitzer, L. and Wheelon, A. D., "Oblateness perturbation of elliptical satellite orbits," J. Appl. Phys. 28, 279 (1957).

[18]Blitzer, L., "Apsidal motion of an IGY satellite orbit," J. Appl. Phys. 28, 1362 (1959).

[19]Altman, S. P., "A spacecraft display system for orbital transfer and rendezvous," IAS Paper 63-14 (January 1963).

[20]Altman, S. P. and Pistiner, J. S., "Analysis of the orbital transfer problem in three-dimensional space," UA SCTM 154 (March 15, 1963).

[21]Neilson, W. A. (editor-in-chief), Webster's New International Dictionary of the English Language (G. & C. Merriam Co., Publishers, Springfield, Mass., 1958).

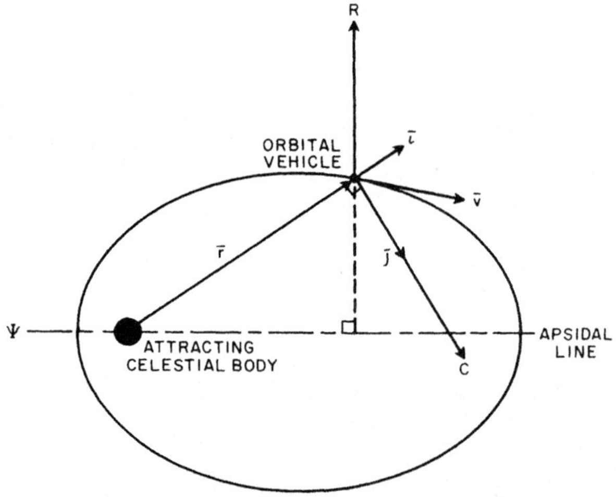

Fig. 1 Hodograph coordinates in two-dimensional space.

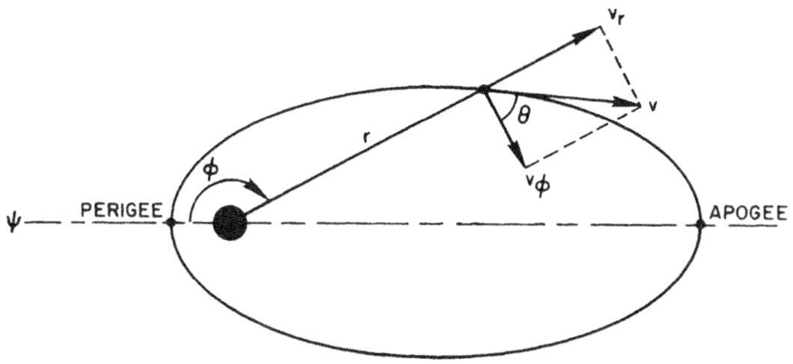

(a)

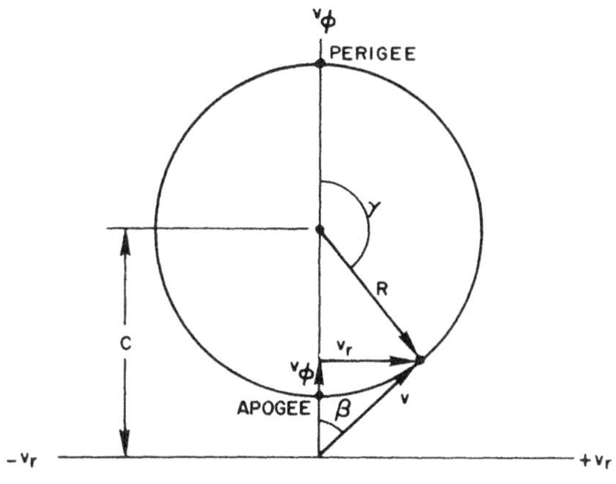

(b)

Fig. 2 Planar orbit and its hodograph.

a) Orbit in space

b) Orbit hodograph

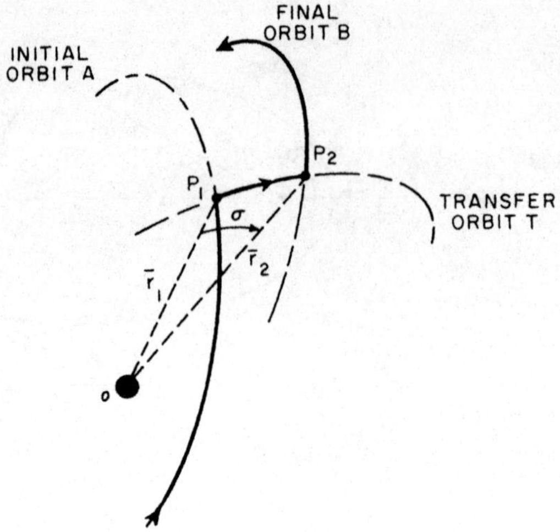

Fig. 3 Schematic diagram of three-dimensional orbital transfer.

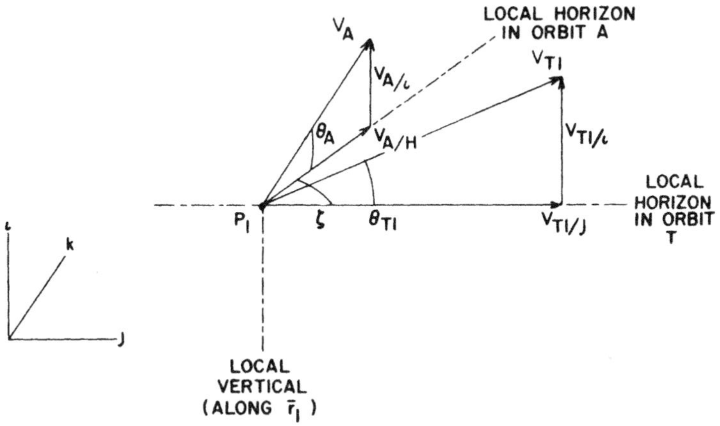

(a)

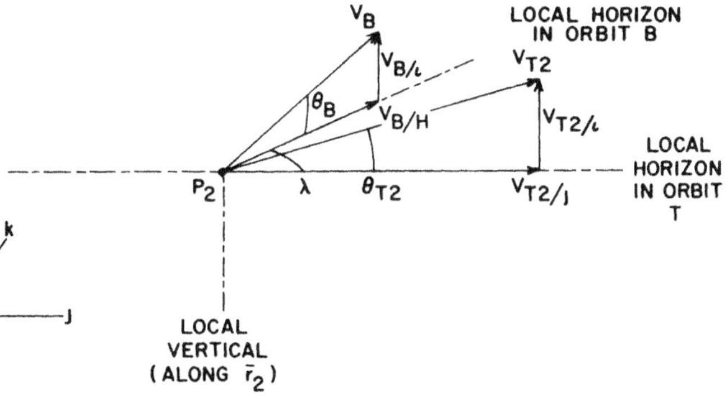

(b)

Fig. 4 Schematic diagrams of end-point velocity vector components.

a) Velocity vectors at initial transfer point

b) Velocity vectors at final transfer point

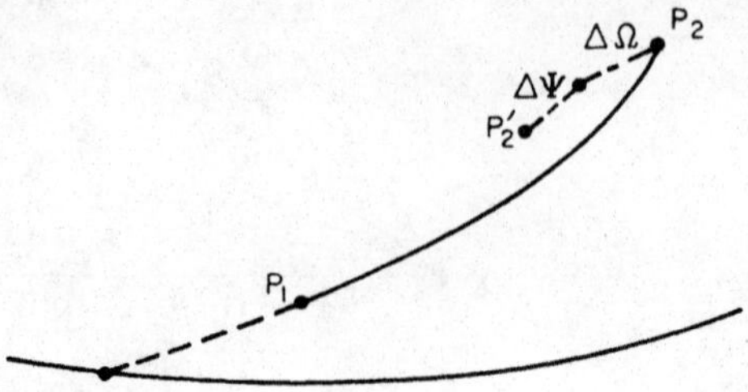

Fig. 5 Oblateness effects upon the apparent transfer terminal point in space.

VI.

ORBIT DETERMINATION

POSITION AND VELOCITY ESTIMATES BY TRAJECTORY RECTIFICATION

Otto R. Spies*

Burroughs Corporation, Paoli, Pa.

Abstract

This paper presents a global method of estimating the true position and velocity of a moving body at some preselected instant, from a succession of its observed, not too widely spaced positions (measured directly by radar or determined indirectly in some other way). The method utilizes fully the a priori knowledge of the integral equations of motion of the body, and may be used in problems of trajectory determination, prediction, or retrodiction, and also in an inverse mode, for refinement of these equations.

Introduction

The age-old problem of determination of orbits of celestial bodies has acquired a new complexion with the injection into the earth's toposphere (≡ aerospace) and beyond of bodies of tellurian origin and the concomitant development of new observational means.

Within the context of celestial mechanics, the problem of orbit determination from relatively few, optically measured position angles over some period of time is solved by the well-known improved classical methods using Keplerian orbits as first approximants. However, the problem of determining the trajectory of a ballistic missile, or an artificial earth satellite, on the basis of positional radar data differs from the former in two respects:

Presented as Preprint 63-427 at the AIAA Astrodynamics Conference, New Haven, Conn., August 19-21, 1963.

*Senior Mathematician, Burroughs Defense and Space Group, Paoli Research Laboratory.

1) In the earth's toposphere, the gravitational field is sensibly not point-symmetric, so that a Keplerian orbit does not necessarily approximate this trajectory closely.

2) Radar measures, less accurately to be sure, both range and position angles, and at relatively many successive instants. The same holds true, mutatis mutandis, of the trajectory of a space vehicle in the combined gravitational fields of several celestial bodies, and of successive positions determined in situ by an on-board space-navigational system.

The method of trajectory rectification described in Sec. I follows quite naturally from the requirements brought about by the changed circumstances:

1) Rather than fitting to the observed quantities the explicit solution of a simplified trajectory problem, in order to estimate parameters of that solution, it is desirable to fit the observed quantities into the equation specifying the original trajectory problem and to estimate parameters of that problem.

2) Rather than singling out a few of the inaccurately measured positions, it is desirable to base the parameter estimates on all observed quantities jointly.

Both requirements are met by writing the problem in the form of a vector integral equation of motion incorporating two vector parameters at a preselected instant. If the time-spacing of the successively measured positions (which are, in fact, sampled values of a specific solution of this integral equation) is not excessively wide, the indefinite twofold vector integral in this equation can be approximated closely by an appropriately weighted vector sum of the associated gravitational forces. In view of the smoothness of the gravitational field and the trajectory, as well as of the smoothing effect of integration, the accuracy of these weighted sums is likely to exceed that of the measured quantities.

The "rectified" position at a sampling instant is obtained as the vector difference between the measured position and the value of the weighted sum at that instant, and would, in the absence of any sources of indeterminacy, lie on a straight world-line tangent to the true trajectory at the preselected instant. All rectified positions are then used jointly in estimating the two vector parameters by the method of least squares.

Sec. II contains a discussion of a number of generalizations and possible modifications and extensions of the rectification method together with its potential applications to trajectory problems in other fields.

A brief summary of the relations between the accuracies of the observed and rectified positions and of the estimated position and velocity at the preselected instant is given in Sec. III.

A list of definitions of the symbols used will be found in Sec. IV.

I. Description of Method

To exhibit the salient features of the method, we consider the following idealized process.

A particle of unit mass moves in a known gravitational force field defined in an inertial frame. Its observed inertial positions $\bar{P}_k = \bar{R}(t_k) + \delta\bar{P}_k$ at a succession of n+1 precisely known instants t_k are determined with accuracies specified by covariance tensors $\bar{\bar{s}}_k(k = 0, 1, \cdots, n)$. (This set of integer values of the index k will not be mentioned explicitly hereinafter.) Our objective is to use these data to obtain estimated values $\hat{\bar{P}}_o$ and $\hat{\dot{\bar{P}}}_o$ of the particle's true position $\bar{R}_o$ and velocity $\dot{\bar{R}}_o$ at the initial instant t_o.

To extract from the positional information the largest possible amount of knowledge pertinent to this objective, we must utilize as fully as we can our a priori knowledge of the process in which this information is acquired:

We know that the particle's trajectory $\bar{R}(t)$ in the inertial frame is described by the solution of an initial-value problem. This problem can be specified in several conceptually and operationally different ways:

1) An ordinary vector differential equation

$$\ddot{\bar{R}}(t) = \bar{F}[\bar{R}(t)] \qquad (t_o \leq t \leq t_\Omega) \tag{1}$$

with the subsidiary initial conditions

$$\dot{\bar{R}}(t_o) \equiv \dot{\bar{R}}_o \ [= \text{vector const}] \tag{2}$$

$$\bar{R}(t_o) \equiv \bar{R}_o \ [= \text{vector const}] \tag{3}$$

2) A vector integro-differential equation

$$\dot{\bar{R}}(t) = \dot{\bar{R}}_o + \int_{t'=t_o}^{t} \bar{F}[\bar{R}(t')]dt' \qquad (t_o \leq t \leq t_\Omega) \tag{4}$$

with the subsidiary initial condition

$$\bar{R}(t_o) \equiv \bar{R}_o \tag{5}$$

3) A vector integral equation

$$\bar{R}(t) = \bar{R}_o + \dot{\bar{R}}_o(t - t_o) + \int_{t''=t_o}^{t} \int_{t'=t_o}^{t''} \bar{F}[\bar{R}(t')]dt'dt'' \qquad (t_o \leq t \leq t_\Omega) \tag{6}$$

In Eqs. (1), (4), and (6), we use the abbreviation

$$\bar{F}[\bar{R}(t)] \equiv -\mu\nabla V(\bar{R})_{\bar{R}=\bar{R}(t)} \tag{7}$$

where $V(\bar{R})$ is the gravitational potential, $\nabla \equiv \partial/\partial\bar{R}$ is the "del" vector operator, and μ is a known physical constant, so that the vector field $\bar{F}(\bar{R})$ has the physical dimension: force per unit mass.

The time interval $(t_o \leq t \leq t_\Omega)$ defining the domain of variation of the independent variable t to which the specified initial-value problem is confined will be taken here as extending between the initial and the final instants of observation, so that $t_\Omega \equiv t_n$. (This time interval will not be mentioned explicitly in the equations to follow.)

Eq. (1) specifies that, at any instant t at which the particle's position $\bar{R}(t)$ is at the point $\bar{R}$, the particle's acceleration $\ddot{\bar{R}}(t)$ is equal to the force per unit mass $\bar{F}[\bar{R}(t)]$ at the point-instant $(\bar{R}, t)$.

Eq. (4) specifies the particle's velocity function $\dot{\bar{R}}(t)$ as the vector sum of its velocity $\dot{\bar{R}}_o$ at the initial instant t_o and of the total velocity increment between the instants t_o and t which, as we know, is equal to the total impulse imparted to the particle by the continuous succession of gravitational forces acting on it as it describes its trajectory $\bar{R}(t)$ between these instants.

Eq. (6) specifies the trajectory $\bar{R}(t)$ as the vector sum of a rectilinear "unforced" trajectory $\bar{R}_o + \dot{\bar{R}}_o(t - t_o)$, which is tangent to $\bar{R}(t)$ at the initial point-instant $(\bar{R}_o, t_o)$, and of the total displacement between the instants t_o and t, which is equal to what may be called the "cumulative impulse" of the above-mentioned forces between these instants.

Adopting the point of view of functional analysis, we can derive Eq. (6) with the aid of singular functions, as follows: The trajectory $\bar{R}(t)$ can be represented by the convolution

$$\bar{R}(t) = \int_{t'=t_o}^{t} U_o(t - t')\, \bar{R}(t')dt' \tag{8}$$

Rewriting Eq. (8) by repeated application of the product rule of integration, we obtain successively other essentially equivalent, but operationally different representations of $\bar{R}(t)$, namely,

$$\left.\begin{aligned} \bar{R}(t) &= \bar{R}_o + \int_{t'=t_o}^{t} U_{-1}(t - t')\dot{\bar{R}}(t')dt' \\ &= \bar{R}_o + \dot{\bar{R}}_o(t - t_o) + \int_{t'=t_o}^{t} U_{-2}(t - t')\ddot{\bar{R}}(t')dt' \\ &= \bar{R}_o + \dot{\bar{R}}_o(t - t_o) + \ddot{\bar{R}}_o \frac{1}{2}(t - t_o)^2 \\ &\quad + \int_{t'=t_o}^{t} U_{-3}(t - t')\dddot{\bar{R}}(t')dt' \\ &\quad \vdots \end{aligned}\right\} \tag{9}$$

Substitution for $\ddot{\bar{R}}(t)$ from Eq. (1) in the second of Eqs. (9) yields the representation

$$\bar{R}(t) = R_o + \dot{\bar{R}}_o(t - t_o) + \int_{t'=t_o}^{t} U_{-2}(t - t')\bar{F}[\bar{R}(t')]dt' \tag{10}$$

A term-by-term comparison of Eq. (6) with Eq. (10) shows that

$$\int_{t''=t_o}^{t} \int_{t'=t_o}^{t''} \bar{F}[\bar{R}(t')]dt'\, dt'' \equiv \int_{t'=t_o}^{t} U_{-2}(t - t')\bar{F}[\bar{R}(t')]dt' \tag{11}$$

that is, that the value of the twofold integral at any instant $t \geq t_o$ may be viewed as a continuous weighted sum of all values of the operand $\bar{F}[\bar{R}(t')]$ of the integral operator up to that instant, the magnitude of the weights increasing linearly with the time-distance from that instant.

In keeping with the global nature of the rectification method, we shall use, in the following the more concise and congenial convolutions, which may readily be transcribed into their respective discrete algebraic counterparts.

The time succession of the particle's observed positions $\bar{P}_k$ ($k = 0, 1, \cdots, n$) is a sequence of somewhat inaccurate and, generally, nonuniformly spaced sampled values of a specific solution $\bar{R}(t)$ of Eq. (10), which may, by employing Eq. (10) in an inverse mode, be used to obtain estimates $\hat{\bar{P}}_o$ and $\hat{\dot{\bar{P}}}_o$, of the unknown specific values of the initial vectors R_o and $\dot{\bar{R}}_o$.

To this end, we first rewrite Eq. (10) in the form

$$\bar{R}^o(t) \equiv \bar{R}(t) - \int_{t'=t_o}^{t} U_{-2}(t-t')\,\bar{F}[\bar{R}(t')]dt' = \bar{R}_o + \dot{\bar{R}}_o(t-t_o) \qquad (12)$$

defining the target's rectified trajectory $\bar{R}^o$ fulcro ($\bar{R}_o$, t_o) as the vector difference of its true trajectory R(t) and its forced displacement from the initial instant t_o, as represented by the convolution of the force-function $\bar{F}[\bar{R}(t)]$, and showing that this rectified trajectory coincides at every instant t with the unforced trajectory $\bar{R}_o + \dot{\bar{R}}_o(t - t_o)$.

Then, we replace the following in Eq. (12):

1) the continuous time t by the succession of n + 1 discrete sampling instants t_k;

2) the continuous trajectory $\bar{R}(t)$ by the succession of n + 1 discrete observed positions $\bar{P}_k \equiv \bar{R}_k + \delta\bar{P}_k$;

3) the continuous force function $\bar{F}[\bar{R}(t)]$ by the succession of n + 1 discrete forces $\bar{\Phi}_k \equiv \bar{F}[\bar{P}_k]$; and

4) the convolution by a correspondingly weighted summation.

We thus obtain a system of n + 1 linear algebraic near-equations

$$\bar{P}_k^o \equiv \bar{P}_k - \sum_{\lambda=0}^{k} t_{k\lambda}\,\bar{\Phi}_\lambda \cong \hat{\bar{P}}_o + \hat{\dot{\bar{P}}}_o(t_k - t_o) \qquad (13)$$

in the two unknown vectors $\hat{\bar{P}}_o$ and $\hat{\dot{\bar{P}}}_o$ in which the scalar coefficients $t_{k\lambda}$ ($\lambda = 0, 1, \cdots, k$) depend in a known manner on the sampling instants t_k only.

Now, both the gravitational force field $\bar{F}(\bar{R})$ and the particle's trajectory $\bar{R}(t)$ are very smooth functions of their respective arguments. Therefore, the values of the n + 1 convolutions will, on the average, be approximated closely by the values of the n + 1 random weighted sums, unless the spacing of the sampling instants t_k is inordinately wide.

Note that the system of n + 1 vector near-equations (13) establishes the relationship existing between the two unknown vectors $\hat{\bar{P}}_o$ and $\hat{\dot{\bar{P}}}_o$, and every one of the 2(n + 1) known vectors $\bar{P}_k$ and $\bar{\Phi}_k$. Therefore, in a joint solution of these n + 1 near-equations, every one of the n + 1 positions $\bar{P}_k$ observed at the sampling instants t_k contributes to the estimated values of the two true initial vectors $\bar{R}_o$ and $\dot{\bar{R}}_o$ at the single instant t_o.

Evidently, taking any two distinct rectified positions $\bar{P}^o_\lambda$ and $\bar{P}^o_k$ (λ, k = 0, 1, ..., n; $\lambda < k$), we can form a system of two linear vector equations

$$\begin{bmatrix} 1 & t_\lambda - t_o \\ 1 & t_k - t_o \end{bmatrix} \begin{bmatrix} \hat{\bar{P}}_o \\ \hat{\dot{\bar{P}}}_o \end{bmatrix} = \begin{bmatrix} \bar{P}^o_\lambda \\ \bar{P}^o_k \end{bmatrix} \tag{14}$$

whose solution may be written directly as

$$\begin{bmatrix} \hat{\bar{P}}_o \\ \hat{\dot{\bar{P}}}_o \end{bmatrix} = \frac{1}{t_k - t_\lambda} \begin{bmatrix} t_k - t_o & -(t_\lambda - t_o) \\ -1 & 1 \end{bmatrix} \begin{bmatrix} \bar{P}^o_\lambda \\ \bar{P}^o_k \end{bmatrix} \tag{15}$$

The accuracy of the estimated initial position $\hat{\bar{P}}_o$ is of the same order as that of the rectified positions used, whereas that of the estimated initial velocity $\hat{\dot{\bar{P}}}_o$ increases with the time-distance $t_k - t_\lambda$ between these rectified positions.

The use of Eq. (15) is well justified when the accuracy and time-spacing of the observed positions $\bar{P}_k$ ensures an acceptable accuracy of the consequent rectified positions. On the other hand, it provides a simple means for roughly estimating the accuracy of the latter by mutual comparison of values of $\hat{\bar{P}}_o$ and $\hat{\dot{\bar{P}}}_o$ computed with the aid of Eq. (15) for a number of different pairs ($\bar{P}^o_\lambda$, $\bar{P}^o_k$) of relatively widely spaced rectified positions.

In the same circumstances, we may obtain estimates of $\hat{\bar{P}}_o$ by identifying the observed and the estimated positions $\bar{P}_o$ and $\hat{\bar{P}}_o$, say, and by subtracting the first of the near-equations (13) from each of the n others, to obtain

$$\bar{P}^o_k - \bar{P}_o \equiv (\bar{P}_k - \bar{P}_o) - \sum_{\lambda=1}^{k} t_{k\lambda}\, \bar{\Phi}_\lambda \cong \hat{\dot{\bar{P}}}_o (t_k - t_o) \qquad k \neq 0 \tag{16}$$

This system of n linear vector equations in a single vector unknown may now be solved for $\hat{\dot{\bar{P}}}_o$ by the method of least squares. The result is:

$$\hat{\bar{P}}_o = \sum_{k=1}^{n} (t_k - t_o)[(\bar{P}_k - \bar{P}_o) - \sum_{\lambda=1}^{k} t_{k\lambda} \bar{\Phi}_\lambda] / \sum_{k=1}^{n} (t_k - t_o)^2 \qquad (17)$$

In general, both estimated values $\hat{\bar{P}}_o$ and $\hat{\dot{\bar{P}}}_o$ must be determined by solving the n + 1 vector near-equations (13) for both by the method of least squares. In doing this, one can use schemes of different degrees of elaboration, depending on the desired accuracy of the estimates and on the affordable computational effort.

We summarize here the formally correct, most elaborate, and computationally most laborious scheme, from which the others may be derived by simplifying assumptions that may be warranted in individual applications of the rectification method.

We define the n + 1 vector residuals as

$$\Delta\bar{P}_k^o \equiv \hat{P}_o + \hat{\dot{P}}_o(t_k - t_o) - \bar{P}_k^o \qquad (18)$$

where P_k^o is the rectified position at the instant t_k, as defined in Eq. (13), and we minimize, with respect to $\hat{P}_o$ and $\hat{\dot{P}}_o$, the accuracy-weighted quadratic form, which may be written, using matrix notation, as

$$Q_n^o(\Delta\bar{P}_k^o, \Delta\bar{P}_\lambda^o) \equiv \frac{1}{2} \underline{\Delta\bar{P}^o}^t \underline{\underline{W}}_n^o \underline{\Delta\bar{P}^o} \equiv \frac{1}{2} \sum_{k=0}^{n} \sum_{\lambda=0}^{n} \underline{\Delta P}_\lambda^{o\,t} \underline{\underline{w}}_{k\lambda}^o \underline{\Delta P}_k^o \qquad (19)$$

The joint accuracy matrix $\underline{\underline{W}}_n^o$ is derived in Sec. III from the n + 1 covariance matrices $\underline{\underline{s}}_k$ characterizing the positional errors $\delta\bar{P}_k$.

Differentiating Eq. (22) partially with respect to $\hat{P}_o$ and $\hat{\dot{P}}_o$, equating either result to the zero vector $\underline{0}$, and collecting terms in $\hat{\underline{P}}_o$ and $\hat{\underline{\dot{P}}}_o$, we obtain the system of two vector equations (≡ the "normal equations") for the two vector unknowns in the form

$$\begin{bmatrix} \underline{\underline{w}}_{n(11)}^o & \underline{\underline{w}}_{n(12)}^o \\ \underline{\underline{w}}_{n(12)}^o & \underline{\underline{w}}_{n(22)}^o \end{bmatrix} \begin{bmatrix} \hat{\underline{P}}_o \\ \hat{\underline{\dot{P}}}_o \end{bmatrix} = \begin{bmatrix} [\underline{\underline{I}}]^t \underline{\underline{W}}_n^o \underline{\bar{P}}^o \\ [I(t - t_o)]^t \underline{\underline{W}}_n^o \underline{\bar{P}}^o \end{bmatrix} = \begin{bmatrix} \sum_{k=0}^{n} \sum_{\lambda=0}^{n} \underline{\underline{w}}_{k\lambda}^o \underline{P}_k^o \\ \sum_{k=0}^{n} \sum_{\lambda=0}^{n} (t_k - t_o) \times \underline{\underline{w}}_{k\lambda}^o \underline{P}_\lambda^o \end{bmatrix} \qquad (20)$$

where we use the abbreviations

$$
\begin{aligned}
\underline{\underline{w}}^{o}_{n(11)} &\equiv \underline{[\underline{\underline{I}}]^{t}}\, \underline{\underline{W}}^{o}_{n}\, \underline{[\underline{\underline{I}}]} = \sum_{k=0}^{n} \sum_{\lambda=0}^{n} \underline{\underline{w}}^{o}_{k\lambda} \\
\underline{\underline{w}}^{o}_{n(12)} &\equiv \underline{[\underline{\underline{I}}]^{t}}\, \underline{\underline{W}}^{o}_{n}\, \underline{[\underline{\underline{I}}(t - t_o)]} = \sum_{k=0}^{n} \sum_{\lambda=0}^{n} (t_k - t_o)\, \underline{\underline{w}}^{o}_{k\lambda} \\
\underline{\underline{w}}^{o}_{n(21)} &\equiv \underline{[\underline{\underline{I}}(t - t_o)]^{t}}\, \underline{\underline{W}}^{o}_{n}\, \underline{[\underline{\underline{I}}]} = \sum_{k=0}^{n} \sum_{\lambda=0}^{n} (t_\lambda - t_o)\, \underline{\underline{w}}^{o}_{k\lambda} = \underline{\underline{w}}^{o}_{n(12)} \qquad (21) \\
\underline{\underline{w}}^{o}_{n(22)} &\equiv \underline{[\underline{\underline{I}}(t - t_o)]^{t}}\, \underline{\underline{W}}^{o}_{n}\, \underline{[\underline{\underline{I}}(t - t_o)]} = \sum_{k=0}^{n} \sum_{\lambda=0}^{n} (t_k - t_o)(t_\lambda - t_o)\, \underline{\underline{w}}^{o}_{k\lambda}
\end{aligned}
$$

The solution of Eq. (20) may then be written in the correspondingly partitioned form

$$
\begin{bmatrix} \hat{\underline{P}}_o \\ \\ \hat{\underline{P}}_o \end{bmatrix} = \begin{bmatrix} \underline{\underline{m}}^{o}_{n(11)} & \underline{\underline{m}}^{o}_{n(12)} \\ \\ \underline{\underline{m}}^{o}_{n(12)} & \underline{\underline{m}}^{o}_{n(22)} \end{bmatrix} \begin{bmatrix} \displaystyle\sum_{k=0}^{n} \sum_{\lambda=0}^{n} \underline{\underline{w}}^{o}_{k\lambda}\, \underline{P}^{o}_{k} \\ \\ \displaystyle\sum_{k=0}^{n} \sum_{\lambda=0}^{n} (t_\lambda - t_o)\, \underline{\underline{w}}^{o}_{k\lambda}\, \underline{P}^{o}_{k} \end{bmatrix} \qquad (22)
$$

where

$$
\begin{aligned}
\underline{\underline{m}}^{o}_{n(11)} &\equiv \left[\underline{\underline{w}}^{o}_{n(11)} - \underline{\underline{w}}^{o}_{n(12)}\, \underline{\underline{w}}^{o-1}_{n(22)}\, \underline{\underline{w}}^{o}_{n(12)} \right]^{-1} \\
\underline{\underline{m}}^{o}_{n(12)} &\equiv \left[\underline{\underline{w}}^{o}_{n(21)} - \underline{\underline{w}}^{o}_{n(22)}\, \underline{\underline{w}}^{o-1}_{n(12)}\, \underline{\underline{w}}^{o}_{n(11)} \right]^{-1} \\
\underline{\underline{m}}^{o}_{n(21)} &\equiv \left[\underline{\underline{w}}^{o}_{n(12)} - \underline{\underline{w}}^{o}_{n(11)}\, \underline{\underline{w}}^{o-1}_{n(21)}\, \underline{\underline{w}}^{o}_{n(22)} \right]^{-1} \qquad (23) \\
&= \underline{\underline{m}}^{o}_{n(12)} \\
\underline{\underline{m}}_{n(22)} &\equiv \left[\underline{\underline{w}}^{o}_{n(22)} - \underline{\underline{w}}^{o}_{n(21)}\, \underline{\underline{w}}^{o-1}_{n(11)}\, \underline{\underline{w}}^{o}_{n(12)} \right]^{-1}
\end{aligned}
$$

It should be noted that, although formally correct, Eqs. (22) and (23) are impractical for computational effort. It is far more advantageous to use the joint accuracy matrix $\underline{\underline{W}}^{o}_{n}$ in the factored form $\underline{\underline{T}}^{o,-1}_{n}\, \underline{\underline{S}}^{-1}_{n}\, \underline{\underline{T}}^{o,-1t}_{n}$ [see Eq. (43)] arising in its derivation. In fact, by factoring the diagonal joint accuracy matrix $\underline{\underline{S}}_n^{-1}$ of the n + 1 positional errors $\delta\bar{P}_k$ into an upper triangular matrix and a lower triangular matrix (its transpose), $\underline{\underline{W}}_n^o$ will be so factored, permitting full computational advantage to be taken of its symmetry.

Carrying out the matrix multiplications in Eq. (23), and collecting terms in the rectified positions $\underline{P}_k^o$, we obtain

$$\hat{\underline{P}}_o = \sum_{k=0}^{n}\left[\underline{\underline{m}}^o_{n(11)}\sum_{\lambda=0}^{n}\underline{\underline{w}}^o_{k\lambda} + \underline{\underline{m}}^o_{n(12)}\sum_{\lambda=0}^{n}(t_\lambda - t_o)\,\underline{\underline{w}}^o_{k\lambda}\right]\underline{P}^o_k$$

$$\hat{\dot{\underline{P}}}_o = \sum_{k=0}^{n}\left[\underline{\underline{m}}^o_{n(12)}\sum_{\lambda=0}^{n}\underline{\underline{w}}^o_{k\lambda} + \underline{\underline{m}}^o_{n(22)}\sum_{\lambda=0}^{n}(t_\lambda - t_o)\,\underline{\underline{w}}^o_{k\lambda}\right]\underline{P}^o_k \tag{24}$$

showing more perspicuously either estimated vector, $\hat{\underline{P}}_o$ or $\hat{\dot{\underline{P}}}_o$, as a weighted sum of the n + 1 rectified-position vectors $\underline{P}_k^o$, with the aid of which the joint accuracy matrix $\underline{\underline{S}}_{n\,\Delta}^{-1}$ of the estimated-position and estimated-velocity errors $\delta\hat{\bar{P}}_o$ and $\delta\hat{\dot{\bar{P}}}_o$ may readily be determined. The derivation of this matrix is given in Sec. III.

A direct measure of the "goodness of fit" of the estimated unforced trajectory $\hat{\bar{P}}_o + \hat{\dot{\bar{P}}}_o(t - t_o)$ is provided by the minimum value of the quadratic form $Q_n^o(\Delta\bar{P}_k^o,\ \Delta\bar{P}_\lambda^o)$ defined in Eq. (19) assumed by it when $\hat{\bar{P}}_o$ and $\hat{\dot{\bar{P}}}_o$ have the values specified by Eqs. (24).

It is likely to prove preferable, first, to obtain relatively rough estimates of $\hat{\bar{P}}_o$ and $\hat{\dot{\bar{P}}}_o$ by a computationally simple least-squares scheme and, then, to refine them by a more elaborate scheme or by iterated substitution into the system of n + 1 near-equations (13).

II. Scope of Method

The rectification method described in Sec. I with the aid of a simple illustrative case can be used in a variety of modifications, extensions, and generalizations, some of which will be considered briefly in the following:

1) By writing the vector integral equation of motion in the "reverse" form

$$\bar{R}(t) = \bar{R}_\Omega + \dot{\bar{R}}_\Omega(t - t_\Omega) + \int_{t'=t_\Omega}^{t} U_{-2}(t' - t)\,\bar{F}[\bar{R}(t')]\,dt' \qquad (t_o \le t \le t_\Omega) \tag{25}$$

we can estimate the particle's final position $\bar{R}_\Omega$ and velocity $\dot{\bar{R}}_\Omega$ in the same manner as we estimated its initial position $\bar{P}_o$ and velocity $\dot{\bar{P}}_o$ by the use of the integral equation in the "forward" form given in Eq. (10). In fact, this estimation problem is the mirror image of the illustrative problem, so that all results derived for the latter hold true here, with the sense of time reversed.

2) The reverse integral equation (25) may be used to estimate the particle's present (≡ "current final") true position $\bar{R}(t_\omega)$ and velocity $\dot{\bar{R}}(t_\omega)$ at the present instant t_ω $(t_o \leq t_\omega < \infty)$ on the basis of all of its positions observed at the (current) past instants $(t_o \leq t \leq t_\omega)$. This form of estimation is typical of navigational problems.

3) It should be clear that the reverse integral equation can also be used when the present true position $\bar{R}(t_\omega)$ and velocity $\dot{\bar{R}}(t_\omega)$ are to be estimated on the basis of the particle's positions observed within the time interval between a current initial and the present (≡ current final) instant.

This "sliding" mode of application of Eq. (25) is effectively realized in the preceding mode whenever the accuracy of the observed positions is relatively low, so that the accuracy weights of rectified positions at relatively remote past instants become so small as to render these terms negligible.

4) A more general form of the vector integral equation of motion, which includes the forward and reverse integral equations (10) and (25) as special cases, is

$$\bar{R}(t) = \bar{R}_p + \dot{\bar{R}}_p(t - t_p) + \int_{t'=t_p}^{t} [U_{-1}(t_p - t')\,U_{-2}(t' - t) + U_{-1}(t' - t_p)\,U_{-2}(t - t')] \times \bar{F}[\bar{R}(t')]dt' \qquad (t_o \leq t \leq t_\Omega) \tag{26}$$

With the aid of Eq. (26), we can estimate a particle's true position $\bar{R}_p$ and velocity $\dot{\bar{R}}_p$ at any preselected "pivotal" instant t_p $(t_o \leq t_p \leq t_\Omega)$. The system of n + 1 algebraic vector near-equations accordingly has the form

$$\bar{P}_k^p \equiv \bar{P}_k - U_{-1}(t_p - t_k) \sum_{j=0}^{p-k} t_{k(p-j)} \bar{\Phi}_{p-j} - U_{-1}(t_k - t_p) \sum_{\lambda=0}^{k-p} t_{k(p+\lambda)} \times \bar{\Phi}_{p+\lambda} \cong \bar{P}_p + \dot{\bar{P}}_p(t_k - t_p) \tag{27}$$

The structure of these pivotal integral and algebraic equations is exhibited very clearly by the $(n + 1) \times (n + 1)$ matrices $\underline{\underline{T}}_n^p$ and $\underline{\underline{T}}_n^{pt}$ of the associated system of $n + 1$ error-vector equations derived in Sec. III, one of which is shown in Eq. (28) for illustration for the special case when $p = 2$ and $n = 5$.

$$\underline{\underline{T}}_5^{2t} \equiv \begin{bmatrix} I - t_{oo}v_o & -t_{o1}v_1 & -t_{o2}v_2 & 0 & 0 & 0 \\ 0 & I - t_{11}v_1 & -t_{12}v_2 & 0 & 0 & 0 \\ 0 & 0 & I - t_{22}v_2 & 0 & 0 & 0 \\ 0 & 0 & -t_{32}v_2 & I - t_{33}v_3 & 0 & 0 \\ 0 & 0 & -t_{42}v_2 & -t_{43}v_3 & I - t_{44}v_4 & 0 \\ 0 & 0 & -t_{52}v_2 & -t_{53}v_3 & -t_{54}v_4 & I - t_{55}v_5 \end{bmatrix} \quad (28)$$

[For brevity, we have omitted the double underscores on the 3×3 Jacobian matrices $\underline{\underline{v}}_k$ $(k = 1, \cdots, 6)$, the 3×3 unit matrices $\underline{\underline{I}}$, and the 3×3 null matrices $\underline{\underline{0}}$.]

The pivotal role of the pivotal instant t_p from which the reverse and the forward integration start is directly evident from the matrix shown in Eq. (28). It is also evident that the results derived for the illustrative special case will hold true in the general case as well, when the $(n + 1) \times (n + 1)$ triangular matrices $\underline{\underline{T}}_n^{ot}$ and $\underline{\underline{T}}_n^{o}$ derived in Sec. III are replaced by the $(n + 1) \times (n + 1)$ bitriangular matrices $\underline{\underline{T}}_n^{pt}$ and $\underline{\underline{T}}_n^{p}$, respectively.

Still another fact emerges from Eq. (28): the number of nonzero matrix off-diagonal elements of the bitriangular matrices varies with p, being largest $[n(n + 1)/2]$ when $p = 0$ or $p = n$. This number attains the smallest value $[n(n + 2) + 4]/4$ when $p = n/2$ and n is even, being slightly larger when n is odd. It follows that, in determining the $n + 1$ rectified positions $\overline{P}_k^p$ fulcro $(\overline{R}_p, t_p)$ from the $n + 1$ observed positions $\overline{P}_k$, the least computational labor will be required when the particle's true position $\overline{R}_p$ and velocity $\dot{\overline{R}}_p$ are estimated at that instant t_p, whose index p equals n/2 (if n is even) or equals (n ± 1)/2 (if n is odd).

It appears likely that, with a given succession of observed positions $\bar{P}_k$, the accuracy of the estimated pivotal position $\bar{P}_p$ and velocity $\dot{\bar{P}}_p$ of the particle will be highest for the same value of the index p. A verification of this conjecture, using the procedure described in Sec. III, has, as yet, not been undertaken.

5) A slight modification of Eqs. (31) is required, when the preselected pivotal instant t_p does not coincide with one but lies between two of the n + 1 sampling instants t_k. This will be the case when the pivotal instant must lie on a standard time grid of uniformly spaced instants employed in the subsequent use of the estimated pivotal vectors $\bar{P}_p$ and $\dot{\bar{P}}_p$, as noted by the author in an unpublished Burroughs report, 1962.

6) One may use vector integral equations of motion in a form appropriate for two-point boundary problems to estimate values of their parameters at two distinct joint-instants. For example, the discrete counterpart of the vector integral equation

$$\bar{R}(t) = \frac{1}{t_\Omega - t_o}\Big\{\bar{R}_o(t_\Omega - t) + \bar{R}_\Omega(t - t_o) + (t_\Omega - t_o)\int_{t'=t_o}^{t} U_{-2}(t - t')\,\bar{F}[\bar{R}(t')]dt' - (t - t_o)\int_{t'=t_o}^{t_\Omega} U_{-2}(t - t')\,\bar{F}[\bar{R}(t')]dt'\Big\} \tag{29}$$

may be used to determine the estimated values $\hat{\bar{P}}_o$ and $\hat{\bar{P}}_\Omega$ of the particle's true initial and final positions $\bar{R}_o$ and $\bar{R}_\Omega$, respectively. In this case, the n + 1 rectified positions $\bar{P}_k^{o\Omega}$ would, in the absence of all sources of indeterminacy, lie on a straight world-line passing through the point-instants (R_o, t_o) and (R_Ω, t_Ω).

7) Evidently, the twofold integration with respect to time (or the equivalent convolution with the unit-ramp function of time) can be carried out as directly as in the illustrative case, when the gravitational force-field varies with time in a known manner. In this case, the gravitational potential is of the more general form $V(\bar{R}, t)$, so that the integrand becomes

$$\bar{F}[\bar{R}(t), t] \equiv - \mu \quad V(\bar{R}, t)_{\bar{R}=\bar{R}(t)} \tag{30}$$

8) In cases where the force acting on the particle is proportional to its velocity $\dot{\bar{R}}(t)$, the factor of proportionality being a known function of time c(t), in the first integration of both sides of the modified differential equation of motion with respect to time, the integral of the expression $c(t)\dot{\bar{R}}(t)$ may be rewritten, by using the product rule of integration, as

$$\int_{t'=t_p}^{t} c(t')\dot{\bar{R}}(t')dt' = c(t)\bar{R}(t) - c_p\bar{R}_p - \int_{t'=t_p}^{t} \dot{c}(t')\bar{R}(t')dt' \tag{31}$$

where $c_p \equiv c(t_p)$, and $\dot{c}(t)$ is the time-derivative of c(t).

Hence, knowledge of the particle's velocity is not required for its evaluation nor, a fortiori, for that of the twofold integral. This is true also in the special case where $c(t) \equiv c$ (= const); then, the time derivative of c is identically equal to zero, and the right side of Eq. (31) reduces to $c[\bar{R}(t) - \bar{R}_p]$.

9) When a force acting on the particle varies with time in a known manner, its twofold integral (or the equivalent convolution integral) can be evaluated independently of the particle's position $\bar{R}(t)$ or velocity $\dot{\bar{R}}(t)$.

10) In view of the foregoing, the rectification method can readily be applied in cases where the particle's motion is specified with respect to a noninertial frame of reference, in which the gravitational field of force may vary with time, and the particle is subject to thrust or to a velocity-dependent force whose variation with time is known.

11) Eq. (12) and its discrete approximant Eq. (13), used in estimating a particle's pivotal state of motion from sampled values of the particle's trajectory--that is, from sampled values of a specific solution of these equations--are global equations of motion of the particle by which its trajectory can be determined from its pivotal state of motion. The solution of integral equations of motion by the method of successive approximation is discussed in a general manner in Spies.[1]

It is possible to combine the two modes of application of Eq. (13) in order to obtain estimates of positions of the particle as some preselected instants jointly with the estimate of its pivotal state of motion ($\bar{R}_p$, $\dot{\bar{R}}_p$). This may be found desirable when the time-spacing of some sampling instants is unduly wide. In this case, one would select one or a few fictitious sampling instants $t_{k'}$, lying between the widely spaced sampling instants, and would include the forces $\bar{\Phi}_{k'}$ corresponding to the tentatively assumed positions $\bar{P}_{k'}$ of the particle as first approximants of its true positions $\bar{R}_{k'}$ at these instants in the definitions of the rectified positions $\bar{P}_k^p$ and $\bar{P}_{k'}^p$, as given by Eq. (27). Defining the residuals $\Delta\bar{P}_k^p$ and $\Delta\bar{P}_{k'}^p$, as in Eq. (18) of the illustrative case, one would form the enlarged weighted quadratic form as in Eq. (19), in which the additional residuals $\Delta\bar{P}_{k'}^p$ would be given some confidence weight, and which one would minimize with respect to $\hat{\bar{P}}_p$ and $\hat{\bar{P}}_p$, as well as with respect to the unknown forces $\bar{\Phi}_{k'}$. The estimated values of these forces $\bar{\Phi}_{k'}$, given by the solutions of the enlarged system of normal equations so obtained, would then be used to compute the particle's estimated positions $\bar{P}_{k'}$, which will, as a rule, differ from the respective first approximants. If these position differences do not significantly deviate from the mean positional error $\delta\bar{P}_{k'}$ of the potentially observed positions at the instants $t_{k'}$, these estimated positions $\bar{P}_{k'}$, as well as the concomitant estimated pivotal state of motion ($\hat{\bar{P}}_p$, $\hat{\bar{P}}_p$), would be accepted. If they do significantly deviate, these deviations are used to determine the second approximants of the particle's positions at the instants $t_{k'}$, and the modified system of normal equations solved for the second estimates $\bar{\Phi}_{k'}$ of the value of the forces, and so on, until the position difference is no longer significant.

12) In many cases, it may be advantageous to obtain first, by the simplest form of the rectification method, relatively inaccurate estimates of the particle's position, or both position and velocity, at several rather widely spaced pivotal instants and, then, to use these jointly as first approximants in the solution of the appropriately formulated integral equation of motion as indicated in Spies.[1]

13) In fitting a straight world-line to the n + 1 rectified positions $\bar{P}_k^p$ by the least-squares method, we may enhance the accuracy of the estimated pivotal position $\hat{\bar{P}}_p$ and velocity $\hat{\bar{P}}_p$ by using: a) additional observed quantities differing from the original ones, as redundant (generalized) coordinates; and b) additional known functional relations between the observed and the estimated quantities, as subsidiary conditions. An example of redundant coordinates is provided by range rates of the particle measured by radar at the

same n + 1 sampling instants t_k at which its n + 1 observed positions P_k are measured. The author described their use in an unpublished paper presented at a 1961 Meeting of SIAM.

If the particle's motion is conservative, one may use the subsidiary condition that the values of the energy function (≡ Hamiltonian function expressed in terms of position and velocity coordinates, instead of positions and conjugate momenta) of any two (or more) estimated states of motion ($\hat{P}_p$, $\hat{\hat{P}}_p$) and ($\hat{P}_{p'}$, $\hat{\hat{P}}_{p'}$) (p, p′ = 0, 1, · · ·, n; p ≠ p′) be mutually equal. A more elegant way of effectively attaining this result is to minimize the difference between these two energy values, weighted by a high relative-importance weight. In an axisymmetric gravitational field, the particle's angular momentum about the axis of symmetry is constant. In such a field, one could use this relationship as an additional subsidiary constraint on any two (or more) estimated states of motion.

14) The form of Eqs. (27) and their locally linearized counterparts, as displayed by the illustrative bitriangular matrix $\underline{\underline{T}}_5^{2t}$ shown in Eq. (28), strongly suggests that, both in the process of estimation and in the process of accuracy determination described in Secs. I and III, respectively, the computational effort may be greatly reduced by the use of recursive methods of summation and matrix inversion. In fact, the use of such methods of computation is imperative in all "current" modes of application of the rectification method. This practically highly important topic is the subject of a number of unpublished memoranda of Bell Telephone Laboratories, referred to in, among others, the very informative paper by Levine,[2] to which the author's attention was kindly drawn at the 1961 Annual Fall Meeting of the Society for Industrial and Applied Mathematics, after presentation of his paper containing a rather elementary application of the rectification method.

15) In the present version of the method, time is viewed as an exactly known determinate independent variable. In the general version, measured instants of time would, like the measured positions in space, be viewed as random variables τ_k characterized by a known variance σ_τ. Correspondingly, in addition to the two pivotal vectors $\hat{P}_p$ and $\hat{\hat{P}}_p$, the pivotal instant τ_p would have to be estimated.

16) A source of indeterminacy of the process of estimation is the incorrectness (or, perhaps better, insufficient

correctness) of the specific form of the force function of the vector differential and integral equations of motion used.

If the integral equation is insufficiently correct, the accuracy of values of the pivotal position $\hat{P}_p$ and velocity $\hat{\dot{P}}_p$, estimated on the basis of actual observed positions $\bar{P}_k$, will be significantly inferior (as determined, for example, by the χ^2-test) to their nominal accuracy, derived with the aid of the (locally linearized) discrete counterpart of that integral equation by the procedure described in Sec. III, from the known accuracy tensors $\underline{\underline{s}}_k$ at these positions.

A succession of rectified positions, to which a smooth curved world-line can be fitted significantly better than a straight one, indicates a systematic error in the force function of the integral equation used which can be corrected. This inverse mode of application of the rectification method is likely to have many features in common with pertinent existing methods.

17) It should be evident that the field of application of the rectification method is not confined to problems of motion of a particle in physical space. In principle, this method can be used to estimate specific values of sets of "state variables" at some preselected instants of any process obeying known generalized equations of motion whose solutions may be viewed as trajectories of a "state point" in an n-dimensional state space.

III. Accuracy Relations

Positional Error, $\delta\bar{P}_k$

At each of the n + 1 sampling instants t_k, the particle's observed position $\bar{P}_k$ relative to an inertial frame is displaced from its true inertial position $\bar{R}_k$, because of the indeterminacy of the observational process, of which $\bar{P}_k$ is an output. The displacement $\delta\bar{P}_k = \bar{P}_k - \bar{R}_k$ is the positional error at $\bar{R}_k$, a random vector that may be characterized probabilistically by a trivariate conditional Gaussian probability distribution centered on $\bar{R}_k$ and specified uniquely by the covariance tensor $\bar{\bar{s}}_k$. This tensor may be represented graphically by the "instantaneous mean-error ellipsoid," centered on $\bar{R}_k$ at the instant t_k, which may be defined by specifying the orientation of its three mutually orthogonal principal axes (≡ eigenvectors) $\bar{1}_{\upsilon k}$, $\bar{1}_{\alpha k}$, $\bar{1}_{\epsilon k}$, with respect to the inertial frame, as well as the lengths (≡ eigenvalues) of the respective half-axes, the principal "standard deviations" (≡ principal mean errors) $\sigma_{\upsilon k}$, $\sigma_{\alpha k}$, $\sigma_{\epsilon k}$. The accuracy tensor $\bar{\bar{u}}_k \equiv \bar{\bar{s}}_k^{-1}$, the inverse of $\bar{\bar{s}}_k$, may be represented by the "instantaneous reciprocal-mean-error or instantaneous accuracy ellipsoid," centered on $\bar{R}_k$ at the instant t_k, whose principal axes coincide with those of $\bar{\bar{s}}_k$ and the lengths of the respective half-axes of which are the reciprocals $1/\sigma_{\upsilon k}$, $1/\sigma_{\alpha k}$, $1/\sigma_{\epsilon k}$, of those of $\bar{\bar{s}}_k$.

The most congenial manner of specification of vectors (≡ first-order tensors) and second-order tensors in some adopted Cartesian inertial frame, OXYZ, is in the form of matrices whose elements are their respective components in that frame.

In general, the inertial components $s_{\mu\nu k} = s_{\nu\mu k}$ (μ, ν, = X, Y, Z) of s_k will not be directly known a priori but will, in most cases, be derivable from the a priori knowledge of the observational process. In those cases when this is not possible, one will fall back on whatever less-detailed accuracy specification is obtainable.

A familiar example of such indirect determination of these components is provided by the case when a ground-based radar unit measures a succession of positions $\bar{r}_k$ of an artificial earth satellite relative to its own noninertial ground frame cxyz fixed to the spinning earth. In this case, the accuracy of the measured values ρ, θ, ψ of the particle's true radar coordinates (range r, azimuth Θ, and elevation Ψ), referring to the ground frame at whose origin c the radar unit is located, is specified by the respective mean errors (≡ standard deviations) σ_r, σ_θ, and σ_ψ.

Gravitational-Force Error, $\delta\bar{\varphi}_k$

Given the force function $\bar{F}(\bar{R})$, specifying the space-variation of the gravitational force $\bar{F}$ in the inertial frame, and given the covariance matrix $\underline{\underline{s}}_k$ characterizing the positional error $\delta\bar{P}_k$ at $\bar{R}_k$, we wish to find the covariance matrix $\underline{\underline{g}}_k$ characterizing the consequent gravitational-force error $\delta\bar{\varphi}_k$ at $\bar{R}_k$.

The gravitational-force error vector $\delta\bar{\varphi}_k$ is defined as the vector difference between the observed force $\bar{\Phi}_k$ and the true force $\bar{F}_k$ at $\bar{R}_k$:

$$\delta\bar{\varphi}_k \equiv \bar{\Phi}_k - \bar{F}_k = \bar{F}(\bar{R}_k + \delta\bar{P}_k) - \bar{F}(\bar{R}_k) \tag{32}$$

The Taylor series expansion of $\bar{F}(\bar{R}_k + \delta\bar{P}_k)$ at the particle's true position $\bar{R}_k$ with respect to the positional error vector $\delta\bar{P}_k$ truncated after the linear term is

$$\bar{F}(\bar{R}_k + \delta\bar{P}_k) \cong \bar{F}(\bar{R}_k) + \left(\frac{\partial\bar{F}}{\partial\bar{R}}\right)_{R=R_k} \delta\bar{P}_k \tag{33}$$

where $(\partial\bar{F}/\partial\bar{R})_{\bar{R}=\bar{R}_k} = -\mu(\partial\,[\nabla V(\bar{R})]/\partial\bar{R})_{\bar{R}=\bar{R}_k}$ is the (symmetrical) Jacobian tensor $\bar{\bar{v}}_k$ at $\bar{R}_k$, which is represented in terms of its inertial components by the Jacobian matrix

$$\underline{\underline{v}}_k \equiv -\mu \begin{bmatrix} \dfrac{\partial^2 V}{\partial X^2} & \dfrac{\partial^2 V}{\partial X \partial Y} & \dfrac{\partial^2 V}{\partial X \partial Z} \\ \dfrac{\partial^2 V}{\partial Y \partial X} & \dfrac{\partial^2 V}{\partial Y^2} & \dfrac{\partial^2 V}{\partial Y \partial Z} \\ \dfrac{\partial^2 V}{\partial Z \partial X} & \dfrac{\partial^2 V}{\partial Z \partial Y} & \dfrac{\partial^2 V}{\partial Z^2} \end{bmatrix}_{\substack{X=X_k \\ Y=Y_k \\ Z=Z_k}} \tag{34}$$

A comparison of Eqs. (32) and (33) shows that

$$\delta\bar{\varphi}_k \cong \bar{\bar{v}}_k\, \delta\bar{P}_k \tag{35}$$

so that, conversely,

$$\delta\bar{P}_k \cong \bar{\bar{v}}_k^{-1}\, \delta\bar{\varphi}_k \tag{36}$$

where $\bar{\bar{v}}_k^{-1}$ is the inverse of $\bar{\bar{v}}_k$, or, using matrix notation,

$$\underline{\delta P}_k \cong \underline{v}_k^{-1}\, \underline{\delta\Phi}_k \quad \therefore \quad \underline{\delta P}_k^t \cong \underline{\delta\Phi}_k^t\, \underline{\underline{v}}_k^{-1t} \tag{37}$$

Recalling the definition of the positional accuracy matrix $\underline{\underline{u}}_k$ at position $\bar{R}_k$, we obtain, using Eq. (37), the gravitational-force accuracy matrix in terms of the inertial components of $\delta\bar{\varphi}_k$ as

$$\underline{\underline{g}}_k^{-1} \equiv \underline{\underline{v}}_k^{-1t}\, \underline{\underline{u}}_k\, \underline{\underline{v}}_k^{-1} \tag{38}$$

The desired covariance matrix of $\delta\bar{\varphi}_k$ at $\bar{R}_k$ is then given by the inverse of Eq. (38):

$$\underline{\underline{g}}_k \equiv \underline{\underline{v}}_k\, \underline{\underline{s}}_k\, \underline{\underline{v}}_k^t \tag{39}$$

Rectified-Position Error, $\delta \bar{P}_k^o$

The rectified position $\bar{P}_k^o$ at the instant t_k (k = 0, 1, ···, n) is defined in Eq. (13). Being a linear function of the k + 2 random vectors, $\bar{P}_k$ and $\bar{\Phi}_\lambda$ (λ = 0, 1, ···, k), $\bar{P}_k^o$ is itself a random vector whose error vector $\delta \bar{P}_k^o$ is the same linear function of the respective k + 2 error vectors $\delta \bar{P}_k$ and $\delta \bar{\varphi}_\lambda$ (λ = 0, 1, ···, k). Thus, we have

$$\delta \bar{P}_k^o = \delta \bar{P}_k - \sum_{\lambda=0}^{k} t_{k\lambda}\, \delta \bar{\varphi}_\lambda \tag{40}$$

or, taking the linear relationships between the error vectors $\delta \bar{\varphi}_\lambda$ and $\delta \bar{P}_\lambda$ given by Eqs. (35) into account, we also have

$$\delta \bar{P}_k^o = \delta \bar{P}_k - \sum_{\lambda=0}^{k} t_{k\lambda}\, \bar{\bar{v}}_\lambda\, \delta \bar{P}_\lambda \tag{41}$$

Eqs. (13) or (41) show that the rectified positions $\bar{P}_k^o$ or their respective error vectors $\delta \bar{P}_k^o$ are not mutually independent random variables, as each contains a partial weighted sum of functions of observed positions or a partial weighted sum of positional errors, also contained by its respective predecessors. Hence, to provide a complete probabilistic characterization of their joint random behavior, we must, reverting to matrix notation, determine the joint covariance matrix $\underline{\underline{S}}_n^o$ of all n + 1 rectified-position error vectors $\underline{\delta P}_k^o$ from the joint covariance matrix $\underline{\underline{S}}_n$ of all n + 1 positional errors $\underline{\delta P}_k$.

In most cases, all n + 1 positional errors $\underline{\delta P}_k$ may rightfully be viewed as mutually independent random vector variables, so that their joint covariance matrix $\underline{\underline{S}}_n$ will be a diagonal (n + 1) × (n + 1) matrix whose n + 1 diagonal elements are the covariance matrices $\underline{\underline{s}}_k$ of the n + 1 positional error vectors $\underline{\delta P}_k$ whereas the remaining n × (n + 1) off-diagonal elements are 3 × 3 null matrices $\underline{\underline{0}}$. Thus, in these cases, $\underline{\underline{S}}_n$ will be of the form

$$\underline{\underline{S}}_n \equiv \begin{bmatrix} \underline{\underline{s}}_o & 0 & 0 \dots & 0 & \dots & 0 \\ 0 & \underline{\underline{s}}_1 & 0 \dots & 0 & \dots & 0 \\ \vdots & & & \vdots & & \vdots \\ 0 & 0 & 0 \dots & \underline{\underline{s}}_k & \dots & 0 \\ & & & & & \end{bmatrix} \tag{42}$$

$$\begin{bmatrix} \vdots & & & \vdots & & \vdots \\ 0 & 0 & 0 \ldots & 0 & \ldots & \underline{\underline{s}}_n \end{bmatrix} \tag{42}$$

The system of n + 1 linear vector equations 41 may likewise be written as a single matrix equation:

$$\underline{\delta \bar{P}^o} = \underline{\underline{T}}_n^{ot} \, \underline{\delta \bar{P}} \tag{43}$$

The joint covariance matrix $\underline{\underline{S}}_n^o$ is now given by matrix product

$$\underline{\underline{S}}_n^o = \underline{\underline{T}}_n^{ot} \, \underline{\underline{S}}_n \, \underline{\underline{T}}_n^o \tag{44}$$

The inverse $\underline{\underline{S}}_n^{o,-1} \equiv \underline{\underline{W}}_n^o$ of $\underline{\underline{S}}_n^o$ is the joint accuracy matrix of the n + 1 rectified positions $\bar{P}_k^o$ used in forming the quadratic form Q_n^o of the n + 1 residuals $\Delta\bar{P}_k^o$. It may be expressed in factored form as

$$\underline{\underline{W}}_n^o = \underline{\underline{T}}_n^{o,-1} \, \underline{\underline{S}}_n^{-1} \, \underline{\underline{T}}_n^{o,-1t} \tag{45}$$

where $\underline{\underline{T}}_n^{o,-1}$ is the inverse of the upper triangular matrix $\underline{\underline{T}}_n^o$, and $\underline{\underline{T}}_n^{o,-1t}$ is the transpose of the former. Here, $\underline{\underline{S}}_n^{-1}$ is the inverse of the diagonal covariance matrix $\underline{\underline{S}}_n$, of which each of the n + 1 diagonal elements $\underline{\underline{s}}_k^{-1}$ is the inverse of the corresponding diagonal element $\underline{\underline{s}}_k$ of $\underline{\underline{S}}_n$.

Since the inversion of a triangular matrix is less laborious than that of a square (symmetric) matrix, it is computationally advantageous to invert $\underline{\underline{T}}_n^o$, $\underline{\underline{T}}_n^{ot}$, and $\underline{\underline{S}}_n$ first, and then to carry out the indicated matrix multiplications.

The accuracy of the rectified positions depends, of course, not only on that of the observed positions $\bar{P}_k$ but also on the accuracy of the approximate representations of the twofold integrals by weighted sums, as well as on the accuracy of the numerical computations of the latter.

It is justifiable in some cases to characterize the computational error by an isotropic covariance tensor, to be added to the individual terms of the joint covariance matrix $\underline{\underline{S}}_n^o$ of the n + 1 rectified positions $\bar{P}_k^o$ defined in Eq. (44). The corresponding joint accuracy matrix $\underline{\underline{W}}_n^o$ would be the inverse of the so-modified $\underline{\underline{S}}_n^o$. This question has not been investigated in the present study.

Estimated-Position and Estimated-Velocity Errors, δP_o, $\delta \dot{P}_o$

The linear relationships between the estimated initial position $\hat{P}_o$ and velocity $\hat{\dot{P}}_o$, on the one hand, and the n + 1 rectified positions $\bar{P}_k^o$ on the other, exhibited by Eqs. (24), hold true also between the respective errors $\delta\hat{P}_o$, $\delta\hat{\dot{P}}_o$, and $\delta\bar{P}_k^o$. Accordingly, we have, using matrix notation

$$\underline{\delta\hat{P}_o} = \sum_{k=0}^{n}\left[\underline{\underline{m}}_{n(11)}^{o} \sum_{\lambda=0}^{n} \underline{\underline{w}}_{k\lambda}^{o} + \underline{\underline{m}}_{n(12)}^{o} \sum_{\lambda=0}^{n} (t_\lambda - t_o)\underline{\underline{w}}_{k\lambda}^{o}\right]\underline{\delta\bar{P}_k^o}$$
$$\underline{\delta\hat{\dot{P}}_o} = \sum_{k=0}^{n}\left[\underline{\underline{m}}_{n(12)}^{o} \sum_{\lambda=0}^{n} \underline{\underline{w}}_{k\lambda}^{o} + \underline{\underline{m}}_{n(22)} \sum_{\lambda=0}^{n} (t_\lambda - t_o)\underline{\underline{w}}_{k\lambda}^{o}\right]\underline{\delta\bar{P}_k^o} \tag{46}$$

Introducing the abbreviations

$$\hat{\underline{\underline{m}}}_{nk}^{o} \equiv \sum_{\lambda=0}^{n}\left[\underline{\underline{m}}_{n(11)}^{o} + \underline{\underline{m}}_{n(12)}^{o}(t_\lambda - t_o)\right] \underline{\underline{w}}_{k\lambda}^{o}$$
$$\hat{\dot{\underline{\underline{m}}}}_{nk}^{o} \equiv \sum_{\lambda=0}^{n}\left[\underline{\underline{m}}_{n(12)}^{o} + \underline{\underline{m}}_{n(22)}^{o}(t_\lambda - t_o)\right] \underline{\underline{w}}_{k\lambda}^{o} \tag{47}$$

we may rewrite either of Eqs. (46) in the scalar-product form:

$$\underline{\delta\hat{P}_o} = \underline{\left[\hat{\underline{\underline{m}}}_{n}^{o}\right]}^{t}\, \underline{\delta\bar{P}^o}$$
$$\underline{\delta\hat{\dot{P}}_o} = \underline{\left[\hat{\dot{\underline{\underline{m}}}}_{n}^{o}\right]}^{t}\, \underline{\delta\bar{P}^o} \tag{48}$$

We know that the variance of a weighted sum of mutually independent random variables is the quadratically weighted sum of the variances of these variables. To make use of this well-known result, we use Eq. (45) to express $\underline{\delta\hat{P}_o}$ and $\underline{\delta\hat{\dot{P}}_o}$ in Eqs. (48) as linear functions of the n + 1 mutually independent positional errors $\underline{\delta P_k}$, characterized by their "diagonal" joint covariance matrix $\underline{\underline{S}}_n$, defined in Eq. (42). We also write both matrix equations so obtained as a single matrix equation

$$\begin{bmatrix}\underline{\delta\hat{P}_o}\\ \underline{\delta\hat{\dot{P}}_o}\end{bmatrix} = \left[\underline{\left[\hat{\underline{\underline{m}}}_{n}^{o}\right]}\;\underline{\left[\hat{\dot{\underline{\underline{m}}}}_{n}^{o}\right]}\right]^{t} \underline{\underline{T}}_{n}^{ot}\, \underline{\delta\bar{P}} \tag{49}$$

The joint covariance matrix $\hat{\underline{\underline{S}}}_n$ of $\underline{\delta \hat{P}_o}$ and $\underline{\delta \hat{P}_o}$ may now be written directly as

$$\hat{\underline{\underline{S}}}_n \equiv \underline{\underline{T}}_n^o \begin{bmatrix} \left[\hat{\underline{\underline{m}}}_n^o\right] \\ \left[\hat{\underline{\underline{m}}}_n^o\right] \end{bmatrix} \underline{\underline{S}}_n \begin{bmatrix} \left[\hat{\underline{\underline{m}}}_n^o\right] \\ \left[\hat{\underline{\underline{m}}}_n^o\right] \end{bmatrix}^t \underline{\underline{T}}_n^{ot} \tag{50}$$

Carrying out the matrix multiplications indicated in Eq. (50), we obtain

$$\hat{\underline{\underline{S}}}_n \equiv \begin{bmatrix} \hat{\underline{\underline{s}}}_{n(11)} & \hat{\underline{\underline{s}}}_{n(12)} \\ \hat{\underline{\underline{s}}}_{n(21)} & \hat{\underline{\underline{s}}}_{n(22)} \end{bmatrix} \tag{51}$$

where

$$\begin{aligned}
\hat{\underline{\underline{s}}}_{n(11)} &\equiv \underline{\underline{T}}_n^o \left[\hat{\underline{\underline{m}}}_n^o\right] \underline{\underline{S}}_n \left[\hat{\underline{\underline{m}}}_n^o\right]^t T_n^{ot} \\
&= \text{variance matrix of } \underline{\delta \hat{P}_o} \\
\hat{\underline{\underline{s}}}_{n(12)} &\equiv \underline{\underline{T}}_n^o \left[\hat{\underline{\underline{m}}}_n^o\right] \underline{\underline{S}}_n \left[\hat{\underline{\underline{m}}}_n^o\right]^t \underline{\underline{T}}_n^{ot} \\
&= \text{covariance matrix of } \underline{\delta \hat{P}_o} \text{ and } \underline{\delta \hat{P}_o} \\
\hat{\underline{\underline{s}}}_{n(21)} &\equiv \underline{\underline{T}}_n^o \left[\hat{\underline{\underline{m}}}_n^o\right] \underline{\underline{S}}_n \left[\hat{\underline{\underline{m}}}_n^o\right]^t \underline{\underline{T}}_n^{ot} = \hat{\underline{\underline{s}}}_{n(12)} \\
\hat{\underline{\underline{s}}}_{n(22)} &\equiv \underline{\underline{T}}_n^o \left[\hat{\underline{\underline{m}}}_n^o\right] \underline{\underline{S}}_n \left[\hat{\underline{\underline{m}}}_n^o\right]^t \underline{\underline{T}}_n^{ot} \\
&= \text{variance matrix of } \underline{\delta \hat{P}_o}
\end{aligned} \tag{52}$$

IV. Notation

$\overline{R}$ = position vector referring to an inertial frame

t = time

$\overline{R}(t)$ = trajectory of particle in inertial frame

$V(\overline{R})$ = gravitational potential in inertial frame

$\overline{F}(\overline{R}) = -\mu \nabla V(\overline{R})$ = gravitational force field in inertial frame

t_k = instant of observation

$\overline{R}_k \equiv \overline{R}(t_k)$ = particle's true inertial position at instant t_k

$\overline{F}_k \equiv \overline{F}(\overline{R}_k)$ = true gravitational force acting on particle at its true position $\overline{R}_k$

t_p $[t_o < t_p < t_\Omega]$ = pivotal instant

t_o = initial instant

$t_\Omega \equiv t_n$ = final instant

t_ω = present ($\equiv$ current final) instant

$\overline{R}_o \equiv \overline{R}(t_o)$ = particle's true inertial position at initial instant t_o

$\dot{\overline{R}}_o \equiv \dot{\overline{R}}(t_o)$ = particle's true inertial velocity at initial instant t_o

$\overline{P}_k$ = particle's "observed" inertial position at instant t_k

$\overline{\Phi}_k \equiv \overline{F}(\overline{P}_k)$ = "observed" gravitational force at particle's "observed" position $\overline{P}_k$

$\delta\overline{P}_k \equiv \overline{P}_k - \overline{R}_k$ = positional error at particle's true position $\overline{R}_k$

$\delta\overline{\varphi}_k \cong \overline{\Phi}_k - \overline{F}_k$ = error in gravitational force at particle's true position $\overline{R}_k$

$$U_o(\xi) = \infty \quad (\xi = 0) \qquad = \text{unit impulse function}$$
$$= 0 \quad (\xi \neq 0) \qquad (\equiv \text{Dirac's delta function})$$

$$U_{-1}(\xi) = \int_{\xi'=0}^{\xi} U_o(\xi')d\xi' = 0 \quad (\xi < 0) \qquad = \text{unit step function}$$
$$= \frac{1}{2} \quad (\xi = 0)$$
$$= 1 \quad (\xi > 0)$$

$$U_{-2}(\xi) = \int_{\xi'=0}^{\xi} U_{-1}(\xi')d\xi' = 0 \quad (\xi < 0) \qquad = \text{unit ramp function}$$
$$= \xi \quad (\xi \geq 0)$$

$$U_{-3}(\xi) = \int_{\xi'=0}^{\xi} U_{-2}(\xi')d\xi' = \int_{\xi'=0}^{\xi} U_{-2}(\xi - \xi')U_{-1}(\xi')d\xi'$$
$$= 0 \quad (\xi < 0)$$
$$= \frac{1}{2}\xi^2 \quad (\xi \geq 0)$$

$\underline{\Delta P}^o$ = $(n + 1) \times 1$ column vector of 3×1 column vectors $\underline{\Delta P}_k^o$

$\underline{\Delta P}_k^o$ = 3×1 column vector of inertial components of vector residual $\underline{\Delta P}_k^o$

$\underline{\underline{W}}_n^o \equiv \underline{\underline{T}}_n^{o,-1} \underline{\underline{S}}_n^{-1} \underline{\underline{T}}_n^{o,-1t} \equiv \underline{\underline{S}}_n^{o,-1}$ = $(n+1) \times (n+1)$ matrix of coaccuracy matrices $\underline{\underline{w}}_{k\lambda}$; inverse of joint covariance matrix $\underline{\underline{S}}_n^o$ of $n+1$ rectified positions $\underline{P}_k^o$

$\underline{\underline{w}}_{k\lambda}^o$ = 3×3 matrix of inertial components of "coaccuracy tensor" $\underline{\underline{w}}_{k\lambda}^o$ of rectified positions $\underline{P}_k^o$ and $\underline{P}_\lambda^o$

$[\underline{\underline{I}}]$ = $(n+1) \times 1$ column vector of 3×3 unit matrices $\underline{\underline{I}}$

$\underline{[\hat{P}_o + \hat{\dot{P}}_o(t - t_o) - \overline{P}_k^o]}$ = $(n+1) \times 1$ column vector of weighted sums of 3×1 column vectors $\underline{\hat{P}}_o$, $\underline{\hat{\dot{P}}}_o$, $\underline{P}_k^o$

$\underline{\hat{P}}_o$, $\underline{\hat{\dot{P}}}_o$, and $\underline{P}_k^o$ = 3×1 column vectors of inertial components of estimated initial position vector $\hat{\overline{P}}_o$, initial velocity vector $\hat{\dot{\overline{P}}}_o$, and rectified-position vector $\overline{P}_k^o$

$\underline{\underline{u}}_k \equiv \underline{\underline{s}}_k^{-1}$ = positional accuracy matrix at $\overline{R}_k$; inverse of $\underline{\underline{s}}_k$

$\underline{\underline{s}}_k$ = covariance matrix of positional error $\delta\overline{P}_k$ at $\overline{R}_k$

$\underline{\delta\overline{P}^o}$ = $(n+1) \times 1$ column vector of 3×1 column vectors $\underline{\delta P}_k^o$

$\underline{\delta P}_k^o$ = 3×1 column vector of inertial components of rectified-position error $\delta\overline{P}_k^o$

$\underline{\delta\overline{P}}$ = $(n+1) \times 1$ column vector of 3×1 column vectors $\underline{\delta P}_k$

$\underline{\delta P}_k$ = 3×1 column vector of inertial components of positional error $\delta\overline{P}_k$ at $\overline{R}_k$

$\underline{[\hat{\underline{\underline{m}}}_n^o]}$ = $(n+1) \times 1$ column vector of 3×3 matrices $\hat{\underline{\underline{m}}}_{nk}^o$ in first of Eqs. (47)

$\underline{[\hat{\dot{\underline{\underline{m}}}}_n^o]}$ = $(n+1) \times 1$ column vector of 3×3 matrices $\hat{\dot{\underline{\underline{m}}}}_{nk}^o$ in second of Eqs. (47)

$\underline{[[\underline{\underline{m}}_n^o][\underline{\underline{m}}_n^o]]}$ = $(n+1) \times 2$ matrix of 3×3 matrices $\hat{\underline{\underline{m}}}_{nk}^o$ and $\hat{\dot{\underline{\underline{m}}}}_{nk}^o$

V. Conclusion

The rectification method of estimating parameters specifying the state of motion of a particle from a succession of observed positions, described in the present paper, is based on concepts of functional analysis underlying the theories of integral equations, integral representations, linear operators, and the like. By using an approximate formulation of the original problem, rather than a solution of an approximate problem, the rectification method enables one to estimate, from the observational data, the values of parameters required for a complete specification of the original problem, the specific solution of which may then be determined by numerical methods.

In contrast to the inherently deterministic, local, differential methods, this global, integral method is inherently probabilistic. Whereas the former are well suited for evaluation of accurate, although perhaps scanty, observational data, the latter is especially suited for evaluation in inaccurate, but redundant, data. Hence, the author feels that the rectification method complements rather than competes with the well-proven differential methods.

The detailed study of its modifications and extensions, some of which are discussed, all too briefly, in Sec. II, is a most intriguing future task.

References

[1] Spies, O. R., "Solution of integral equations of motion by successive approximation," Institute of Electrical and Electronic Engineers, Aerospace Ground Support Conference (August 4-9, 1963).

[2] Levine, N., "A new technique for increasing the flexibility of recursive least squares data smoothing," Bell System Tech. J. 40, 821-840 (1961).

SELF-CONTAINED ORBIT DETERMINATION TECHNIQUES

Robert H. Gersten* and Z. E. Schwarzbein*

North American Aviation, Inc., Downey, Calif.

Abstract

The importance of self-contained orbit determination techniques (with all measurements and computations made on-board) has greatly increased with the approach of the era of flight to our celestial neighbors. This paper presents an extensive treatment of self-contained techniques, any of which can be applied by space vehicles on geocentric, selenocentric, heliocentric, or planetocentric orbits. These methods utilize observations of landmarks of the principal body (under the assumption that the body-centered latitude and longitude of these landmarks as well as the rotation rate of the principal body are known) or stellar observations (as described in the text) to determine the orientation of the orbit plane in inertial space. Once the orientation of the orbit plane is known, three more measurements are utilized to uniquely determine the two-body orbit. Orbit determination from the following combinations of measurements are considered: 1) three measurements of either the altitude, radial velocity, or time rate of change of true anomaly and the associated times; 2) two measurements of any one of the preceding quantities, the first of which is taken simultaneously with the measurement of one of the other quantities, e.g., a simultaneous measurement of altitude and range-rate along with a later measurement of altitude, and the associated times. A primary advantage of these methods is that, although an additional measurement is sometimes required to remove ambiguities, the solution is always obtained in closed

Presented as Preprint 63-431 at the AIAA Astrodynamics Conference, New Haven, Conn., August 19-21, 1963, when authors were with Northrop Space Laboratories.

*Research Specialist, Space and Information Systems Division.

form and without iteration. It should be pointed out that altitude and radial velocity can be obtained directly from measurements of range and range-rate along the local vertical of the (assumed spherical) principal body. Likewise, time rate of change of true anomaly is obtained from successive determinations of the direction of the local vertical.

Nomenclature

Scalars

A = $(\dot{r}_2 - \dot{r}_1 \cos \Delta v_{12}) \csc \Delta v_{12}$, auxiliary quantity used with $\dot{v}_1$, $\dot{r}_1$, $\dot{r}_2$

a, b = auxiliary quantities used to analyze multiple solutions with $\dot{r}_1$, $\dot{v}_1$, $\dot{v}_2$

e = orbital eccentricity

p = parameter or semilatus rectum

P = $(p)^{1/2}$

$\overline{P}$ = $(p/\mu)^{1/2}$

$\tilde{P}$ = $\overline{P}^{1/2}$ = $(p/\mu)^{1/4}$

$\hat{P}$ = auxiliary quantity used to analyze multiple solution with $\dot{r}_1$, $\dot{v}_1$, $\dot{v}_2$

P^* = value of P associated with maximum r_2 (for fixed r_1 and $\dot{r}_1$) used with $\dot{r}_1$, r_1, r_2

r = separation of vehicle from dynamical center

$\dot{r}$ = radial component of velocity

t = time

v = true anomaly

Δv = increment in true anomaly between measurements

$\dot{v}$ = derivative of v with respect to time

Scalars

ϕ = sublatitude of vehicle

λ = sublongitude of vehicle

$\dot{\theta}$ = rotation rate of principal body

μ = $g R^2$ where g is the gravitational attraction at the surface of the principal body and R is its radius

β, α = azimuth and elevation angles, respectively, of the vehicle in the x_1, y_1, z_1 frame

δ, γ = azimuth and elevation angles, respectively, of known stars in the x, y, z frame

Vectors

$\underline{i}, \underline{j}, \underline{k}$ = unit vectors in the x, y, z directions discussed in the text, respectively

$\underline{r}$ (x, y, z) = position vector of the vehicle

$\dot{\underline{r}}$ ($\dot{x}, \dot{y}, \dot{z}$) = velocity vector of the vehicle

$\underline{S}_1, \underline{t}_1$ = unit vectors directed from the vehicle to two known stars

$\underline{S}_1, \underline{S}_2, \underline{S}_3$ = unit vectors along the x_1 y_1 z_1 directions discussed in the text, respectively

$\underline{U}$ = unit vector along $\underline{r}$

$\underline{V}$ = unit vector in the orbit plane directed so that $\underline{U}$, $\underline{V}$, $\underline{W}$ form a right-handed set

$\underline{W}$ = unit vector orthogonal to the orbit plane in the direction of the moment of momentum vector

Subscripts

1, 2, 3	=	quantities associated with the first, second, and third observations, respectively
j	=	2, 3 preceding
k	=	1, 2, 3 preceding
a, b	=	used to differentiate between multiple solutions in the $\dot{r}_1$, r_1, r_2 and $\dot{r}_1$, $\dot{v}_1$, $\dot{v}_2$ cases
S1, t1	=	quantities associated with two known stars
x, y, z	=	x, y, z components, respectively
x_1, y_1, z_1	=	x_1, y_1, z_1 components, respectively

General Remarks

1. Orientation of the Orbit Plane

The following inertial coordinate system is adopted.[1] The origin of coordinates is taken as the geometric center of the principal body with the x and y axes lying in the equator plane and the z axis directed toward the North Pole. The x axis is coincident with the intersection of a reference prime meridian with the equator plane at the time of the first observation. The sense of the y axis is such that x, y, z form a right-handed set (see Fig. 1).

If local cartography is utilized, the unit vectors $\underline{U}_1$, $\underline{U}_2$, $\underline{U}_3$ directed from the origin to the satellite at the observation times are given by

$$\underline{U}_1 = \cos\varphi_1 \cos\lambda_1 \underline{i} + \cos\varphi_1 \sin\lambda_1 \underline{j} + \sin\varphi_1 \underline{k} \tag{1a}$$

$$\underline{U}_j = \cos\varphi_j \cos\left[\lambda_j + \dot{\theta}\,(t_j - t_1)\right] \underline{i} \tag{1b}$$

$$+ \cos\varphi_j \sin\left[\lambda_j + \dot{\theta}\,(t_j - t_1)\right] \underline{j} + \sin\varphi_j \underline{k}$$

$$j = 2, 3$$

where $\underline{i}$, $\underline{j}$, $\underline{k}$ are unit vectors in the x, y, z directions, respectively, $\dot{\theta}$ is the rotation rate of the principal body, ϕ and λ represent the vehicular sublatitude and sublongitude, respectively, t is the observation time, and the subscripts 1, 2, and 3 correspond to the first, second, and third observations. For this analysis the longitude is measured as positive in the direction of the attracting body's rotation with 0° longitude located along the prime meridian. The increments in true anomalies Δv_{1j} (j = 2, 3) between the preceding vectors is given by the dot product

$$\begin{aligned} \underline{U}_1 \cdot \underline{U}_j &= \cos (v_j - v_1) = \cos \Delta v_{1j} \\ &= \cos \varphi_1 \cos \varphi_j \cos \lambda_1 \cos [\lambda_j + \dot{\theta} (t_j - t_1)] + \\ &\quad \cos \varphi_1 \cos \varphi_j \sin \lambda_1 \sin [\lambda_j + \dot{\theta} (t_j - t_1)] + \\ &\quad \sin \varphi_1 \sin \varphi_j \qquad (2) \\ &\quad j = 2, 3 \end{aligned}$$

If an inertial platform monitored by star trackers is utilized, Δv_{1j} is obtained as follows. The unit vectors $\underline{U}_k$ (k = 1, 2, 3) are defined coincident with the radial direction with sense directed away from the dynamical center and with origin in the vehicle. Further, the coordinates of $\underline{U}_k$ relative to the inertial x_1, y_1, z_1 frame established below are given by the measured azimuth β_k and elevation angle α_k in this frame (see Fig. 2). Let us further assume that the preceding inertial system is defined by unit vectors $\underline{S}_1$ and $\underline{t}_1$ directed toward two known stars. Thus, the coordinates of $\underline{S}_1$ and $\underline{t}_1$ relative to the inertial frame x, y, z with origin at the dynamical center are

$$\underline{S}_1 = \cos\gamma_{S1}\cos\delta_{S1}\,\underline{i} + \cos\gamma_{S1}\sin\delta_{S1}\,\underline{j} + \sin\gamma_{S1}\,\underline{k} \quad (3a)$$

$$\underline{t}_1 = \cos\gamma_{t1}\cos\delta_{t1}\,\underline{i} + \cos\gamma_{t1}\sin\delta_{t1}\,\underline{j} + \sin\gamma_{t1}\,\underline{k} \quad (3b)$$

where δ and γ are the azimuth and elevation angles, respectively, of the two stars in the x, y, z frame. The x, y, z components of $\underline{S}_1$ and $\underline{t}_1$ are computed from ephemerides of the two stars and of the x, y, z system in a common inertial frame. The separation between vehicle and stars is sufficiently large that $\underline{S}_1$ and $\underline{t}_1$ can be assumed to be constant throughout the flight. A unit vector $\underline{S}_3$ orthogonal to the $\underline{S}_1$, $\underline{t}_1$ plane is obtained from

$$\underline{S}_3 = \underline{S}_1 \times \underline{t}_1 \big/ |\underline{S}_1 \times \underline{t}_1| \quad (3c)$$

Likewise, a unit vector $\underline{S}_2$ normal to the $\underline{S}_1$, $\underline{S}_3$ plane is given by

$$\underline{S}_2 = \underline{S}_3 \times \underline{S}_1 \quad (3d)$$

Let us now define the x_1, y_1, z_1 axes as lying along the $\underline{S}_1$, $\underline{S}_2$, and $\underline{S}_3$ vectors, respectively. In this manner, x_1, y_1, z_1 form a right-handed set. Hence, $\underline{U}_k$ can be written relative to the (inertially fixed) x_1, y_1, z_1 system as

$$\underline{U}_k = U_{x_1 k}\,\underline{S}_1 + U_{y_1 k}\,\underline{S}_2 + U_{z_1 k}\,\underline{S}_3 \quad (4)$$

where, in terms of the measured angles,

$$U_{x_1 k} = \cos \alpha_k \cos \beta_k \tag{4a}$$

$$U_{y_1 k} = \cos \alpha_k \sin \beta_k \tag{4b}$$

$$U_{z_1 k} = \sin \alpha_k \tag{4c}$$

$$k = 1, 2, 3$$

The components of this vector in the x, y, z system are given by

$$\underline{U}_k = U_{xk}\, \underline{i} + U_{yk}\, \underline{j} + U_{zk}\, \underline{k} \tag{5}$$

where

$$U_{xk} = \underline{U}_k \cdot \underline{i} \tag{5a}$$

$$U_{yk} = \underline{U}_k \cdot \underline{j} \tag{5b}$$

$$U_{zk} = \underline{U}_k \cdot \underline{k} \tag{5c}$$

$$k = 1, 2, 3$$

The increment in true anomaly Δv_{1j} is obtained from

$$\underline{U}_1 \cdot \underline{U}_j = \cos (v_j - v_1) = \cos \Delta v_{1j} \tag{6}$$

$$j = 2, 3$$

For this analysis, the assumption is made that $0 < \Delta v < \pi$. Thus, Δv_{1j} may be assigned to the first or second quadrant and uniquely determined from Eq. (2) or (6).

2. Basic Formulas

Combinations of the following formulas will appear throughout the several methods:

$$r_k = \frac{p}{1 + e \cos (v_1 + \Delta v_{1k})} \tag{7a}$$

$$\dot{r}_k = (\mu / p)^{\frac{1}{2}} e \sin (v_1 + \Delta v_{1k}) \tag{7b}$$

$$\dot{v}_k = (\mu / p^3)^{\frac{1}{2}} [1 + e \cos (v_1 + \Delta v_{1k})]^2 \tag{7c}$$

$$k = 1, 2, 3 \text{ where } \Delta v_{11} = 0$$

where e is the orbital eccentricity, p is the parameter (semilatus rectum), μ is the product of the universal gravitational constant and the mass of the attracting body, $\dot{r}$ is the radial velocity, r is the vehicular separation from the geometrical center of the principal body, t is the observation time, $\dot{v}$ is the time rate of change of true anomaly, and the subscripts k = 1, 2, 3 correspond, respectively, to the first, second, and third observations. Since the nonsphericity of the principal body and local anomalies are neglected in this analysis, the measured range-rate (along the local vertical) may be set equal to $\dot{r}$ and the sum of the measured range (along the local vertical) and the (assumed constant) radius of the principal body is taken as r. Likewise, $\dot{v}$ is given by the angular rate determined from successive measurements of increments of true anomaly obtained by a horizon scanner and the corresponding times. This determination is valid if the attracting body is assumed to be spherically symmetric.

The following section is concerned with the determination of the semilatus rectum p, orbital eccentricity e, and true anomaly v_1, from various combinations of r, $\dot{r}$, and $\dot{v}$ obtained from measured data in the manner indicated previously. It is assumed in all cases that the increments in true anomalies between measurements have been obtained from Eq. (2) or (6) prior to commencing the computations. The case of orbit determination from three r's, the associated $\underline{U}$'s being obtained as in Eq. (1) or (5), is identical to the classical Gibbsian method, which is analyzed in Refs. 2

and 3 and therefore is not presented here. As can be seen from Eq. (7b), although two (or more) measurements of $\dot{r}$ yield discrete values of the quantity $e/p^{1/2}$, it is impossible to obtain either the eccentricity or semilatus rectum separately. Thus, the case of orbit determination from measurements of $\dot{r}$ only is impossible of solution and is not discussed. Having obtained p, e, and v_1, the last section considers the determination of the initial position and velocity vectors as functions of the preceding quantities.

Specific Cases

1. One Altitude and Two Radial Velocities (r_1, $\dot{r}_1$, $\dot{r}_2$)

The true anomaly at the first observation is obtained, after some manipulation, from Eq. (7b) for k = 1 and 2 as

$$\tan v_1 = \frac{\dot{r}_1}{(\dot{r}_2 - \dot{r}_1 \cos \Delta v_{12}) \csc \Delta v_{12}} \tag{8}$$

In assigning the proper quadrant to v_1, the signs of the numerator and denominator are equal, respectively, to the signs of $\sin v_1$ and $\cos v_1$. From Eqs. (7a) and (7b) for k = 1 we obtain

$$e \cos v_1 = \frac{p}{r_1} - 1 \tag{9a}$$

$$e \sin v_1 = \left(\frac{p}{\mu}\right)^{\frac{1}{2}} \dot{r}_1 \tag{9b}$$

The preceding yields

$$\frac{\tan v_1}{r_1} P^2 - \frac{\dot{r}_1}{\mu^{\frac{1}{2}}} P - \tan v_1 = 0 \tag{10}$$

where

$$P \triangleq p^{\frac{1}{2}} \tag{10a}$$

The solution of the quadratic Eq. (10) is

$$P = \frac{\dot{r}_1/\mu^{1/2} \pm \left[(\dot{r}_1^{\,2}/\mu) + (4 \tan^2 v_1 / r_1)\right]^{1/2}}{2\,(\tan v_1 / r_1)} \tag{11}$$

where the + sign is utilized when v_1 lies in the first or third quadrant and the - sign when v_1 lies in the second or fourth quadrant. Of course, upon solution of Eq. (11), the semilatus rectum is obtained directly from Eq. (10a). The eccentricity e results from taking the square root of the sum of the squares of Eqs. (9a) and (9b).

2. One Radial Velocity and Two Altitudes ($\dot{r}_1$, r_1, r_2)

Substituting Eqs. (9a) and (9b) into Eq. (7a) for k = 2 yields

$$\left(\frac{\cos \Delta v_{12}}{r_1} - \frac{1}{r_2}\right) P^2 - \frac{\dot{r}_1 \sin \Delta v_{12}}{\mu^{\frac{1}{2}}} P + (1 - \cos \Delta v_{12}) = 0 \tag{12}$$

where again

$$P \triangleq p^{\frac{1}{2}} \tag{12a}$$

The solution to the quadratic equation (12) is

$$P = \frac{\dfrac{\dot{r}_1 \sin \Delta v_{12}}{\mu^{\frac{1}{2}}} \pm \left[\dfrac{\dot{r}_1^{\,2} \sin^2 \Delta v_{12}}{\mu}\right.}{2\,\dfrac{r_2 \cos \Delta v_{12} - r_1}{r_1 r_2}}$$

$$\frac{\left. -4\,\dfrac{r_2 \cos \Delta v_{12} - r_1}{r_1 r_2}\,(1 - \cos \Delta v_{12})\right]^{\frac{1}{2}}}{2\,\dfrac{r_2 \cos \Delta v_{12} - r_1}{r_1 r_2}} \tag{13}$$

When $r_2 \cos \Delta v_{12} < r_1$, the negative sign is utilized in Eq. (13). When $r_2 \cos \Delta v_{12} = r_1$, P is given from Eq. (12) as $P = \mu^{1/2} (1 - \cos \Delta v_{12}) / (\dot{r}_1 \sin \Delta v_{12})$. The case when $r_2 \cos \Delta v_{12} > r_1$ requires further analysis as illustrated in Appendix A. Of course, once P is computed, p can be obtained from Eq. (12a). One obtains v_1 from

$$\tan v_1 = (p/\mu)^{1/2} \dot{r}_1 / (p/r_1 - 1) \tag{14}$$

which results from Eqs. (9a) and (9b). The signs of the numerator and denominator in Eq. (14) are equal, respectively, to the signs of $\sin v_1$ and $\cos v_1$. The eccentricity is obtained from Eqs. (9a) and (9b), as in case 1.

3. One Time Rate of Change of True Anomaly and Two Radial Velocities ($\dot{v}_1$, $\dot{r}_1$, $\dot{r}_2$)

The true anomaly corresponding to the first observation results directly from Eq. (8). Then from Eq. (7b) for k = 1 and 2 and Eq. (7c) for k = 1, the following cubic equation is obtained:

$$\mu \dot{v}_1 \bar{P}^3 - A^2 \bar{P}^2 - 2A \bar{P} - 1 = 0 \tag{15}$$

where

$$\bar{P} = (p/\mu)^{1/2} \tag{15a}$$

$$A = (\dot{r}_2 - \dot{r}_1 \cos \Delta v_{12}) \csc \Delta v_{12} \tag{15b}$$

Proof that Eq. (15) yields one and only one positive real root of $\bar{P}$ for conic section orbits is presented in Appendix B. Of course, $\bar{P}$ is utilized to give p from Eq. (15a). The eccentricity is then obtained from[1]

$$e = \left\{ \frac{p}{\mu} \left[\dot{r}_1^{\,2} + \left(\frac{\dot{r}_2 - \dot{r}_1 \cos \Delta v_{12}}{\sin \Delta v_{12}} \right)^2 \right] \right\}^{\frac{1}{2}} \tag{16}$$

4. One Radial Velocity and Two Time Rates of Change of True Anomaly ($\dot{r}_1$, $\dot{v}_1$, $\dot{v}_2$)

Dividing Eq. (7c) for k = 1 by Eq. (7c) for k = 2, and taking the square root of the resultant, yields

$$\left(\frac{\dot{v}_1}{\dot{v}_2}\right)^{\frac{1}{2}} = \frac{1 + e \cos v_1}{1+e \cos v_1 \cos \Delta v_{12} - e \sin v_1 \sin \Delta v_{12}} \tag{17}$$

Substituting Eq. (7b) for k = 1 into the preceding gives

$$1 + e \cos v_1 = \frac{(\dot{v}_1/\dot{v}_2)^{1/2} \left[1 - \dot{r}_1 \sin \Delta v_{12} \bar{P} - \cos \Delta v_{12}\right]}{1 - (\dot{v}_1/\dot{v}_2)^{1/2} \cos \Delta v_{12}} \tag{18}$$

where

$$\bar{P} = (p/\mu)^{1/2} \tag{18a}$$

Substituting Eq. (18) into Eq. (7c) for k = 1, one obtains, after some manipulation,

$$\mu \dot{v}_2 \left[1 - \left(\frac{\dot{v}_1}{\dot{v}_2}\right)^{\frac{1}{2}} \cos \Delta v_{12}\right]^2 \bar{P}^3 - \dot{r}_1^{\,2} \sin^2 \Delta v_{12} \bar{P}^2$$

$$+ 2 \dot{r}_1 \sin \Delta v_{12} (1 - \cos \Delta v_{12}) \bar{P}$$

$$- (1 - \cos \Delta v_{12})^2 = 0 \tag{19}$$

A discussion of the utilization of Eq. (19) to yield the correct $\bar{P}$ is contained in Appendix C. Having $\bar{P}$, one obtains v_1 and e from the following relationships obtained from Eqs. (9b) and (18)

$$e \sin v_1 = \bar{P} \dot{r}_1 \tag{20a}$$

$$e \cos v_1 = \frac{\left(\frac{\dot{v}_1}{\dot{v}_2}\right)^{\frac{1}{2}} [1 - \dot{r}_1 \sin \Delta v_{12} \bar{P}] - 1}{1 - \left(\frac{\dot{v}_1}{\dot{v}_2}\right)^{\frac{1}{2}} \cos \Delta v_{12}} \tag{20b}$$

Of course, p is obtained directly from Eq. (18a).

5. One Time Rate of Change of True Anomaly and Two Altitudes ($\dot{v}_1$, r_1, r_2)

The semilatus rectum p results directly from

$$p = \frac{r_1^4 \dot{v}_1^2}{\mu} \tag{21}$$

which is an alternate form of Eq. (7c) for k = 1. One obtains v_1 and e, after some manipulation of Eq. (7a), for k = 1 and 2 from

$$\tan v_1 = \frac{1}{\sin \Delta v_{12}} \frac{\left[p\left(\frac{\cos \Delta v_{12}}{r_1} - \frac{1}{r_2}\right)\right.}{\frac{p}{r_1} - 1} \frac{\left. + (1 - \cos \Delta v_{12})\right]}{\frac{p}{r_1} - 1} \tag{22}$$

and

$$e = \left\{ \left[\frac{p}{r_1} - 1\right]^2 + \left[p\left(\frac{\cos \Delta v_{12}}{r_1} - \frac{1}{r_2}\right) + (1 - \cos \Delta v_{12})\right]^2 \frac{1}{\sin^2 \Delta v_{12}} \right\}^{\frac{1}{2}} \tag{23}$$

where the quadrant of v_1 is determined from the sign of the numerator and denominator of Eq. (22).

6. One Altitude and Two Time Rates of Change of True Anomaly (r_1, $\dot{v}_1$, $\dot{v}_2$)

The semilatus rectum p, is obtained directly from Eq. (21). The values of v_1 and e are then obtained after some manipulation of Eq. (7a) for k = 1 and Eq. (7c) for k = 2 from

$$\tan v_1 = \frac{1}{\sin \Delta v_{12}} \frac{\left[\frac{p}{r_1}\left\{\cos \Delta v_{12} - \left(\frac{\dot{v}_2}{\dot{v}_1}\right)^{\frac{1}{2}}\right\}\right.}{\frac{p}{r_1} - 1} \frac{\left. + (1 - \cos \Delta v_{12})\right]}{\frac{p}{r_1} - 1} \tag{24}$$

$$e = \left| \left\{\frac{p}{r_1} - 1\right\}^2 + \left\{\frac{p}{r_1}\left[\cos \Delta v_{12} - \left(\frac{\dot{v}_2}{\dot{v}_1}\right)^{\frac{1}{2}}\right] + (1 - \cos \Delta v_{12})\right\}^2 \frac{1}{\sin^2 \Delta v_{12}} \right|^{\frac{1}{2}} \tag{25}$$

where the quadrant of v_1 is determined from the sign of the numerator and denominator of Eq. (24).

7. Three Time Rates of Change of True Anomaly ($\dot{v}_1$, $\dot{v}_2$, $\dot{v}_3$)

Equations (7c) for k = 1, 2, and 3 are utilized to give

$$(\dot{v}_1)^{\frac{1}{2}} (1 + e \cos v_1 \cos \Delta v_{12} - e \sin v_1 \sin \Delta v_{12}) = \dot{v}_2^{\frac{1}{2}} (1 + e \cos v_1) \tag{26a}$$

and

$$(\dot{v}_1)^{\frac{1}{2}} (1 + e \cos v_1 \cos \Delta v_{13} - e \sin v_1 \sin \Delta v_{13})$$
$$= \dot{v}_3^{\frac{1}{2}} (1 + e \cos v_1). \tag{26b}$$

From Eqs. (26a) and (26b) one obtains

$$e \cos v_1 = \tilde{Q}/\tilde{S} \tag{27a}$$

$$e \sin v_1 = \tilde{R}/\tilde{S} \tag{27b}$$

where

$$\tilde{Q} = \dot{v}_2^{\frac{1}{2}} \sin \Delta v_{13} - \dot{v}_3^{\frac{1}{2}} \sin \Delta v_{12} + \dot{v}_1^{\frac{1}{2}} (\sin \Delta v_{12} - \sin \Delta v_{13})$$

$$\tilde{R} = -(\dot{v}_3^{\frac{1}{2}} - \dot{v}_1^{\frac{1}{2}}) \cos \Delta v_{12} + (\dot{v}_2^{\frac{1}{2}} - \dot{v}_1^{\frac{1}{2}}) \cos \Delta v_{13} - (\dot{v}_2^{\frac{1}{2}} - \dot{v}_3^{\frac{1}{2}})$$

$$\tilde{S} = \dot{v}_1^{\frac{1}{2}} \sin (\Delta v_{13} - \Delta v_{12}) + \dot{v}_3^{\frac{1}{2}} \sin \Delta v_{12} - \dot{v}_2^{\frac{1}{2}} \sin \Delta v_{13}$$

The values of v_1 and e are obtained from Eqs. (27); the former by setting tan v_1 equal to the quotient of Eq. (27b) over Eq. (27a) and determining the quadrant from the sign of the numerator and denominator and the latter by taking the square root of the sum of the squares of these equations. Finally, having obtained e cos v_1 from Eq. (27a), the parameter p is computed from Eq. (7c) for $k = 1$, rewritten in the form

$$p = \left[\frac{\mu (1 + e \cos v_1)^4}{\dot{v}_1^2} \right]^{1/3} \tag{28}$$

<u>Determination of Initial Position and Velocity Vectors:</u> One now has sufficient information to determine the position and velocity at the first observation time in the inertial coordinate system selected. This can be accomplished in two steps. The first step involves determination of the triad of

unit vectors $\underline{U}_1$, $\underline{V}_1$, $\underline{W}$, where $\underline{U}_1$ is directed from the origin of coordinates to the vehicle at the first observation time, $\underline{W}$ is normal to the orbit plane in the direction of the moment of momentum vector, and $\underline{V}_1$ lies in the orbit plane such that $\underline{U}_1$, $\underline{V}_1$, $\underline{W}$ form a right-handed set. The unit vector orthogonal to the orbital plane $\underline{W}$ is given by

$$\underline{W} = \underline{U}_1 \times \underline{U}_2 \, / \, |\underline{U}_1 \times \underline{U}_2| \tag{29}$$

From $\underline{W}$ and $\underline{U}_1$, the unit vector $\underline{V}_1$ is obtained by

$$\underline{V}_1 = \underline{W} \times \underline{U}_1 \tag{30}$$

The second step consists of calculating the separation of the vehicle from the dynamical center, r_1, the transverse component of velocity, $(r\,\dot{v})_1$, and the radial velocity $\dot{r}_1$, at the time of the first observation from Eqs. (7) for k = 1. Of course, any of the preceding which are observed need not be computed.

Using the preceding values, the position vector $\underline{r}_1$ and velocity vector $\dot{\underline{r}}_1$ at the time of the first observation are given by

$$\underline{r}_1 = r_1 \, \underline{U}_1 \tag{31a}$$

and

$$\dot{\underline{r}}_1 = \dot{r}_1 \, \underline{U}_1 + (r\,\dot{v})_1 \, \underline{V}_1 \tag{31b}$$

The last two equations uniquely define the two-body orbit in inertial space.

Conclusions

All seven cases considered yield closed-form, no iteration determinations of the position and velocity vectors in inertial space at the initial observation time. For five of these cases, only three of the measured quantities are required to yield unambiguous solutions. The cases ($\dot{r}_1$, r_1, r_2, and $\dot{r}_1$, $\dot{v}_1$, $\dot{v}_2$) require a fourth measurement to avoid ambiguity as discussed in the text.

Appendix A: Analysis of Multiple Solutions for One Radial Velocity and Two Altitudes ($\dot{r}_1$, r_1, r_2)

Differentiating Eq. (12) with respect to P for fixed values of r_1 and $\dot{r}_1$ yields

$$\frac{P^2}{r_2^2} \frac{d\, r_2}{d\, P} = \frac{2}{P} (1 - \cos \Delta v_{12}) - \frac{\dot{r}_1 \sin \Delta v_{12}}{\mu^{\frac{1}{2}}} \tag{A1}$$

or

$$\frac{d\, r_2}{d\, P} = \frac{r_2^2}{P^3 \mu^{\frac{1}{2}}} [2 \mu^{\frac{1}{2}} (1 - \cos \Delta v_{12}) - \dot{r}_1 \sin \Delta v_{12} P] \tag{A2}$$

Equation (A2) gives the variation of r_2 with respect to P. This variation is always positive for values of $\dot{r}_1 \leq 0$. However, this constraint corresponds to the condition $r_2 \cos \Delta v_{12} < r_1$, which already has been shown to yield a unique value of P. For $\dot{r}_1 > 0$, $d\, r_2/d\, P$ is positive, negative, or zero, depending upon the value of P. The maximum value of r_2 [associated with the zero value of Eq. (A2)] corresponds to $P = P^*$ where

$$P^* = \frac{2 \mu^{\frac{1}{2}} (1 - \cos \Delta v_{12})}{\dot{r}_1 \sin \Delta v_{12}} = \frac{2 \mu^{\frac{1}{2}}}{\dot{r}_1} \tan \frac{\Delta v_{12}}{2} \tag{A3}$$

A schematic of Eq. (12) is shown in Fig. 3 for a lunar satellite.

It can be seen from examination of Fig. 3 that when $r_2 \cos \Delta v_{12} > r_1$, there are, in general, two values of P corresponding to each value of r_2. This ambiguity can be resolved if $\dot{r}_2$ is measured by substituting the values of p, e, and v_1 corresponding to each of the P's obtained into Eq. (7b) for k = 2 and selecting the value yielding the observed $\dot{r}_2$. The uniqueness of this solution is proved below. Extending Eq. (9b) to the second observation time, and remembering Eq. (10a),

$$e \sin (v_1 + \Delta v) = (p/\mu)^{\frac{1}{2}} \dot{r}_2 \tag{A4}$$

or

$$P \dot{r}_2 = \mu^{\frac{1}{2}} \{e \sin v_1 \cos \Delta v + e \cos v_1 \sin \Delta v\} \tag{A5}$$

Substituting Eqs. (9a) and (9b) for each of the P's obtained in Eq. (13) (designated P_a and P_b) into Eq. (A5) yields

$$P_i \dot{r}_2 = P_i \dot{r}_1 \cos \Delta v_{12} + \mu^{\frac{1}{2}} \left(\frac{P_i^2}{r_1} - 1 \right) \sin \Delta v_{12} \tag{A6}$$

$$i = a \text{ and } b$$

If each of the P's is associated with the same value of $\dot{r}_2$, then

$$\frac{P_a}{P_b} = \frac{P_a \dot{r}_1 \cos \Delta v_{12} + \mu^{\frac{1}{2}} \left(\frac{P_a^2}{r_1} - 1 \right) \sin \Delta v_{12}}{P_b \dot{r}_1 \cos \Delta v_{12} + \mu^{\frac{1}{2}} \left(\frac{P_b^2}{r_1} - 1 \right) \sin \Delta v_{12}} \tag{A7}$$

Since Eq. (A7) is satisfied if and only if $P_a = P_b$, there can be one and only one value of P yielding the observed $\dot{r}_2$. Another alternative that can be utilized when $\dot{r}_2$ is measured is to compute P directly from

$$P = \frac{\dfrac{\dot{r}_2 \sin \Delta v_{12}}{\mu^{\frac{1}{2}}} \pm \left[\dfrac{\dot{r}_2^2 \sin^2 \Delta v_{12}}{\mu} + 4 \dfrac{r_2 - r_1 \cos \Delta v_{12}}{r_1 r_2} (1 - \cos \Delta v_{12}) \right]^{\frac{1}{2}}}{2 \dfrac{r_2 - r_1 \cos \Delta v_{12}}{r_1 r_2}} \tag{A8}$$

Equation (A8), which is derived in a manner similar to Eq. (13), is valid when the positive sign is utilized throughout the region $r_2 > r_1 \cos \Delta v_{12}$, which contains the region in question $r_2 \cos \Delta v_{12} > r_1$.

Appendix B: Proof of the Uniqueness of the Solution for One Time Rate of Change of True Anomaly and Two Radial Velocities ($\dot{v}_1$, $\dot{r}_1$, $\dot{r}_2$)

From Eq. (15)

$$\mu \dot{v}_1 = (1 + A \bar{P})^2 / \bar{P}^3 \tag{B1}$$

Differentiating the preceding equation with respect to $\bar{P}$ for fixed $\dot{r}_1$ and $\dot{r}_2$ yields

$$\frac{d(\mu \dot{v}_1)}{d \bar{P}} = - \frac{(1 + A \bar{P})(3 + A \bar{P})}{\bar{P}^4} \tag{B2}$$

But, from Eq. (7b) for k = 1 and 2 and Eq. (15b),

$$e \cos v_1 = A \bar{P} \tag{B3}$$

Substituting Eq. (B3) into Eq. (B2) gives

$$\frac{d(\mu \dot{v}_1)}{d \bar{P}} = - \frac{(1 + e \cos v_1)(3 + e \cos v_1)}{\bar{P}^4} \tag{B4}$$

Since $(1 + e \cos v_1) \geq 0$ for all conic sections, $\dot{v}_1$ increases (decreases) monotonically as $\bar{P}$ decreases (increases), and there is one and only one $\bar{P}$ corresponding to each $\dot{v}_1$. This value can be obtained by iterating on Eq. (15) or solving the cubic equation directly, discarding roots of $\bar{P}$ yielding values of $A \bar{P} < -1$.

Appendix C: Analysis of Multiple Solutions for One Radial Velocity and Two Time Rates of Change of True Anomaly ($\dot{r}_1$, $\dot{v}_1$, $\dot{v}_2$)

Equation (19) may be transformed into

$$\hat{P}^3 + a \hat{P} + b = 0 \tag{C1}$$

where

$$\hat{P} = \bar{P} + \frac{\dot{r}_1^{\,2} \sin^2 \Delta v_{12}}{3 \mu \dot{v}_2 \left[1 - \left(\frac{\dot{v}_1}{\dot{v}_2}\right)^{\frac{1}{2}} \cos \Delta v_{12}\right]^2} \qquad \text{(C2)}$$

$$a = \frac{\dot{r}_1 \sin \Delta v_{12}}{3 \mu^2 \dot{v}_2^{\,2} \left[1 - \left(\frac{\dot{v}_1}{\dot{v}_2}\right)^{\frac{1}{2}} \cos \Delta v_{12}\right]^4}$$

$$\left\{6 \mu \dot{v}_2 (1 - \cos \Delta v_{12}) \left[1 - \left(\frac{\dot{v}_1}{\dot{v}_2}\right)^{\frac{1}{2}} \cos \Delta v_{12}\right]^2 - \dot{r}_1^{\,3} \sin^3 \Delta v_{12}\right\} \qquad \text{(C3)}$$

$$b = \frac{1}{27\mu^3 \dot{v}_2^{\,3} \left[1 - \left(\frac{\dot{v}_1}{\dot{v}_2}\right)^{\frac{1}{2}} \cos\Delta v_{12}\right]^6} \left\{- 2\dot{r}_1^{\,6} \sin^6 \Delta v_{12} + \right.$$

$$18\dot{r}_1^{\,3} \sin^3 \Delta v_{12} \mu \dot{v}_2 (1-\cos\Delta v_{12}) \left[1 - \left(\frac{\dot{v}_1}{\dot{v}_2}\right)^{\frac{1}{2}} \cos\Delta v_{12}\right]^2$$

$$\left. -27(1-\cos\Delta v_{12})^2 \mu^2 \dot{v}_2^{\,2} \left[1 - \left(\frac{\dot{v}_1}{\dot{v}_2}\right)^{\frac{1}{2}} \cos\Delta v_{12}\right]^4 \right\} \qquad \text{(C4)}$$

Equations (C3) and (C4) yield

$$\frac{b^2}{4} + \frac{a^3}{27} = \frac{(1 - \cos \Delta v_{12})^3}{\mu^3 \dot{v}_2^{\,3} \left[1 - \left(\frac{\dot{v}_1}{\dot{v}_2}\right)^{\frac{1}{2}} \cos \Delta v_{12}\right]^6}$$

$$\left\{\frac{1}{4} (1 - \cos \Delta v_{12}) \mu \dot{v}_2 \left[1 - \left(\frac{\dot{v}_1}{\dot{v}_2}\right)^{\frac{1}{2}} \cos \Delta v_{12}\right]^2\right.$$

$$\left. - \frac{1}{27} \dot{r}_1^{\,3} \sin^3 \Delta v_{12}\right\} \qquad \text{(C5)}$$

There are one or three real roots of, $\hat{P}$, depending upon whether Eq. (C5) is greater or less than zero. Since the leading term on the right-hand side is always positive, there is one (or three) real root if the terms within the parentheses are greater (or less) than zero.

Equation (19) can be regarded as a quadratic equation in $\dot{r}_1$ with solution

$$\dot{r}_1 = \frac{(1-\cos\Delta v_{12}) \pm (\mu\dot{v}_2)^{\frac{1}{2}} \left[1 - (\dot{v}_1/\dot{v}_2)^{\frac{1}{2}} \cos\Delta v_{12}\right] \tilde{P}^3}{\sin \Delta v_{12}\ \tilde{P}^2} \qquad \text{(C6)}$$

where

$$\tilde{P} = \bar{P}^{-\frac{1}{2}} = (p/\mu)^{\frac{1}{4}} \qquad \text{(C6a)}$$

A schematic of Eq. (C6) illustrating the regions of one and three roots is presented in Fig. 4.

The ambiguity associated with more than one positive real value of $\bar{P}$ can be resolved by measuring $\dot{r}_2$. One then substitutes each of the values of $\bar{P}$ and its associated v_1 [as determined from Eqs. (20)] into the relationship

$$\dot{r}_2 = e \sin (v_1 + \Delta v_{12})/\bar{P} \qquad \text{(C7)}$$

The $\bar{P}$ value yielding the measured $\dot{r}_2$ is the one accepted.

The uniqueness of this solution is proved below. From Eq. (C7), two values of $\bar{P}$, designated $\bar{P}_a$ and $\bar{P}_b$, yielding the same $\dot{r}_2$ must be related by

$$\frac{\bar{P}_a}{\bar{P}_b} = \frac{e_a \sin v_{1a} \cos \Delta v_{12} + e_a \cos v_{1a} \sin \Delta v_{12}}{e_b \sin v_{1b} \cos \Delta v_{12} + e_b \cos v_{1b} \sin \Delta v_{12}} \qquad \text{(C8)}$$

where the subscripts a and b indicate quantities associated with $\bar{P}_a$ and $\bar{P}_b$, respectively. Substituting Eqs. (20a) and (20b) into Eq. (C8) yields

$$\bar{P}_a / \bar{P}_b = (A_a + B_a)/(A_b + B_b) \qquad \text{(C9)}$$

where

$$A_i = \bar{P}_i \dot{r}_1 \cos \Delta v_{12} \left[1 - (\dot{v}_1 / \dot{v}_2)^{\frac{1}{2}} \cos \Delta v_{12} \right]$$

$$B_i = \left\{ (\dot{v}_1 / \dot{v}_2)^{\frac{1}{2}} \left[1 - \dot{r}_1 \sin \Delta v_{12} \bar{P}_i \right] - 1 \right\} \sin \Delta v_{12}$$

$$i = a, b$$

Since the relationship in Eq. (C9) is satisfied only if $\bar{P}_a = \bar{P}_b$, it is evident that one and only one $\bar{P}$ can yield the measured $\dot{r}_2$.

References

[1]Schwarzbein, Z. E. and Gersten, R. H., "Preliminary orbit determination for a moon satellite from range-rate data and lunar cartography and/or stellar observations," AIAA J. 1, 467-469 (February 1963).

[2]Herrick, S., Astrodynamics (D. Van Nostrand Co., Inc., Princeton, N. J., in publication).

[3]Baker, R. M. L., Jr. and Makemson, M. W., An Introduction to Astrodynamics (Academic Press, New York, 1960), pp. 119-120.

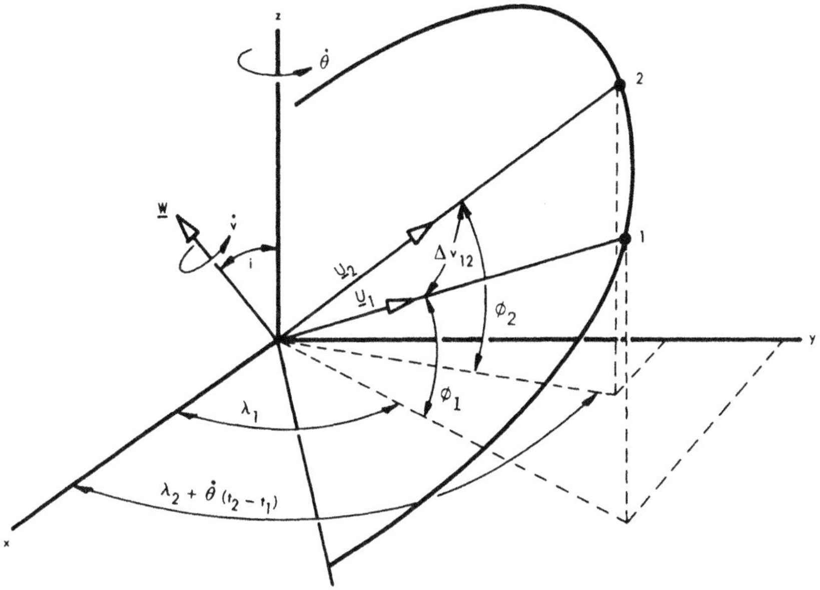

Fig. 1 Inertial x, y, z coordinate system.

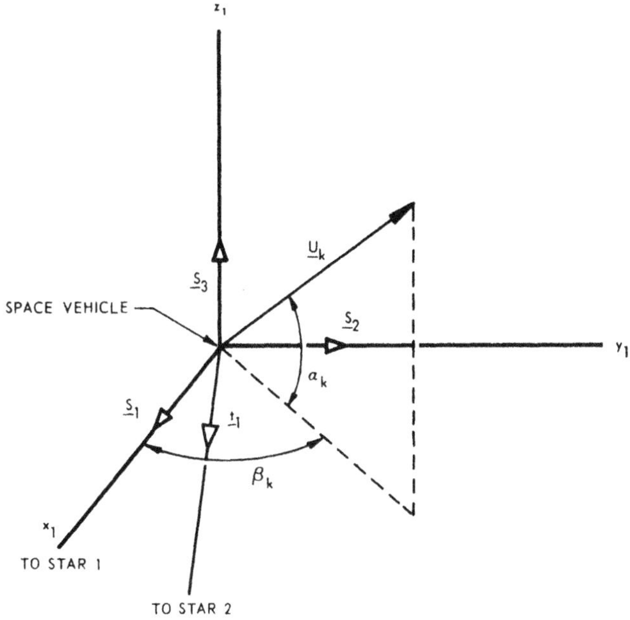

Fig. 2 Inertial x_1, y_1, z_1 coordinate system.

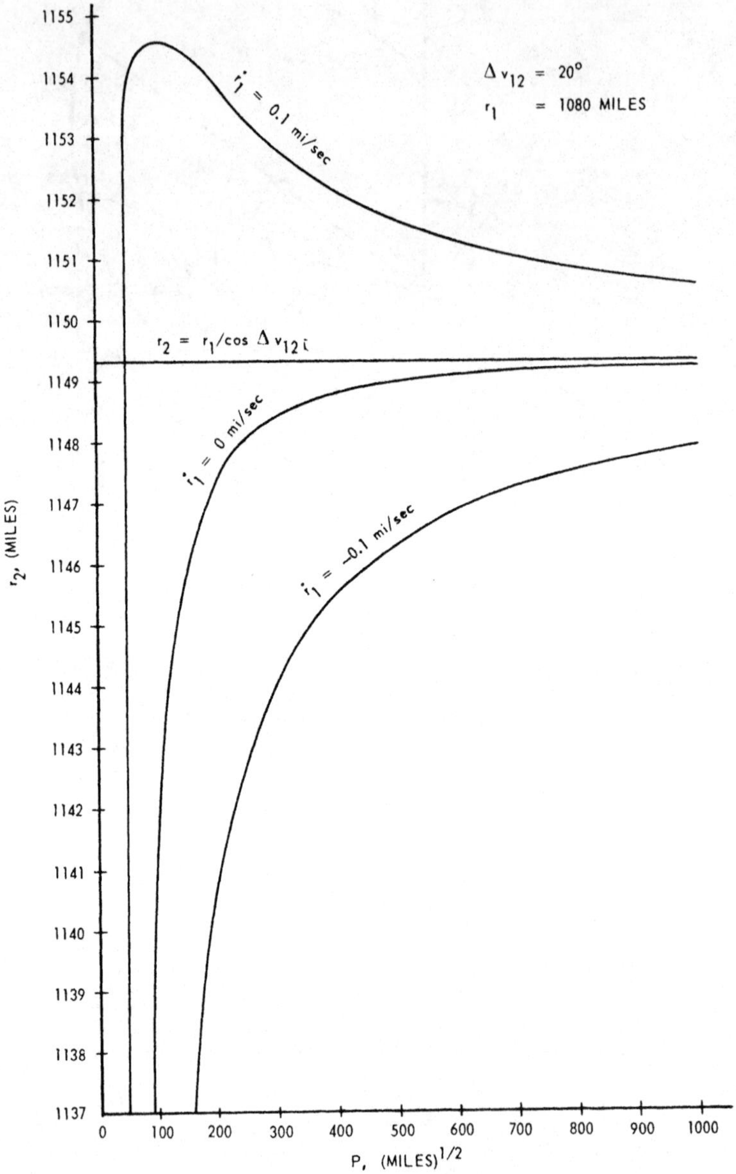

Fig. 3 Variation of r_2 vs P for fixed Δv_{12} and r_1.

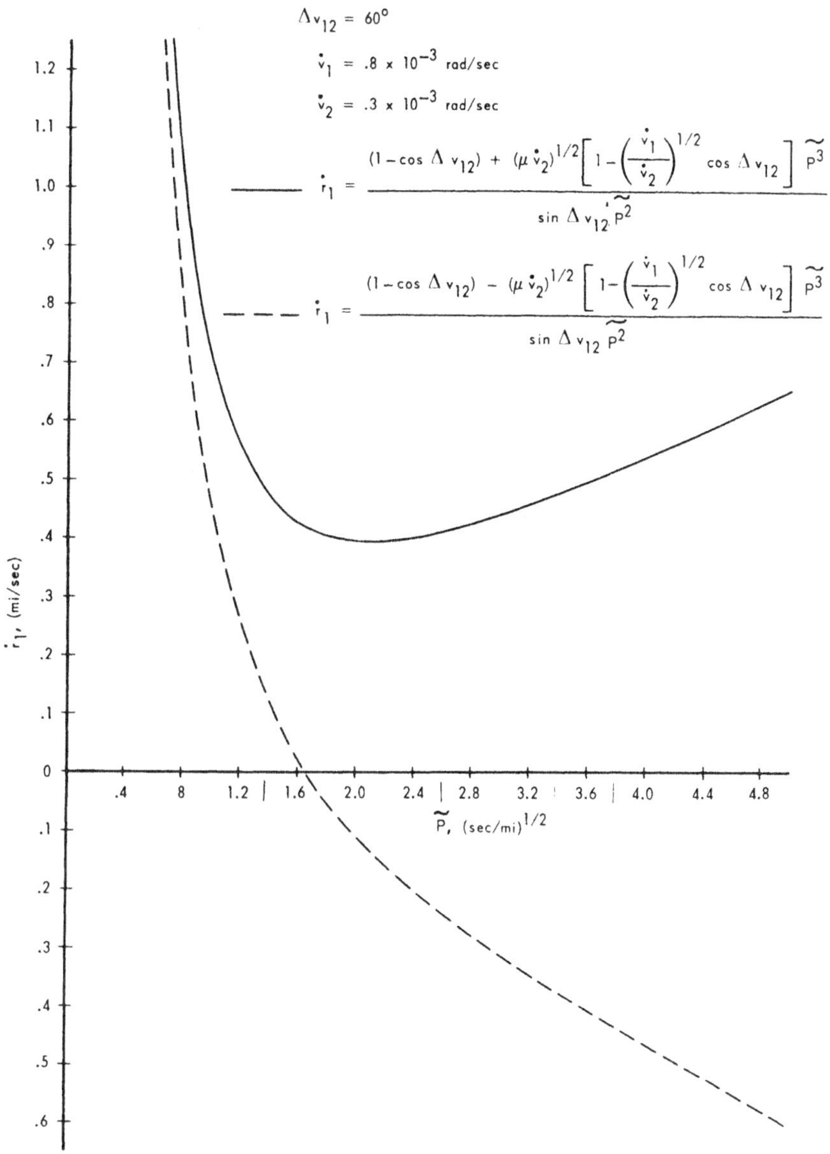

Fig. 4 Variation of $\dot{r}_1$ vs $\tilde{P}$ for fixed Δv_{12}, $\dot{v}_1$, and $\dot{v}_2$.

LINEARIZED DRAG ANALYSIS AND ORBIT DETERMINATION

Kurt Forster* and Robert M. L. Baker Jr.†
University of California, Los Angeles, Calif.

Abstract

Difficulties to properly account for atmospheric drag in orbit determination and prediction present themselves in two ways. First, the atmospheric density varies not only with respect to position but also with respect to time. Second, it is difficult to develop a sufficiently comprehensive theory in analytical form. A number of approaches have been successfully applied to the final satellite decay problem, but only a few theories have been developed for the case of predicting ahead a fraction or a few revolutions. In this paper the second of these two problems is attacked for orbits of small eccentricity taking into account not only a reference density but also a scale height. By linearization of the dynamical equations analytical expressions, valid for a radial range of two scale heights, are developed for the perturbation due to drag. Being in closed form in terms of elementary functions, these expressions can be used in various ways to study the influence of small eccentricity and drag on other quantities. As a specific application to satellite orbit determination partial derivatives are developed that could be used in connection with a differential correction procedure for the improvement of both reference density and scale height.

Linearized Equations

The nonlinear problem of the motion of a satellite in an atmosphere which decreases exponentially with altitude has been the subject of outstanding papers by King-Hele[1], Sterne[2], and others. The motion of a satellite in an orbit of small eccentricity was also variously considered, notably by Perkins[3], Lawden[4], and Karrenberg et al.[5,6] We consider a spherical

Presented as Preprint 63-428 at the AIAA Astrodynamics Conference, New Haven, Conn., August 19-21, 1963.

*Professor of Engineering.

†Lecturer, Department of Engineering; also Head, Astrodynamics Research Center, Lockheed-California Company, Los Angeles, California.

planet with a stationary atmosphere of density $\rho(r)$ and the motion of a satellite of mass m in that atmosphere. The satellite is subject to a drag force $\frac{1}{2}C_D A\rho V^2$, with A the cross-sectional area and V the speed; if the satellite is small compared with the mean free path, the value for C_D is close to 2. Let the satellite also be subject to a thrust with components F_r and F_v in the radial and normal directions, respectively. Then the equations of motion in polar coordinates become

$$m\left(\ddot{r} - r\dot{v}^2 + \frac{\mu}{r^2}\right) = F_r(t) - \tfrac{1}{2}C_D A\dot{r}V\rho(r) \tag{1}$$

$$m(r\ddot{v} + 2\dot{r}\dot{v}) = F_V(t) - \tfrac{1}{2}C_D Ar\dot{v}V\rho(r) \tag{2}$$

where μ/r^2 is the local gravitational force per unit mass.

The nonlinear character of Eqs. (1) and (2) is usually stressed in the literature. It is the purpose of this paper to make use of the fact that a linearization of the equations of motion for a near-circular orbit yields a set of linear differential equations with constant coefficients. As a consequence, the linearized solution can be expressed in terms of elementary functions and in closed form. From these expressions some basic information about the essential character of the motion can be obtained which would be difficult to extract from results of machine computations. Such solutions can be useful, for instance, for the interpretation of localized drag measurements and in the study of orbit-sustaining techniques, as well as for certain rendezvous problems that will presumably be concerned with orbits that are nearly circular.

The motion of a satellite that is in a nearly circular orbit was, of course, variously considered in the literature. In Ref. 3 the differential equation for small deviations Δr from a circular orbit of radius r under the action of a constant tangentially directed thrust, without consideration of drag is formulated. It is stated that the sinewave oscillations about a mean path damp out with time, and the velocity along this mean path is found to be proportional to time. It is difficult to see, though, how the oscillations about the mean path <u>can</u> damp out in a nondissipative system, and Lawden[4], who carried the calculation to the next order, found that the oscillations increase steadily in amplitude, eventually becoming comparable with the nonoscillatory terms.

Karrenberg, Levin, and Lewis[6] used Perkin's results of Ref. 3 (motion for constant tangential thrust), treating the drag as a negative constant thrust in an atmosphere of uniform density. They thereby found the same type of motion as just mentioned: velocity proportional to time, with a sinusoidal oscillation of constant amplitude superimposed. Inasmuch as Perkins' expressions were derived for takeoff (at t=0) from an exactly circular orbit, these initial conditions also carry over into Ref. 6; conclusions on the atmospheric density are applicable provided that the satellite was, in fact, in an exactly circular orbit at the initial observation. For a satellite that is subject to drag, however, such an initial state is not realized, and, as we shall see later, an initial eccentricity even as small as a few thousandths gives rise to terms that are just as large as the drag perturbation. Conclusions drawn from the drag perturbation as suggested in Ref. 6 may then easily lead to erroneous values for the atmospheric density.

It seems, in summary, that the motion of a satellite in a near-circular orbit was variously studied, particularly for constant thrust. The expressions resulting for constant thrust were then used to represent the motion of the satellite in an atmosphere of constant density. A general treatment of the linearized problem, including drag in an atmosphere of varying density, for arbitrary initial conditions (injection), and for arbitrary thrust given as a function of time, is not known to the present authors. Such treatment of the problem seems desirable and we shall present it here for a plane orbit, while an extension to three dimensions may be presented at a later date.

Suppose that at time t = 0 the satellite was observed to have angular velocity $\dot{v}(o)$ at distance r_o with respect to the earth's center. We define an angular velocity at epoch n_o, which is in general not equal to $\dot{v}(o)$, by the relation

$$r_o n_o^2 = \mu / r_o^2 \tag{3}$$

and, instead of r and v, we introduce the unknown functions $\varepsilon(t)$ and $\varphi(t)$ by writing

$$r(t) = r_o \left[1 + \varepsilon(t)\right] \tag{4}$$

$$\dot{v}(t) = n_o \left[1 + \varphi(t)\right] \tag{5}$$

and we linearize Eqs. (1) and (2) for ε and φ by neglecting ε^2, φ^2, and their products as compared with ε and φ for the case $\varepsilon(t)$ and $\varphi(t) << 1$.

For example, for the velocity V, we obtain, from $V^2 = \dot{r}^2 + r^2\,\dot{v}^2$,

$$V^2 = r_o^{\,2} n_o^{\,2} \left[1 + 2\varepsilon + 2\varphi + (\dot{\varepsilon}^2/n_o^{\,2}) + \ldots \right] \tag{6}$$

where the dots stand for terms of order ε^2, φ^2, etc.; thus the velocity to the first order in ε and φ becomes

$$V = r_o n_o (1 + \varepsilon + \varphi) \tag{7}$$

Let the atmospheric density at r_o be ρ_o, and let its variation be given by

$$\rho(r) = \rho_o \exp\left[-K(r-r_o)/r_o \right] \tag{8}$$

where we have written $K = r_o/H_{SH}$ in terms of the scale height H_{SH}; Eq. (4) then gives the following equation for the variation of density, neglecting terms of the order ε^2 and smaller:

$$\rho(r) = \rho_o (1 - K\varepsilon) \tag{9}$$

For brevity, we write $\beta = \frac{1}{2} C_D A \rho/m$, and hence, from Eq. (9),

$$\beta = \beta_o (1 - K\varepsilon) \tag{10}$$

The quantity (βV), which occurs in the equations of motion, becomes

$$\beta V = r_o n_o \beta_o \left[1 + \varphi + \varepsilon (1 - K) \right] \tag{11}$$

When Eqs.(4) and (5) are introduced into Eqs. (1) and (2) and all variable terms are developed to the first power, the terms given by Eq. (3) can be canceled on both sides; the following equations of motion result:

$$\ddot{\varepsilon} + \dot{\varepsilon}\beta_o r_o n_o - n_o^{\,2}(3\varepsilon + 2\varphi) = n_o F_1(t) \tag{12}$$

$$\dot{\varphi} + 2\dot{\varepsilon} + \beta_o r_o n_o \left[2\varphi + \varepsilon(2 - K) \right] = F_2(t) - \beta_o r_o n_o \quad (13)$$

For convenience, Eq. (1) was divided by $m\ r_o$ and Eq. (2) by $m\ r_o n_o$; hence, F_1 and F_2 are F_r and F_v, respectively, divided by $(m\ r_o n_o)$.

The equations of motion, Eqs. (12) and (13), are linear equations with constant coefficients. The initial conditions for ε and φ at t=0 are

$$\varepsilon(0) = 0 \qquad \dot{\varepsilon}(0) = \dot{r}(0)/r_o = An_o \qquad A \ll 1 \quad (14)$$

$$\varphi(0) = \left[\dot{v}(0)/n_o\right] - 1 = B \qquad B \ll 1 \quad (15)$$

with A and B constants assumed to be known from observation. The quantity n_o was made a factor of A to render A dimensionless.

The osculating orbital element at t=0, the eccentricity e, semilatus rectum p, and semimajor axis a can be expressed in terms of A and B by using Eqs. (14, 15, and 3) in standard formulas[7]:

$$p = r_o(1 + B)^2$$

$$e^2 = A^2(1 + B)^2 + B^2(2 + B)^2$$

$$a = r_o(1 + B)^2/(1 - e^2)$$

As an illustration of the solution for the force-free satellite subject to drag, we shall apply Eqs. (12) and (13) to the motion of a satellite in a stationary atmosphere, with $F_1 = F_2 = 0$.

For brevity let us write α for the dimensionless quantity $\beta_o r_o$; thus we wish to find a solution to

$$\ddot{\varepsilon} + \dot{\varepsilon}\alpha n_o^2(3\varepsilon + 2\varphi) = 0 \quad (16)$$

$$\dot{\varphi} + 2\dot{\varepsilon} + \alpha n_o \left[2\varphi + \varepsilon(2 - K) \right] = -\alpha n_o \tag{17}$$

subject to initial conditions given by Eqs. (14) and (15). For a system of equations of this type, the Laplace-transform method is particularly suited. Let us designate the transforms of ε and φ by $\overline{\varepsilon}(s)$ and $\overline{\varphi}(s)$, with s the transform variable. The Eqs. (14, 16, and 17) are transformed into the following:

$$\overline{\varepsilon}(s^2 + s\alpha n_o - 3n_o^2) - \overline{2\varphi} n_o^2 = An_o \tag{18}$$

$$\overline{\varepsilon}\left[2s + \alpha n_o(2 - K) \right] + \overline{\varphi}(s + 2\alpha n_o) = B - (\alpha n_o/s) \tag{19}$$

which are two algebraic equations to be solved for $\overline{\varepsilon}$ and $\overline{\varphi}$. Let

$$\Delta = s(s^3 + C_1 s^2 + C_2 s + C_3) \tag{20}$$

$$C_1 = 3\alpha n_o; \quad C_2 = n_o^2(1+2\alpha^2); \quad C_3 = -2\alpha n_o^3(1+K)$$

$$\Delta_1(s) = An_o s^2 + A_1 s + A_2 \tag{21}$$

$$A_1 = 2n_o^2(\alpha A + B)$$

$$A_2 = -2\alpha n_o^3$$

$$\Delta_2(s) = Bs^3 + B_1 s^2 + B_2 s + B_3 \tag{22}$$

$$B_1 = n_o(\alpha B - \alpha - 2A); \quad B_3 = 3\alpha n_o^3$$

$$B_2 = -n_o^2\left[3B + \alpha^2 + \alpha A(2 - K) \right]$$

The transforms of ε and of φ are then given by the ratio of the determinants:

$$\overline{\varepsilon} = \Delta_1(s)/\Delta(s) \qquad \overline{\varphi} = \Delta_2(s)/\Delta(s) \tag{23}$$

For the inversion of Eqs. (23), the roots of the equation $\Delta = 0$ are of primary importance. One root is $s_o = 0$; the roots of the cubic equation can easily be found exactly by Cardano's formula; to indicate their real or complex nature, we write them in the form

$$s_1 = \lambda_1, \quad s_2 = \lambda_2 + \lambda_o j, \quad s_3 = \lambda_2 - \lambda_o j \tag{24}$$

or, alternatively, we write Δ in the form

$$\Delta = s(s - \lambda_1)(s^2 - 2\lambda_2 s + \lambda_o^2 + \lambda_2^2) \tag{25}$$

with λ_1, λ_2, and λ_3 being real numbers. We shall show that λ_1 is positive and λ_2 negative in the present example.

Our knowledge of the simple poles of Eqs. (23) given by Eqs. (24)and (25) directly provides the general information that the radial and perpendicular perturbations ε and φ each have three terms, namely, 1) a constant term due to the pole $s_o = 0$; 2) an exponentially increasing term $e^{+\lambda_1 t}$ due to the pole $s_1 = \lambda_1$; and 3) a sinusoidal term of frequency λ_o, with amplitude decreasing exponentially as $e^{\lambda_2 t}$. It remains to evaluate the constant coefficients by which these terms are multiplied; this can be done with the aid of tables of transforms[8] or by expanding Eqs. (23) into partial fractions. We chose the latter course to obtain the following results.

From the roots $\lambda_{0,1,2}$ of Eq. (24), the following numbers are computed:

$$N_o = (\lambda_2 - \lambda_1)^2 + \lambda_o^2 \qquad N_1 = \lambda_2^2 - \lambda_1\lambda_2 - \lambda_o^2$$

$$N_2 = \lambda_o(2\lambda_2 - \lambda_1) \qquad N = N_1^2 + N_2^2 = N_o(\lambda_2^2 + \lambda_o^2)$$

$$M_1 = \Delta_1(\lambda_2) - An_o\lambda_o^2 \qquad M_2 = \lambda_o(2An_o\lambda_2 + A_1)$$

Then, if we abbreviate

$$c_o = \Delta_1(\lambda_1) \qquad c_1 = N_1M_1 + N_2M_2$$

$$c_2 = N_1M_2 - N_2M_1$$

the function $\varepsilon(t)$ becomes

$$\varepsilon(t) = \frac{c_o}{\lambda_1 N_o}(e^{\lambda_1 t} - 1) + \frac{e^{\lambda_2 t}}{\lambda_o N_o}\left[c_1 \sin\lambda_o t + c_2(\cos\lambda_o t - 1)\right] \quad (26)$$

The expressions for the roots, though easily computed from Cardano's formula for numerically given values of A and B, are too cumbersome to use explicitely for arbitrary A and B. It is instructive, however, to have explicit expressions for the coefficients in Eq. (26). To this end, α being a small number, we develop the roots of Eq. (2) (in Cardano's formula) in terms of α and obtain, to the second order in α,

$$\lambda_1 = 2\alpha n_o(K+1) \qquad \lambda_2 = -\alpha n_o(K + \tfrac{5}{2})$$

$$\lambda_o = n_o(1+\delta\alpha^2) \qquad \text{with } \delta = \tfrac{9}{2}(K + \tfrac{3}{2})^2 - \tfrac{1}{2} \quad (27)$$

We note that α occurs in the roots, not by itself, but multiplied with (K+1), $(K + \frac{3}{2})$, etc. The quantity K, being roughly the earth's radius over scale height, has a numerical value of the order 100 and is thus a large quantity.

With the roots given in Eq. (27), we may calculate the coefficients of Eq. (26) in terms of K and α to the first order; $\varepsilon(t)$ is then explicitely

$$\varepsilon(t) \approx \left[2B - (K+1)^{-1} + 2A\,\alpha(K+2)\right](e^{2\alpha n_o(K+1)t} - 1)$$
$$+ e^{-\alpha n_o(K+\frac{5}{2})t}\left\{\left[A + 2\alpha\left(1 - 2B(K+1)\right)\right]\sin n_o(1+\delta\alpha^2)t\right.$$
$$\left. - 2\left[B + \alpha A(K+2)\right]\left[\cos n_o(1+\delta\alpha^2)t - 1\right]\right\} \quad (28)$$

It is apparent from Eq. (28) that the initial conditions, A, B, enter into the expressions for the motion to the same order as the perturbing influence of drag; this is seen, for instance, by the simultaneous presence of the quantities 2B and $(K+1)^{-1}$ in the first bracket in Eq. (28). The eccentricity of the initial orbit is given in terms of A and B, as just mentioned,

$$e^2 = A^2(1 + B)^2 + B^2(2 + B)^2$$

which is approximately equal to $(A^2 + 4B^2)$, inasmuch as A and B are both much less than one. With K of the order of 100, it follows from Eq. (28) that an eccentricity even as small as a few thousandths affects the motion to the same order as the drag perturbation. This may also be seen from the presence of the initial value A together with the quantity (2α), which is a measure of the density, in the second bracket in Eq. (28). In an estimate of the perturbing effects of drag and conclusions on the density from expressions calculated for $A=B=0$, these facts must be carefully considered.

Some improvements in the accuracy and in the limitations of applicability of Eqs. (26) and (28) can be obtained by the following considerations, which apply equally to the similar expressions to be derived for $\varphi(t)$ in the sequel. Inaccuracies stemming from the linearization of the problem enter Eqs. (26) and (28) in two ways:

1) The motion in the initial osculating orbit $(\alpha = 0)$, which would be the orbit without drag, is approximated by the terms

$$A \sin n_o t - 2B(\cos n_o t - 1) \tag{29}$$

Therefore, we form the expression

$$D(t) = \varepsilon(t) - A \sin n_o t + 2B(\cos n_o t - 1) \tag{30}$$

which represents the linearized perturbation due to drag of the initial orbit. By superimposing D(t) onto the ellipse calculated for initial values A and B in the standard way, we can reduce the inaccuracy incurred by the process of linearization.
2) The validity of the perturbation that is caused by drag $(\alpha \neq 0)$ is limited by the inaccuracy inherent in the linearization of the differential equations; the most important contribution to this inaccuracy stems from that quantity that changes most rapidly with position; this is the density, as given in Eq. (8). Thus the extent of the validity of Eq. (9) in place of Eq. (8) largely determines the validity of the linearized approximation of the perturbation due to drag. This limits A and B to values at most of the same order as (1/K) in Eq. (9).

As far as perturbations by drag are concerned, the foregoing condition expresses quantitatively what is meant by the term "near-circular" orbit. It must be noted that within these limits the magnitudes of A, B, α, and K are of equal importance for the determination of position or, conversely, for a determination of α, and hence of the atmospheric density, as well as scale height from a series of observations.

The foregoing considerations limit the total radial variation to about one scale height. At altitudes of 200 to 300 km, for instance, this limits the variations to about 25%.

Since the density varies greatly with position, the tangent-approximation of Eq. (9) to Eq. (8) may not be the best linear approximation for our purposes.

If the initial excursions ε are of the order of a scale height, an expression can be used which is more accurate for large values of the density at perigee than for the small and therefore less important values at apogee; this expression extends the range of applicability of the linearized formulas. We can find values from Eq. (29) for the initial maximum of ε, which we denote by ε_1, and for the minimum ε_2, which are sufficiently accurate for our present purposes:

$$\varepsilon_{1,2} = 2B\left[1 \pm \left(1 + \frac{A^2}{4B^2}\right)^{\frac{1}{2}}\right] \tag{31}$$

Let us denote the approximate densities at apogee (for ε_1) and at perigee (for ε_2, which is negative) by ρ_1 and ρ_2, respectively; then we replace ρ_o in Eq. (9) by

$$\rho_x = \frac{\rho_2}{1 + \left(|\varepsilon_2| / \varepsilon_1\right)} \tag{32}$$

and we obtain the following expression in place of Eq. (9) for the case where the initial values A and B are of the order of (1/K):

$$\rho(r) = \rho_x\left[1 - (\varepsilon/\varepsilon_1)\right] \tag{9'}$$

Equation (9') is a linear interpolation that is quite accurate near perigee, but that neglects the density near apogee. Since the drag decreases by an order of magnitude over the distance of two scale heights, Eq. (9') will yield better accuracy, under the stated conditions than Eq. (9).

By a partial-fractions expansion, using Eq. (25), we obtain $\varphi(t)$ by inversion of Eq. (23). From the roots $\lambda_{0,1,2}$, the following additional numbers are computed:

$$P_1 = \Delta_2(\lambda_2) - \lambda_o^2(3B\lambda_2 + B_1)$$

$$P_2 = \lambda_o(3\lambda_2^2 - \lambda_o^2)B + 2\lambda_2 B_1 + B_2$$

$$d_o = \Delta_2(\lambda_1), \qquad d_1 = N_1P_1 + N_2P_2, \qquad d_2 = N_1P_2 - N_2P_1$$

whence

$$\begin{aligned}\varphi(t) &= B + d_o(e^{\lambda_1 t} - 1)/\lambda_1 N_o \\ &+ e^{\lambda_2 t}\left[(d_1 \sin\lambda_o t + d_2 \cos\lambda_o t) - d_2\right]/\lambda_o N\end{aligned} \tag{33}$$

The expression for the initial osculating orbit

$$4B \cos n_o t - 2A \sin n_o t - 3B \tag{34}$$

subtracted from $\varphi(t)$ yields the drag perturbation of $\dot{v}$:

$$\Phi(t) = \varphi(t) - 4B \cos n_o t + 2A \sin n_o t + 3B \tag{35}$$

Inclusion of Thrust

In consequence of the linearity of Eqs. (12) and (13), the inclusion of thrust functions F_1 and F_2 involves merely the addition of another term to the solutions given by Eqs. (26) and (33).

Let $\bar{F}_1(s)$ and $\bar{F}_2(s)$ be the transforms of $F_1(t)$ and $F_2(t)$, respectively; then, to include the thrust, we have to insert $\left[A + \bar{F}_1(s)\right]$ in place of A, and $\left[B + \bar{F}_2(s)\right]$ in place of B on the right-hand side of Eqs. (18) and (19); this follows directly from Eqs. (12) and (13). Making this substitution in the determinants that lead from Eqs. (18) and (19) to Eq. (23), we obtain the following additional terms representing the thrust, on the right-hand side of Eq. (23):

$$\bar{\varepsilon}_F = \frac{n_o(s+2\alpha n_o)\,\bar{F}_1(s)+2n_o^2\,\bar{F}_2(s)}{\Delta'(s)} \tag{36}$$

$$\bar{\varphi}_F = \frac{-n_o\left[2s+\alpha n_o(2-K)\right]\,\bar{F}_1(s)+(s^2+s\alpha n_o-3n_o^2)\,\bar{F}_2(s)}{\Delta'(s)} \tag{37}$$

where $\Delta'(s)$ is the cubic expression in Eq. (20) or, explicitly,

$$\Delta'(s) = s^3 + C_1 s^2 + C_2 s + C_3$$

For a given thrust F(t), the inverse transforms of Eqs. (36) and (37) may be obtained either from tables of transforms or directly, using Eq. (24). In this manner, the inverse transforms of Eqs. (36) and (37), added to the solutions of Eqs. (26) and (33), give explicit expressions for the entire problem for an arbitrary given thrust program, including drag.

Differential Correction

In order to make the linearized solution of use in orbit determination, it is necessary to recast the equations in a three-dimensional form. Following the $\underset{\sim}{U}_o\, \underset{\sim}{V}_o$ system developed by Herrick and presented on p. 126 ff. of Ref. 7, we have

$$\underset{\sim}{r} = \underset{\sim}{U}_o\left[r(t)\right] \cos\left[\int_o^\tau (\dot{v} + v')d\tau\right]$$

$$+ \underset{\sim}{V}_o\left[r(t)\right] \sin\left[\int_o^\tau (\dot{v} + v')d\tau\right] \tag{38}$$

where $\underset{\sim}{U}_o$ and $\underset{\sim}{V}_o$ are, respectively, unit vectors in the direction of the satellite and perpendicular to that direction in the orbit plane at the time of epoch t_o, $\tau = k_e(t - t_o)$, r is defined by Eqs. (4) and (30), $\dot{v}$ = unperturbed angular rate, and v' is defined by Eq. (35). The integral

$$\int_o^\tau (\dot{v} + v')d\tau$$

involves the "in-path" perturbation due to drag and must be evaluated with special care, since most of the drag perturbation occurs in this direction. For the small eccentricity orbit under consideration here,

$$\int_{t_o}^{t} \dot{v}dt = n(t - t_o) + 2e \sin n(t - t_o)$$

$$+ \frac{5}{4} e^2 \sin 2n(t - t_o) + . . . \tag{39}$$

where $n = \mu\ /a^{3/2}$.

The integral of Eq. (35) is simply the sum of the following integrals:

$$\left(4B - \frac{d_o}{\lambda_1 N_o} - \frac{d_2}{\lambda_o N}\right)(t - t_o) + \frac{d_o}{\lambda_1{}^2 N_o}(e^{\lambda_1 t} - e^{\lambda_1 t_o})$$

$$+ \frac{1}{\lambda_o N} \frac{e^{\lambda_2 t}}{(\lambda_2{}^2 + \lambda_o{}^2)} \left[(d_1\lambda_2 + d_2\lambda_o)\sin\lambda_o t + (d_2\lambda_2 - d_1\lambda_o\cos\lambda_o t\right]$$

$$- \frac{1}{\lambda_o N} \frac{e^{\lambda_2 t_o}}{(\lambda_2{}^2 + \lambda_o{}^2)} \left[(d_1\lambda_2 + d_2\lambda_o)\sin\lambda_o t_o + (d_2\lambda_2 - d_1\lambda_o)\cos\lambda_o t_o\right]$$

$$+ \frac{4B}{n_o}(\sin n_o t_o - \sin n_o t) + \frac{2A}{n_o}(\cos n_o t_o - \cos n_o t) \tag{40}$$

The differential correction formulas follow directly from Eq. (38), i.e.,

$$\Delta x = r(t) \cos\left[\int_o^\tau (\dot{v} + v')d\tau\right]\Delta y_o$$

$$+ r(t) \sin\left[\int_0^{\tau} (\dot{v} + v')d\tau\right]\Delta \underset{\sim}{V}_o$$

$$+ \left[\underset{\sim}{U}_o \cos \int_0^{\tau} (\dot{v}+v')d\tau + \underset{\sim}{V}_o \sin \int_0^{\tau} (\dot{v}+v')d\tau\right] \Delta r(t)$$

$$+ \left[\underset{\sim}{V}_o r(t)\cos \int_0^{\tau} (\dot{v}+v')d\tau - \underset{\sim}{U}_o r(t)\sin \int_0^{\tau} (\dot{v}+v')d\tau\right]$$

$$\Delta \int_0^{\tau} (\dot{v}+v')d\tau \tag{41}$$

The $\Delta \underset{\sim}{U}_o$, $\Delta \underset{\sim}{V}_o$, $\Delta r(t)$, and $\Delta \int_0^{\tau} (\dot{v} + v')d\tau$ can be attributed to inaccuracy in the reference density ρ_o, inaccuracy in the scale height H_{SH}, inaccuracy in the harmonics of the earth's gravitational field J_i's, inaccuracy in the location of the topos λ_T, φ_T, etc. In order to serve the purpose of the present analysis, we shall assume that $\Delta \underset{\sim}{\chi}$ arises entirely from poorly determined $\underset{\sim}{r}_o$, $\underset{\sim}{\dot{r}}_o$, ρ_o, and H_{SH}. This does not mean that the other inaccuracies in the mathematical model describing the satellite's motion and observations of this motion could not also be included, but merely that we wish to focus our attention on these eight parameters.

A simple analytical treatment leads to the fact that, since $\underset{\sim}{U}_o \triangleq \underset{\sim}{r}_o/r_o$, the following holds:

$$\Delta \underset{\sim}{U}_o = \Delta \underset{\sim}{r}_o/r_o - (\underset{\sim}{r}_o/r_o^2)\Delta r_o \tag{42}$$

$$\underset{\sim}{r}_o = x_o \underset{\sim}{I} + y_o \underset{\sim}{J} + z_o \underset{\sim}{K}$$

$$r_o^2 = x_o^2 + y_o^2 + z_o^2$$

$$\Delta \underset{\sim}{r}_o = \underset{\sim}{I}\,\Delta x_o + \underset{\sim}{J}\,\Delta y_o + \underset{\sim}{K}\,\Delta z_o$$

$$r_o \Delta r_o = x_o \Delta x_o + y_o \Delta y_o + z_o \Delta z_o$$

In a similar fashion (for a nearly circular orbit),

$$\Delta \underset{\sim}{V}_o = \frac{\Delta \dot{\underset{\sim}{r}}_o}{\dot{s}_o} - \frac{\underset{\sim}{r}_o}{\dot{s}_o^{\,2}}\,\Delta \dot{s}_o \tag{43}$$

$$\dot{\underset{\sim}{r}}_o = \dot{x}_o \underset{\sim}{I} + \dot{y}_o \underset{\sim}{J} + \dot{z}_o \underset{\sim}{K}$$

$$\dot{s}_o^{\,2} = \dot{x}_o^{\,2} + \dot{y}_o^{\,2} + \dot{z}_o^{\,2}$$

$$\Delta \dot{\underset{\sim}{r}}_o = \underset{\sim}{I}\Delta \dot{x}_o + \underset{\sim}{J}\Delta \dot{y}_o + \underset{\sim}{K}\Delta \dot{z}_o$$

$$s_o \Delta s_o = \dot{x}_o \Delta \dot{x}_o + \dot{y}_o \Delta \dot{y}_o + \dot{z}_o \Delta \dot{z}_o$$

The explicit evaluation of $\Delta\, \varepsilon(t)$ as a function of $\Delta \rho_o$ would require the differentiation of Eq. (26) with respect to α and with respect to K. Since the coefficients (for instance, $c_o/\lambda_1 N_o$) depend on α and K, the explicit differentiation leads to very cumbersome expressions. Since the coefficients themselves, however, are simple functions of these parameters, it is proposed that their differentials be generated numerically by evaluation for slightly different ρ_o's and H_{SH}'s. Such numerical evaluation of differences may equally well be carried out for the value of $\Delta \varepsilon(t)$ (for a desired fixed t) or for

$$\Delta \left[\int_o^{\tau} (\dot{v} + v')dt \right]$$

The foregoing procedure, being based on analytical expressions, offers advantages over similar procedures in use today in which the functions of improvements or corrections to $\underset{\sim}{r}_o$, $\dot{\underset{\sim}{r}}_o$, ρ_o, and H_{SH} are obtained in a completely numerical fashion by means of nine simultaneous numerical integrations (with associated representations) of the satellite orbit.

References

[1] King-Hele, D. G. and Cook, G. E., "Review of earth satellite orbital studies," Royal Aircraft Establishment GW no. 25 (July 1961); also "The contraction of satellite orbits under the influence of air drag," Proc. Roy. Soc. (London) A 257, 224-249 (1962).

[2] Sterne, T. E., "An atmospheric model, and some remarks on the inference of density from the orbit of a close earth satellite," Astron. J. 63,81-87 (1958).

[3] Perkins, F. M., "Flight mechanism of low-thrust spacecraft," J. Aero/Space Sci. 26, 291-297 (1959).

[4] Lawden, D. F., "Optimal escape from a circular orbit," Astronaut.Acta. 4, 218-233 (1958).

[5] Geyling, F. T., "Drag displacements and decay of near-circular satellite orbits," Congr. Int. Astronaut. Fed., Stockholm, Sweden (1960).

[6] Karrenberg, H. K., Levin, E., and Lewis, D. H., "Variation of satellite position with uncertainties in the mean atmospheric density," ARS J 32, 576-582 (1962).

[7] Baker, R. M. L., Jr. and Makemson, M. W., An Introduction to Astrodynamics (Academic Press, New York, 1960), pp. 113.

[8] See, for instance, Nixon, F. E., Handbook of Laplace Transformation Tables (Prentice-Hall, Inc., Englewood Cliffs, New Jersey, 1960).

ORBIT DETERMINATION IN THE PRESENCE OF SYSTEMATIC ERRORS

Alfons J. Claus*

Bell Telephone Laboratories, Inc., Whippany, N. J.

ABSTRACT

A method of orbit refinement from tracking data is presented. The data are assumed to contain random errors as well as errors that can be expressed in some known functional form but involving unknown constants. These assumptions are shown to lead to a smoothing technique capable of producing orbital elements with a minimum sensitivity to systematic errors. The formulas are derived explicitly in the case where the only systematic errors present are constant throughout the visibility zone of the pertinent observer. The hypothetical case in which no random errors appear is treated in some detail. One possible way of implementing the method on an electronic computer is discussed briefly and indicates that only minor modifications are required of more conventional computer routines where the presence of systematic errors is ignored completely.

INTRODUCTION

In processing radar data for orbit determination, one usually has to cope with basically two types of errors. Errors of one kind vary randomly from one observation to the next, whereas other errors exhibit some sort of systematic behavior. Orbit refinement methods can be designed to varying degrees of sophistication in order to increase the amount of information extracted from the available data. A brief investigation readily reveals, however, that the optimum processing of non-random error components rapidly leads to quite tedious algebra and, in spite of its optimality, results in a rather impractical computational technique. One is consequently forced to the introduction of simplifying assumptions. The challenge then consists of introducing enough simplifications to make the method practically attractive but yet retaining the

---

Received by AIAA June 1, 1964.

*Member of the Technical Staff, Analytical and Aerospace Mechanics Department.

necessary essential features to produce a scheme as highly optimal as possible. This paper constitutes an effort in that direction.

Observational errors are assumed to contain basically two different components. One component varies randomly from one data point to the next, whereas the other component can be expressed in some known functional form but involving unknown constants. Such a systematic error may, for instance, arise when gathering data with large antennas where the deformation of the structure depends mainly on the elevation angle.

Maximum likelihood estimates of the orbital elements are computed assuming normal error distributions. The normal assumption may be removed provided that, instead of using the maximum likelihood principle, one estimates orbital elements by minimizing their resultant variances. The two disciplines can be shown to lead ultimately to the same technique for the reduction of orbital data.

The smoothing formulas presented here are found to be only slightly more complicated than the ones used in some standard orbit refinement methods where the existence of bias errors is ignored completely. On the other hand, they lead to quite encouraging results. In fact, one of the most important features of the ensuing method lies in the property that it minimizes the sensitivity of the computed orbit to systematic errors in the data. To be more specific, if radar data were merely corrupted by a systematic error as described earlier, correct orbital elements could be obtained in spite of the fact that the problem remains a six dimensional one.

For the sake of simplicity, the smoothing formulas are explicitly derived only for systematic errors that are constant in time, henceforth referred to as bias errors. It is shown that a computer implementation of the present formulation causes little additional complexity to usual orbit reduction programs. A computer run, involving 50 data points, requires less than 5000 additional arithmetic operations, whereas additional storage requirements range from 150 to 450 locations depending on the desired flexibility for weighting the data.

DERIVATION OF THE SMOOTHING FORMULA

As stated in the Introduction, the derivation of the formulas will be limited to the case of observational errors consisting of a component varying randomly from one data point to the other and a component that is constant within the visibility zone (henceforth referred to as "pass") of the relevant tracking station. However, it is assumed that bias errors from different stations or different passes are statistically independent. If the passes in question originated from different stations, the assumption of uncorrelated bias errors would certainly seem logical. On the other hand, if the passes involve the same tracking device, matters tend to be somewhat more complicated.

First of all, one will recognize that, at least for adequately surveyed land stations, errors resulting from station location uncertainties are usually less important than the ones resulting from instrument bias. For instance, a typical error of 100 ft in station location causes an error of approximately 0.0002° in angular data for a range of 5000 miles, whereas an instrument bias of 0.01° is not uncommon. This is the reason we have chosen not to include station location errors in our subsequent analysis although they can, of course, be accounted for at the expense of having to use more complicated smoothing formulas.

This leaves us with bias errors resulting mainly from instrument imperfection. It appears to be rather difficult to estimate the variation of such errors with time, since they depend on many extraneous circumstances. Obviously, the true situation lies somewhere between complete statistical independence and exact functional dependence. A realistic choice somewhere between these two extreme cases would be severely influenced by the length of the time interval between the relevant passes. For simplicity's sake, we have taken the assumption of bias errors varying randomly from one pass to the next one as the starting point for the analysis.

To illustrate the underlying principles, we shall consider the case of n data points, each point consisting of a reading for range, a reading for the azimuth angle, and a reading for the elevation angle. It should be noted that, with minor modifications, this includes the case of only angular data by setting the weights of the range measurements equal to zero.

In the presence of normally distributed observational errors, the maximum likelihood principle is known to lead to the formula[1,2]

$$\hat{\mu} = \alpha + (A^{-1} + J' \Phi^{-1} J)^{-1} \left[J' \Phi^{-1} J (\bar{\mu} - \alpha) + J' \Phi^{-1} \Delta\lambda \right] \qquad [1]$$

where $\hat{\mu}$ denotes the refined orbital elements, α represents estimates of the orbital elements with covariance matrix A, J is the matrix of partial derivatives of the observations with respect to the elements, Φ is the covariance matrix of the observations, and $\Delta\lambda$ is the difference between observed data and the corresponding simulated data, the latter being computed from elements $\bar{\mu}$. Primes indicate transposition. In all applications in this paper, we shall assume that no prior data have been processed so that, in effect, $A^{-1} = 0$. Eq. 1 then reduces to

$$\hat{\mu} = \bar{\mu} + (J' \Phi^{-1} J)^{-1} J' \Phi^{-1} \Delta\lambda \qquad [2]$$

The covariance matrix Φ must represent the random error component as well as the bias errors and is readily shown to be

$$\Phi = \Psi + N \Lambda N' \qquad [3]$$

where Ψ is the covariance matrix of the random component of the total observational error, Λ represents the covariance matrix of the bias component, and N is the matrix of bias sensitivities, that is, the matrix relating increments of observational errors resulting from small increments of the bias errors.

As indicated earlier, the case treated in some detail pertains to n sets of observations, each set consisting of three concurrent readings: one for range R, one for azimuth angle A, and one for elevation angle E. We shall assume Ψ to be of the form

$$\Psi = \left(W_i^{-1} \delta_{ij} \right) \qquad (i, j = 1, 2, 3, \ldots, n) \qquad [4]$$

where the W_i $(i = 1,2,3,\ldots,n)$ denote 3×3 matrices and δ_{ij} is the Kronecker delta. This is equivalent to assuming that random errors corresponding to concurrent readings may be correlated, but no correlation is considered between different triplets R, A, E. In the vast majority of the cases, all W_i are diagonal.

Since bias errors are assumed to originate solely from instrument imperfection, the matrix N can be written as

$$N = \begin{bmatrix} I \\ I \\ \vdots \\ I \end{bmatrix} \qquad [5]$$

consisting of one column of n 3×3 identity matrices. Let us define

$$J = \begin{bmatrix} J_1 \\ J_2 \\ \vdots \\ J_n \end{bmatrix} \quad \text{and} \quad \Delta\lambda = \begin{bmatrix} \Delta\lambda_1 \\ \Delta\lambda_2 \\ \vdots \\ \Delta\lambda_n \end{bmatrix} \qquad [6]$$

where

$$J_i = \frac{\partial(R_i, A_i, E_i)}{\partial\mu} \quad \text{and} \quad \Delta\lambda_i = \begin{bmatrix} \Delta R_i \\ \Delta A_i \\ \Delta E_i \end{bmatrix} \quad (i = 1, 2, 3, \cdots, n) \qquad [7]$$

All elements of the $3n \times 3n$ matrix Φ are, in general, different from zero. Thus, a straightforward application of Eq. 2 is very impractical for various reasons. An obvious one is the necessary inversion of a large-size matrix. Other reasons will become apparent later. In order to transform the formula, we will make use of the matrix identity

$$(\Psi + N\Lambda N')^{-1} = \Psi^{-1} - \Psi^{-1}N(\Lambda^{-1} + N'\Psi^{-1}N)^{-1}N'\Psi^{-1} \qquad [8]$$

Thus

$$J'\Phi^{-1}J = J'\Psi^{-1}J - J'\Psi^{-1}N(\Lambda^{-1} + N'\Psi^{-1}N)^{-1}N'\Psi^{-1}J \qquad [9]$$

and

$$J'\Phi^{-1}\Delta\lambda = J'\Psi^{-1}\Delta\lambda - J'\Psi^{-1}N(\Lambda^{-1} + N'\Psi^{-1}N)^{-1}N'\Psi^{-1}\Delta\lambda \qquad [10]$$

Using Eqs. 4-6 for Ψ, N, J, and $\Delta\lambda$, Eqs. 9 and 10 can readily be transformed to

$$J'\Phi^{-1}J = \sum_{i=1}^{n} J_i' W_i J_i - K(\Lambda^{-1} + P)^{-1}K' \qquad [11]$$

$$J'\Phi^{-1}\Delta\lambda = \sum_{i=1}^{n} J_i' W_i \Delta\lambda_i - K(\Lambda^{-1} + P)^{-1}\sum_{i=1}^{n} W_i \Delta\lambda_i \qquad [12]$$

where

$$K = \sum_{i=1}^{n} J_i' W_i \quad \text{and} \quad P = \sum_{i=1}^{n} W_i \qquad [13]$$

Inversion of the 6×6 matrix appearing on the right-hand side of Eq. 11 provides the covariance matrix of the computed orbital elements, whereas Eqs. 11 and 12 together with Eq. 2 establish the iterative procedure for refinement of the orbit.

For implementation of the method on an electronic computer, the differences $\Delta\lambda_i$ should be computed from $\Delta\lambda_i = H_i(x_i - \bar{x}_i)$, where x_i, $\bar{x}_i$ are, respectively, inertial coordinates corresponding to the ith data point as computed from R_i, A_i, E_i and from the initial estimate $\bar{\mu}$ of the orbital elements (note that x_i and $\bar{x}_i$ represent three by one matrices); H_i is given by

$$H_i = \frac{\partial(R_i, A_i, E_i)}{\partial x_i}$$

It may be worthwhile to investigate what additional computing time and storage requirements are involved when comparing the present formulation to a method where bias errors are neglected completely. The following algorithm seems to be quite expedient.

The quantities W_i, H_i, x_i are computed (if not already available) and stored for $i = 1, 2, \ldots, n$. The following computations are then carried out at each data point: 1) compute $G_i = \partial x_i / \partial \mu$; 2) compute $J_i = H_i G_i$; 3) compute $W_i J_i$ and accumulate; 4) compute $J_i' W_i J_i$ and accumulate; 5) compute Δx_i; 6) compute $\Delta\lambda_i = H_i \Delta x_i$; 7) compute $W_i \Delta\lambda_i$ and accumulate; and 8) compute $J_i' W_i \Delta\lambda_i$ and accumulate. The matrix $(\Lambda^{-1} + P)^{-1}$ depends directly on the input and is computed in a straightforward fashion.

A count of the additional computations to be performed in the case where bias errors are accounted for leads to the following approximate formula for the increase in computing time:

$$\delta T \approx [9n(7r-4) + 200]\delta t$$

where n is the number of data points to be processed, r is the number of iterations allowed, and δt is the average duration of an arithmetic operation. Similarly, storage requirements are approximately increased by ΔM and $\Delta M = 9n$ in the case where each triplet R, A, E is weighted differently, $\Delta M = 6n$ in the case where each triplet is weighted differently but with no correlation between individual readings, and $\Delta M = 3n$ in the case where all data points are weighted equally but no correlation between readings is assumed. For instance, with typically 50 data points and two iterations, the number of required slots would approximately increase by 150 in the last case, the one quite commonly used in practice. Additional computing time would amount to roughly 0.05 sec on an IBM 7094 computer.

DISCUSSION

It may be of interest to investigate special cases of Eqs. 11, 12, and 2. First of all, if no bias errors are present, that is, if $\Lambda = 0$, the expressions reduce to the usual smoothing formulas for uncorrelated observational errors. This is directly apparent from Eqs. 11 and 12, where the matrix $(\Lambda^{-1} + P)^{-1} = \Lambda(\Lambda + P^{-1})^{-1} P^{-1}$ is then equal to the null matrix.

More interesting is the case where no random errors are present, i.e., $\Psi = 0$. It should be noted that Eq. 2 without Eqs. 11 and 12 is then meaningless, since the covariance matrix $N\Lambda N'$ of the observations is singular, its rank being at the most equal to three in the case considered in Sec. 2. The smoothing formula and covariance matrix of the refined elements must be obtained by a suitable limiting process in Eqs. 11 and 12.

Let $W_i = I/\varepsilon$ $(i = 1,2,\ldots,n)$, where I is a 3×3 unit matrix and ε is a positive number. We can let the elements of Ψ become arbitrarily small by letting ε approach zero. If we define the matrix $S(\varepsilon)$ as

$$S(\varepsilon) = \sum_{i=1}^{n} J_i' J_i - \frac{1}{n}\left(\sum_{i=1}^{n} J_i'\right)\Lambda\left(\Lambda + \frac{\varepsilon}{n} I\right)^{-1}\left(\sum_{i=1}^{n} J_i\right)$$

then the matrix $J'\Phi^{-1}J$ can be written as

$$J'\Phi^{-1}J = \frac{1}{\varepsilon} S(\varepsilon) \quad [14]$$

Also,

$$J'\Phi^{-1}\Delta\lambda = \frac{1}{\varepsilon}\left[\sum_{i=1}^{n} J_i'\Delta\lambda_i - \frac{1}{n}\left(\sum_{i=1}^{n} J_i'\right)\Lambda\left(\Lambda + \frac{\varepsilon}{n} I\right)^{-1}\left(\sum_{i=1}^{n} \Delta\lambda_i\right)\right]$$

Finally, the covariance matrix of the refined elements becomes

$$\operatorname{cov}(\hat{\mu}) = \varepsilon S^{-1}(\varepsilon) \quad [15]$$

and the smoothing formula is

$$\hat{\mu} = \bar{\mu} + S^{-1}(\varepsilon)\left[\sum_{i=1}^{n} J_i'\Delta\lambda_i - \frac{1}{n}\left(\sum_{i=1}^{n} J_i'\right)\Lambda\left(\Lambda + \frac{\varepsilon}{n} I\right)^{-1}\left(\sum_{i=1}^{n} \Delta\lambda_i\right)\right] \quad [16]$$

Assuming $S(o)$ to be nonsingular, it follows from Eq. 15 that

$$\lim_{\varepsilon \to 0} \operatorname{cov}(\hat{\mu}) = 0 \quad [17]$$

If the formulation is statistically consistent, Eq. 17 indicates that it should be possible to compute exact orbital elements from biased data according to the preceding formalism and in spite of the fact that the bias errors are not explicitly solved for. That this is indeed the case can be demonstrated as follows. Obviously, the entire argument hinges on the nonsingularity of S(o). This matter will be discussed later in this section.

Let us denote the correct orbital elements by μ and the correct data by λ. Suppose that an estimate $\bar{\mu}$ of μ is available from which the simulated observations $\bar{\lambda}$ are computed. The available data are corrupted only by a constant bias error and can thus be represented by $\lambda+s$, where s is given by

$$s = \begin{pmatrix} s_1 \\ s_1 \\ \vdots \\ s_1 \end{pmatrix}$$

consisting of n identical vectors s_1. Thus, the data can be written as

$$\begin{pmatrix} \lambda_1 + s_1 \\ \lambda_2 + s_1 \\ \cdots \\ \lambda_n + s_1 \end{pmatrix}$$

and the increments $\Delta\lambda_i$ (i = 1,2,...n) are $\Delta\lambda_i = \lambda_i + s_1 - \bar{\lambda}_i$. If $\hat{\mu}$ are the refined elements, Eq. 16 becomes, with $\varepsilon = 0$,

$$\hat{\mu} = \bar{\mu} + S^{-1}(o)\left[\sum_{i=1}^{n} J_i'(s_1+\lambda_i-\bar{\lambda}_i) - \frac{1}{n}\left(\sum_{i=1}^{n} J_i'\right)\sum_{i=1}^{n}(s_1+\lambda_i-\bar{\lambda}_i)\right]$$

or

$$\hat{\mu} = \bar{\mu} + S^{-1}(o)\left[\sum_{i=1}^{n} J_i'(\lambda_i-\bar{\lambda}_i) - \frac{1}{n}\left(\sum_{i=1}^{n} J_i'\right)\sum_{i=1}^{n}(\lambda_i-\bar{\lambda}_i)\right] \qquad [18]$$

It is thus already evident that the magnitude of the bias error s_1 is irrelevant. Furthermore, to first order we have $\lambda_i - \bar{\lambda}_i = J_i(\mu-\bar{\mu})$. Putting this into Eq. 18, we obtain

$$\hat{\mu} = \bar{\mu} + S^{-1}(o)\left[\sum_{i=1}^{n} J_i' J_i - \frac{1}{n}\left(\sum_{i=1}^{n} J_i'\right)\left(\sum_{i=1}^{n} J_i\right)\right](\mu - \bar{\mu})$$

or, observing that

$$S(o) = \sum_{i=1}^{n} J_i' J_i - \frac{1}{n}\left(\sum_{i=1}^{n} J_i'\right)\left(\sum_{i=1}^{n} J_i\right)$$

we have $\hat{\mu} = \mu$. Hence, in the presence of bias, but without random errors, the procedure will adjust any initial estimates of the elements to their correct values provided, of course, convergence is guaranteed. This is in agreement with the earlier conclusion that the covariance matrix of the refined elements is equal to the null matrix in the absence of random errors.

In the arguments just presented, the random errors were suppressed by letting their covariance matrix Ψ approach the null matrix. The same consequences could be demonstrated by letting the number of data points increase indefinitely but maintaining a certain bias. It then becomes evident that the present method should be preferable to a method ignoring bias errors completely, especially in the case where the amount of data is such that random errors are essentially "averaged out" as happens quite frequently in practice. This is all the more true with tracking devices with a small mean-squared error compared to possible magnitudes of bias errors.

We mentioned earlier that the matrix S(o) is of crucial importance. Indeed, if S(o) turns out to be singular, no correct elements can be obtained. Intuitively, it is clear that two points in inertial space cannot possibly yield correct elements, if any error whatsoever is present. In terms of the foregoing analysis, this implies that the matrix S(o) must be singular when only two triplets R, A, E are available. In the latter case, S(o) can be written as

$$S(o) = \frac{1}{2}\left(J_1' - J_2'\right)\left(J_1 - J_2\right) \qquad [19]$$

It follows from Eq. 19 that, at the most, S(o) can be of rank three and must therefore indeed be singular since S(o) is a 6×6 matrix. If three or more triplets R, A, E are available, S(o) will, in general, be nonsingular. As soon as this happens, the method becomes insensitive to bias errors if enough iterations are allowed.

It is to be anticipated that the singular behavior of S(o) is closely related to the fact that observations, offset by a constant bias, may not exactly fit any orbit whatsoever. Presumably this is, then, the feature by which bias errors are detected and ultimately corrected for. Of course, it may very well be possible to produce cases where S(o) always remains singular for any number of processed observations. These cases are rather exceptional and of minor practical interest. Geometrically, it means that an orbit can always be passed exactly through the biased data points. Hence, for such peculiar geometries there exists some increment $d\mu$ of the orbital elements for which all resulting increments $d\lambda_i$ ($i = 1, 2, \ldots n$) of the observations are equal to, say, $d\lambda_1$. This can be expressed as

$$d\lambda_1 = J_i \, d\mu \qquad (i = 1, 2, \cdots, n) \tag{20}$$

Summing Eq. 20 over all points and dividing by n yields

$$d\lambda_1 = \frac{1}{n}\left(\sum_{i=1}^{n} J_i\right) d\mu \tag{21}$$

It follows from Eqs. 20 and 21 that

$$\left[J_i - \frac{1}{n}\left(\sum_{i=1}^{n} J_i\right)\right] d\mu = 0$$

Premultiplying by J_i' and summing over all points gives

$$\left[\sum_{i=1}^{n} J_i' J_i - \frac{1}{n}\left(\sum_{i=1}^{n} J_i'\right)\left(\sum_{i=1}^{n} J_i\right)\right] d\mu = 0$$

or

$$S(o)\, d\mu = o \qquad [22]$$

Since the increment dμ is nonzero, it follows that the matrix S(o) must be singular, as was indeed expected.

The previous discussion has touched on rather fundamental aspects of the present smoothing scheme. From a practical point of view, it is clear that the limited accuracy of computers may cause difficulties for geometries sufficiently close to the singular ones referred to earlier. This is an immediate consequence of the fact that the inversion of S(o) becomes increasingly more difficult the better the prevailing geometry resembles a singular case. The growing inability of bias detection could result in the presence of "inbred" bias errors if a tracking device must acquire completely on the basis of its own predictions. Of course, a method neglecting bias errors altogether is expected to be much more vulnerable to the latter shortcoming than the one presented here.

EXAMPLES

Many of the conclusions reached in the previous section can be illustrated using rather simple examples. As a first example, consider the problem of determining the radius of a circle centered at the origin of a Cartesian system Oxy (Fig. 1), given a number of points on the circle but corrupted with a bias error s. Assume ρ to be the exact value of the radius and let the points A and B be given, i.e., AA' = BB' = s. The two observations λ_1 and λ_2 can then be taken as λ_1 = AA" and λ_2 = BB". Thus

$$J_i = \frac{\partial \lambda_i}{\partial \rho} = \frac{\bar{\rho}}{\lambda_i} = J_i' \qquad (i = 1,2)$$

where $\bar{\rho}$ is an estimate of the radius. Furthermore, we have

$$\lambda_i = s + \sqrt{\rho^2 - \alpha_i^2} \qquad (i = 1,2)$$

and

$$\Delta\lambda_i = s + \sqrt{\rho^2 - \alpha_i^2} - \sqrt{\bar{\rho}^2 - \alpha_i^2} \approx s + \frac{\bar{\rho}}{\lambda_i}(\rho - \bar{\rho}) \qquad [23]$$

If only one data point is used, it is intuitively obvious that no bias errors can be detected. This is in agreement with the fact that, in the latter case,

$$S(o) = \left(\frac{\bar{\rho}}{\lambda_1}\right)^2 - \frac{1}{1}\left(\frac{\bar{\rho}}{\lambda_1}\right)\left(\frac{\bar{\rho}}{\lambda_1}\right) = 0$$

If both observations are used, we obtain

$$S(o) = \frac{\bar{\rho}^2}{2}\left(\frac{1}{\lambda_1} - \frac{1}{\lambda_2}\right)^2$$

If $\lambda_1 = \lambda_2$, S(o) is singular. However, this would mean that either the two points A and B coincide or they are located symmetrically with respect to the y axis. In both cases, it is geometrically evident that bias detection is impossible. If $\lambda_1 \neq \lambda_2$, as in Fig. 1, S(o) is nonsingular, and the correct radius can be found. Indeed,

$$\hat{\rho} = \bar{\rho} + \frac{2}{\bar{\rho}^2} \cdot \frac{1}{(1/\lambda_1 - 1/\lambda_2)^2}\left[\frac{\bar{\rho}}{\lambda_1}\Delta\lambda_1 + \frac{\bar{\rho}}{\lambda_2}\Delta\lambda_2 - \frac{1}{2}\left(\frac{\bar{\rho}}{\lambda_1} + \frac{\bar{\rho}}{\lambda_2}\right)(\Delta\lambda_1 + \Delta\lambda_2)\right]$$

Using Eq. 23 for $\Delta\lambda_1$, we obtain

$$\hat{\rho} = \bar{\rho} + \frac{2}{(1/\lambda_1 - 1/\lambda_2)^2}\left[\frac{1}{\lambda_1^2} + \frac{1}{\lambda_2^2} - \frac{1}{2}\left(\frac{1}{\lambda_1} + \frac{1}{\lambda_2}\right)^2\right](\rho - \bar{\rho}) = \rho$$

The second example pertains more to the problem of orbit determination. A point is moving at a uniform speed along a rectilinear trajectory, taken as the x axis in the Cartesian Oxy system in Fig. 2. The motion is exactly given by $x(t) = x_o + v_o t$, where x_o, v_o can be regarded as a set of state variables of which estimates $\bar{x}_o$ and $\bar{v}_o$ are assumed to be known. Assume further that, at the times t_1, t_2, and t_3 respectively, three range measurements λ_1, λ_2, λ_3 from point T are available, all of which are in error by the same amount s. Thus, the measurements can be expressed as

$$\lambda_i = \sqrt{h^2 + (x_0 + v_0 t_i)^2} + s \qquad (i = 1,2,3)$$

We will again apply Eq. 16 with $\varepsilon = 0$. The matrix J_i can be written as $(i = 1,2,3)$

$$J_i = (\alpha_i \quad \alpha_i t_i) \qquad [24]$$

where

$$\alpha_i = \frac{x_0 + v_0 t_i}{\sqrt{h^2 + (x_0 + v_0 t_i)^2}} \qquad [25]$$

The matrix S(o) becomes

$$S(o) = \begin{bmatrix} p_{11} & p_{12} \\ p_{21} & p_{22} \end{bmatrix}$$

in which

$$p_{11} = \sum_1^3 \alpha_i^2 - \frac{1}{3}\left(\sum_1^3 \alpha_i\right)^2$$

$$p_{12} = p_{21} = \sum_1^3 \alpha_i^2 t_i - \frac{1}{3}\left(\sum_1^3 \alpha_i\right)\left(\sum_1^3 \alpha_i t_i\right)$$

$$p_{22} = \sum_1^3 \alpha_i^2 t_i^2 - \frac{1}{3}\left(\sum_1^3 \alpha_i t_i\right)^2$$

The determinant $|S(o)|$ of the matrix S(o) may be put in the form

$$|S(o)| = \frac{1}{3}\left[\alpha_1 t_1 (\alpha_2 - \alpha_3) + \alpha_2 t_2 (\alpha_3 - \alpha_1) + \alpha_3 t_3 (\alpha_1 - \alpha_2)\right]^2 \quad [26]$$

Some interesting conclusions follow quite readily from Eq. 26. First, if any two of the three observations coincide in time, the determinant $|S(o)|$ vanishes. Indeed, if the first two observations (for example) occur at time t_1, it follows that $\alpha_1 = \alpha_2$, and it may easily be verified that $|S(o)|$ then reduces to zero. The situation is, in effect, equivalent to the case of two observations, that is, the same number of observations as the number of state variables to be determined. Under these conditions, it is to be anticipated that no bias detection is possible as was indeed confirmed by the resulting zero value of $|S(o)|$.

Second, if the observing station (point T) coincides with the origin O, we have $\alpha_1 = \alpha_2 = \alpha_3 = 1$, and again we obtain a zero value for $|S(o)|$. This then corresponds to a singular geometry briefly discussed at the end of the previous section. It should be clear that, in the latter case, the state variable x_0 cannot be obtained exactly.

If $|S(o)| \neq 0$, the inverse matrix $S^{-1}(o)$ is given by

$$S^{-1}(o) = \frac{1}{|S(o)|}\begin{bmatrix} P_{22} & -P_{12} \\ -P_{21} & P_{11} \end{bmatrix}$$

Furthermore, the residuals $\Delta\lambda_i$ $(i = 1,2,3)$ in the data can be written as

$$\Delta\lambda_i = \sqrt{h^2 + (x_0 + v_0 t_i)^2} + s - \sqrt{h^2 + (\bar{x}_0 + \bar{v}_0 t_i)^2}$$

or, to a first approximation,

$$\Delta\lambda_i = \alpha_i (x_0 - \bar{x}_0) + \alpha_i t_i (v_0 - \bar{v}_0) + s \quad [27]$$

where α_1, α_2, and α_3 are given by Eq. 25. Using Eqs. 24 and 27 for J_i and $\Delta\lambda_i$, respectively, it is readily shown that

$$\sum_{1}^{3} J_i' \Delta\lambda_i - \frac{1}{3}\left(\sum_{1}^{3} J_i'\right)\left(\sum_{1}^{3} \Delta\lambda_i\right) = \begin{bmatrix} p_{11} \\ p_{12} \end{bmatrix} (x_o - \bar{x}_o) + \begin{bmatrix} p_{12} \\ p_{22} \end{bmatrix} (v_o - \bar{v}_o)$$

Application of Eq. 16 with $\varepsilon = 0$ yields

$$\begin{bmatrix} \hat{x}_o \\ \hat{v}_o \end{bmatrix} = \begin{bmatrix} \bar{x}_o \\ \bar{v}_o \end{bmatrix} + \frac{1}{p_{11}p_{22} - p_{12}^2} \begin{bmatrix} p_{22} & -p_{12} \\ -p_{21} & p_{11} \end{bmatrix} \times \left\{ \begin{bmatrix} p_{11} \\ p_{12} \end{bmatrix} (x_o - \bar{x}_o) + \begin{bmatrix} p_{12} \\ p_{22} \end{bmatrix} (v_o - \bar{v}_o) \right\}$$

Hence,

$$\begin{bmatrix} \hat{x}_o \\ \hat{v}_o \end{bmatrix} = \begin{bmatrix} \bar{x}_o \\ \bar{v}_o \end{bmatrix} + \begin{bmatrix} 1 \\ 0 \end{bmatrix} (x_o - \bar{x}_o) + \begin{bmatrix} 0 \\ 1 \end{bmatrix} (v_o - \bar{v}_o)$$

or $\hat{x}_o = x_o$, $\hat{v}_o = v_o$, and the estimates $\bar{x}_o$, $\bar{v}_o$ of the state variables have thus been adjusted to their correct values x_o and v_o.

CONCLUSIONS

A method of orbit refinement has been presented whose chief merit consists of its ability to obtain orbital elements having a minimum sensitivity to systematic errors in the observations. In particular, in the hypothetical case that the only errors present can be put in a form merely involving unknown constants, correct orbital elements are computed provided enough observations are processed.

In certain special cases largely of academic interest, the method may fail to yield any orbital elements whatsoever. This situation is characterized by the appearance of a singular matrix. It has been indicated that this phenomenon should not be regarded as a basic shortcoming of the method but is rather inherent to the prevailing geometry. In such a case, bias errors cannot effectively be eliminated by any other method. The presented formulas are still applicable. However, the covariance matrix of the bias component is now to be put equal to the null matrix, thereby completely ignoring the presence of any systematic errors. Of course, the orbit so computed should not be expected to remain insensitive to whatever systematic errors are present.

REFERENCES

1 Blackman, R. B., "Methods of orbit refinement," Bell System Tech. J. 43, 885-909 (1964).

2 Swerling, P., "First order error propagation in a stagewise smoothing procedure for satellite observations," J. Astronaut. Sci. 6, 46-52 (Autumn 1959).

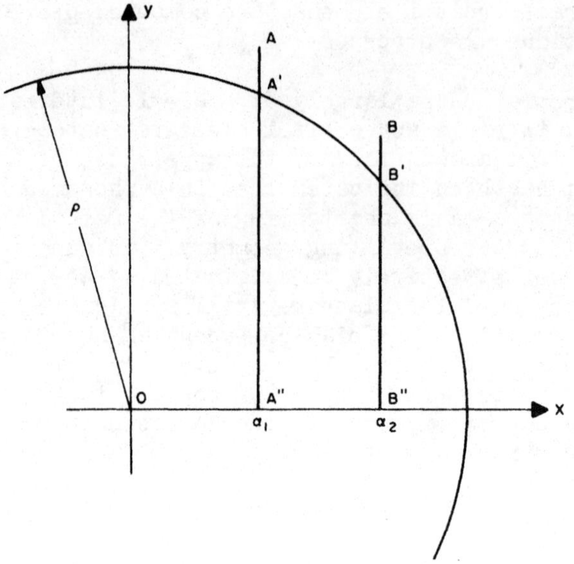

Fig. 1 Computation of the radius of a circle from biased ordinates of two points.

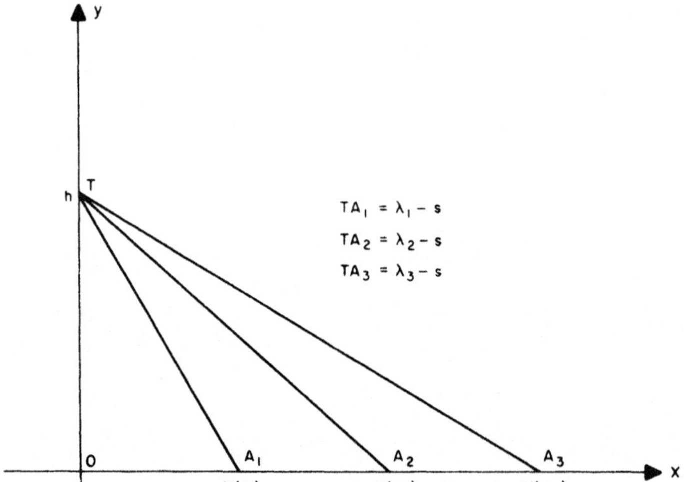

Fig. 2 Computation of position and velocity from biased range measurements.

CONTRIBUTORS TO VOLUME 14

Printed in Dunstable, United Kingdom

91898120R00429